Food Packaging Science and Technology

Food Packaging Science and Technology

DONG SUN LEE

KIT L. YAM

LUCIANO PIERGIOVANNI

CRC Press
Taylor & Francis Group
Boca Raton   London   New York

CRC Press is an imprint of the
Taylor & Francis Group, an **Informa** business

CRC Press
Taylor & Francis Group
6000 Broken Sound Parkway NW, Suite 300
Boca Raton, FL 33487-2742

© 2008 by Taylor & Francis Group, LLC
CRC Press is an imprint of Taylor & Francis Group, an Informa business

No claim to original U.S. Government works
Printed in the United States of America on acid-free paper
10 9 8 7 6 5 4 3 2 1

International Standard Book Number-13: 978-0-8247-2779-6 (Hardcover)

Visit the Taylor & Francis Web site at
http://www.taylorandfrancis.com

and the CRC Press Web site at
http://www.crcpress.com

Table of Contents

Chapter 1: Overview of Food Packaging Systems.....................1
1.1   Introduction 1
1.2   Science and Technology of Food Packaging.................................... 2
        1.2.1   Food Packaging Science···3
        1.2.2   Food Packaging Technology ·······································3
1.3   Socioeconomic Needs 4
1.4   Packaging Functions.. 5
1.5   Packaging Environments 6
1.6   Food Packaging Systems 7
        1.6.1   Levels of Packaging ···7
        1.6.2   Package Forms ···8
        1.6.3   Primary Food Packaging System···································8
1.7   Tables for Analyzing Food Packaging Systems 9
        1.7.1   Functions/Socioeconomics Table··································9
        1.7.2   Functions/Technologies Table····································· 10
        1.7.3   Functions/Environments Table···································· 11
1.8   Food Package Development 13
Discussion Questions and Problems 14
Bibliography 15

Part One: Packaging Material Science

Chapter 2: Chemical Structures and Properties of Packaging Materials...19
2.1   Introduction 19
2.2   Chemical Constituents.. 20
2.3   Chemical Bonding... 21
        2.3.1   Ionic Bonding ·· 21
        2.3.2   Metallic Bonds ·· 23
        2.3.3   Covalent Bond Model ·· 25
2.4   Intermolecular Forces.. 27
        2.4.1   Ion-Dipole Forces··· 28
        2.4.2   Dipole-Dipole Forces ·· 29
        2.4.3   Hydrogen Bonding ·· 29
        2.4.4   Dispersion Forces ··· 30
        2.4.5   Consequences of Intermolecular Interaction: Cohesion, Adhesion, and Surface Tension ··· 30
2.5   Spatial Arrangements 32
        2.5.1   Tacticity·· 32
        2.5.2   Crystalline vs. Amorphous State ································· 34

2.6 Chemical Reactivity and Susceptibility of Packaging 35
    2.6.1 Oxidation ···36
    2.6.2 Biodegradation and Biodeterioration ···························38
    2.6.3 Chemical Resistance, Etching, and Weathering ············40
Discussion Questions and Problems 41
Bibliography 41

Chapter 3: Physical Properties of Packaging Materials...........43

3.1 Introduction 43
3.2 Thermal Properties 44
    3.2.1 Thermal Conductivity (k) ·······································45
    3.2.2 Heat Capacity (Cp) ···46
    3.2.3 Thermal Expansion (linear and volumetric) ················47
    3.2.4 Tolerable Thermal Range ··48
    3.2.5 Transition Temperatures (Tm, Tg) ····························48
    3.2.6 Heat of Combustion (Qc) ··50
3.3 Electromagnetic Properties...51
    3.3.1 Refractive Index (n)···52
    3.3.2 Transparency (%T)··53
    3.3.3 Transmittance/Absorption Spectra in UV, VIS, and IR ····55
    3.3.4 Haze···58
    3.3.5 Gloss···59
    3.3.6 Behavior to Ionizing Radiations ·······························60
    3.3.7 Behavior to Microwaves ···62
3.4 Mechanical Properties 63
    3.4.1 Density and Related Properties ·································63
    3.4.2 Coefficient of Friction ···64
    3.4.3 Strength Properties: Tensile, Tear, Burst, and Creep ······66
    3.4.4 Response to Dynamic Stresses: Impact Resistance and Cushioning ··72
Discussion Questions and Problems 75
Bibliography 76

Chapter 4: Permeation of Gas and Vapor................…..........79

4.1 Introduction 79
4.2 Basic Concepts of Permeation..80
    4.2.1 Mechanism of Gas Transport through Permeation ·········81
    4.2.2 Diffusion of Permeant ···82
    4.2.3 Adsorption and Desorption of Permeant ····················82
4.3 Theoretical Analysis of Permeation...83
    4.3.1 Mass Balance Analysis···83
    4.3.2 Concentration Profile within Film at Steady State ·········85
    4.3.3 Derivation of Permeation Rate Equation ····················85
    4.3.4 Physical Meanings of Permeation Rate Equation···········86
4.4 Terminology and Units for Permeation 89

4.4.1 Transmission Rate 90
4.4.2 Permeance 91
4.4.3 Permeability 91
4.5 Permeability of Food Packaging Polymers............................92
4.6 Factors Governing Permeation..96
4.6.1 Nature of Polymer 96
4.6.2 Nature of Permeant 97
4.6.3 Ambient Environment 97
4.7 Measurements for Permeation Properties.............................98
4.7.1 Basic Design Principles..98
4.7.2 Isostatic Method 99
4.7.3 Quasi-Isostatic Method..100
4.7.4 Measuring Permeation Rate of Finished Packages................101
4.7.5 Another Method for Estimating Permeation of Plastic Containers ··102
4.7.6 Gravimetric Method 104
4.8 Gas Transport through Leaks 105
Discussion Questions and Problems 106
Bibliography 108

Chapter 5: Migration and Food-Package Interactions.............109
5.1 Introduction 109
5.2 Phenomenal Description of Migration Process 110
5.2.1 Phenomenal Description of Migration Process 111
5.2.2 Kinetic and Thermodynamic Approach 114
5.3 Migration Issues in Food Packaging...............................116
5.3.1 Chemicals from Plastics 117
5.3.2 Recycled Plastics...117
5.3.3 Microwave Susceptor 120
5.4 Flavor Scalping and Sorption 121
5.5 Migration Testing...122
5.5.1 General Principles 123
5.5.2 Food Simulants...124
5.5.3 Analytical Techniques 124
5.6 Predictive Migration Models......................................125
5.6.1 Models for Simplified Systems 125
5.6.2 Estimation of Diffusion Coefficient and Partition Coefficient ·······128
5.6.3 Modeling for Worst Case Scenario 130
5.7 Regulatory Considerations 132
5.7.1 Commons and Differences between Regulations ···················133
5.7.2 Sensory Tainting ··136
Discussion Questions and Problems 137
Bibliography 138

Chapter 6: Food Packaging Polymers.....................................141
6.1   Basic Concepts of Polymers...142
     6.1.1   Chemical Structure ···142
     6.1.2   Molecular Shape ··143
     6.1.3   Thermoplastics versus Thermosets ····················144
     6.1.4   Homopolymers versus Copolymers ··················144
     6.1.5   Polymer Blends ···145
6.2   Polymerization Reactions..145
     6.2.1   Addition Polymerization ·····································145
     6.2.2   Condensation Polymerization ····························146
6.3   Plastics versus Polymers..147
     6.3.1   Production of Polymers and Plastic Resins ·······147
     6.3.2   Advantages and Disadvantages of Plastics ·······148
6.4   Composition/Processing/Morphology/Properties Relationships 148
6.5   Characteristics of Packaging Polymers 149
     6.5.1   Molecular Weight ···149
     6.5.2   Chain Entanglement ···151
     6.5.3   Summation of Intermolecular Forces between Polymer Chains······151
     6.5.4   Polymer Morphology ···152
6.6   Food Packaging Polymers..154
     6.6.1   Polyethylene (PE) ···155
     6.6.2   Polypropylene (PP) ··157
     6.6.3   Polystyrene (PS) ···157
     6.6.4   Polyvinyl Chloride (PVC) ···································158
     6.6.5   Polyethylene Terephthalate (PET) ·····················158
     6.6.6   Polyvinylidene Chloride (PVDC)·······················159
     6.6.7   Ethylene Vinyl Alcohol Copolymer (EVOH) ····160
     6.6.8   Ionomer ··160
     6.6.9   Ethylene Vinyl Acetate (EVA) Copolymer ·······161
     6.6.10 Polyamides (Nylons)···162
     6.6.11 Polycarbonate (PC) ··162
     6.6.12 Edible Coatings and Films ··································163
     6.6.13 Recycling Symbols···163
6.7   Polymer Processing..163
     6.7.1   Extrusion ··164
     6.7.2   Coextrusion ··164
     6.7.3   Cast Film Extrusion···165
     6.7.4   Blown Film Extrusion ···166
     6.7.5   Injection Molding ··168
     6.7.6   Blow Molding ··168
     6.7.7   Extrusion Coating··171
     6.7.8   Extrusion Lamination ··172
     6.7.9   Adhesive Lamination ···173
     6.7.10 Thermoforming ···174

6.7.11 Vacuum Metallization ·· 174
Discussion Questions and Problems 175
Bibliography 176

Chapter 7: Glass Packaging.............................177
7.1   Introduction 177
7.2   Chemical Structure 178
7.3   Glass Properties.. 180
      7.3.1   Mechanical Property·· 180
      7.3.2   Thermal Property·· 182
      7.3.3   Electromagnetic Property ································ 183
      7.3.4   Chemical Inertness ······································ 184
7.4   Glass Containers Manufacturing 185
      7.4.1   Glass Making·· 185
      7.4.2   Container Manufacturing ································ 188
      7.4.3   Post Blowing Operations································ 189
7.5   Glass Container Strengthening Treatments 192
7.6   Use of Glass Containers in Food Packaging.................. 193
7.7   Other Ceramic Containers 194
Discussion Questions and Problems 195
Bibliography 195

Chapter 8: Metal Packaging.............................197
8.1   Introduction 197
8.2   Aluminum.. 198
      8.2.1   Aluminum Foil ·· 201
8.3   Coated Steels.. 203
      8.3.1   Tinplate ·· 204
      8.3.2   Tin Free Steels··· 207
      8.3.3   Polymer Coated Steels··································· 208
8.4   Stainless Steel.. 209
8.5   Metal Corrosion.. 210
      8.5.1   Basic Corrosion Theory·································· 210
      8.5.2   Corrosion of Tinplate ··································· 215
      8.5.3   Corrosion of TFS··· 218
      8.5.4   Corrosion of Aluminum ································· 218
      8.5.5   Corrosion of Stainless Steel ··························· 218
      8.5.6   Microbiologically Induced Corrosion ················· 219
8.6   Metal Container Manufacturing 220
      8.6.1   Three-piece Can ··· 221
      8.6.2   Two-piece Can - D&I···································· 224
      8.6.3   Two-piece Can - DRD···································· 225
      8.6.4   Can Ends ··· 226
      8.6.5   Double Seaming ·· 228

    8.6.6 Kegs and Drums ···229
    8.6.7 Trays··230
    8.6.8 Collapsible Tubes and Aerosol Containers ·············230
8.7 Protective Lacquers 232
    8.7.1 Types of Coatings···232
    8.7.2 Methods of Coating Application ····························236
8.8 Recent Developments and Trends 237
Discussion Questions and Problems 238
Bibliography 239

Chapter 9: Cellulosic Packaging 243
9.1 Introduction 243
9.2 Cellulose Fiber-Morphology 245
9.3 Cellulose Fiber-Chemistry.......................................247
    9.3.1 Lignin ··247
    9.3.2 Hemicelluloses ···247
    9.3.3 Cellulose···249
9.4 Paper and Paperboard Production..............................251
    9.4.1 Pulping Technology ·······································251
    9.4.2 Papermaking···253
    9.4.3 Converting and Special Paper Products·············257
9.5 Paper Bags and Wrappings.......................................259
9.6 Corrugated Board and Boxes....................................261
9.7 Folding Cartons and Set-up Boxes 264
9.8 Composite Cans and Fiber Drums.............................266
9.9 Cartons for Liquids..267
9.10 Molded Cellulose 270
9.11 Cellophane...270
9.12 Other Quasi Cellulosic Materials 272
    9.12.1 Wood ··272
    9.12.2 Cork···273
    9.12.3 Nonwovens···273
Discussion Questions and Problems 274
Bibliography 274

Part Two: Packaging Technologies

Chapter 10: End-of-Line Operations 279
10.1 Introduction 279
10.2 Printing..279
    10.2.1 Image Printed ··280
    10.2.2 Conventional Printing Methods····························281
    10.2.3 Digital and Novel Printing Methods·····················288
10.3 Label and Labeling...291

    10.3.1 Label Types and Materials ···292
    10.3.2 Labeling Process ··294
10.4 Coding..296
    10.4.1 Code Imprinting ···296
    10.4.2 Bar Code··297
10.5 Sealing of Plastic Surfaces 301
    10.5.1 Heat Sealing ··301
    10.5.2 Cold Seal ···307
    10.5.3 Seal Strength Test···308
10.6 Case Study: Finding Sealing Conditions for an LLDPE Film 309
Discussion Questions and Problems 311
Bibliography 312

Chapter 11: Food Packaging Operations and Technology.....313
11.1 Food Packaging Line...313
11.2 Filling of Liquid and Wet Food Products............................317
    11.2.1 Filling to Predetermined Level ···318
    11.2.2 Filling to Predetermined Volume ·····································321
11.3 Filling of Dry Solid Foods...322
    11.3.1 Filling by Count ···323
    11.3.2 Filling by Volume ··324
    11.3.3 Filling by Weight ···326
11.4 Closure and Closing Operation..327
    11.4.1 Closures···327
    11.4.2 Closing Operation ···333
    11.4.3 Package Integrity Testing ···333
11.5 Methods of Wrapping and Bagging.....................................336
    11.5.1 Wrapping···336
    11.5.2 Bagging ··337
11.6 Form-Fill-Seal 339
    11.6.1 Vertical Form-Fill-Seal ··340
    11.6.2 Horizontal Form-Fill-Seal ··342
11.7 Various Forms of Contact and Contour Packaging 344
    11.7.1 Skin Packaging ···345
    11.7.2 Blister Packaging···345
    11.7.3 Shrink Packaging··346
    11.7.4 Stretch Wrapping··348
11.8 Case Studies 349
    11.8.1 Performance of a Piston-type Filler····································349
    11.8.2 Pressure Differential System for Nondestructive Leak Detection····351
Discussion Questions and Problems 352
Bibliography 353

Chapter 12: Thermally Preserved Food Packaging: Retortable and Aseptic..357
12.1 Introduction 357
12.2 Thermal Destruction of Microorganisms and Food Quality......................358
    12.2.1 Kinetics of Microbial Inactivation··358
    12.2.2 Kinetics in Thermal Degradation of Food Quality ··················362
12.3 Basics of Thermal Processing Design 362
    12.3.1 Selection of Processing Conditions ···································362
    12.3.2 Determination of Thermal Process Condition ····················365
12.4 Hot Filling 367
12.5 In-container Pasteurization and Sterilization...................................368
    12.5.1 Heat Penetration into Packaged Foods ···························369
    12.5.2 Pasteurization ···371
    12.5.3 Sterilization for Shelf-Stable Low Acid Foods ··················372
    12.5.4 Containers for heat preserved foods ····························376
12.6 Aseptic Packaging 379
    12.6.1 Continuous Heat Processing of Fluid-Based Foods ···············381
    12.6.2 Sterilization of Packages and Food Contact Surfaces ·············384
    12.6.3 Aseptic Filling and Packaging System ·························386
    12.6.4 Modified Version of Aseptic Packaging ·························390
12.7 Case Study: Design of a Thermal-Processed Tray-Set Containing High- and Low-Acid Foods...392
Discussion Questions and Problems 393
Bibliography 394

Chapter 13: Vacuum/Modified Atmosphere Packaging..........397
13.1 Basic Principles 397
    13.1.1 Gases for MAP ···399
    13.1.2 Packaging Operations for MAP Foods···························404
    13.1.3 Safety of MAP foods··406
13.2 Nonrespiring Products..410
13.3 Respiring Products...412
13.4 Case Study: Design of Modified Atmosphere Package for Blueberries......420
Discussion Questions and Problems 422
Bibliography 422

Chapter 14: Microwavable Packaging..................................425
14.1 Microwaves and Microwave Oven..425
14.2 Microwave/Food/Packaging Interactions 426
    14.2.1 Microwave/Heat Conversion Mechanism ······················427
    14.2.2 Dielectric Properties ··428
    14.2.3 Penetration Depth ··429
    14.2.4 Mathematical Equations and Models ···························430
14.3 Challenges in Microwave Heating of Foods....................................431

14.3.1 Nonuniform Heating ···432
14.3.2 Lack of Browning and Crisping ································433
14.3.3 Variation in Microwave Ovens ·······························434
14.3.4 Meeting the Challenges ···434
14.4 Microwavable Packaging Materials 435
14.4.1 Microwave Transparent Materials··························435
14.4.2 Microwave Reflective Materials ·····························436
14.4.3 Microwave Absorbent Materials ····························437
14.5 Interactive Microwave Food Packages............................439
14.5.1 Surface Heating Packages ·································439
14.5.2 Field Modification Packages ·······························440
14.5.3 Steam Cooking Packages ·································441
14.6 Case Study: Effect of Metal Shielding on Microwave Heating Uniformity 442
Discussion Questions and Problems 443
Bibliography 443

Chapter 15: Active and Intelligent Packaging........................445

15.1 Introduction..445
15.2 Active Packaging – Absorbing System 448
15.2.1 Oxygen Absorbers··449
15.2.2 Moisture Absorbers ··451
15.2.3 Carbon Dioxide Absorbers·································452
15.2.4 Ethylene Absorbers ···453
15.2.5 Other Absorbers or Removers ····························453
15.3 Active Packaging – Releasing System 454
15.3.1 Carbon Dioxide Emitters····································456
15.3.2 Antimicrobial Packaging Systems ·····················456
15.3.3 Antioxidant Releasers··459
15.3.4 Other Releasers ··460
15.4 Active Packaging – Other System................................460
15.4.1 Self-Heating Systems ···460
15.4.2 Self-Cooling Systems ···461
15.4.3 Selective Permeation Devices and Others ···········461
15.5 Intelligent Packaging Framework 462
15.6 Smart Packaging Devices..464
15.6.1 Time Temperature Integrator or Indicator···········465
15.6.2 Leak, Gas, or Other Indicators ·························466
15.6.3 Freshness Indicators ···467
15.6.4 Bar Code··468
15.6.5 Electronic Identification Tags ····························468
15.7 Legislative and Human Behavior Issues 470
15.8 Case Study: Intelligent Microwave Oven with Barcode Reader..............471
Discussion Questions and Problems 472
Bibliography 473

Part Three: Packaging Food Science

Chapter 16: Shelf Life of Packaged Food Products...............479
16.1 Basic Concepts 480
    16.1.1 Definitions of Shelf Life··480
    16.1.2 Factors Affecting Shelf Life of Packaged Foods··············482
16.2 Food Factors Affecting Shelf Life...483
    16.2.1 Food Deterioration Modes···483
    16.2.2 Package Dependent versus Product Dependent Deteriorations·······485
    16.2.3 Quality Indexes and Critical Limits································486
    16.2.4 Sensory Quality ···487
    16.2.5 Microbial Count···488
    16.2.6 Losses in Nutrients and Pigments·································490
    16.2.7 Production of Undesirable Components···························492
    16.2.8 Physical Changes and Processes···································492
16.3 Kinetics of Food Deterioration...493
    16.3.1 Chemical Kinetics: Reaction Order and Rate Constant···········493
    16.3.2 Microbial Growth Model···497
    16.3.3 Other Kinetic Model Associated with Physical Changes ············498
    16.3.4 Temperature Dependence: Arrhenius and Shelf Life Plots ············499
    16.3.5 Moisture Dependence···502
    16.3.6 Oxygen Dependence··506
16.4 Environmental Factors Affecting Shelf Life 507
    16.4.1 Ambient Environment ··507
    16.4.2 Physical Environment···510
    16.4.3 Human Environment··512
16.5 Package Factors Affecting Shelf Life..512
    16.5.1 Packaging Parameters Important to Shelf Life ···················512
    16.5.2 Permeation versus Reaction Controlled Shelf Life················513
    16.5.3 Package Interactions ···514
16.6 Shelf Life Studies..515
    16.6.1 Testing Under Normal Conditions·································515
    16.6.2 Testing Under Accelerated Conditions···························516
    16.6.3 Procedures for Shelf Life Studies ·································517
16.7 Shelf Life Models..521
    16.7.1 Shelf Life of Chemical and Microbial Deteriorations ············522
    16.7.2 Shelf Life Models of Constant H2O and O2 Driving Forces ·········524
    16.7.3 Shelf Life Model of Variable H2O Driving Force·················527
    16.7.4 Shelf Life Model of Variable O2 Driving Force ·················531
    16.7.5 Package/Food Compatibility Dependent Shelf Life Model·········533
16.8 Case Study: Shelf Life of Potato Chip with Two Interacting Quality Deterioration Mechanisms 534
Discussion Questions and Problems 536
Bibliography 540

Chapter 17: Food Products Stability and Packaging Requirements...........................543

17.1 Introduction 543
17.2 Cereals and Bakery Products........................545
    17.2.1 Cereal Grains and Flours··545
    17.2.2 Ready-to-Eat Breakfast Cereals and Snacks ·····························547
    17.2.3 Fresh and Dried Pasta Products································548
    17.2.4 Fresh Bakery Products································550
17.3 Meat and fish products 553
    17.3.1 Fresh Meat and Poultry ································553
    17.3.2 Processed Meat Products································556
    17.3.3 Fish Products ································557
17.4 Dairy Products........................559
    17.4.1 Pasteurized and UHT Sterilized Milk and Cream ·····························559
    17.4.2 Dried Milk Products································561
    17.4.3 Cheese ································562
    17.4.4 Fermented Milks ································564
17.5 Confectionery Products 565
    17.5.1 Chocolate Products································566
    17.5.2 Hard Boiled Sweets ································567
    17.5.3 Toffees and Other Confectioneries································568
17.6 Fats and Oils........................570
17.7 Drinks........................571
    17.7.1 Fruit Juices ································571
    17.7.2 Soft Drinks ································573
    17.7.3 Beer ································574
    17.7.4 Wine ································575
17.8 Fresh Fruits and Vegetables 576
17.9 Frozen Foods........................581
Discussion Questions and Problems 585
Bibliography 585

Part Four: Packaging Sociology

Chapter 18: Sustainable Packaging........................595

18.1 Introduction 595
18.2 Sustainable Packaging 596
    18.2.1 Basic Concepts ································596
    18.2.2 Development of Sustainable Packaging ································597
18.3 Environmental Issues Relating to Packaging........................598
    18.3.1 Solid Wastes································598
    18.3.2 Hazardous Compounds································598
    18.3.3 Ozone Depletion································598
18.4 Packaging Waste Management........................599
    18.4.1 Reduction ································599

18.4.2 Reuse ···599
18.4.3 Recycling··600
18.4.4 Composting ··601
18.4.5 Incineration···601
18.4.6 Landfill ··602
18.5 Life Cycle Assessment 602
    18.5.1 Scope of LCA Studies ··································602
    18.5.2 Methodology of LCA ····································603
    18.5.3 LCA Studies ··604
18.6 Degradable Packaging Polymers 604
    18.6.1 Biodegradable Packaging ·······················605
    18.6.2 Photodegradable Packaging ···················606
Discussion Questions and Problems 607
Bibliography 607

Chapter 19: Sociological and Legislative Considerations......609
19.1 Introduction 609
19.2 Tamper Evident Packaging.......................................609
    19.2.1 Conventional Techniques Commonly Used ···········611
    19.2.2 Innovative Techniques ·····························613
19.3 Product Liability..614
19.4 Labeling Information..616
Discussion Questions and Problems 618
Bibliography 618

Index...621

Preface

For many years, we have been teaching food packaging at undergraduate and graduate levels, conducting research in a wide range of food packaging areas, and participating in the development of commercial products and systems relating to food packaging—this book is based on our collective experiences. We hope that students and professionals interested in food packaging will find this book informative and useful.

Food packaging is partly art and partly science—both are important for the development of effective and efficient food packages or packaging systems. Nevertheless, we observe that many commercial food packages are developed largely by empirical or trial-and-error methods. The practice of packaging art alone, with little consideration of the relevant science, frequently leads to ineffective designs and poor products, especially when dealing with complex food packaging systems. In this book, we attempt to strike a healthy balance between art and science, hoping to equip the reader with the necessary knowledge and tools to tackle food packaging problems.

This book is unique in several aspects. Most noticeably is its richness in illustrations, examples, discussion questions, and case studies, which are useful learning tools but are either absent or few in quantity in other packaging books. Another unique aspect is the breath and depth of materials, covering basic principles and technologies as well as more advanced or current topics such as active packaging, intelligent packaging, and sustainable packaging. Also, this book is unique because its authors are from Asia, United States, and Europe, who have local and global perspectives about food packaging.

Following an overview of food packaging by Chapter 1, the remaining chapters are organized into four parts:

- Part 1 is composed of eight chapters related to packaging materials science. Chapters 2-3 review the basic concepts of chemical and physical properties most relevant to packaging materials. Chapters 4-5 discuss gas permeation and migration, which are important transport phenomena to packaging materials. Chapters 6-9 provide detailed information about the four basic types of packaging materials: plastics, glass, metal, and cellulosic materials.

- Part 2 is composed of six chapters related to packaging technologies. Chapters 10-13 cover traditional technologies such as end-of-line operations, food packaging operations and technologies, canning and aseptic packaging, and vacuum/modified atmosphere packaging. Chapter 14-15 cover more advanced technologies including microwavable packaging, active packaging, and intelligent packaging.

- Part 3 is composed of two chapters related to packaging food science. Chapter 16 discusses the shelf life of packaged food products, while

Chapter 17 discusses the elements of packaging, distribution, and shelf lives of various food categories.

- Part 4 is composed of two chapters related to packaging sociology. Chapter 18 addresses sustainable packaging and related issues, while Chapter 19 addresses sociological and legislative considerations pertaining to food packaging.

Solutions of some example and case studies can be found and downloaded as Mathcad® programs or Excel® sheets in the website of this book (http://food.kyungnam.ac.kr/foodpack/).

Dong Sun Lee
Kit L. Yam
Luciano Piergiovanni

Abstract

Food packaging is partly art and partly science—both are important for the development of effective and efficient food packages or packaging systems. Nevertheless, many commercial food packages are developed largely by empirical or trial-and-error methods. The practice of packaging art alone, with little consideration of the relevant science, frequently leads to ineffective designs and poor products, especially when dealing with complex food packaging systems. In this book, the authors attempt to strike a healthy balance between art and science, hoping to equip the reader with the necessary knowledge and tools to tackle food packaging problems.

This book is unique in several aspects. Most noticeably is its richness in examples, discussion questions, and case studies, which are useful learning tools but are either absent or few in quantity in other packaging books. Another unique aspect is the breath of materials, covering not only the basic principles and technologies, but also the more advanced topics such as active packaging, intelligent packaging, and sustainable packaging. Also, this book is unique because its authors are from Asia, United States, and Europe, who have local and global perspectives about food packaging.

Acknowledgements

Much of the materials are derived from our lectures, conference presentations, and consulting assignments. We thank many of our students, colleagues, and friends who have either directly or indirectly contribute to this book; we especially thank our talented students Stefano Farris, Aishwarya Balasubramanian, Jae Deok Jang, Ki-Eun Lee, and Il Seo who helped edit the manuscript and prepare some of the artwork. We also thank the publishers who kindly granted permission for reproducing some figures and tables. Most importantly, we thank our beloved wives (Eun-Mee, Aileen, and Mariangela) and families for their support and patience during the writing of this book.

Dong Sun Lee
Kit L. Yam
Luciano Piergiovanni

Biographies

Dong Sun Lee is full professor of food engineering and packaging in the Department of Food Science and Biotechnology at Kyungnam University, Korea. He earned a B.S. degree in food technology from Seoul National University and M.S. and Ph.D. degrees in food engineering from Yonsei University. He was as a researcher at Food Research Institute of Agriculture and Fishery Development Corporation in Korea, as well as a visiting scientist at Massachusetts Institute of Technology (MIT) and Rutgers University. He also serves on the editorial board of Packaging Technology and Science. His research areas include modified atmosphere packaging, active packaging, and intelligent packaging for fresh produce, fermented foods, and prepared meals.

Kit L. Yam is full professor of food engineering and packaging in the Department of Food Science at Rutgers University, United States. He earned B.S., M.S., and Ph.D. degrees in chemical engineering from Michigan State University. He teaches university level courses and professional continuing education courses and has been a frequent invited speaker at technical conferences. He also holds two other professorships overseas, serves on the editorial boards of several journals, and serves as a consultant for food and packaging companies. His research projects are focused on shelf life of packaged foods, controlled release packaging, and intelligent packaging, many of which supported by competitive grants from USDA, DoD, NSF, NASA, and the industry.

Luciano Piergiovanni is full professor and academic program director of food science and technology in the Department of Food Science and Microbiology (diSTAM) at the State University of Milan, Italy. He has been visiting professor at the University of San Paolo (Brasil) and the University of Santafè de Bogotà (Colombia). He serves on several editorial boards and has published over 200 journal articles, conference proceedings, and technical papers. He is president of the Italian Scientific Group for Food Packaging (GSICA) and coordinator of several research projects sponsored by National Institutions (CNR, MIUR) and the European Union. His research is focused on modified atmosphere packaging of perishable foods, modeling and prediction of shelf life of foods in flexible packaging, and validation of new packaging materials and techniques.

Chapter 1

Overview of Food Packaging Systems

1.1	Introduction	1
1.2	Science and Technology of Food Packaging	2
	1.2.1 Food Packaging Science	3
	1.2.2 Food Packaging Technology	3
1.3	Socioeconomic Needs	4
1.4	Packaging Functions	5
1.5	Packaging Environments	6
1.6	Food Packaging Systems	7
	1.6.1 Levels of Packaging	7
	1.6.2 Package Forms	8
	1.6.3 Primary Food Packaging System	8
1.7	Tables for Analyzing Food Packaging Systems	9
	1.7.1 Functions/Socioeconomics Table	9
	1.7.2 Functions/Technologies Table	10
	1.7.3 Functions/Environments Table	11
1.8	Food Package Development	13
	Discussion Questions and Problems	14
	Bibliography	15

1.1 Introduction

To the average consumer, food packaging is simply the enclosure of a food product in a plastic pouch, a metal can, or a glass bottle. To a scientist or engineer, however, food packaging is a rather technical matter—it is a coordinated system designed for the efficient delivery of high quality, safe food products throughout every phase of the supply chain, from raw material production to food manufacture, packing, retail, wholesale, consumer use, disposal, and recycling or other means of resource recovery.

The scope of food packaging is very broad. It encompasses technical activities such as machinery design, graphic design, package development, package manufacture, shelf life testing, distribution, and marketing. It deals with various types and forms of food packages including metal cans, glass containers, paper cartons, plastic containers, and pouches. It involves the participation of packaging technologists, scientists and engineers, packaging material suppliers, packaging converters, packaging machinery manufacturers, food processors, food retailers, and regulatory agencies.

This chapter presents an overview of food packaging. Section 1.2 introduces food packaging as a scientific discipline that applies the principles of materials science, food science, information science, and socioeconomics to develop useful technologies for the society. Section 1.3 describes the socioeconomic needs that are responsible for the market-pull for food packaging technologies. Section 1.4 describes four packaging functions, and Section 1.5 describes three packaging environments. Section 1.6 introduces the important concept that food packaging is a system consisting of not only the package, but also the food and the environment. The packaging system may involve up to four levels of packaging that may be optimized to provide the most efficient solution. Section 1.7 introduces three tables for analyzing food packaging systems. Finally, Section 1.8 describes a typical food packaging development process which may serve as a guide for designing food packaging systems.

1.2 Science and Technology of Food Packaging

Food packaging science is a discipline which applies the principles from four major areas of science (materials science, food science, information science, and socioeconomics) to understand the properties of packaging materials, the packaging requirements of foods, the packaging system, etc. Examples of those principles are kinetics of food deterioration, mass transport phenomena, and stress-strain relationship. Food packaging technology is a set of practical solutions for delivering high quality and safe food products to the consumer in an efficient manner. Examples of food packaging technology are modified atmosphere packaging, microwavable packaging, and aseptic packaging.

Figure 1.1 illustrates the concept of technology push and market pull important to food packaging. Technology push means that new products or innovations are created as a result of advances in materials science, food science, and information science, and then those new products or innovations are pushed forward to seek market acceptance. Market pull means that market needs are created by the dynamics of socioeconomics, and then technology is sought to provide solutions to satisfy those needs.

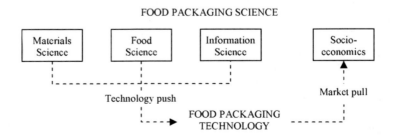

Figure 1.1 Dynamics of packaging science and technology

In the food packaging industry, market pull is typically the initiating force followed by technology push. For example, to satisfy the demand of the consumer for convenience, microwavable food packages were developed. To address the concern of plastic wastes, the technology of plastics recycling was developed in the 1980s.

1.2.1 Food Packaging Science

Historically, materials science and food science have been two major contributors for food packaging development. Materials science is important to understand the mechanical strength, barrier properties, appearance, and other physical and chemical properties of paper, glass, metal, polymer, and composite materials which are important to the manufacture and performance of food packages. Food science is important to understand the deterioration kinetics of foods (such as microbial growth, lipid oxidation, moisture gain or loss) which governs the shelf life of food.

In recent decades, socioeconomics surrounding packaging has evolved to become another important area of food packaging science. This area involves the collection and analysis of data to study the impact of food packaging on the changing society and economy. Socioeconomic problems such as those relating to environmental protection, product liability, and sustainability are constantly demanding specific and prompt actions from the packaging community.

More recently, information science is also evolving to become yet another important area for food packaging. Advances in information science are radically changing the lifestyles and conventional values of the society. The use of powerful hardware and software to enhance the efficiency of food packaging systems has become commonplace. Intelligent packaging [1], a relatively new concept for applying information technologies to enhance food quality and safety, is also gaining appreciation.

The Japanese Packaging Institute has published a proposal for packaging in the 21st century [2] which suggests that packaging science is a discipline consisting of packaging sociology, packaging materials science, and packaging technology. The proposal also recommends specific steps and measures for enterprises, universities, government agencies, and consumers to make packaging more useful and environmentally friendly.

1.2.2 Food Packaging Technology

Food packaging technology is a set of science-based solutions to address specific food packaging needs. Examples are tamper evident packaging, modified atmosphere packaging, aseptic packaging, and microwavable packaging which are aimed at enhancing safety, quality, and convenience for the consumer. Packaging technologies are also important for improving the efficiency of package manufacture, distribution, retail display, and waste disposal for the industry.

Innovative technologies such as antimicrobial packaging, controlled release packaging, nanotechnology, biosensors, and radio frequency identification (RFID) have attracted much interest from the packaging community in recent years. Since the development of a new technology is typically costly, it should be justified carefully based on its ability of this technology to enhance certain packaging functions to meet certain socioeconomic needs.

1.3 Socioeconomic Needs

The food packaging industry is largely driven by market pull to satisfy the needs of the society and economy. Hence the viability of any new packaging technology depends on its ability to fill these socioeconomic needs. Below are some major socioeconomic needs that drive the food packaging industry.

- *Consumer lifestyle.* This is the need that drives innovations in convenient packaged foods. In recent years, consumer lifestyle has been influenced greatly by the aging population, increasing number of smaller families, single-person households, and dual-income families. As a result, the consumer is increasingly demanding food products that are convenient, good tasting, safe, wholesome, and nutritious. This has also created opportunities for innovative food packaging to satisfy target consumers of diverse demographics. For instance, packaging has played an important role in the development of convenient food products such as on-the-go snacks, microwavable foods, and refrigerated meals (also known as meal solutions).

- *Value.* This need, defined as benefits/cost ratio, is driven by the consumer. Higher benefits may be achieved by enhancing the functions of packaging to satisfy the unmet needs of the consumer. Lower cost may be achieved by using less expensive packaging materials, using high-speed machines to increase productivity, and using more compact package designs to reduce distribution cost. For instance, material cost may be reduced by replacing a thicker monolayer material with a thinner multilayer material, production speed may be increased by replacing double seamed containers with heat sealed containers, and distribution cost may be reduced by replacing heavy glass containers with plastic containers. In the efforts of reducing cost, it is important that product quality and safety are not significantly compromised.

- *Profits.* This need is driven by food companies to maintain or grow their businesses. To earn profits, food companies frequently rely on packaging innovations to meet the ever-changing market needs. Profits also spark intense competition in the packaging industry to fill the market needs. For example, beverages may be packaged in different forms, including glass bottles, plastic bottles, aluminium cans, and stand-up pouches—and this provides a battleground for packaging materials suppliers to capture the market through optimization or innovation.

- *Food safety and biosecurity.* This is the need that drives innovations in protective food packaging. Each year in the United States alone, foodborne diseases cause approximately 76 million illnesses, 325000 hospitalizations, and 5000 deaths [3]. Microbial contamination is a major cause of foodborne illness that can occur during harvesting, processing, distributing, handling, store display, and food preparation. After the tragic event of September 11, food bioterrorism (e.g., deliberate contamination of commercial food products) has become a serious public threat. Packaging can effectively protect against microbial contamination and product tampering. Innovative food packaging such as antimicrobial packaging, advanced package integrity inspection systems, tamper-evident packaging, and biochemical sensors are increasingly sought to provide enhanced food safety and biosecurity.

- *Food packaging regulations.* This is the need that drives research and development relating to food packaging safety issues such as the migration of unwanted compounds from package to food (especially during situations such as microwaving) and the use of recycled materials in food packages. The purpose of food packaging regulations is to protect the consumer against unacceptable levels of food contamination by packaging components. This has also led to a continued interest in studying the migration of packaging components under various conditions and in developing sophisticated analytical methods to detect volatile compounds at lower concentrations.

- *Environmental concerns.* This is the need that drives innovations in environmentally friendly packaging. As landfill sites are diminishing, packaging waste disposal has become a major public concern in many developed countries. There is a growing pressure for the society to favor food packages that use less material and are easy to reuse, recycle, or incinerate. There is also a continued quest for biobased and biodegradable packaging materials that have good mechanical and barrier properties.

1.4 Packaging Functions

A food package must serve one or more functions to justify its existence. Traditionally, food packaging has four basic functions: protection, convenience, communication, and containment. A package design may be evaluated based on how well the package performs the required functions in a cost effective manner.

- *Protection.* Protecting the food from physical damage, physiochemical deterioration, microbial spoilage, and product tampering is probably the most important function of packaging. Without proper protection, the food may become unappetizing, less nutritious, and unsafe to consume.

The required packaging protection depends on the stability and fragility of the food, the desired shelf life of the food package, and the distribution environment. Good package integrity is also required to protect against loss of hermetic condition and microbial penetration.

Generally, the protection function of packaging is limited to foods whose shelf lives are controlled by environmental factors relating to physical damages, humidity, oxygen, light, and to some extent, temperature. Packaging is usually not effective for protecting foods whose shelf lives are controlled by internal factors. For example, consider a tuna sandwich in which moisture may migrate internally, from the tuna to the bread, causing the bread to become soggy and unacceptable to the consumer. The internal moisture migration is driven by the difference in water activities between the tuna and the bread, not by the relative humidity outside the sandwich. Thus enclosing the sandwich with a package will not solve the problem; perhaps placing an edible coating between the tuna and the bread may provide an acceptable solution.

- *Convenience.* This is an important function to satisfy the busy consumer lifestyle. Examples of convenient food packages are ready-to-eat meals, heat-and-eat meals, and self-heating packages. Examples of convenient features are easy opening, resealability, and microwavability. Innovations are constantly sought to provide more convenience without sacrificing quality or increasing cost.

- *Communication.* The function of communicating is important to create brand identity and influence consumer buying decisions. The package communicates with the consumer through written texts, brand logo, and graphics. In many countries, nutritional facts such as calories, fat, cholesterol, and carbohydrate are required on all food packages.

  The communication function is also important for facilitating distribution and retail checkouts. The barcode has virtually become an integral part of every commercial food package. Besides the barcode, there are other package devices, such as time-temperature indicator and radio frequency identification (RFID), which enable the package to communicate more effectively for the purpose of ensuring food quality and safety. As discussed in Chapter 15, intelligent packaging is a new technology that pushes the communication function to a higher level.

- *Containment.* Containing the food product is the most basic function of packaging. The requirement for containment depends on the size, weight, form, and shape of the enclosed food; for example, a solid food has different requirements from a liquid food. The containment function is also closely related to the rigidity of the package.

1.5 Packaging Environments

The package typically functions under three environments:

- *Physical environment.* This environment is concerned with the physical conditions that a package may encounter during its life cycle. For example, the package may have to withstand the harsh temperature/pressure conditions during retorting, as well as shock and vibration, falls and bumps, crushing from stacking, and attack from insects and rodents during storage and distribution. To protect against this environment, the package must have adequate mechanical strength and thermal stability, as well as other properties depending on the situation.

- *Ambient environment.* This environment is concerned with the oxygen, moisture, odors, molds, bacteria, light, and heat which are ubiquitous during storage and distribution. To protect against the negative impact of the ambient environment, the package must have an adequate level of barrier properties depending on the requirements of the food.

- *Human environment.* This complex environment is concerned with the human aspects of packaging, such as the user-friendliness, liking/disliking, and safety of a package to the consumer. Many socioeconomic issues such as the impact of packaging on the environment, packaging related legislations and regulations, and packaging related litigations are largely influenced by humans. Human perception, vision, dexterity, and language are also included in this environment.

Understanding these three environments and the nature of the food product is important for specifying the packaging requirements.

1.6 Food Packaging Systems

Food packaging is a system that involves certain physical components and operations. The major physical components are the food, the package, and the environment; the major operations are the manufacture, the distribution, and the disposal of the food package. In designing a food packaging system, these physical components and operations must be integrated efficiently to prevent overpackaging or underpackaging.

1.6.1 Levels of Packaging

A food packaging system may involve up to four levels of packaging:

- The primary package is usually a single unit purchased by the consumer at the retail store. Examples are a bottle of milk, a box of chocolates, a bag of potato chips, a can of ham, etc. Since the primary package is in direct contact with the food, it is in the best position to protect and promote the food product.

- The secondary package is usually a corrugated fiberboard box that contains a number of primary packages. The simplest type of secondary package is probably a plastic ring that holds several cans or bottles. The major role of the secondary package is to facilitate handling of multiple primary packages in the retail store.

■   The tertiary package is one that holds a number of secondary packages. An example is a stretched-wrapped pallet of corrugated fiberboard boxes. Its major role is to facilitate handling in the warehouse and retail store.

■   The quaternary package is one that holds a number of tertiary packages. An example is a large metal container holding several pallets to be placed inside a truck, train, or ship. Its major role is to facilitate long distance distribution.

These different levels should be optimized to provide the most efficient packaging solution. For example, the protection of a food product may be allocated to the primary packaging and the secondary packaging in a judicious manner so that the protection requirement is met while the cost is minimized.

1.6.2 Package Forms

Packaging may be classified into three forms: flexible packaging, semirigid packaging, and rigid packaging. These forms of packaging provide a wide selection of choices for various foods. A flexible package, such as a pouch or a bag, is one whose shape or contour is significantly affected when filled and sealed with the enclosed product. Due to the pliancy of the package wall, the internal pressure of a flexible package is approximately the same as the external pressure, and the package may be inflated or deflated according to the external conditions. A semirigid package, such as a plastic tray or container, is one whose shape or contour is not significantly affected when filled and sealed with the enclosed product, but it can be deformed with finger pressure (about 10 psi or 60 kPa). A rigid package, such as a metal can or a glass jar, is one that does not change shape or break unless excessive external force is applied.

All three forms of packaging have both advantages and disadvantages. For instance, a flexible pouch can easily accommodate foods of irregular shapes, but its closure integrity may not be as good as that of a rigid container.

1.6.3 Primary Food Packaging System

This book deals mostly with the primary package, since this level of packaging has the greatest influence on food quality and safety. Figure 1.2 shows that the primary food packaging system has four components: the food, the internal environment, the package, and the external environment. A good understanding of the interactions between these components is necessary for designing the food package.

Interactions between the food and the internal environment are the most important factors, since the rate of food deterioration under the conditions of the internal environment frequently governs the shelf life of the food package. Typically the internal environment contains a headspace filled with air. To extend the shelf life of oxygen or moisture sensitive foods, the headspace may be flushed with nitrogen or carbon dioxide, by a technique known as gas flushing or modified atmosphere packaging, to remove oxygen and moisture in the headspace. Sometimes the headspace volume is minimized by shrink wrapping or vacuum packing the food with a flexible packaging material.

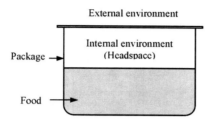

Figure 1.2 Primary food packaging

The package is an enclosure of the food, which protects the food against the negative impact of physical abuses and deteriorative atmospheric effects from the external environment. For example, the package often serves as a barrier to retard the ingress of moisture or oxygen from the external environment to the internal environment, thereby extending the shelf life of foods that are sensitive to moisture and oxygen.

Interactions between the food and the package may be desirable or undesirable. Examples of desirable food/package interactions include oxygen scavenging films to absorb residual oxygen in food, antimicrobial films to release antimicrobials for inhibiting microbial growth at food surface, and permeability films to allow optimal exchange of O_2 and CO_2 in modified atmosphere packaging of fresh produce. Examples of undesirable food/package interactions include migration of undesirable packaging components to the food and migration of flavor compounds from the food to the package (also known as flavor scalping).

1.7 Tables for Analyzing Food Packaging Systems

Based on the above discussions, the following tables may be constructed to provide an overview of the relationships between packaging functions and socioeconomic needs, technology, and packaging environments. These tables are particularly useful during the early stage of the package development process for facilitating brainstorming and identifying key areas for further consideration.

1.7.1 Functions/Socioeconomics Table

In this table, packaging functions and socioeconomic needs are arranged in columns and rows. For a particular situation, the socioeconomic needs are identified and prioritized, not necessary the same as shown in Table 1.1. Each cell in the table is then considered to determine its relevance and importance. Using this table, a product development team may conclude that, for example, food safety and saving time are most important to the consumer and to the success of a new product. The team will then focus on enhancing the protection and convenience functions of the package to satisfy these needs.

Table 1.1 Functions/socioeconomics table

		PACKAGING FUNCTIONS			
		Containment	Protection	Convenience	Communication
SOCIOECONOMICS	Consumer lifestyle				
	Values				
	Profits				
	Safety/biosecurity				
	Regulations				
	Environmental				

1.7.2 Functions/Technologies Table

In this table, packaging functions and technologies are arranged in columns and rows. This table is useful for evaluating the strengths and weaknesses of new technologies, based on their impact on the packaging functions. New technologies may also be introduced to improve operational efficiency or to reduce cost, preferably without compromising the existing packaging functions.

Table 1.2 Functions/technologies table

		PACKAGING FUNCTIONS			
		Containment	Protection	Convenience	Communication
TECHNOLOGIES	Active packaging				
	Bio-based materials				
	Barrier materials				
	Intelligent packaging				
	Advanced machinery				

Table 1.2 lists some technologies that have received much attention in recent years. For example, active packaging (such as oxygen scavenger and antimicrobial film) is for enhancing the protection function, and intelligent

packaging (such as time-temperature indicator and RFID tags) is for enhancing the communication function of the package.

While new technologies are aimed at achieving certain benefits, they sometimes also create unintended problems. For example, replacing glass with plastics may provide the benefits of versatility and cost effectiveness, but it also creates concerns of undesired compounds from the plastics migrating to the food. Replacing synthetic plastics with biobased materials can help minimize packaging waste to the environment, but most biobased materials have relatively poor mechanical properties.

1.7.3 Functions/Environments Table

In this table, packaging functions and environments are arranged in columns and rows. This table is useful for identifying certain packaging features, operations, devices, or considerations that are important for the package to function under certain environments.

Table 1.3 Functions/environments table

		PACKAGING FUNCTIONS			
		Containment	Protection	Convenience	Communication
ENVIRONMENTS	Physical				
	Ambient				
	Human				

For many food packaging systems, it is possible to associate items to most of the cells in Table 1.3. Although the selection and interpretation of the cell items involve somewhat subjective choices, the functions/environments table helps the product development team to consider all aspects of the food packaging system in a systematic manner. Shown below are examples of items that may be associated with the cells. With some imagination, other helpful items may also be associated.

- *Containment/physical.* The size and type of package determine how much and what kind of food can be contained and how much physical abuses the package can withstand.

- *Containment/ambient.* The size and type of package determine how much and what kind of food can be contained and the ability of the package to protect against oxygen, moisture, or light from the ambient environment.

- *Containment/human.* The size and type of package determine how much and what kind of food can be contained and ease of use of the package to humans.

- *Protection/physical.* Mechanical strength and seal strength are important for protecting the product from physical abuses.

- *Protection/ambient.* Barrier properties and seal integrity are important for protecting the food product from the adverse effects of oxygen, moisture, microbes, and light in the ambient environment.

- *Protection/human.* Tamper evident packaging is useful to protect the consumer from product tampering. Regulations are necessary to protect from unsafe packages harmful to the consumer.

- *Convenience/physical.* Efficiently bundled packages offer convenience to the manufacturer during storage and distribution.

- *Convenience/ambient.* Aseptic packages offer the convenience of storing food products at ambient environment instead of refrigerated conditions.

- *Convenience/human.* Microwavable packages and easy-to-open packages offer convenience to the consumer.

- *Communication/physical.* Barcodes can facilitate communication of information through the physical distribution of the product in the supply chain.

- *Communication/ambient.* Time-temperature indicators and biosensors may be attached to the package to communicate the conditions of the ambient environment.

- *Communication/human.* Nutrition labels, instructions, and graphics on the package can provide useful information to the consumer.

To take a step further, relative values (from 0 to 100) may be assigned to the cells in Figure 1.3 [4]. A 3-dimensional chart may then be generated using the functions, environments, and relative values as x, y, z coordinates. The determination of these relative values is a process that frequently requires subjective judgment. A hypothetical example of this chart is shown in Figure 1.3. The chart provides a bird's eye view of the functions/environments relationships and is useful for prioritizing the cells and identifying areas of strength and weakness of the food packaging system.

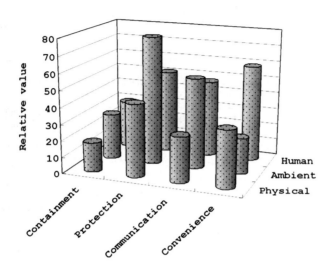

Figure 1.3 Three-dimensional plot of functions/environments

1.8 Food Package Development

The food packaging development process involves many considerations and activities. The socioeconomic pulling forces, different levels of packaging, different forms of packaging, and functions/environments matrix are all important considerations. Typical activities involved in food package development are shown in Table 1.4.

While designing a food packaging system is a complicated process, using the concepts and tables presented in this chapter will provide a good start for this process. In later chapters, more detailed information about the science and technology of food packaging will be presented.

Table 1.4 A typical food packaging development process

Step 1 Determine product/package requirements	Step 2 Select package materials and equipment	Step 3 Evaluate prototype packages	Step 4 Test packaging system in market
Food product • Formulation • Processing • Stability to light, O_2. and H_2O Marketing • Desired shelf life • Distribution chain • Cost constraints • Launch timing • Product positioning • Package size Production • Existing equipment • Plant capability • Cost per unit calculations	Work with suppliers of packaging materials and machines to identify options Determine cost and availability Compare stock versus custom package Work closely with product development to better understand ingredients and possible issues relating to packaging Regulation compliance	Conduct shelf life testing to determine quality retention at ambient and elevated temperatures Conduct distribution testing to determine robustness and integrity of package Conduct product/package interaction testing Evaluate feasibility of scale-up at final production facility	Produce final food packages Confirm all major requirements are met Retrieve packages from market for evaluation and confirmation Monitor consumer feedback during test market or national rollout Refine package/process design if necessary
Marketing group typically leads discussions based upon consumer data. while technical group provides input on feasibility.	Timing is critical to package design. Identify potential issues that may delay timeline.	Scale-up package at production facility is important prior to initial run to identify potential process issues.	Final package is tested as part of a concept fulfillment test to confirm consumer acceptance of product and package.

Discussion Questions and Problems

1. Estimate the sizes of the food packaging industry ($/year) in your country and the world. The exercise is aimed at providing a better understanding of the scope and impact of the food packaging industry. When appropriate, state the assumptions used in obtaining these estimates.

2. Discuss how the physical environment, ambient environment, and human environment individually affect the internal environment of the primary package in Figure 1.2.

3. Discuss how the food, package, and environment affect the shelf life of a packaged food.

4. Select a food package from the supermarket and analyze its strengths and weaknesses. Then write a report to include a description of the food product

and the package, a functions/environments table, and a discussion of the analysis. Photos or diagrams of the food package will be helpful.

5. Your marketing group would like to develop a refrigerated meal containing items such as chicken and pasta. Describe how you would develop a packaging system for this product.

Bibliography

1. KL Yam, PT Takhistov, J Miltz. Intelligent packaging: concepts and applications. J Food Sci 70:R1-10, 2005.
2. Anonymous. A proposal for packaging in 21st century: realization of packaging contributing to a sustainable society. Tokyo, Japan Packaging Institute, 2000.
3. PS Mead, L Slutsker, V Dietz, LF McCaig, JS Bresee, C Shapiro, PM Griffin, RV Tauxe. Food-related illness and death in the United States. Emerg Infect Dis 5(5):607-625, 1999.
4. HE Lockhart. A paradigm for packaging. Packag Technol Sci 10:237-252, 1997.

16

Part One

Packaging Material Science

Food packaging materials consist of a family of heterogeneous materials including plastic, glass, metal, and paper with a broad range of performance characteristics. A good understanding of the science governing the behavior of these materials is essential for material selection, specification, package design, cost control, and waste management.

Although foods and packages are made of distinctively different materials, the same chemical or physical properties are often used to describe both. For example, oxidation sensitivity may be used to describe food stability and package stability, thermal conductivity and heat capacity may be used to describe thermal processing of food and heat sealing of package, and compression strength may be used to describe food texture and package deformability. Moreover, many operations involved in food production and package production are governed by the same scientific principles, and thus knowledge from food production may sometimes be applied to packaging production and vice versa.

The first part of this book is composed of eight chapters devoted to reviewing the basic concepts of materials science, with the aim of demonstrating their applications to food packaging. Chapters 2-3 discuss the chemical and physical properties most relevant to the functions of packaging materials. Chapters 4-5 discuss gas permeation and migration, which are important transport phenomena for packaging materials. Chapters 6-9 provide detailed information about the four basic types of packaging materials: plastics, glass, metal, and cellulosic materials.

Chapter 2

Chemical Structures and Properties of Packaging Materials

2.1 Introduction 19
2.2 Chemical Constituents 20
2.3 Chemical Bonding 21
    2.3.1 Ionic Bonding 21
    2.3.2 Metallic Bonds 23
    2.3.3 Covalent Bond Model 25
2.4 Intermolecular Forces 27
    2.4.1 Ion-Dipole Forces 28
    2.4.2 Dipole-Dipole Forces 29
    2.4.3 Hydrogen Bonding 29
    2.4.4 Dispersion Forces 30
    2.4.5 Consequences of Intermolecular Interaction: Cohesion, Adhesion, and Surface Tension 30
2.5 Spatial Arrangements 32
    2.5.1 Tacticity 32
    2.5.2 Crystalline vs. Amorphous State 34
2.6 Chemical Reactivity and Susceptibility of Packaging 35
    2.6.1 Oxidation 36
    2.6.2 Biodegradation and Biodeterioration 38
    2.6.3 Chemical Resistance, Etching, and Weathering 40
Discussion Questions and Problems 41
Bibliography 41

2.1 Introduction

The theme of this chapter is that the properties of materials are closely related to their chemical structures, and thus understanding the structure/property relationships is important to the study of food packaging. More specifically, the properties of packaging materials depend on the atomic and molecular structures at four levels: (1) *chemical constituents*, what atoms are composed in the material, (2) *chemical bonding*, what forces hold the atoms together to form a molecule, (3) *intermolecular forces*, what forces attract molecules together to form a material, and (4) *spatial arrangements*, how molecules are arranged in three dimensional space. While each level has distinct characteristics, the four levels are closely related, with each level affecting the sequential levels in determining the observed material properties.

In a large sense, the first two levels of structures—chemical constituents and chemical bonding—are responsible for the chemical properties of the materials. Chemical properties determine the sensitivity of a material to chemical changes, which involve breaking chemical bonds to transform part or all of the original material into new substances. Examples of chemical changes important to food packaging include oxidation, corrosion of metal containers, and incineration of packaging wastes.

On the other hand, the last two levels of structures—intermolecular forces and spatial arrangement—are responsible for the physical properties of materials. Physical properties determine the physical behavior of a material under conditions such as heat, pressure, and concentration gradient that do not involve breaking chemical bonds and forming new substances. Examples of physical properties important to food packaging include gas permeation through package walls, migration of volatile compounds from package to food, and shock and vibration during distribution, which will be discussed in Chapters 4, 5, and 3, respectively.

2.2 Chemical Constituents

The first level of structure considers the chemical constituents, the atoms composed in the packaging material. These atoms determine what type of chemical bonds can be formed and what chemical properties can be expected.

In general, food packages are constructed of four basic types of materials (plastics, paper, glass, and metals) with chemical constituents shown in Table 2.1. It is important to note that plastics and paper are carbon based organic materials while glass and metal are inorganic materials, since this categorization has profound implications on their properties, sources of raw material, manufacturing technologies, and economic value. Like most organic materials, plastics and paper are inherently lighter, weaker, and more susceptible to chemical reactions than inorganic materials such as glass and metals. Plastics or paper are also more likely to interact with food, since they are all organic materials.

Table 2.1 Chemical constituents, chemical bonds, and intermolecular forces in packaging materials

Material	Commonly found atoms	Chemical bond	Intermolecular forces
Plastics	Major: C, H Minor: O, N, Cl, etc.	C-C backbone, ionic	van der Waals forces
Paper	C, H, O	Covalent	van der Waals forces
Glass	Major: Si, O Minor: Na, K, Ca, Al, etc.	Covalent, ionic	Ionic solid
Metals	Major: Fe, Al Minor: Cr, Sn, etc.	Metallic	Coulombic forces

While most packages are made of single materials, some packages are made of two or more different materials to provide the necessary functionalities for certain applications. For example, aseptic cartons for milk and juices are made of a laminate consisting of 70% paper, 6% aluminum (Al), and 24% polyethylene (PE). The typical structure of the laminate is PE/paper/PE/Al/PE, in which the paper layer provides strength and stiffness, the thin aluminum foil layer provides gas and light barrier; the polyethylene layers provide inner and outer protection, heat sealing, and binding between paper and aluminum.

2.3 Chemical Bonding

The second level of structure is chemical bonding which determines how atoms are bound together to form a molecule. A chemical bond is formed when the outermost electrons from two atoms interact with each other to achieve a lower and more stable energy state. These interactions sometimes involve the transferring of electrons to form ionic bonds, or sometimes involve the sharing of electrons to form covalent bonds or metallic bonds, depending on the electronegativities of the atoms (i.e., their abilities to attract electrons). Ionic, covalent, and metallic bonds are known as primary bonds, which are characterized by strong bond strengths as shown in Table 2.2. General classification of the solid packaging materials based on primary bonds can relate their properties to the chemical bonding and intermolecular interactions (Table 2.3).

Table 2.2 Typical bond strengths of chemical bonding and intermolecular forces

	Interaction type	Bond energy ($kJ \ mol^{-1}$)
Chemical bonding	Covalent bonds	200 - 800
	Metallic bonds	100 - 400
	Ionic bonds	40 - 500
Intermolecular forces	Hydrogen bonds	4 – 40
	Dipole-dipole forces	0.15 –15
	Ion-dipole	5 – 60
	Dispersion forces	0.4 - 4

2.3.1 Ionic Bonding

An ionic bond is formed when two atoms come together to transfer one or more electrons between each other. For example, a sodium atom will transfer an electron to a chloride atom to yield a Na^+ ion and a Cl^- ion:

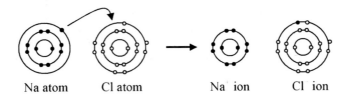

Na atom   Cl atom   Na⁺ ion   Cl⁻ ion

The electrostatic attraction binding the oppositely charged ions together is known as the ionic bond.

Table 2.3 The main properties of the different materials related to their chemical bonding

Property	Ionic bond materials	Covalent bond materials	Metallic bond materials
Melting and boiling points	High melting and boiling points because of strong attractions between atoms (ions)	Relatively low thermal transition temperatures due to weak attractive forces between molecules	High melting and boiling points due to attraction of metal ions with the delocalized electrons
Electrical and thermal conductivity	No electric conductivity and poor thermal conductivity due to absence of mobile charged particles	The electron sharing makes the electrons tightly bound to atoms, thus no charge available for conductivity	High conductivity because of easy transmission of vibration energy through the 'electron cloud'; metallic luster coming from free electrons
Mechanical properties (hardness and brittleness)	Generally hard because the ions are strongly bound to the lattice; also often brittle because of strong repulsion between equal charges at close proximity at the distortion	Generally soft because the intermolecular bonds are weak and easily displaced	Dense, hard, and strong with tightly packed atoms in the lattice; malleable and ductile as the distortion is possible without disrupting the inter-atomic bond

Formation of an ionic bond occurs between two atoms of greatly different electronegativities. Typically, one atom is a metal (such as Li, Na, Al, Fe) because of its tendency to donate electrons, and the other atom is a nonmetal (such as O, F, S, Cl) because of its tendency to accept electrons. Ionic materials are strong, hard, brittle, resistant to aggressive chemicals, and have a high melting point (Table 2.3).

Compared to covalent and metallic bonds, ionic bonds are less commonly found in packaging materials. Nevertheless, ionic compounds are found in protective layers for packaging applications, such as alumina (Al_2O_3) on foil and tin oxide (SnO_x) on tin plate. Ionic bonds are also found in ionomers, which are copolymers containing both ionic and nonionic repeat units. Shown below is the chemical structure of Surlyn®, an ionomer polymer containing a small portion of methacrylic acid partially substituted with Na in its carboxyl group:

$$
\left[\begin{array}{cc} H & H \\ | & | \\ -C - C \\ | & | \\ H & H \end{array}\right]_x
\left[\begin{array}{cc} H & CH_3 \\ | & | \\ -C - C \\ | & | \\ H & \\ & C=O \\ & | \\ & OH \end{array}\right]_y
\left[\begin{array}{cc} H & CH_3 \\ | & | \\ -C - C \\ | & | \\ H & \\ & C=O \\ & | \\ & O^-Na^+ \end{array}\right]_z
$$

The ionic groups in an ionomer may interact each other to form a cluster with ionic bonds as shown in Figure 2.1. However, this type of bonding may be seen on the borderline between intramolecular bonding and intermolecular interaction. Ionomer is used as sealant between a plastic layer and a foil, in packaging applications where excellent heat sealability and hot tack are required.

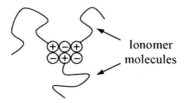
Ionomer molecules

Figure 2.1 Ionic interactions in ionomer molecules

2.3.2 Metallic Bonds

Metallic bonds are found in aluminum foil, metal containers, and other metal containing packaging materials. Atoms in metal typically have low electronegativities and are willing to share electrons freely, not only with neighbor atoms, but also with all other atoms in the lattice. For example, Figure 2.2 shows the formation of metallic bonding, when aluminum atoms lose their outer electrons, become positively charged Al^{+3} ions, and those electrons in turn form a 'sea of electrons' to surround the regularly spaced Al^{+3} ions. It is important to note that the electrons in metallic bonds are 'delocalized' or free, meaning that they are nondirectional and free to move in all directions in the lattice, unlike the electrons in covalent bonds which are bound to specific atoms and locations.

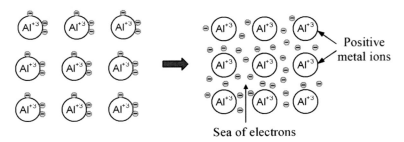

Positive metal ions

Sea of electrons

Figure 2.2 Formation of metallic bonding

Metallic bonds are largely responsible for the strength, malleability, and ductility of metals. Malleability is the ability to be hammered into various shapes, ductility is the ability to be drawn out into thin wires, while the shape of the material is changed without cracking or breaking in both cases. Therefore, a metallic material such as an aluminum sheet may be rolled into a very thin aluminum foil or an aluminum disk may be hammered into cups used for the popular soda cans.

To explain why metallic bonded materials are malleable and ductile, Figure 2.3 shows several layers of metal ions held closely together:

Layers of ions before sliding   Layers of ions after sliding

Figure 2.3 Sliding between layers of metal ions

Although strong, metallic bonds are not bound to specific metal ions. Hence, forces from hammering or stretching can cause layers of metal ions to slide over each another. As the metal ions move, they also attract the delocalized electrons to their new locations, while the sea of electrons is not broken and still holds the metal ions together.

Metallic bonds are also responsible for high electrical and thermal conductivities of metals (Table 2.3). The high electrical conductivity may be explained by the ability of the delocalized electrons to flow freely in the lattice of positive metal ions when a voltage is applied. The high thermal conductivity may be explained by the ability of metal ions to absorb heat energy by vibrating faster; since the metal ions are closely packed in the lattice, the faster vibrating ions can collide with their neighboring ions to transfer the heat energy.

2.3.3 Covalent Bond Model

Covalent bonds are the most abundant chemical bond found in plastics, paper, and glass. A covalent bond is formed when two atoms come together to share their outermost electrons. Typically these atoms have high and similar electronegativities, meaning that they have strong affinity to electrons and are willing to share them with their partner atoms. These atoms are from nonmetal elements (such as C, O, N, Cl) located on the right side of the periodic table. As an example, shown below are two oxygen atoms sharing two pairs of outermost electron to form an oxygen molecule joined by a double covalent bond:

Two oxygen atoms   One oxygen molecule

Although small molecules such as O_2 and H_2O are important to food products, the molecules in packaging materials are much larger and more complex. For example, polyethylene, the most frequently used packaging polymer shown below, is a molecule consisting of thousands of C-C and C-H covalent bonds:

$$
\sim\!\!\left[\begin{array}{cc} H & H \\ | & | \\ C-C \\ | & | \\ H & H \end{array}\right]\!\!\left[\begin{array}{cc} H & H \\ | & | \\ C-C \\ | & | \\ H & H \end{array}\right]\!\!\left[\begin{array}{cc} H & H \\ | & | \\ C-C \\ | & | \\ H & H \end{array}\right]\!\!\left[\begin{array}{cc} H & H \\ | & | \\ C-C \\ | & | \\ H & H \end{array}\right]\!\!\sim
$$

For a polymer molecule, the degree of polymerization determines the number of covalent bonds and thus molecular weight, which affect the property and performance of the material. Significance of molecular weight and its distribution in plastic material is dealt with in Chapter 6. While C-C and C-H bonds are the most common, C=C, C=O, O-H, and other covalent bonds are also frequently found in plastics and paper. The most common covalent bond in glass is Si-O.

An important attribute of covalent bond is bond polarity, which is determined by the difference in electronegativities between the atoms joined by the bond. When the difference is zero or negligible, the electrons are shared equally between atoms and a nonpolar covalent bond such as C-C is formed. When the difference is small but significant, the electrons are no longer shared equally between atoms and a polar covalent bond such as C=O is formed. When the difference is large, the electrons are no longer shared but are transferred from one atom to another and an ionic bond is formed.

Bond polarity can greatly affect the polarity of the molecule and thus the properties of the material. If all its bonds are nonpolar, the molecule is also nonpolar. If some of its bonds are polar, the molecule could be polar or nonpolar, depending on the symmetry of the molecule. This concept is illustrated by considering the covalent bonds in two common packaging polymers, polyvinyl chloride (PVC) and polyethylene (PE):

$$\left[\begin{array}{cc} H & Cl \\ | & | \\ C & - & C \\ | & | \\ H & H \end{array}\right]_n \qquad \left[\begin{array}{cc} H & H \\ | & | \\ C & - & C \\ | & | \\ H & H \end{array}\right]_n$$

PVC (polar)   PE (nonpolar)

Since C-H and C-Cl are both polar bonds, PVC is a polar molecule. But PE is a nonpolar molecule, because its C-H bonds are arranged symmetrically in such a manner that the dipoles cancel out resulting in no net dipole. The more polar the molecule, the stronger are the intermolecular forces, and the stronger intermolecular forces accounts for the reason why PVC has higher mechanical and barrier properties than PE.

Covalent bonds may be broken when packaging materials such as plastics and paper are exposed to severe conditions such as aggressive chemicals, high temperature, and high shear. High temperature, for example, is encountered during heating of food packages in the microwave oven (especially when a heating device known as susceptor is used in the package) or in the conventional oven. High temperature and high shear are frequently encountered during the processing of polymers such as extruding polymer resins to form plastic films or containers. A major concern associated with the breakage of chemical bonds is possible formation of undesirable oligomer compounds which can migrate from package to food.

Covalent bonds are typically stronger than metallic and ionic bonds (Table 2.2). The amount of energy required to break a covalent bond between two linked atoms is known as bond dissociation energy. Some covalent bond energies relating to packaging materials are shown in Table 2.4.

Table 2.4 The strength of some selected covalent chemical bonds at 25°C

Bond	Bond energy (kJ mol^{-1})	Bond	Bond energy (kJ mol^{-1})
C- Cl	338	O – H	463
C – F	565	C – C	348
O – O	146	C = C	612
Si – O	466	C – N	305
C – H	412	C – O	360
N – H	388	C = O	743

It needs to be mentioned that a molecule may have more than one type of chemical bond. For example, ionomer is a packaging polymer which consists of mostly of covalent bonds and a small but significant number of ionic bonds (Na^- or Zn^+ ions) to its side chains. Its covalent-ionic nature provides ionomer with unique properties (heat seal strength, abrasion resistance, and melt elasticity) not found in polymers with only covalent bonds.

2.4 Intermolecular Forces

The third level of structure considers the intermolecular forces that attract molecules together. While atoms are bond together tightly by strong covalent bonds to form a molecule, a group of molecules are held together loosely by weak intermolecular forces. Typically covalent bonds are 10-100 times stronger than intermolecular forces. Intermolecular forces can be cohesive between like molecules in surface tension or adhesive between unlike molecules.

Plastics and paper are composed of polymer molecules, and their physical properties are greatly influenced by the intermolecular forces between these molecules. When a PE film is punctured or melted, what are broken are only the intermolecular forces holding the PE molecules, whereas the much stronger covalent bonds holding the C and H atoms remain unaffected. The weak intermolecular forces are largely responsible for the low strength and low melting point of these materials (Table 2.3).

A plastic or a paper material is composed of many molecules holding together by intermolecular forces, known also as secondary bonds, which are much weaker than primary bonds (Table 2.2).

On the other hand, glass and metal are not composed of a collection of molecules. A glass or metal material is composed of a single giant molecule, in which all the atoms are bound together by primary bonds to form a network. Glass is a network of Si-O bonds extending throughout the lattice, and metal is a network of metallic ions held in the sea of electrons. Hence, cutting a metal container involves breaking metallic bonds, and shattering a glass bottle involves breaking covalent bonds. Metal and glass are inherently stronger and stiffer than plastics and paper, since metallic bonds and covalent bonds are much stronger than intermolecular forces.

Intermolecular forces, often called van der Waals forces, correspond to a combination of different ways of attraction among molecules according to their dipole moment (dipole-dipole, dipole induced dipole, dispersion forces) as shown in Figure 2.4. Dispersion forces are the weakest intermolecular force (one hundredth-one thousandth the strength of a covalent bond), hydrogen bonds are the strongest intermolecular force (about one-tenth the strength of a covalent bond): dispersion forces < dipole-dipole interactions < hydrogen bonds. Interactions between ionic moieties in ionomer may act as a stronger bonding to form a molecular cluster (Figure 2.1). These intermolecular forces, which weakly bind the molecules in the solid state of several materials, are responsible for the low melting and low boiling point of the organic materials with covalent bonds (Table 2.3).

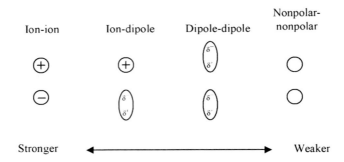

Figure 2.4 Types of intermolecular forces

2.4.1 Ion-Dipole Forces

An ion-dipole force is an attractive force that results from the electrostatic attraction between an ion and a neutral molecule that has a dipole. Attractive electrostatic forces exist between charged ion and a polar molecule. They are most commonly found in solutions. They are mainly responsible for the dissolution of ionic substances in polar solvents such aqueous salt solution.

2.4.2 Dipole-Dipole Forces

Dipole-dipole forces are intermolecular forces between polar molecules: there are attractive forces between the positive end of one polar molecule and the negative end of another polar molecule. They have a significant effect only when the molecules involved are close together (touching or almost touching). Dipole-dipole forces have strengths that range from 0.15 to 15 kJ mol^{-1}, being much weaker than ionic or covalent bonds (Table 2.2).

Consider the chemical structures of two polar molecules, polyvinyl chloride (PVC) and ethylene vinyl alcohol copolymer (EVOH):

PVC   EVOH

For PVC, each C-H bond is polar since carbon is more electronegative than hydrogen, and the C-Cl bond is polar since chlorine is more electronegative than either carbon or hydrogen. Each hydrogen atom will take on a partial positive charge and the chlorine atom will take on a partial negative charge resulting in a net dipole since the dipoles will not cancel out owing to the difference in electronegativities of carbon, hydrogen, and chlorine. For EVOH, vinyl alcohol part is polar since OH group is more electronegative than hydrogen. When these molecular chains are in close proximity, there will occur the dipole-dipole attractions.

2.4.3 Hydrogen Bonding

Hydrogen bonding is a special type of dipole-dipole interaction. It occurs when a hydrogen atom is covalently bonded to a small highly electronegative atom such as nitrogen or oxygen. The result is a dipolar molecule. The hydrogen atom has a partial positive charge and can interact with another highly electronegative atom in an adjacent molecule (again N, O, or F). This results in a stabilizing interaction that binds the two molecules together. Hydrogen bonds vary in their bond strength from about 4 to 40 kJ mol^{-1} being much weaker than typical covalent bonds (Table 2.2), but are stronger than dipole-dipole or dispersion forces.

Hydrogen bonding, responsible for the particular properties of water, is also very important for many organic solid materials, and it is a useful concept in the description of the behavior of many packaging materials like plastics and cellulosics.

2.4.4 Dispersion Forces

These intermolecular forces exist between nonpolar molecules. They are forces between temporary dipole molecules as a result of positive nuclei of one molecule attracting the electrons of another molecule. The dispersion force is the weakest intermolecular force. This force is sometimes called an induced dipole-induced dipole attraction. Dispersion forces play an important role in plastic and biopolymer structure. The term hydrophobic interaction is employed to describe interaction between nonpolar groups in macromolecules in aqueous or polar solution.

2.4.5 Consequences of Intermolecular Interaction: Cohesion, Adhesion, and Surface Tension

Intermolecular forces affect many applications of packaging materials in converting process and final usage. The attraction forces between same or similar molecules are called cohesion; those between different molecules are called adhesive forces. As discussed above, cohesive forces in plastic and paper materials explain their strengths under tensile, tearing, or puncturing stress. The cohesive force or strength depends on both the chemical nature and the physical state of a material. Adhesion plays important role in lamination, coating, and printing. Particularly adhesive in lamination and cold sealing is used to develop maximum adhesive force between the adherend and the adhesive: two surfaces are held together by interfacial adhesive force, while cohesive force plays a role of strengthening and supporting the solidified adhesive itself in the seal or attachment. Both cohesive and adhesive forces are important in understanding the behavior of packaging materials in coextrusion, lamination, printing, sealing, and other converting and processing operations. Cohesive and adhesive forces can vary greatly in strength depending on application: usually cohesive force is much higher than adhesive force while in adhesive seal layer the latter is higher than the former.

One direct measurement of intermolecular cohesive force in liquid is surface tension, which is defined as Gibbs free energy assigned to surface area. The surface tension results from cohesive forces between molecules inside the material which has an exposed surface. Whereas the molecules down in the liquid are shared with all the surrounding molecules in the cohesive attraction, those on the surface have no neighboring molecules above. As a consequence of these unbalanced forces, they exhibit stronger attractive forces toward their nearest molecules at the surface: expansion of this surface area requires energy to counteract the cohesive force, which is called surface tension. Surface tension of energy per surface area is typically measured in dyne cm^{-1} or $N\ m^{-1}$ for liquid phases where the work done by cohesive forces corresponds to new surface formed. But for solids, whose stronger intermolecular bonds do not permit such deformation, a more precise thermodynamic discussion is needed for measurement, which is beyond the scope of this book but is available in specific literature [1, 2].

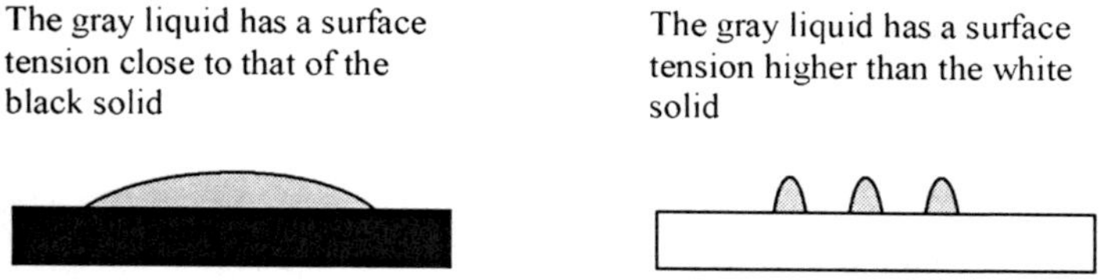

The gray liquid has a surface tension close to that of the black solid

The gray liquid has a surface tension higher than the white solid

Figure 2.5 Wetting of liquid on solid surface in relevant packaging operations

In case of two distinct phases, e.g., one solid and one liquid (e.g., in coating or printing the surface of a packaging material) or two solid surfaces in close contact with each other (e.g., during the sliding of one surface over another), the energy to increase the interface area is termed as interfacial tension. Interfacial tension is useful in explaining several common behaviors of liquid on solid surface, such as wettability, capillarity, and drop formation (Figure 2.5). When the intermolecular adhesive forces between liquid and the solid surface are stronger than the cohesive forces, wetting of the surface and upward meniscus in the capillary occur. If on the contrary the cohesive forces are stronger than the adhesive forces, the liquid beads-up and does not wet the solid surface.

The generated interfacial tension between two different phases is an important parameter for understanding different chemical and physical properties of the materials in packaging operations and functions, such as friction (see 3.4.2 in Chapter 3), adhesion, cleaning ability, fogging, and printing. A liquid phase can wet completely a solid surface (leading to a large surface covered by a liquid layer and not to several separated small drops) only if its surface tension is lower than that of the solid (Figure 2.5); the generally low surface tension values of plastic material in comparison with water solutions and also with several organic solvents, make it quite difficult to run important operations like printing, laminating, and coating, requiring the intimate contact between the plastics and liquids (Table 2.5). Therefore, numerous surface treatments (corona discharge, chemical etching, plasma treatment, mechanical abrasion) are applied to enhance the surface tension of plastic surfaces. The other way to promote wetting is to lower the surface tension of the liquid by surface active agents. The uses of antifogging and surfactant additives are other examples controlling the energies at surface of packaging materials in practical applications. It also needs to be mentioned that surface tension decreases with temperature: hot solvent molecules more readily wet, reach, and clean the solid surfaces than cold ones.

Table 2.5 Surface tension of some packaging and reference materials

Material	Surface tension at 25°C (dyne cm^{-1})
Polytetrafluoroethylene	20
Polyethylene	31
Polyethylene terephthalate	43
Cellophane	44
Polyamide (nylon 6.6)	46
Water	73
Glass (Pyrex)	170
Iron	1100

2.5 Spatial Arrangements

The fourth level is to consider spatial arrangements, how molecules are arranged in the material. In addition to chemical bonds, the spatial arrangement of atoms in the molecule and the molecular weight are also important factors determining the chemical and physical properties, particularly for polymeric materials including plastics and paper. The fine structure of material resulting from chemical composition, molecular arrangement, and crystalline degree is referred to as morphology.

2.5.1 Tacticity

The atomic arrangement in three dimensional molecular configuration is called stereochemistry or tacticity [3]. The way the molecular or ionic units are geometrically organized, distinguishes between crystalline and amorphous solids: the former, having symmetric, periodical, and well ordered structural organization, the latter being asymmetric, not periodical and unsettled (Figure 2.6). For example, methyl groups in polypropylene (PP) may be positioned in three different ways: isotactic PP has all the methyl groups on the same side of the chain, syndiotactic PP has the substitute group alternately above and below the chain plane, and atactic PP is with random sequence (Figure 2.7). The tacticity or steroregularity is important in determining the ability of the solid material to crystallize: regular atomic arrangement in the molecular structure makes the molecules easily pack together into rigid crystals. Therefore isotactic and syndiotactic PPs are crystalline whereas atactic is amorphous.

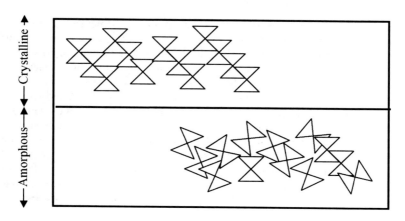

Figure 2.6 A schematic representation of the crystalline and amorphous
spatial configuration of solid constituents

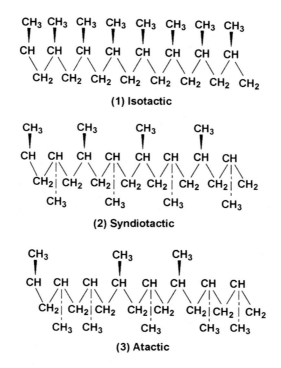

Figure 2.7 Different tacticities in polypropylene

2.5.2 Crystalline vs. Amorphous State

Crystalline structural organization of solids applies to structures composed of repeating identical units, orderly disposed with a fixed geometric relation between each other. The framework derived from such a symmetric organization of units may be more or less complete and homogeneous (with vacancy or impurities), but it always has the lowest energy, and therefore is the most favored one. Stereochemistry, kinetics of solidification from a liquid state, and presence of crystallization promoters are factors determining degree of crystallinity. Performances of the packaging materials are greatly affected by the crystalline or amorphous structure (Table 2.6). Glass is typically noncrystalline, amorphous material. Ceramic material may be in different degrees of crystallinity depending on the chemical composition and manufacturing conditions. Most plastic materials are composed of crystalline and amorphous regions. It is possible to modulate their degree of crystallinity by processing of the packaging materials.

Table 2.6 The crystalline and amorphous state for general categories of materials used for packaging

Material	Spatial organization
Glassy	Amorphous
Ceramic	Crystalline/amorphous
Cellulosic	Crystalline (pure cellulose)
Metallic	Crystalline
Plastic	Amorphous/crystalline

Important consequences of the crystalline state are high density, good mechanical properties, reduced free volume, and low gas transmission throughout. Some solids can crystallize in different forms (this behavior is called polymorphism) and each form has its definite set of chemical and physical properties; this, for instance, is an important point in understanding different resistance to corrosion or mechanical strength of different types of stainless steel or aluminum alloys. The potential of the crystalline structural organization, in establishing several and even unusual properties, is well exemplified by the fullurenes, discovered in the mid '80s, which show carbon atoms arranged in hollow spheres (that can be possibly filled with functional molecules) or in the very recent 'nanotubes' or 'nanogears' which appeared in the most advanced metallic nano-technologies [4].

Solid structure lacking orderly arrangement is of an amorphous state. Amorphous materials like glass do not have a definite melting point as crystalline materials do. The structural organization of amorphous materials can be explained well by that of liquids, which are deprived of any order or symmetric organization. Amorphous solids also behave like solidified liquids, having properties of flowing, ductility, and transparency. Amorphous solid is formed when the molecular chains have little orientation and regularity throughout the bulk structure.

Most plastic materials are semi-crystalline and the controlled combination of crystalline and amorphous structures can attain a material with advantageous properties of strength and stiffness. Generally increase in crystallinity causes increased density, increased tensile and compression strength, increased gas and moisture barrier, increased sealing temperature, decreased transparency, decreased tearing resistance, decreased impact strength, decreased toughness, and decreased elongation.

2.6 Chemical Reactivity and Susceptibility of Packaging

Food packaging materials have been used and adopted for suitable applications in consideration of their benefits and limitations in functions provided by their chemical nature and properties. The chemical properties refer to the measurable phenomena to which a modification of the chemical structure is related, and are involved very usefully and importantly in protective role of food packaging. The chemical behavior of food packaging materials needs to be considered in selecting their food packaging applications. It needs to be mentioned again that chemical and other properties are governed by what happens at the atomic and molecular levels. Understanding the underlying principle governing the material features helps to select the best choice desired among the different options of the packaging materials. For example, polyethylene consisting of linear chains with little side branches shows higher density and lower gas permeability than that of branched chains.

Examples of chemical properties important to food packaging include stability to oxidation, combustion, and degradation, particularly for plastics and paper; resistance to corrosion for aluminum, stainless steel, and tin plate; and resistance to aggressive chemicals. In Table 2.7 most of the important chemical properties for food packaging materials are listed. Some of the most important and pertinent chemical properties are measured or compared according to the official standards or well established methods which make possible an unequivocal description of important features for food packaging materials. However, many other behaviors/features of the materials are seldom defined clearly, are not associated with easily measurable properties, and thus are often evaluated empirically or by subjective comparisons.

Table 2.7 Chemical properties important to food packaging materials

Behavior	Main materials involved	When it might be important
Oxidation	Plastics, some cellulosic	Weathering, aging, storage, combustion
Burning/flame	Cellulosic, plastics	Waste treatment, material identification
Corrosion	Metals	Weathering, storage, food contact
Biodegradation	Cellulosic, plastics	Waste treatment, food contact
Biodeterioration	Cellulosic, plastics, metals	Waste treatment, food contact, storage
Chemical resistance	All	Food contact, handling
Etching/leaching	All	Food contact

2.6.1 Oxidation

Packaging materials are oxidized when exposed to atmospheric oxygen. Too fast oxidation rate of packaging material endangers its function of food packaging in many ways. The term oxidation can be defined in many different ways. In origin the word was derived from the observation that almost all the chemical elements are able to react with oxygen, forming compounds called oxides. The most rigorous way of explaining what oxidation is, however, refers to the loss of electrons from an atom or a molecule. The electrons lost by what is oxidized are not destroyed but collected by other chemicals, which are reduced. Therefore oxidation always happens in a reaction coupled with reduction. Some phenomenal description of oxidation-reduction reactions notable in packaging is given below.

- *Combustion or burning.* Oxidation reactions are usually exothermic and combustion is probably the most common example of oxidation. It is also the process which converts the potential energy of material into heat and light, and may be a way to destroy the organic packaging materials after their use. Their potential energy is released as heat during the combustion process (see 3.2.6 in Chapter 3). For simplicity, the combustion of polypropylene can be written as:

$$2 (C_3H_6)_n + 9n\ O_2 \rightarrow 6n\ H_2O + 6n\ CO_2 + Heat \qquad (2.1)$$

As polypropylene burns in air, its carbon atoms are oxidized to form carbon dioxide combining with oxygen, and oxygen molecules are reduced by the hydrogen atoms producing water and heat energy corresponding to the potential energy of that polymer. Sometimes observing behavior of the material with flame may help to identify the material. When materials (particularly the plastics) burn, the flame's color, the smoke produced, the burnt edge appearance, and the combustion products may be used to identify and recognize the type and nature of material (Table 2.8) [5].

Table 2.8 The burning behavior of some polymers

Polymer	Flame	Smoke
Polyethylene/polypropylene	Mostly yellow with blue inside	Weak smoke, paraffin odor
Polyvinyl chloride	Bright yellow	Soot type, HCl odor
Polyamide	Bluish with yellow edge	White smokes, horn odor
Polyester	Bright	Sweet, provoking
Polystyrene	Yellow, bright, lambent	Weak odor of plastic
Polymethyl methacrylate	Bright	Fruit odor
Cellophane	Pale orange, charred edges	Paper, wood odor

- *Bleaching.* Bleaching agents are used to chemically remove color from the packaging materials such as papers and plastics. Their action is somehow the same phenomenon of progressive color fading which happens when these products are exposed to sun and air. The chemical bleachers are largely used in manufacturing the cellulose packaging materials, where they can generate also some environmental problems. Most of commercial bleachers are oxidizing agents, such as sodium hypochlorite (NaOCl), other chlorine-based mixtures, and hydrogen peroxide (H_2O_2). Their decolorizing action is mainly due to the removal of the electrons activated by visible light. The hypochlorite ion, for instance, is reduced resulting in a basic solution as it accepts electrons from the colored material:

$$Na^+ + OCl^- + 2e^- + H_2O \text{-------->} Na^+ + Cl^- + 2\,OH^- \tag{2.2}$$

A great number of environmental restrictions set during the 1990s led to a progressive termination in use of most traditional chlorine based bleaching agents, and fully bleached pulp can nowadays be produced with modern bleaching process, either elemental chlorine free (ECF) or totally chlorine free (TCF) [6].

- *Corrosion.* This phenomenon is so important and so peculiar to the metallic materials that the topic will be more deeply approached further in Chapter 8 dedicated to metals.

Also corrosion may be defined as an oxidation-reduction process. All the metals used as food packaging material (mainly aluminum, stainless steel, tinplate) suffer from the risk of corrosion resulting from interactions both with the external environment (e.g., moist and salty air) and with the food contact (e.g., acidic fruit or vegetable juices). Great commitment of packaging scientists and technologists is devoted to reducing the possible damages of the corrosion phenomena, selecting the best alloy for food contact (those showing the highest resistance to corrosion), or characterizing the most useful coating for metal packaging in order to prevent the phenomenon. Organic coatings are the common choice for corrosion protection of metal packaging, mainly because they offer a good barrier to the passage of oxygen and water, which are responsible for corrosion [7].

The metal's corrosion is a serious problem mainly because the oxide formed does not firmly adhere to the surface of the metal and flakes off easily causing structural weakness, possible container perforation, and even potential food contamination. For instance if iron is exposed to air in presence of moisture, the oxidation reactions (corrosion) completed through several complex steps lead to the formation of rust ($Fe_2O_3.xH_2O$) which is incoherent and tends to flake off:

$$4Fe + 3O_2 + 2xH_2O \rightarrow 2Fe_2O_3\ xH_2O \qquad (2.3)$$

2.6.2 Biodegradation and Biodeterioration

Biodegradation, a topic of increasing interest, may be defined as the complete biochemical decomposition of organic molecules by microorganisms, or the biologically-catalyzed decrease in the complexity of chemicals. Frequently but not necessarily, it involves mineralization, i.e., the conversion of carbon, nitrogen, sulfur, and phosphorous content of the organic compounds to inorganic products.

Once, this kind of biodegradation behavior was considered unallowable and carefully avoided for food contact materials, for which high inertia and low reactivity appeared more desired. The growing problem of waste management, however, makes now biodegradation property desirable for food packaging material, and it becomes important to have a choice of biodegradable packaging and evaluate biodegradability objectively. The extent of biodegradation is usually measured in the production of inorganic CO_2 from the test material inoculated with a mixed population of microorganisms: percentage to the theoretically producible maximum CO_2 amount is used as degree of biodegradation (Figure 2.8) [8]. According to the European harmonized norm EN 13432, a material is considered compostable if at least 90 % of it biodegrades in a specified test (EN 14046 or ISO 14855) within 6 months.

From a theoretical point of view, all the naturally occurred materials (e.g., cellulose and derivatives) may undergo a biodegradation whereas the synthetic materials cannot. The synthetic materials show chemical inertia, which is called 'recalcitrance or recalcitrancy', due to the high average molecular weight (500 daltons might be considered a cut-off for living organisms), the presence of end groups incompatible with enzymatic attack, their hydrophobic nature, the presence of chemicals (additive or residuals) unfavorable for bacterial growth, and the steric hindrance from branched side chains. These characteristics mostly apply to plastic polymers, but may be modified by production technologies or additives (Figure 2.8).

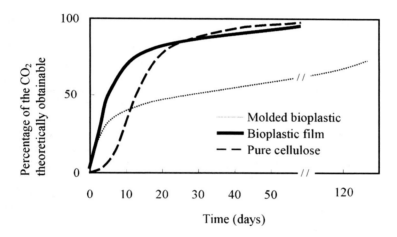

Figure 2.8 Aerobic biodegradation under controlled composting conditions (ISO 14855) of two bioplastics and pure cellulose

However, synthetic or inorganic materials may also be deteriorated or affected by microbial action. For example, corrosion of metal surface can be induced by microorganisms (see Section 8.5.6). Some degree of microbial damage or impairment of the material is termed biodeterioration, which applies mostly to relatively stable and durable materials.

Another phenomenon, which may be included in biodeterioration, is the formation of a biofilm. Microbial biofilms develop on all surfaces in contact with aqueous environments and are composed of immobilized cells embedded in an organic polymer matrix. Being absorptive and porous, they may contain solutes, heavy metals, and inorganic particles in addition to cellular constituents. The interfacial conditions of biofilm are rather diverse and different from that of the bulk medium in pH, oxygen, and other chemical species dissolved. The biofilms usually show a resistance to antimicrobial substances, enhancing the possibility of bacteria proliferation. In some circumstances, the presence of microorganisms within biofilms at material surfaces can cause deep corrosion (microbiologically triggered corrosion) and other degradations. The surface characteristics of a material determine the adhesion of microbial cells on it: roughness, electrostatic charges, and surface tension can influence the first adhesion of microorganisms and thus may also affect the hygienic conditions of glass, stainless steel, aluminum, and plastic surfaces as well as possible deterioration phenomena for paper, board, plastics, and even metal [9, 10].

2.6.3 Chemical Resistance, Etching, and Weathering

The generic term, chemical resistance refers to the capacity of a material to maintain its fundamental characteristics when exposed to some chemically aggressive substances. Etching and leaching are also general words often used to indicate the removal of matter from the surfaces of packaging materials by dissolution or scraping. Weathering is the chemical (also physical) decomposition of a material on exposure to atmospheric conditions. Such practical properties or performances are measured by controlled exposure of standardized sample material to standardized contact conditions, which are sometimes defined as abuse test conditions. The changes of weight (absorption), dimension (absorption, interaction), mechanical properties, and appearance (possible chemical changes) are measured objectively or evaluated subjectively after the time of exposure. Behaviors to oil, solvent, and some chemical substances in contact are the most interesting chemical resistance properties evaluated for packaging materials: a few examples of such tests are listed in Table 2.9.

Table 2.9 A selection of standards for chemical resistance

Document	Topic
ASTM D 543	Chemical resistance (effects on weight, dimension and appearance during time of exposure to a standard list of chemicals)
ASTM D 570	Water absorption (weight gain on water immersion)
ASTM D 1693	Stress cracking (polyethylene samples with flaws are exposed to oils or other chemicals, and number of failures is counted)
ASTMD 2651	Stress cracking
ISO 5634	Grease penetration (time to oil penetration under pressure)
TAPPI 454	Grease penetration (time to oil penetration under pressure)
ASTM F 119	Grease penetration (time to oil penetration under pressure)

Discussion Questions and Problems

1. When a plastic film is cut, what kind of bonding is broken?

2. If we can increase the crystallinity of a specific material, the phenomena of light transmission throughout are increased or decreased? Explain why.

3. Is the metal corrosion a chemical or physical phenomenon?

4. When a liquid easily wets a solid surface, does it mean that its surface tension is lower than that of the solid?

5. How does the degree of crystallinity affect the properties of the plastic packaging material? Explain by using examples of high density polyethylene and low density polyethylene which are different in crystallinity.

6. If a material is polar, does it mean that its atomic constituents have similar or different electronegativity?

7. During a can's fabrication, tinplate and aluminum are bent and deformed very deep without any breaking. How can this feature of metals be explained in terms of chemical structure?

8. What is the role of ionic bond or interaction in ionomer?

9. Please consult a reference book for the properties and use of isotactic, syndiotactic, and atactic PPs in packaging applications. Discuss them with reference to their tacticity.

10. Biodegradation is different from biodeterioration. Explain the two phenomena.

11. Is oxidation a phenomenon in which oxygen is always involved? Explain.

Bibliography

1. AW Adamson. Physical Chemistry of Surfaces. 5th ed. New York: John Wiley & Sons, 1990, pp. 260-293.
2. JN Israelachvili. Intermolecular and Surface Forces. 2nd ed. London: Academic Press, 1992, pp. 201-204.
3. RJ Hernandez, SEM Selke, JD Culter. Plastics Packaging. Munchi, Germany: Hanser Publishers, 2000, pp. 13-40.
4. PJF Harris. Carbon Nanotubes and Related Structures. Cambridge, UK: Cambridge University Press, 1999, pp. 16-60.
5. VV Tumanov, AA Berlin, NA Khalturinskii. Study of burning out of polymers. Polymer Science USSR 20(12):3121-3129, 1978.
6. F Lopez, MJ Daz, ME Eugenio, J Ariza, A Rodrguez, L Jimenez. Optimization of hydrogen peroxide in totally chlorine free bleaching of

cellulose pulp from olive tree residues. Bioresource Technol 87(3):255-261, 2003.

7. NS Sangaj, VC Malshe. Permeability of polymers in protective organic coatings. Prog Org Coat 50(1):28-39, 2004.

8. NS Battersby. The ISO headspace CO_2 biodegradation test. Chemosphere 34(8):1813-1822, 1997.

9. B Carpenter, O Cerf. Biofilms and their consequences, with particular reference to hygiene in the food industry. J Appl Bacteriol 75:499-511, 1993.

10. IC Blackman, JF Frank. Growth of *Listeria monocytogenes* as a biofilm on various food-processing surfaces. J Food Protect 59:827-831, 1996.

Chapter 3

Physical Properties of Packaging Materials

3.1   Introduction 43
3.2   Thermal Properties 44
      3.2.1 Thermal Conductivity (k) ··45
      3.2.2 Heat Capacity (C_p)··46
      3.2.3 Thermal Expansion (linear and volumetric) ·······································47
      3.2.4 Tolerable Thermal Range ···48
      3.2.5 Transition Temperatures (T_m, T_g)···48
      3.2.6 Heat of Combustion (Q_c) ··50
3.3   Electromagnetic Properties...51
      3.3.1 Refractive Index (n)···52
      3.3.2 Transparency (%T) ···53
      3.3.3 Transmittance/Absorption Spectra in UV, VIS, and IR ·····················55
      3.3.4 Haze···58
      3.3.5 Gloss ···59
      3.3.6 Behavior to Ionizing Radiations ···60
      3.3.7 Behavior to Microwaves···62
3.4   Mechanical Properties 63
      3.4.1 Density and Related Properties···63
      3.4.2 Coefficient of Friction ···64
      3.4.3 Strength Properties: Tensile, Tear, Burst, and Creep ·························66
      3.4.4 Response to Dynamic Stresses: Impact Resistance and Cushioning ···72
Discussion Questions and Problems 75
Bibliography 76

3.1 Introduction

Physical properties of a material describe its behavior under physical stress or treatment, and do not involve the modification or change of its chemical structure. Since physical changes are frequently encountered during manufacture and usage of food packages, a good knowledge on the physical properties of the packaging materials is useful for controlling the package fabrication process and designing suitable food package. Physical behavior of the materials under appropriate conditions also enables one to differentiate apparently similar materials and to identify a certain type of packaging material in question. Hence understanding the physical properties is a fundamental part of packaging materials science.

Physical properties of solids are usually well defined with scientific meaning and quantified precisely in measurable unit, in contrast to the chemical properties that often rely on empirical evaluation or indirect measurements. Diverse physical properties of the packaging materials are classified into thermal, electromagnetic, mechanical, and diffusional ones as done with foods and beverages [1]. Table 3.1 is a nonexhaustive list of physical properties in four classes; all of them contribute to the complete description of the materials for food packaging applications; many of them appear in the technical sheets and dossiers which accompany the shipments of packaging materials. This chapter describes the physical properties of packaging materials with reference to their use in food packaging, while diffusional properties relating to permeation and migration are deferred to Chapters 4 and 5, respectively.

Table 3.1 A nonexhaustive list of the physical properties of food packaging materials

Thermal	Electromagnetic[a]	Mechanical	Diffusional[b]
Describe the behavior of a material to heat	*Describe the behavior of a material to electromagnetic radiation*	*Describe the behavior of a material to mechanical forces*	*Describe the behavior of mass transport phenomena*
Thermal conductivity	Refractive index	Coefficient of friction	Solubility
Thermal capacity	Transparency/absorption/ reflectivity	Mechanical strength	Permeability
Specific heat	Gloss		Diffusion
Heat of formation	Haze	Density	
Heat of combustion	Behavior to microwave	Shock resistance	
Transition temperatures	Behavior to ionizing radiation	Elongation	
Sealing properties			
Thermal strain			

[a]Also called as 'electric' and 'optical' properties; [b]include sorption properties.

3.2 Thermal Properties

At times food packaging scientists or technologists need to know how the packaging material behaves when exposed to the conditions of losing or acquiring heat. Heat transport phenomena occur during the thermal treatment (pasteurization, sterilization of packed food) or long term storage with temperature variations as well as during the package fabrication (soldering, thermoforming, molding, etc.). In many cases, our interest focuses on the behavior of the material as heat exchanger, i.e., we are interested in evaluating how the heat flows across the material; sometimes the material property change under heat is also a matter of our concern. Generally speaking, the thermal properties of a material describe the heat transport behavior and the change in material state under application of thermal energy.

3.2.1 Thermal Conductivity (k)

Thermal conductivity is the tendency of a solid body to exchange heat by conduction which is the main mechanism of heat transfer through a packaging material. Thermal conductivity usually stands for the easiness of thermal flow between two points inside the solid, due to a temperature gradient; thus it means the rate of heat transfer through unit area of unit thickness for unit temperature difference as appearing in the Fourier's Law as:

$$\frac{dQ}{dt} = -kA \frac{dT}{dx} \tag{3.1}$$

where Q is heat, t is time, A is the surface area, T is temperature, x is the distance in thermal flow direction, and k is the thermal conductivity. Some common units of thermal conductivity are kcal m^{-1} h^{-1} $°C^{-1}$, J m^{-1} s^{-1} $°C^{-1}$, or W m^{-1} $°C^{-1}$. The negative sign in the equation denotes that heat flows from a point of high temperature to a point of low temperature.

Table 3.2 Typical thermal conductivity values of some materials used for packaging

Material	Thermal conductivity (W m^{-1} $°C^{-1}$)
Aluminum (25°C)	237.00
Stainless steel	54.00
Glass (30°C)	0.96
Low density polyethylene (LDPE, density = 920 kg m^{-3})	0.48
Expanded LDPE (density = 43 kg m^{-3})	0.053
Wood (oak)	0.14
Cork	0.04

The packaging materials of different thermal conductivity may affect the heating or cooling rate in heat processes such as pasteurization or sterilization of the packaged food. As evident from Table 3.2, metallic materials have high thermal conductivity due to the electron mobility in metallic bonds. Aluminum, usually used in a thin layer, shows very high rate of thermal energy transfer. The organic materials, such as plastics and wood products, have low thermal conductivity due to their atomic and intermolecular bonds. Different applications would dictate different structure requirements of the packaging materials for the desired thermal conductivity in their pertinent performance.

From the fact that still air (at about 20°C) has a thermal conductivity of 0.02 W m^{-1} °C^{-1}, it can be easily understood why corrugated board and expanded polystyrene are, at the same thickness, better insulating materials than solid board and homogeneous plastics, respectively. The former materials have a large amount of air in their structure, containing few free electrons available for energy transfer.

3.2.2 Heat Capacity (C$_p$)

When heat is transferred to or from an object, its temperature increases or decreases. The relationship between the transferred heat (Q) and the change in temperature (ΔT) has the relationship:

$$Q = C_p \, \Delta T \qquad\qquad (3.2)$$

The proportionality constant in Equation 3.2 is called the thermal or heat capacity (C$_p$). The heat capacity, therefore, measures the amount of heat necessary to produce a unit change of temperature in a body.

The most common units used for heat capacity are kcal °C^{-1} or J K^{-1}. If the capacity is referred to for unit mass (kcal kg^{-1} °C^{-1}) and not for a body, it is often called specific heat and some examples of pertinent values are shown in Table 3.3.

Table 3.3 Typical specific heat of some packaging materials

Material	Specific heat (kJ kg^{-1} °C^{-1})
Stainless steel	0.46
Glass	0.84
Aluminum	0.90
LDPE	2.30
Wood	2.72

Once again, it is possible to note that metallic and inorganic materials react to the thermal energy input more sensitively than organic materials: the former materials increase their temperature with a less amount of energy than the latter materials. In general, an amorphous solid has a larger heat capacity than a crystalline one, thus the heat capacity of semicrystalline materials (plastics, cellulosics) changes with the distribution between amorphous and crystalline phase fractions. A larger heat capacity is also a measure of better thermal insulation, which can be provided by the organic materials protecting sensitive foods under heat-shock conditions.

3.2.3 Thermal Expansion (linear and volumetric)

The coefficients of linear expansion (α) and volume expansion (β) express the relative changes in length or volume, respectively, for a temperature change at constant pressure according to the following forms:

$$L = L_o (1+\alpha\, \Delta T) \text{ or } \alpha = \frac{L - L_o}{L_o \Delta T} \tag{3.3}$$

$$V = V_o (1+\beta\, \Delta T) \text{ or } \beta = \frac{V - V_o}{V_o \Delta T} \tag{3.4}$$

where L and V are, respectively, length and volume attained with temperature change of ΔT while L_o and V_o are those before temperature change. The dimensions of both α and β are the inverse of temperature.

Even if these properties are seldom reported in the technical sheets of packaging materials, it is worthy to note that a change in packaging dimensions (due to temperature change) may have dramatic effects. The linear expansion or shrinking of belts assuring a load during a shipment, for instance, may cause serious economic consequences; the expansion of glass jars in a constrained space can lead to breakage; the repeated exposure of the flexible packages to heat and cold may also cause an expansive phenomenon. In general α is equal to $\beta/3$, and the coefficients depend on the temperature; in Table 3.4, the linear thermal expansion coefficients in the temperature range of 0-100°C are listed for some common packaging materials.

Table 3.4 Linear thermal expansion coefficients of some packaging materials

Material	Thermal expansion α (1/°C)
Polypropylene	110×10^{-6}
Rubber	77×10^{-6}
Aluminum	24×10^{-6}
Polyester	17×10^{-6}
Steel	13×10^{-6}
Glass	6×10^{-6}
Wood	4×10^{-6}
Ceramic	3×10^{-6}

3.2.4 Tolerable Thermal Range

The lowest limit of this range is the temperature at which the material shows the maximum fragility affordable during the commercial distribution; the highest temperature is one that begins to cause a physical distortion of the object (e.g., heat deflection temperature in ASTM D648-06), a lack of performance, or the beginning of a chemical change (browning, oxidation, etc.). This characteristic of performance (also named service temperature), more typical of the final package than of the raw material, may be largely influenced by the method of manufacturing: for instance the degree of crystallization or orientation of plastic objects (see Chapter 6), as well as the presence of some ingredients, like inorganic fillers for plastic, boron or aluminum for glasses, and some alloying for the metals. The limits of tolerable temperature range are given for some materials in Table 3.5.

Table 3.5 Lowest and highest temperatures tolerable for some packaging materials

Material	Lowest temperature (°C)	Highest temperature (°C)
Stainless steel	$\cong$ - 45	600
Glass (boron aluminate)	-*	490
Glass (sodium silicate)	-*	460
Aluminum (anodized)	$\cong$ - 40	350-400
Fluoropolymers	$\cong$ - 100	260
Polyester (crystallized PET)	$\cong$ - 60	220- 230
Cellophane	$\cong$ - 18	180
Polyamide (Nylon 6)	$\cong$ - 10	60-160
Polyester (amorphous PET)	$\cong$ - 60	65-70
High density polyethylene	$\cong$ - 90	60-85
Polyvinyl chloride	$\cong$ - 15	65-85

*Depends on thermal shock.

3.2.5 Transition Temperatures (T_m, T_g)

Generally speaking, these are the temperatures at which a change in the state of substance occurs under standard pressure: thus the well-known concepts of boiling and melting points apply to the definition of transition temperatures. For the packaging materials, the important transition temperatures are usually melting point, glass transition, and the crystallization temperature.

By adjusting the degree of crystallization depending on temperature, it is possible to change dramatically some important features of plastics (diffusional, mechanical, thermal properties), ceramic materials (strength, fragility), and metals (ductility).

The transition temperature corresponding to the phase change from a solid to a liquid state (the melting point, T_m) is important for whatever material, organic or inorganic, and moreover becomes crucial for those intended to take a final form in some steps of package preparation through solidification from a liquid or semi-solid state. Only the materials with a true crystalline organization have sharply defined melting temperatures (T_m), whereas amorphous, noncrystalline substances have no melting point, and soften with heating to reach progressively a fluid form (Table 3.6). This happens both to glasses (the typical amorphous ceramic) and to amorphous polymers, and the phenomenon is well described in a plot of 'stiffness versus temperature' (Figure 3.1).

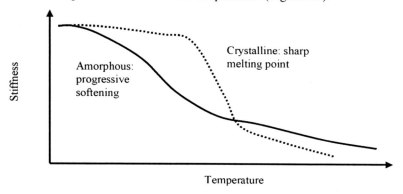

Figure 3.1 Variation of stiffness with increasing temperature for amorphous and crystalline materials

As temperature increases, the behavior of amorphous material differentiates greatly from that of the crystalline one: the former becomes easier to shape with reduction of the stiffness, while the crystalline material is still restricted in the extent of deformation until the melting temperature is reached. As a consequence, for amorphous ceramic materials (e.g., glass, china) and amorphous polymers (e.g., polystyrene, some polyesters), there are no clearly defined melting temperatures, but a range of melting temperatures; these materials are conveniently formed into shape by several different processing technologies.

Below the melting point, there is another transition temperature, the glass transition temperature (T_g), at which a substantial change in molecular chain mobility takes place (Table 3.6); below the T_g the molecules show the minimum of their mobility and a significant decrease of the free volume resulting in maximum fragility and also low diffusion properties of the packaging materials. Therefore around the T_g, the material exhibits an unusual thermal sensitivity. Even if the T_g is theoretically definable for whatever material, it has a special importance for polymers, both synthetic and natural macromolecules. Plastic polymers, the amorphous as well as the semi-crystalline, can be manufactured to attain glass transition temperatures above and below room temperature: those with T_g below the room temperature are rubbery, and those with T_g above the room temperature are glassy at ambient conditions. The behavior of the rubbery state is soft and flexible being similar to a rubber, whereas the glassy polymer shows a more rigid character at the temperature of use.

Table 3.6 Melting and glass transition temperatures for some packaging materials

Material	T_m (°C)	T_g (°C)
Glass	-	1450[a]
Stainless steel	1430[b]	-
Aluminum	660[b]	-
Polyester	267	69
Tin	232[b]	-
Polypropylene	176	-18
Polyvinyl chloride	212	87
Polyamide (Nylon 6.6)	265	50
Cellulose	-	40
Synthetic rubber (1-4 polyisoprene)	-	≅ -70
Polymethyl methacrylate	-	100[a]
Polystyrene atactic	-	94[a]
High density polyethylene	137	-125
Linear low density polyethylene	112	-110

[a]Completely amorphous materials; [b]no glass transition temperatures for metals.

3.2.6 Heat of Combustion (Q_c)

The heat of combustion represents the amount of thermal energy released from the total combustion of a material. It may give an idea of the environmental impact of the different packaging alternatives. Obviously the concept of combustion heat applies only to the organic materials (cellulosics and plastics) for which a thermal destruction has a sense of energy recovery. The heat of combustion is measured by burning the material in a bomb calorimeter pressurized with a large excess of oxygen. Table 3.7 offers a realistic and complete comparison of combustion heat values.

Table 3.7 Heat of combustion for some packaging materials and fuels

Material	Heat of combustion ($MJ\ kg^{-1}$)
Polyester	33
Polypropylene	46
Polyvinyl chloride	20
Cellulose	15-30
Wood	16-20
Fuel oil	43
MSW(municipal solid waste)	25

3.3 Electromagnetic Properties

The energy that propagates through the space or through a material with change in electric and magnetic fields is named electromagnetic radiation or simply radiation. As is well known, this kind of energy is characterized by wavelengths and/or frequency of the disturbance of coupled electric and magnetic fields. The array of known electromagnetic radiation, extends from the cosmic rays (the shortest in wavelengths), through γ and X rays, ultraviolet, visible, and infrared lights, microwaves, to radio waves of the longest wavelengths (Table 3.8). Most of radiation types are applied to package manufacturing and food packaging operations. From aspects of food packaging, it is of our concern to know how and how much these different kinds of energy are absorbed, reflected, or transmitted by the packaging materials.

The knowledge of the material's behavior to electromagnetic radiation, i.e., the evaluation of their electromagnetic properties, is often useful for many different purposes (Table 3.8). For instance, the infrared absorption spectrum permits the identification of synthetic polymers, the transparency, clarity, and color of the plastic films and glasses are an objective description of important aesthetic/optical properties, the gloss of a metallized surface may be used for homogeneity evaluation of a thin layer coating, and the X ray absorption is commonly used to measure the thickness of continuously extruded plastic films.

The optical/electromagnetic properties of packaging materials, most frequently reported in the technical literature, are the refraction index, the transparency, the gloss, the haze, and the absorption/transmission behavior to ultraviolet, visible, infrared, and microwave radiations. Some hints to how they are defined, measured, and used are described successively.

Table 3.8 Electromagnetic radiation as relevant use in packaging applications

Radiation	Frequency (Hz)	Wavelength (nm)	Interest
γ rays	10^{21}-10^{18}	10^{-4}-0.1	Material sterilization
X rays	10^{19}-10^{16}	0.01-10	Thickness measurement
Ultraviolet	10^{16}-10^{15}	10-400	Food protection performance
Visible	10^{15}	400-800	Color, transparency, etc.
Infrared	10^{15}-10^{12}	800-20000	Identification
Microwave	10^{11}-10^{10}	10^4-10^5	Heating, etc.

3.3.1 Refractive Index (n)

According to Law of Refraction or Snell's Law, when a monochromatic light passes from a medium m_1 to another m_2 of higher density, the angle of incidence ϕ_i and the angle of refraction ϕ_r (both measured with respect to the plane normal to the interface as shown in Figure 3.2) are different with the latter being lower, because the velocity of the light in m_1 (v_1) decreases in m_2 (v_2); their ratio (n) is called index of refraction or refractive index:

$$n = \frac{\sin \phi_i}{\sin \phi_r} = \frac{v_1}{v_2} \qquad (3.5)$$

The ratio of the speed of light in a vacuum to the speed of light in a given material is called the absolute index of refraction of the material. The absolute index of refraction in air is about 1.0003, i.e., the light travels slightly slower in air than in a vacuum. The higher the index of refraction for a given material, the slower the light travels through it. The ratio of the absolute indices of refraction for light traveling from any medium to another is called the relative index of refraction (Table 3.9). It is a function of wavelength, temperature, and pressure, and it is normally measured in standardized conditions. The knowledge of refractive index is occasionally useful in special circumstances (like pigment characterization or optical fiber production, for instance), but with regard to packaging materials, this property is generally used for an objective characterization of clearness of glasses and transparent plastics. Some ingredients (e.g., lead's ores) are specifically used in the glass fabrication to increase the index of refraction and thus to improve their clearness. The higher the index of refraction of a transparent material, in fact, the more the incidence light will deviate toward the normality to the interface (see Figure 3.2), thus the clearer will appear the vision. The plastics that better simulate glass for their clearness are those like polycarbonate or polyesters, which show the highest index of refraction.

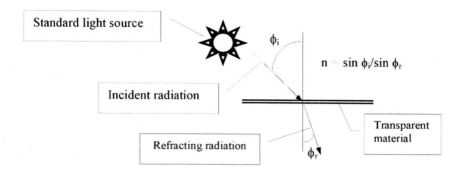

Figure 3.2 Schematic apparatus for determining relative refractive index

Table 3.9 Relative refractive index for some packaging materials and references

Material	Refractive index (20°C, 589.3 nm)
PbO	2.5–2.7
Diamond	2.42
Glass	1.52–1.89
Polyester	1.64
Polypropylene	1.50
Polycarbonate	1.59
Polymethyl methacrylate	1.49
Polyamide (nylon 6.6)	1.57
Water	1.33
Air	1.00

3.3.2 Transparency (%T)

This is typically an aesthetic characteristic of the packaging materials and one of the properties most wanted and appreciated by the consumers. Even if it is hedonistic and somehow subjective perception, the transparency can be accurately measured in standard conditions as the specular transmittance (i.e., the transmitted radiant flux measured includes only that in the same direction as that of the incident flux, with a deviation of an angle less than 0.1°) of an electromagnetic radiation between 540 and 560 nm, central in the visible light spectrum [2]. Actually, some confusion still exists about this optical property, which is sometimes presented as an empirical determination of 'see through clarity'. Nowadays it is easily measured objectively for the transparent materials in standard conditions (Figure 3.3). Table 3.10 gives some typical values.

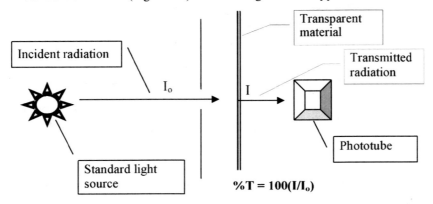

Figure 3.3 Schematic apparatus for transparency determination

Transparency is inversely correlated to the thickness of the material, following Beer's Law:

$$I = I_o \exp(-kcl) \tag{3.6}$$

where I is the intensity of the radiation, whose original intensity is I_o, after passing through a thickness l (cm) of a material with a molar absorptivity of k (cm^{-1} M^{-1}) and a molar concentration of c (M). The ratio I/I_o gives the transmission factor (T) and its percentage (%T) is termed as transmittance providing the basis for transparency definition. The transparency of flexible packaging materials is greatly affected by the degree of crystallinity, homogeneity, and amount of some ingredients (e.g., pigments and inert fillers), as well as the thickness of some coatings like varnishes or very thin metallic layers; therefore, it can be modified, beyond the concept of an optical/hedonistic property, for the protection of light-sensitive foods [3]. A widely used concept related to transparency is the absorbance of a material. The absorbance or optical density (OD) defined as $\log_{10}(I_o/I)$, is related to transmittance by the following:

$$OD = \log_{10}(1/T) \text{ or } OD = \log_{10}(e) \times kcl \tag{3.7}$$

Also note that the absorbance is normally directly proportional to the concentration of absorbing ingredient.

Table 3.10 Transparency of some packaging materials

Material	%T (specular transmittance at 550 nm)
Glass (180 µm)	92.1
Polypropylene biaxially oriented (28 µm)	92.0
Cellophane (20 µm)	92.0
Polyamide (nylon 6.6) biaxially oriented (18 µm)	89.2
Polyvinyl chloride (40 µm)	87.3
Polyamide (Nylon 6.6) cast (50 µm)	82.6
Polypropylene cast (40 µm)	82.3
Polyester (25 µm)	81.8
Polystyrene (38 µm)	43.5
Polytetrafluoroethylene (50 µm)	21.4
Pergamine (45 µm)	8.7
White polypropylene (36 µm)	3.1
Vegetable parchment (78 µm)	1.3

3.3.3 Transmittance/Absorption Spectra in UV, VIS, and IR

The most complete and accurate way to describe the behavior of a transparent packaging material in response to the light irradiation is certainly represented by the transmittance/absorption spectra in the ultraviolet (UV) and visible (VIS) regions of the electromagnetic spectrum. These spectra, recorded generally between 200 and 800 nm in wavelength, permit to quantify the transparency at any wavelength and consequently to know which proportion of the light energy is absorbed by the material. These electromagnetic radiations (UV/VIS) are absorbed mostly by atoms whose electrons are undergoing transitions to higher energy state; the information achieved by the spectra are therefore related to the atomic structure, to the color of the material examined, and finally to the kind and level of protection that a packaging material can give to light-sensitive foods. Electromagnetic properties in UV or visible light are measured by various kinds of spectrophotometer instruments.

UV radiations (wavelength of 200-400 nm) might be dangerous both for foods and for packaging materials. In many food products they activate the mechanisms of photo-oxidation of lipids and can damage other photo-sensitive substances, such as pigments and vitamins; in several plastics and cellulosic materials they can lead to changes in color, gloss, loss of flexibility, and enhancement of migration. Even low levels of UV radiations in natural and artificial lights (incandescence and fluorescence lamps) present a serious problem for various foods during shelf life that should never be underestimated. Recently several researches investigated the effects of light on food stability and the role of packaging in the protection against light [3-5]. Many additives are available today, both for plastics and glasses, to reduce the UV transmission acting in different ways: absorbing UV radiation, emitting less dangerous longer wavelength light by the quenching phenomenon (taking down a UV excited substance to a stable state without any unwanted side reactions), or scavenging the radicals formed by UV irradiation [6]. Figure 3.4 presents the UV spectra of some common packaging materials from which UV cut-off values can be recognized quite obviously.

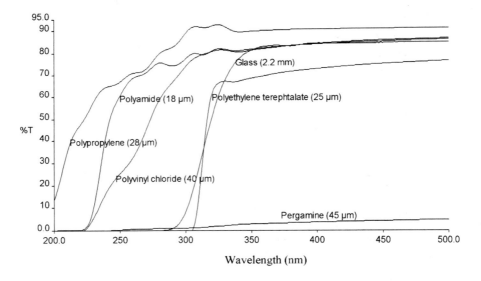

Figure 3.4 UV transmission spectra of some common packaging materials

Also visible light radiations (wavelength of 400-800 nm) may pose a risk against color stability, flavor maintenance, and general appearance of foods during their shelf life even to a lesser extent and also this region of absorption spectra should be considered for protective purposes more than for aesthetic aims such as see-through clarity or color brightness. The coloring of transparent materials can change their visible transmission spectra.

It is worthy to underline that the visible absorption data can be recorded usefully for an objective characterization of their color, both for transparent and nontransparent materials, i.e., as a transmittance or a reflectance spectrum. In order to put under control the important converting operations, for instance printing and decoration of packaging materials, the reflectance spectra of an opaque colored object or material can be measured as essential information about appearance of the final packaging.

Other useful applications of the UV/VIS transmittance/absorption information are represented by the spectra collected on paper sheets and bleached pulps. In several cellulosic products, bleaching additives are used to increase whiteness and to avoid loss of lightness and change into yellowish color during ageing and light exposing; different fluorescent whitening agents are used to preserve the paper quality. Reflectance spectrum recorded in the visible light region can evaluate the paper color change or defect. The presence of the additives in paper products is easily recognized by their specific absorption in the UV region between 320 and 370 nm [7].

The absorption spectrum of a compound in infrared radiation is unique by reflecting its molecular structure and thus is used for identifying the substance. The region of electromagnetic spectrum named 'infrared', includes wavelength between about 800-50000 nm (0.8-50 µm); it is normally divided into regions of 'near', 'medium', and 'far' IR (infrared), and the different locations of an IR spectrum are frequently indicated as 'wavenumbers', corresponding to the reciprocal of the wavelength in centimeters (cm^{-1}). The medium infrared between 3000-25000 nm, where the absorption is mainly due to the vibration levels of the molecules, is the region from which the major information for the identification of organic molecules is obtained, however it is also possible to achieve useful information also in the NIR (near infrared) region.

The techniques for recording infrared spectra have been developed and advanced so that nowadays IR spectra can be obtained from samples in any states of matter, shape, and dimensions, as absorption, transmittance, and reflectance. For example, many aspects of cellulose chemical structure, thus concerning paper and board applications, involve the hydroxyl groups and their ability to generate hydrogen bonds. By means of an IR spectrum which can identify the O-H stretching bands, it is possible to quantify free OH groups and inter- or intra-molecular hydrogen bonds, obtaining useful information about the mechanical properties of the cellulosic sheets. Studying near infrared reflectance spectra makes it possible to easily differentiate cellulosic materials as well as to measure water absorption in paper, a phenomenon that leads to a dramatic decrease of mechanical characteristics [8]. Plastic polymers and their compositional constituents are identified by using an infrared spectrophotometer. Moreover, the study of IR spectra may give useful information concerning how the chemical structure of polymers can change with the effects of light, time, and temperature during their commercial life or in abuse storage tests.

3.3.4 Haze

For transparent materials such as plastic films and glass, haze as the degree of opacity is measured in the percentage of transmitted light which deviates more than 2.5° from the incident radiation by forward scattering in passing through a specimen [9]. Being an aesthetic characteristic of transparent materials, it also represents the homogeneous distribution or the dimension of constituents, which may lead to some light scattering phenomena decreasing transparency. For instance, the agglomerates of crystalline and amorphous region, such as spherulytes typically present in semi-crystalline polymers like polyolefin, are large enough to scatter the visible light, whose wavelengths are similar to their average diameter.

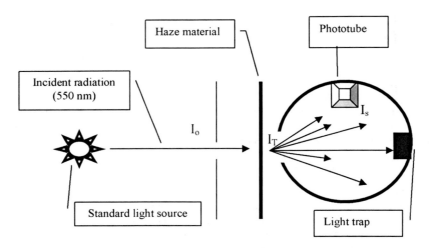

Figure 3.5 Schematic apparatus for haze determination

The haze parameter is almost always included in the specifications of plastic films (in Europe it is included even in anti-dumping conditions of international transactions), and therefore it is important to measure and report accurately the haze of transparent packaging materials. Several instruments dedicated to it are available today, and a schematic drawing of these instruments is given in Figure 3.5. They always incorporate an integrating sphere of a hollow sphere as an attachment of spectrophotometer instrument. Its internal surface is coated by a white reflective substance capable of collecting and reflecting to the photocell all the light transmitted from a specimen of the material, which is placed ahead the accessory. The transmitted light consists of specular-transmitted portion and one scattered through the sample. A light trap is located on the integrating sphere in such a way that just specular-transmitted light portion is absorbed allowing only the scattered light (deviated more than 2.5°) to be measured. Therefore, the measurements of sample transmittance, with and without the light trap, lead to the calculation of the haze as follows:

$$\text{Haze} = 100 \times \frac{I_s}{I_T} \tag{3.8}$$

where I_s is the intensity of scattered light (with the trap light), and I_T the total light transmitted (without the trap light). Table 3.11 shows the haze values of some common clear packaging materials.

It is worth mentioning that the light scattered by a clear packaging material is not included in the information of UV/VIS spectra and transparency. As a part of globally transmitted light, the scattered light portion affects the quality of the packaged food shortening its shelf life. Therefore, consideration of this optical property of packaging materials in food shelf life studies may be helpful to understand the effect of packaging condition on the quality change.

Table 3.11 Haze values for some packaging materials

Material	Haze (%)
Glass	0-0.2
Polystyrene	0.1-3.0
Low density polyethylene (50 μm)	30
Polypropylene (25 μm)	2.50
High density polyethylene (30 μm)	5-6
Polycarbonate	0.5-2.0
Cellophane	3.0
Polyester (15-30 μm)	2.5 – 5.0

3.3.5 Gloss

Gloss, or specular gloss in better meaning, is a particular attribute of a surface indicating objectively the shiny appearance of packaging materials or objects, transparent or opaque [10]. It is defined as the ratio of the luminous flux reflected from a specimen at a specific solid angle to the luminous flux incident at the same angle, i.e., in the specular direction. Therefore the higher the gloss value, the shinier the surface and generally the more appreciated the colored print or the metallized surface of a package. The light source used in the measuring device should not differ significantly from a standard 'CIE Source C', which is similar to that of a shiny day at noon. The drawing of Figure 3.6 shows schematic concept of a glossmeter; angles of 20, 45, 60, and even 85° are recommended for the different applications. This is probably the only electromagnetic property more relevant for aesthetic purposes than for functionality, even if this objective physical property can work as useful information concerning the effectiveness of important converting operations, like decoration, coating, or metallizing. Gloss values are used mainly in smoothness evaluation of colored or uncolored surfaces, and given in Table 3.12 typically for some packaging materials.

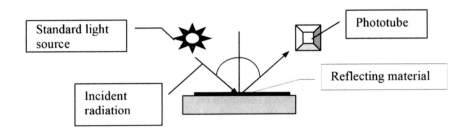

Figure 3.6 Schematic apparatus for gloss determination

Table 3.12 Gloss values for some packaging materials

Material (angle, thickness)	Gloss units
Polyvinyl chloride (20°, 20-50 µm)	120-160
Polyvinylidene chloride (45°, 60 µm)	115
White polypropylene (25 µm)	70
Polypropylene (25 µm)	80
Ionomers (20°)	25

3.3.6 Behavior to Ionizing Radiations

Some electromagnetic radiation may promote ionization and change some properties of the materials through which they pass; this can happen directly from the radiations carrying a charge like accelerated electrons (β radiation), or indirectly from their high energy (X and γ radiation).

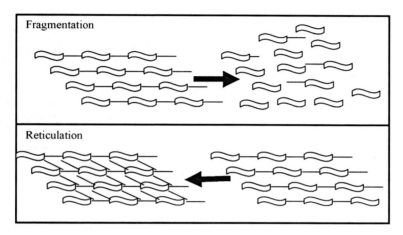

Figure 3.7 Fragmentation and reticulation phenomena in a polymer under the effects of electromagnetic radiation

Table 3.13 Behavior of some packaging materials to ionizing radiation

Material	Prevailing behavior
Polyethylene	Networking
Polystyrene	Networking
Polyester	Networking
Ethylene vinyl acetate	Networking
Polytetrafluoroethylene	Networking
Glass	Networking
Polyvinylidene chloride	Fragmentation
Cellulose	Fragmentation
Polyvinyl chloride	Both phenomena
Cellulose esters	Both phenomena
Polycarbonate	Both phenomena
Polyamide	Both phenomena
Polypropylene	Both phenomena

The consequences of ionization process are quite different with different media, leading to possible fragmentation of macromolecules or, on the contrary, to extensive networking by side chain bonds (Figure 3.7). The former phenomenon (fragmentation) on cell cytoplasmic constituents can be positive in the lethal effect on unwanted microorganisms: in fact γ radiations are used to sterilize packaging materials and also packed foods. Nevertheless, the consequences of the fragmentation in packaging materials are rather negative, since they will lead to intense migration, loss of mechanical resistance, and increased permeability. Otherwise, if the networking phenomenon is prevalent, the ionization treatment results in a general improvement of the performance of packaging materials, because they will become stronger, less permeable, and more inert. Actually, some plastic films as well as some glasses are submitted to ionizing radiation to improve their properties: for plastics the most important feature is generally an increase in gas barrier properties (a permeability decrease) as well as an increase in mechanical performance, whereas for glasses a lesser fragility is achieved. In Table 3.13 the responses to ionizing radiation are provided for some important materials [11].

Table 3.14 Behavior of some packaging materials to microwave irradiation

Material	Behavior	Notes
Metals	Absorbing/reflective	Risk of electric arcs
Glass	Transparent	-
Cellulose	Transparent/absorbing	Heat dissipation
Polyethylene	Transparent	Risk of melting
Polyester	Transparent	-
Polyamide	Transparent	-

3.3.7 Behavior to Microwaves

The electromagnetic radiations with frequencies between 300 MHz and 30 GHz are commonly named microwaves, and their exploitation in different fields of application, both at the industrial level and in the home environment, has grown dramatically in recent years. Food packages are submitted to microwave irradiation in several circumstances, mainly for heating, cooking, or defrosting of foods, and in some rare cases for sealing purposes of flexible materials. The behavior of different materials may be different according to their chemical composition and to the presence of some impurities. Table 3.14 summarizes it in transparent, absorbing, and reflective manner. If the packaging material has residual moisture or contains free ions or polar molecules able to orient according to the direction of propagation of electromagnetic fields, heat dissipation happens to increase the material temperature even to melting. Usually this is not desired for packaging materials, which are generally appreciated for their transparency to the microwave. One significant exception is represented by the susceptor materials: composite packaging materials, generally constituted of a plastic or cellulosic matrix (transparent to microwaves) containing finely dispersed substances sensitive to the electric or to the magnetic field. These substances sensitive to the oscillating magnetic field (like iron, cobalt, oxides, and alloys of these elements) or sensitive to the electric field (like aluminum or carbon black) can increase in temperature quickly and greatly when irradiated by microwaves. The applications of the microwave interactive packaging materials are discussed in detail in Chapter 14.

3.4 Mechanical Properties

All those properties that describe how a material reacts in response to application of forces (particularly its behavior under mechanical loads) are conveniently named mechanical properties. The kind of force involved, its time of application, its magnitude, its direction, its source, etc. permit to define several different properties. Some refer to the resistance of materials to possible stresses (stiffness, fragility, burst strength, resilience, etc.) and others are performance characteristics (elasticity, coefficient of friction, flex resistance, tackiness, cushioning properties, etc.). All of them, however, may be important in the selection of material, design and production of packages, optimization of transport and distribution, and shelf life prediction under some special situations. As mechanical properties are so numerous and different, those widely used by food packaging technologists and researchers will be discussed here with emphasis on the basic concept.

3.4.1 Density and Related Properties

The force which always acts on whatever material and body is its weight, which is the product of its mass and the acceleration of gravity. Thus mass is measured in weight and these two terms are often used interchangeably in our daily life.

As a property strictly related to the mass, density (d) is a way we evaluate its concentration in a material; it is measured by the mass per unit volume and its most common units are g cm^{-3} or kg m^{-3}. In Table 3.15 the densities of several materials used for food packaging application are given; they range in a small interval (about $0.8 - 8.0$ g cm^{-3}), but nevertheless help to identify similar materials and also have important economic consequences. There are several standard methods available for measuring density of the materials of interest (metals, cellulosics, plastics, ceramics).

For some nonhomogeneous materials used for food packaging application bulk density is often used with a definition of mass per total volume of the material consisting of solids and voids: plastic foams, expanded polymers, corrugated boards, and raw material in pellets or granules are often characterized in term of their bulk density. Since bulk densities are generally quite low, the units mostly used are kg m^{-3} (kg m^{-3}/1000 = g cm^{-3}).

Table 3.15 Density of some packaging materials

Material	Mass per unit volume (g cm^{-3})
Steel	8.0
Tinplate	8.0
Aluminum	2.70 - 3.20
Glass	2.40 - 2.80
Cellophane	1.50
Paper	0.75 – 1.15
Polyethylene	0.91 – 0.94
Polyester	1.35 – 1.40
Polyvinyl chloride	1.20 – 1.35
Polyamide (Nylon 6)	1.14
Polypropylene	0.88 – 0.91
Polystyrene	1.04 – 1.08
Wood (balsa)	0.11 - 0.14
Wood (oak)	0.60 – 0.90
Cork	0.22 – 0.26

Related to density is also grammage (G) or basis weight, which is defined as the mass per unit area of surface; generally g m^{-2} is the unit for grammage, a property which is widely used not only for paper and paperboard but also for plastic films, coatings, and varnishes. It is useful for the situation where the mass per unit surface is relevant. It is worth noting the relationship between density and grammage; the ratio of grammage to density gives the thickness (l) or caliper of a material, and is directly provided in μm for their respective standard units (g m^{-2}/g cm^{-3}):

$$G \text{ (g m}^{-2}) / d \text{ (g cm}^{-3}) = l \text{ (μm)} \qquad (3.9)$$

3.4.2 Coefficient of Friction

For many materials used as films or sheets on automatic packaging and printing machinery, the resisting force arising between two surfaces (of the same or different materials) that slide each over is of primary practical importance. This force is named friction, and the coefficient of friction (COF) is a fundamental mechanical property of flexible packaging materials which can often set the speed of automatic operations when the material unrolls from a reel and runs on forming machinery, printing, or laminating equipment. Frictional forces arise in so many different circumstances of food packaging that frictional property is important for every case of material and package. Restricting the force resistant to sliding is often desired, but there are occasions when proper level of friction between adjoining surfaces is required. When several packages are required to be stacked one over another, for instance, any possible slip of each item is unwanted and a resisting force between the surfaces is welcome; on the contrary when several glass bottles have to run quickly close each other along a filling line, frictional forces need to be at minimum to avoid possible breakage and consequent stopping of the operations.

The resisting forces exerting between sliding surfaces have different origins: natural roughness of superficial layers, surface treatments, and processing additives such as lubricants and plasticizers particularly for plastics. All of these different causes have also direct and measurable effects on the surface tension of solid surfaces (see Section 2.4.5).

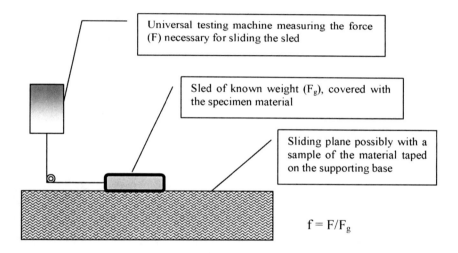

Universal testing machine measuring the force (F) necessary for sliding the sled

Sled of known weight (F_g), covered with the specimen material

Sliding plane possibly with a sample of the material taped on the supporting base

$$f = F/F_g$$

Figure 3.8 Schematic apparatus for determining coefficient of friction

Two kinds of friction, and consequently two types of COF, are generally recognized: static coefficient of friction is the resistance opposing the force required to start the movement of one surface over the other, whereas dynamic or kinetic COF is the resistance which opposes the force to move one surface at a fixed speed. The two COF values are not the same, and it is therefore important to discriminate between them. An accurate definition of coefficient of friction (f) is stated as the ratio of frictional force (F) to the force usually gravitational (F_g) acting perpendicularly onto the two surfaces in contact; consequently both the static COF and kinetic COF are nondimensional property as ratio of two forces. Well established standards [12] describe procedures for determining COF by means of testing machines equipped with a load cell and a constant rate of motion device. The film or sheet material to be tested is wrapped around a sled (a metal block of known weight) connected by a chord to the load cell (Figure 3.8); the sled is leaned on a plate (which might be covered with same test material), where it can slide by being pulled by the motion system, while the resistance is measured and recorded continuously or at intervals. In Table 3.16 typical values of COF for different materials are presented.

Table 3.16 Coefficient of friction (K= kinetic, S= static) of some packaging materials

Material	Coefficient of friction (N/N)
Cellophane on itself (K)	4.50
Aluminum on itself (S)	1.90
Glass on glass (S)	0.9 – 1.0
Oriented polypropylene on itself (K)	0.94
Glass on glass lubricated (S)	0.30 – 0.60
Steel on steel (S)	0.58
Polystyrene on polystyrene (S)	0.50
Paper kraft 67 μm, on itself (K)	0.30
Polyvinyl chloride on itself (K)	0.27
Polyethylene low density on itself (S)	0.20
Wood on wood (S)	0.20
Polyester on polyester (K)	0.45 - 0.38

3.4.3 Strength Properties: Tensile, Tear, Burst, and Creep

In the technical language, mechanical properties of packaging materials are mostly the properties referring to the strength and resistance of materials or finished packages against different and various applied loads. And the strength properties can be understood in terms of rheological science handling with deformation of material. This part of applied science on material properties has been developed, in origin, for protective purposes of nonfood items (electronics, weapons, computers, etc.) and in the second half of last century several works appeared to provide a general theory and standard procedures for evaluating materials in the perspective of reducing shipping and distribution damages of packaged goods [13-15]. More recently this knowledge has been progressively transferred to food packaging science and progressively matured to understand the role of mechanical performance of materials in the design of new forms, shapes, and uses of food packages [16].

Actually, several specific words are used to indicate the strength properties of different packaging materials, sometimes ambiguously, but their correct definitions are stated in standards [17], from which Table 3.17 has been prepared. Strain refers to the change in size or shape of a body in response to a stress applied. Stress is usually expressed by the load divided by the area corresponding to its action against the body. Brittleness, elasticity, tensile strength, and toughness are related to physical durability or resistance of the packaging materials. Ductility and plasticity should be considered in the package manufacture such as cold forming of the metal or plastics. Creep phenomenon is observed in the pressurized package such as carbonated drinks. Relaxation of stress needs to be considered in selection of the proper material for stretch wrapping (see Section 11.7.4).

Table 3.17 Glossary of mechanical behavior

Property	Definition
Brittleness	The inability to experience a significant amount of strain without rupture; the attitude of breaking in the elastic field
Creep	The slow deformation measured under constant stress
Ductility	The ability of a material to be plastically deformed by elongation without rupture
Elasticity	A reversible stress/strain behavior
Load	The value of an applied force; generally measured in N
Plasticity	The ability to be molded and retain a shape for a significant period under finite forces
Relaxation	The attitude to experience, under a constant deformation, a reduction in stress required to sustain that deformation
Strain	The measurement of deformation relative to a reference configuration of length, area, or volume (nondimensional)
Stress	A force per unit area ($N\ mm^{-2}$ or MPa)
Tensile strength	The maximum tensile stress that a material can sustain
Toughness	The energy that a material can absorb before rupturing; it is generally measured as the area under the stress strain curve

To describe objectively and effectively the behavior of a solid material responding to an applied force, it is necessary to classify type, origin, and direction of the force itself. The applied force may be in static or dynamic form. The former is one that acts for length of time in a constant intensity such as the load due to some stacked packages on a shelf or a truck, and the stress of a stretch wrapping. The latter is, on the contrary, quick in their action and their intensity is variable; examples of dynamic stresses are those due to impacts, falls, or vibrations. As far as the origin of the force is concerned, it is possible to discriminate between that internal to the package (such as overpressure of a carbonated beverage or pressure changes during pasteurization and sterilization) and that external to the package (such as over-imposed stacking, knocks, and impact). Considering the direction of the force, two general cases are normally recognized: tensile stresses (e.g., the stretching force applied to wrap a food by a transparent thin film) and compressive stresses (e.g., the crushing forces exerted by industrial equipments to reduce the volume of plastic objects for waste disposal).

In real conditions (packaging, handling, transport, storage, etc.) all the possible combinations of type, origin, and direction of forces are likely to be imposed, but if the aim is to characterize the strength properties of a material, tensile test is mostly performed on material specimens, measuring the stresses and the corresponding strains under a low constant rate of loading (Figure 3.9). Tensile test is easy to standardize and, moreover, solids are generally more sensitive and vulnerable to tensile stresses than to compressive ones. However, when specific performances are requested, compressive and deflection tests are also applied in testing activities.

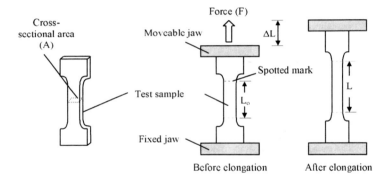

Figure 3.9 Tensile test of a material specimen. A typical sample may be 150 mm long and 20 mm wide, with the center section 10 mm wide and 60 mm long

- **Tensile Strength**

The most convenient way to have a useful description of the strength of a material is to test it by a universal testing machine in a tensile test [18].

During this kind of test (Figure 3.9), specimens of different shape and specified dimension can be used depending on the material. In particular, a straight rectangular strip is generally used for thin films, whereas a dog-bone shape is requested for thicker samples, in which specific neck of central section makes easier the elongation without detachment of the sample from the grips. When the analysis starts, the specimen is stretched between a fixed jaw and a moveable head, which moves with a constant and low rate; during this motion the resulting changes in length (strain) are monitored together with the corresponding stress, leading to a strain-stress plot, from which several useful mechanical parameters are obtained.

In Figure 3.10 an idealized strain-stress curve is presented, showing that useful information is obtained from such a tensile test, covering the start of the deformation up to the breakage of the specimen. From the plot, stress or strain can be read or recorded. Stress value on the ordinate measures the force (F) per unit cross-sectional area (A) of the specimen and is expressed usually in MPa (N mm^{-2}):

$$\text{Stress} = \sigma = \frac{F}{A} \tag{3.10}$$

Strain on the abscissa is dimensionless, being usually expressed by the ratio of lengths having the same unit:

$$\text{Strain} = \varepsilon = \frac{L - L_o}{L_o} = \frac{\Delta L}{L_o} \tag{3.11}$$

where L_o is the original length of the sample and L is the length after the deformation.

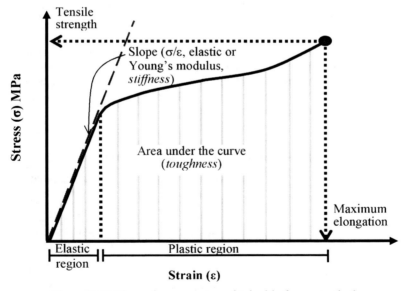

Figure 3.10 The main parameters obtainable from a typical 'stress-strain' curve in a tensile test

The above mentioned curve can be divided in two main regions. The first one is given by the area under the first linear tract of the plot and is called elastic region. Here the material behaves elastically and the deformation is reversible. That means that when the load is released the specimen will recover its original, undeformed configuration, i.e., its original length. The second one is called plastic region and is defined by the area under the remaining part of the plot up to the sample break. In this area, any increases in stress will lead to permanent deformation. Moreover, in this part of the curve the deformation will be relatively large for small, almost negligible increases in the stress.

Useful parameters that can be drawn from such a tensile test are:

- *Elastic modulus* (also called *Young's modulus*) It is graphically represented by the slope of the first rising and linear part of the stress-strain plot. Since it is the ratio of stress to strain, it is also expressed as MPa. In practice, this parameter is useful linking to an important physical properties of plastic films, i.e., the *stiffness*. Specifically, the higher the modulus value of a given material, the higher its stiffness.

- *Area under the curve*. It represents the energy required to break the sample of material. As product between the ordinate and the abscissa values of the stress-strain plot, it can be expressed in MPa (product of the stress, in MPa, and the strain, dimensionless) or in J (product of the applied force, in N, and the deformation underwent by the specimen, in m). The physical property of plastic polymers related to this parameter is the *toughness*.

- *Tensile strength*. It indicates the maximum stress that a specimen is able to sustain before failure. For plastic films, it is given by the y-value of the ultimate point of the stress-strain plot, before breakage takes place. The corresponding value on the x-axis defines the elongation that a given material can sustain without rupture.

Table 3.18 Strength values of some packaging materials in tensile test

Material	Modulus of elasticity (Mpa)	Breaking point (Mpa)
Aluminum	70 000	70 – 210
Tinplate	1 800 000	330 – 740
Glass	70 000	70
Cellulosic (kraft paper)	-	25 – 100
Polypropylene	2 000 - 3 500	35 – 50
Polyester	4 000 - 5 000	170 –270

Another great advantage of a tensile test is that the test makes it possible and easy to measure and compare objective parameters of the mechanical behavior of several different materials. In Table 3.18, some typical values of strength in tensile test for common packaging materials are presented.

- **Tear Strength**

The resistance of packaging materials to tearing forces is important and useful particularly for flat materials. This strength property is frequently measured both for cellulosic and plastic packaging materials, after proper conditioning and often in both the machine and transverse directions. The standardized tests [19, 20] evaluate the force (N) to initiate and/or propagate a tearing of sample as well as the energy (J = N m) absorbed by the specimen during the tear beginning and/or growing from an initiation cut. A universal testing machine is generally used for initial tearing whereas a pendulum impact tester is used to measure the force required to propagate an existing slit.

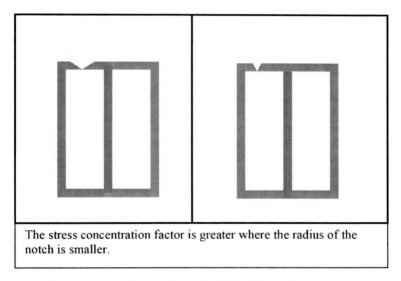

The stress concentration factor is greater where the radius of the notch is smaller.

Figure 3.11 Notches on the seal of flexible pouches act as stress amplifiers and facilitate the propagation of tearing

Initial tearing point is of general interest because the strength of the material is seriously reduced if a notch in its edge or any other physical defect (micro-holes, micro-bubbles, sudden thickness changes, corners, cracks, etc.) is present: they all act as stress concentrators or amplifiers and may produce a local stress even above the material maximum resistance. The amplification of a stress is the magnification of the forces existing at the defect or geometrical discontinuity well above the nominal stress, which is the load divided by the area over which the load is applied. Therefore, a stress concentration factor is calculated as the ratio of local stress to nominal stress; the stress concentration factor depends on the geometry of the defect and, in most general cases of a notch, the smaller is its radius, the greater is the concentration factor. The geometrical discontinuities are dangerous and feared defects in all the packaging production lines, but actually they are also positively exploited in some circumstances such as easy opening devices in several packaging forms (Figure 3.11).

- **Burst Strength**

The resistance to bursting means the pressure value required to obtain the rupture of the material in standard conditions [21]. The pressure is applied at a controlled, slowly increasing rate, and it can be provided by water or compressed air according to specimen under examination. The bursting strength test is common for paper, board, plastic films, and plastic sheets, but glass bottles are also inspected for their internal pressure resistance by means of tests quite similar in principles to those used for flexible packaging materials. Measuring the burst strength of a material specimen or a finished package may be useful for predicting possible performance during the real usage of that item as well as for highlighting hidden or nonvisible defects in the structure (see also Section 10.5.3 in Chapter 10).

- **Creep Resistance**

As already defined (Table 3.17), creep is the slow deformation measured over time under constant stress. Packages and packaging materials, mostly the flexible ones, have several occasions to be challenged by a possible creep deformation. It may happen for packages stacked on a pallet in the warehouse or on a shelf at point of sale (creep by external forces) as well as for containers of pressurized beverage (creep by internal forces). A heat processing also increases the internal pressure of the food containers. Because creep resistance is time dependent, it has some relationship with the commercial life of the package and/or the packaged product; the shelf life of a product might be limited and consequently predicted by the resistance of its packaging material to such slow but constant stresses.

3.4.4 Response to Dynamic Stresses: Impact Resistance and Cushioning

In the strength properties mentioned in the previous section, the stress changes slowly during the testing, and thus we can assume the loading is almost static. In the real situations of handling and transport, the packed foods are also exposed to stresses that are variable in intensity and can be of very short duration in the application; this applies to the cases of impacts, falls, and vibrations, which are likely to occur during the packaging production as well as along the distribution routes. One relevant responsibility in the selection of packaging material, therefore, is to find one that will be resistant against the predictable dynamic stresses; in other words, it might be useful or even essential to evaluate the packaging materials in terms of their cushioning attitudes, in order to provide protection against the possible damages to the food inside, coming from impact and/or vibrations.

Foods products are generally considered not to be so fragile products, but underestimating the sensitivity of foods to these kinds of dynamic stresses may lead to wrong decision. It has been demonstrated that the type of the transport and packaging material as well as the sensitivity of the food item to the particular stress can influence the fruit bruising and consequent (not only aesthetic) defects in product quality [22, 23]. It is necessary to provide adequate protection against possible shocks and vibrations by choosing the reliable form of packaging. Moreover, it is useful to distinguish between external and internal fragility because some food products suffer from the highest internal fragility related to their complex inner structure. For internal damages, the bruises are not immediately evident and may be masked by the surface color and fruit skin [22, 23], or the stresses can injure nonvisible internal constituents or parts of the food product. For example, the complex and delicate internal structure of peppers (with seeds) and eggs (with yolk) may suffer from internal damage if mechanical stresses are strong enough for the organized structure to be shaken and ruptured even under external protection.

- **Impact Resistance**

In real usage situation, the severe shocks result from packages falling onto rigid surfaces or knocking against hard objects. The resistance of different materials against these stresses is measured by several procedures. This property is sometimes called resilience and assumed as the capacity to absorb energy in impulsive form (shock). In some cases, we call the inverse concept of resilience by the term fragility. The most appropriate definition of resilience is the capacity of a material to absorb energy when it is deformed elastically and then to return this energy with removal of the load; in fact it may be measured as the area under the strain-stress curve in the elastic region.

The common ways of evaluating the impact resistance of packaging materials are the notched Izod Impact (IZOD Method) [24] and the free-falling dart method [25]. As shown in Figure 3.12, the notched Izod Impact method measures the resistance of a material to the impact from a swinging pendulum. The specimen is notched to prevent deformation and facilitate fracture and breakage more reproducibly. The figure also shows that the pendulum is at a lower position after breaking the specimen, which is due to energy absorbed by the specimen. The energy for breaking the specimen (J), E_b, is equal to the loss of potential energy:

$$E_b = mg\Delta h \tag{3.12}$$

where m is the mass of the pendulum (kg), g is gravitational acceleration (9.81 m s^{-2}), and Δh is the difference in heights before and after impact (m).

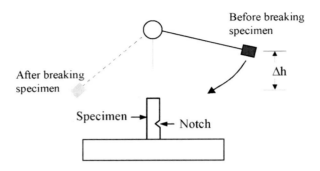

Figure 3.12 Notched Izod impact method

The more recent devices are equipped with sensors able to measure directly in real time the load on the specimen during its rupture. The free-falling dart method (GARDNER impact) is also a common procedure for the evaluation of impact resistance of flat materials. The instruments are equipped with a set of darts that are suspended at predetermined column heights and then dropped onto the specimen. The impact resistance is derived from the dart mass and the drop height that cause the specimen to crack or is given as the weight required to break 50% of a large number of specimens.

- **Cushioning Properties**

If a packaging material or package has to provide the food with protection against dynamic impacts or falls, it must first show its own resistance to the stresses applied (it should not break) and it must also able to absorb enough energy of the dynamic stress to reduce the stress amount reaching the food product. This can be done (i) by reducing the force per unit area, spreading them over the maximum possible surface and avoiding a concentrated load on a single part (but it cannot protect against internal fragility), (ii) by sustaining the product at the strongest part in order to concentrate the stress on the more resistant points (this is practically impossible for foods and does not reduce the internal fragility), and finally (iii) by extending the time of arresting or in the same way extending the distance of arresting, i.e., the length of space from the beginning of the contact to the end of movement. This last solution corresponds to the choice of a cushion material that is compressed on impact and absorbs a part of the energy of dynamically applied stress, producing a work of deformation and reducing the deceleration.

Designing cushion packaging for a delicate item consists of identifying product's fragility factor and dynamic cushioning property of the packaging material. Fragility factor can be determined by means of a fragility tester (the shock machine) which consists of a table where the item is secured and that can move vertically, reproducing any drop height: when the table stops, the deceleration the item receives is measured and the possible damage observed [26]. The ratio of the deceleration of the respective mass to the normal gravitational acceleration at occurrence of the first damage is recorded as fragility factor. Delicate items are supposed to have fragility factors in the range of 15-25, whereas those of rugged wares are over 100. Dynamic cushioning test of packaging material consists of repeated dropping of a specific size platen from a specific height onto a flat cushion sample. The test is repeated using different weight platens, cushion thickness, and drop heights. The maximum shock sustained as the falling platen impacts the cushion sample is measured and recorded as function of static stress which is weight per load bearing area. Parameters to characterize the effectiveness of a cushioning system can be obtained to compare different cushioning systems [27].

Discussion Questions and Problems

1. Haze, gloss, and transparency may be considered as functional properties of flexible materials. Define them and explain their concept with measurement principle.

2. How can the 'coefficient of friction' be defined and why is it important as packaging performance?

3. What is a 'stress concentrator'?

4. Where is the breaking point in a stress/strain curve located?

5. During a tensile test, is the maximum strain always corresponding to the breaking elongation?

6. Explain the relationship between density and grammage.

7. Explain the difference between static and dynamic COF.

8. If in a tensile test the breaking load is 150 N, the width of the sample is 20 mm, and its thickness 35 μm, what is the tensile strength?

9. What meaning does the heat of combustion have in a food packaging aspect?

10. Are the glass transition temperatures of polymers always lower than room temperature? Discuss the different possibilities.

11. Are the effects of ionizing radiations on packaging material useful or negative? Explain why.

12. An expanded polymer is a better insulating material than its homogeneous form. Explain why.

13. May the haze value be higher than 100? Explain the reason for your answer.

Bibliography

1. P Nesvadba, M Houska, W Wolf, V Gekas, D Jarvis, PA Sadd, AI Johns. Database of physical properties of agro-food materials. J Food Eng 61:497-503, 2004.
2. ASTM. Standard test method for transparency of plastic sheeting, in Annual Book of ASTM Standards D1746-70. 1988, American Society for Testing and Materials: Philadephia, PA.
3. L Piergiovanni, S Limbo. The protective effect of film metallization against oxidative deterioration and discoloration of sensitive foods. Packag Technol Sci 17(3):155-164, 2004.
4. JP Wold, A Veberg, A Nilsen, V Iani, P Juzenas, J Moan. The role of naturally occurring chlorophyll and porphyrins in light-induced oxidation of dairy products. A study based on fluorescence spectroscopy and sensory analysis. Int Dairy J 15(4):343-353, 2005.
5. JO Bosset, PU Gallmann, R Sieber. Influence of light transmittance of packaging materials on the shelf life of milk and dairy products - review. In: M Mathlouthi, ed. Food Packaging and Preservation. London, UK: Blackie Academic & Professional, 1994,
6. RJ Hernandez, SEM Selke, JD Culter. Plastics Packaging. Munich, Germany: Hanser Publishers, 2000, pp. 141-142.
7. ML Jao, MI Liao, CT Huang, CC Cheng, SS Chou. Studies on analytical method for determination of fluorescent whitening agents in paper containers for food. J Food Drug Anal 7(1):53-63, 1999.
8. C Bizet, S Desobry, J Hardy. A NIR reflectance method to measure water absorption in paper. In: P Ackermann, M Jagerstad, O T., ed. Food and Packaging Materials - Chemical Interaction. Cambridge, UK: The Royal Society of Chemistry, 1995, pp. 140-145.
9. ASTM. Standard test method for haze and luminous transmittance of transparent plastics, in Annual Book of ASTM Standards D1003-00. 2000, American Society for Testing and Materials: Philadelphia, PA.
10. ASTM. Standard test method for specular gloss of plastic films and solid plastics, in Annual Book of ASTM Standards D2457-03. 2003, American Society for Testing and Materials: Philadelphia, PA.
11. KA Riganakos, WD Koller, DAE Ehlermann, B Bauer, MG Kontominas. Effects of ionizing radiation on properties of monolayer and multilayer flexible food packaging materials. Radiat Phy Chem 54(5):527-540, 1999.
12. ASTM. Standard test method for static and kinetic coefficients of friction of plastic film and sheeting, in Annual Book of ASTM Standards D1894-87. 1987, American Society for Testing and Materials: Philadelphia, PA.
13. RD Mindlin. Dynamics of package cushioning. Bell Sys Tech J 24:361-365, 1945.
14. RE Newton. Theory of Shock Isolation. In: CM Harris, CE Crede, ed. Shock and Vibration Handbook. New York: McGraw-Hill, 1961, pp. 31-31-31-33.

15. HC Blacke. A cushion design method and static stress-peak deceleration curves for selected cushioning materials. East Lansing, Michigan, School of Packaging, Michigan State University, 1963.
16. WE Brown. Plastics in Food Packaging. New York: Marcel Dekker, 1992, pp. 247-291.
17. BS Institute. Glossary of rheological terms, in British Standard BS 5168:1975. 1975, British Standard Institute: London, UK.
18. ASTM. Standard test method for tensile properties of paper and paperboard using constant-rate-of-elongation apparatus, in Annual Book of ASTM Standards D828-97. 2002, American Society for Testing and Materials: Philadelphia, PA.
19. ASTM. Standard test method for initial tear resistance of plastic film and sheeting, in Annual Book of ASTM Standards D1004-94a. 1994, American Society for Testing and Materials: Philadelphia, PA.
20. ASTM. Standard test method for tear-propagation resistance (trouser tear) of plastic film and thin sheeting by a single-tear method, in Annual Book of ASTM Standards D1938-94. 1994, American Society for Testing and Materials: Philadelphia, PA.
21. ASTM. Standard test method for bursting strength of paper, in Annual Book of ASTM Standards D774-97. 2002, American Society for Testing and Materials: Philadelphia, PA.
22. SP Singh, G Burgess, M Xu. Bruising of apples in four different packages using simulated truck vibration. Packag Technol Sci 5(2):145-150, 1992.
23. PJ Vergano, RF Testing, WC Newall. Distinguishing among bruises in peaches caused by impact, vibration, and compression. J Food Qual 14:285-298, 1991.
24. ASTM. Standard test method for determining the Izod pendulum impact resistance of plastics. Philadelphia, PA, American Society for Testing and Materials, 2002.
25. ASTM. Standard test method for impact resistance of plastic film by the free-falling dart method, in Annual Book of ASTM Standards D1709-03. 2003, American Society for Testing and Materials: Philadelphia, PA.
26. ASTM. Standard test method for mechanical-shock fragility of products, in Annual Book of ASTM Standards D3332-99. 2004, American Society for Testing and Materials: Philadelphia, PA.
27. FA Paine. The Packaging User's Handbook. Glasgow, UK: Blackie, 1995, pp. 565-588.

Chapter 4

Permeation of Gas and Vapor

4.1   Introduction 79
4.2   Basic Concepts of Permeation 80
      4.2.1 Mechanism of Gas Transport through Permeation ············· 81
      4.2.2 Diffusion of Permeant ··· 82
      4.2.3 Adsorption and Desorption of Permeant ···························· 82
4.3   Theoretical Analysis of Permeation 83
      4.3.1 Mass Balance Analysis ··· 83
      4.3.2 Concentration Profile within Film at Steady State ············· 85
      4.3.3 Derivation of Permeation Rate Equation ······················· 85
      4.3.4 Physical Meanings of Permeation Rate Equation ··············· 86
4.4   Terminology and Units for Permeation 89
      4.4.1 Transmission Rate ··· 90
      4.4.2 Permeance ··· 91
      4.4.3 Permeability ··· 91
4.5   Permeability of Food Packaging Polymers 92
4.6   Factors Governing Permeation 96
      4.6.1 Nature of Polymer ··· 96
      4.6.2 Nature of Permeant ·· 97
      4.6.3 Ambient Environment ·· 97
4.7   Measurements for Permeation Properties 98
      4.7.1 Basic Design Principles ··· 98
      4.7.2 Isostatic Method ·· 99
      4.7.3 Quasi-Isostatic Method ··· 100
      4.7.4 Measuring Permeation Rate of Finished Packages ············· 101
      4.7.5 Another Method for Estimating Permeation of Plastic Containers···· 102
      4.7.6 Gravimetric Method ·· 104
4.8   Gas Transport through Leaks 105
Discussion Questions and Problems 106
Bibliography 108

4.1 Introduction

The transport of gas or vapor through a food package can greatly influence the keeping quality of the packed food. Oxygen ingress into the package can cause oxidation in lipid foods (especially dehydrated meat, egg, cheese, as well as foods cooked in frying oil) that leads to off-flavors, loss of color and nutrient value. Water vapor infiltrating the package can cause moisture gain leading to sogginess or microbial growth in food, while water vapor escaping from the package can cause moisture loss leading to undesirable textual changes in food.

On the contrary, there are occasions when the transport of gases and vapors is desirable. In modified atmosphere packaging of fresh produce, the exchange of oxygen, carbon dioxide, and water vapor through the package is necessary to accommodate the respiration and transpiration of the still bioactive product and to maintain an optimum gas composition in the package. In packaging of freshly roasted coffee, degassing is necessary to prevent excessive pressure buildup since the product still emits carbon dioxide after packing.

There is a subtle difference between gas and vapor, although these two terms are frequently used interchangeably. A gas refers to low-boiling-point molecules that are noncondensable and exist only in the gaseous state at ambient temperature and atmospheric pressure. 'Permanent gases' describe oxygen, nitrogen, hydrogen, carbon dioxide, and other gases that follow the ideal behavior; for example, they obey Henry's Law and their permeabilities and diffusivities are constant, not concentration dependent. On the other hand, a vapor is condensable at ambient temperature. Condensation of water vapor may cause the problems of fogging on package surface and microbial growth on food surface. Water vapor and organic vapors often deviate from the ideal behavior, and their permeabilities and diffusivities are concentration dependent in hydrophilic polymers such as nylon and ethylene vinyl alcohol (EVOH).

There are two major mechanisms by which gas or vapor may be transported through the package—permeation and leak. Permeation involves the exchange of a gas or vapor (also known as permeant) through a plastic film or package wall. Leak involves the exchange of gas or vapor through pinholes or channel leaks, to be explained further in Section 4.8. Although permeation is the focus of this chapter, it is important to realize that leakers from defective packages are often more influential than permeation in determining the overall transmission rate [1].

In the following sections, we describe the three major steps—adsorption, diffusion, desorption—involved in the permeation process and derive a permeation equation based on this understanding. To characterize the barrier properties of polymeric packaging materials, we introduce the term permeability (also known as permeability coefficient), defined as a product of diffusion coefficient and solubility coefficient. Other permeation properties including transmission rate and permeance are clarified and methods for measuring them are described.

4.2 Basic Concepts of Permeation

Permeation and gas barrier are closely related terms. Permeation is inversely proportional to gas barrier; for example, a package which allows gas to permeate quickly is a package of low gas barrier. To protect foods that are oxygen or moisture sensitive, high gas barrier packages should be used to retard the rate of permeation.

This chapter is limited to permeation in polymeric packaging materials. It is important to mention that all packaging polymers are permeable to gas and vapor to various degrees. These polymers provide a wide range of permeability for different applications. Hence the study of permeation is important to ensure that adequate gas barrier protection is provided for plastic pouches, plastic containers, and other plastic packages. On the other hand, glass and metal packaging materials are not permeable and paper packaging materials (unless coated with barrier layer) are too permeable. Unlike polymers, these materials do not provide an opportunity for the designer to optimize the barrier property for various applications.

4.2.1 Mechanism of Gas Transport through Permeation

Figure 4.1 shows that permeation of a permeant through a polymer film (or a package wall) is driven by concentration gradient in the direction from high to low concentration. The mechanism of permeation consists of three sequential steps: adsorption of the permeant onto the high concentration side of the film surface, diffusion of the permeant across the film, and desorption of the permeant from the low concentration side of the film surface.

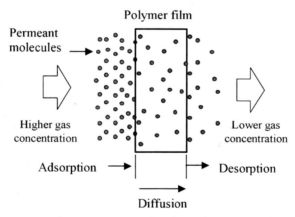

Figure 4.1 Permeation through a polymer film

When measuring permeation rates, experiments are usually set up such that both sides of the film are exposed to the gas phase. In this case, the permeation process involves only the gas and solid phases, where the permeant in one gas phase (left side) enters the solid phase of the polymer film (middle) and exits to another gas phase (right side). In fact, all permeability values in the literature were measured in this manner.

In reality, permeation process may also involve the liquid phase. In situations such as a HDPE milk bottle where the product inside is a liquid, the permeation process involves the permeant in gas phase enters into the solid phase of the package wall and then exits into the liquid phase. Unfortunately, permeation involving liquid phase is seldom studied and thus permeability values for liquid permeation are not available in the literature.

4.2.2 Diffusion of Permeant

Diffusion is the movement of molecules from a region of high concentration to a region of low concentration as a result of intermingling of the molecules due to random thermal agitation. While individual molecules move randomly with no sense of direction, the intermingling of the molecules results in a net movement from high to low concentration—this means more molecules moving from a region of high concentration than those moving from the opposite direction.

The diffusion of a permeant through in a polymer film may be described by Fick's First Law:

$$J = -D \frac{dc}{dx} \qquad (4.1)$$

where J (mol cm^{-2} s^{-1}) is diffusion flux, D (cm^2 s^{-1}) is diffusion coefficient (also called diffusivity), c (mol cm^{-3}) is permeant concentration, x (cm) is distance in the flow direction. Since the increase in concentration is customarily associated with the positive x direction, a negative sign is needed in Equation 4.1 to denote that diffusion is in the negative x direction or toward a lower concentration.

Diffusion flux J is the amount of permeant diffusing per area per time through a plane perpendicular to the flow direction. The term dc/dx is frequently referred to as concentration gradient or driving force of diffusion. When dc/dx = 0, net diffusion ceases—it does not mean the molecules are no longer moving, but the rates of molecular movement on both sides of the film counterbalance each other. Diffusion coefficient D is a measure of the mobility of permeant molecules in the polymer film.

There are two types of behavior for molecular diffusion in polymer. Fickian behavior is characterized by concentration independent diffusion coefficients. This behavior is commonly observed for diffusion of permanent gases in polymers; for example, the diffusion coefficient of oxygen in a polyethylene (PE) film remains the same when oxygen concentration is changed from 21% to 100%. On the other hand, non-Fickian behavior is characterized by concentration dependent diffusion coefficients. This behavior is commonly observed for diffusion of water vapor or organic vapors in polar polymers such as nylon and EVOH, where there exist strong interactions between the permeant and the polymer matrix.

4.2.3 Adsorption and Desorption of Permeant

Adsorption and desorption are related to the solution or sorption behavior of the permeant molecules in the polymer film, which is governed by the relative strengths of interactions between permeant/permeant, permeant/polymer, polymer/polymer.

The simplest or 'ideal' sorption isotherm is expressed by Henry's Law:

$$c_s = S p \qquad (4.2)$$

where c_s is permeant concentration at the solid-phase film surface (mol cm^{-3}), p (atm) is partial pressure of the permeant, S is solubility coefficient (mol cm^{-3} atm^{-1}). If the solubility coefficient S is concentration independent, then Equation 4.2 expresses a linear relationship—this ideal behavior is applicable to the adsorption and desorption of permanent gases at atmospheric pressure, where the permeant/permeant and permeant/polymer interactions are weak compared to the polymer/polymer interactions.

Henry's Law provides a convenient means for estimating c_s once p and S are known. It is difficult to measure the permeant concentration at the film surface (c_s), but it is relatively easy to measure the permeant partial pressure (p) using a gas chromatograph or a portable gas analyzer.

4.3 Theoretical Analysis of Permeation

The individual steps of adsorption, diffusion, desorption in the permeation mechanism may be integrated using the principle of mass balance because permeation involves the transport of mass (i.e., permeant molecules) from one place to another. By applying the mass balance, along with Fick's First Law and Henry's Law, useful equations may be obtained to guide the design of food packages.

4.3.1 Mass Balance Analysis

A mass balance around a thin layer Δx located within a polymer film (Figure 4.2) may be written as:

$$
\begin{array}{ccc}
\text{rate of permeant} & \text{rate of permeant} & \text{rate of permeant} \\
\text{diffusion into the} \quad - & \text{diffusion out of the} \quad = & \text{accumulation in the} \\
\text{plane at x} & \text{plane at x+}\Delta x & \text{layer } \Delta x
\end{array}
$$

$$(4.3)$$

In terms of diffusion flux, the mass balance equation is expressed as:

$$J_x A - J_{x+\Delta x} A = \frac{d(cA\Delta x)}{dt} \qquad (4.4)$$

where c is permeant concentration and A is surface area of the film. $cA\Delta x$ is amount of permeant in the Δx layer at any time t, and the term on the right of Equation 4.4 represents the rate of change of permeant within this layer. Dividing Equation 4.4 by $A\Delta x$ and taking the limit as Δx approaches zero:

$$\lim_{\Delta x \to 0} \frac{J_x - J_{x+\Delta x}}{\Delta x} = -\frac{\partial J}{\partial x} = \frac{\partial c}{\partial t} \qquad (4.5)$$

Substituting J by Equation 4.1 and assuming D is constant:

$$D\frac{\partial^2 c}{\partial x^2} = \frac{\partial c}{\partial t} \qquad (4.6)$$

which is known as Fick's Second Law, an equation useful for describing the unsteady state conditions of the film.

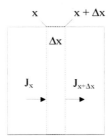

Figure 4.2 Analysis of a thin layer Δx in a polymer film

As shown in Figure 4.3, a film when challenged by a pressure difference between p_A and p_B, will undergo a transient period during which its concentration profile changes with time. Assume the permeant concentration within the film is initially c_L and, at time zero (t_o), constant concentrations c_0 and c_L are imposed on the sides. During the time between t_o and t_s, the concentration profile gradually adjusts itself until steady state is achieved—this transient period, sometimes referred to as the conditioning of the film, is useful for estimating diffusion coefficients (see Equations 4.42 and 4.43 later).

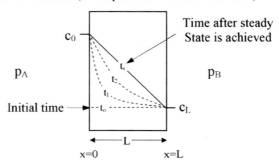

Figure 4.3 Concentration profile of a film as function of time

4.3.2 Concentration Profile within Film at Steady State

At t_s and beyond, steady state is reached when the concentration profile is linear and the diffusion flux is constant across the film. In terms of the mass balance equation, steady state means no permeant accumulation in the film (i.e., the term $\partial c/\partial t = 0$) and thus Equation 4.6 becomes:

$$D\frac{d^2c}{dx^2} = 0 \tag{4.7}$$

If the following conditions are imposed on the film (Figure 4.3):

$$\text{At } x = 0, \, c = c_0$$
$$\text{At } x = L, \, c = c_L \tag{4.8}$$

where L is the film thickness, then the solution becomes:

$$c = c_0 - \left(\frac{c_0 - c_L}{L}\right)x \tag{4.9}$$

which confirms that the concentration profile at steady state (when $t \geq t_s$) is linear as shown in Figure 4.3.

The transient period from t_o to t_s is usually within hours, which is often a much shorter period compared to the product shelf life. Hence, this transient period may sometimes be assumed to be small or negligible. In solving practical problems, it is often adequate to consider only the time period after steady state has been achieved.

4.3.3 Derivation of Permeation Rate Equation

A useful equation to describe the permeation rate at steady state is derived as follows. Recall that the permeation process involves three steps: adsorption, diffusion, and desorption. The derivation begins with describing the diffusion step using Fick's First Law, followed by describing the adsorption and desorption steps using Henry's Law.

According to Fick's First Law, the diffusion rate may be expressed as:

$$\text{Diffusion rate} = JA = -DA\frac{dc}{dx} \tag{4.10}$$

At steady state, the concentration gradient, dc/dx, may be obtained by taking the derivative of Equation 4.9 with respect to x to obtain:

$$\frac{dc}{dx} = \frac{c_L - c_0}{L} \tag{4.11}$$

Steady state also dictates the condition that the diffusion rate is equal to the permeation rate. Accordingly, Equation 4.10 becomes:

$$Q = DA \frac{c_0 - c_L}{L} \qquad (4.12)$$

where Q is the steady state permeation rate.

According to Henry's Law in Equation 4.2, $c_0 = p_A S$ and $c_L = p_B S$. Substituting them into the above equation yields:

$$Q = DA \frac{p_A S - p_B S}{L} = DSA \frac{p_A - p_B}{L} \qquad (4.13)$$

where p_A and p_B are partial pressures on both sides of the film (Figure 4.3). This equation assumes that instant equilibrium is established across the film surface. It is convenient to introduce the term permeability, defined as the product of diffusion coefficient and solubility coefficient, as:

$$\overline{P} = D\ S \qquad (4.14)$$

Hence Equation 4.13 becomes

$$Q = \overline{P}A \frac{p_A - p_B}{L} = \frac{\overline{P}A}{L} \Delta p \qquad (4.15)$$

which is the permeation rate equation at steady state.

4.3.4 Physical Meanings of Permeation Rate Equation

The permeation rate equation, Equation 4.15, succinctly describes the permeation rate as a function of package and environmental variables. The package variables include permeability $\overline{P}$ and package dimensions (A and L). The environmental variables include p_A and p_B; temperature and relative humidity are implicitly included in permeability. $\overline{P}$ is a function of temperature, and for some polymers such as EVOH, also a function of relative humidity.

It is instructive to mention that gas permeation through a polymer film, like water through a pipe or electric current through a resistor, may be described by the empirical relationship:

$$\text{flow rate} = \frac{\text{driving force}}{\text{resistance}} \qquad (4.16)$$

Comparing this relationship with Equation 4.15 indicates that Q is flow rate, pressure difference Δp is driving force for permeation, and:

$$\text{resistance to permeation} = \frac{L}{\overline{P}A} \qquad (4.17)$$

Gas permeation through a film may be compared to water flow through a pipe connecting two water tanks of different heights (Figure 4.4a). Intuitively, the water flow rate is directly proportional to the difference between the water levels (Δh) and inversely proportional to the pipe resistance—this is consistent with Equation 4.17. The pipe resistance increases with the length and friction factor of the pipe but decreases with its cross-sectional area. The similarities are as follows: water flow rate is similar to gas permeation rate, water level difference between the two tanks (Δh) is similar to pressure difference across the film (Δp), cross-sectional area of pipe is similar to surface area of film, length of pipe is similar to thickness of film, and friction factor of pipe is similar to the inverse of permeability of film.

(a)   (b)

Figure 4.4 (a) Flow of water through a pipe, (b) flow of electric current through a resistor

Gas permeation is also similar to the flow of electric current through a resistor (Figure 4.4b). The relationship between electric current I, voltage V, and resistance R across a resistor is described by Ohm's Law:

$$I = \frac{V}{R} \qquad (4.18)$$

which has the same form as Equation 4.17. The total resistance is equal to the sum of individual resistances in series (Figure 4.5a):

$$R_T = R_1 + R_2 + R_3 + \dots \qquad (4.19)$$

$\overline{P}_1$	$\overline{P}_2$	$\overline{P}_3$
L_1	L_2	L_3

$R_1 \quad R_2 \quad R_3$

$\mathsf{\Lambda\!\Lambda\!\Lambda}\!-\!\mathsf{\Lambda\!\Lambda\!\Lambda}\!-\!\mathsf{\Lambda\!\Lambda\!\Lambda}$

(a)   (b)

Figure 4.5 Resistance across (a) resistors in series and (b) a multilayer film

Similarly, the total resistance in a multilayer film is equal to the sum of resistances in individual layers. Since the resistance to permeation for each layer is shown earlier to be $L/(PA)$:

$$\frac{L_T}{\overline{P}_o A} = \frac{L_1}{\overline{P}_1 A} + \frac{L_2}{\overline{P}_2 A} + \frac{L_3}{\overline{P}_3 A} + ... \qquad (4.20)$$

or

$$\frac{L_T}{\overline{P}_o} = \frac{L_1}{\overline{P}_1} + \frac{L_2}{\overline{P}_2} + \frac{L_3}{\overline{P}_3} + ... \qquad (4.21)$$

where L_T is total thickness, $\overline{P}_o$ is overall permeability, and the subscripts 1, 2, 3 refer to the individual layers in the film.

Example 4.1 Overall permeability of a multilayer film

A multilayer film is constructed of 3 mil PP / 0.5 mil PVDC / 3 mil PP. Assume that the oxygen permeabilities for the PP (polypropylene) and PVDC (polyvinylidene chloride) layers are 150 and 1.2 (cm^3·mil)/(100in^2·day·atm), respectively. What is the overall oxygen permeability $\overline{P}_o$ of the film?

$\overline{P}_o$ may be calculated using Equation 4.21:

$$\frac{3 + 0.5 + 3}{\overline{P}_o} = \frac{3}{150} + \frac{0.5}{1.2} + \frac{3}{150}$$

where the calculated $\overline{P}_o$ is 22.4 (cm^3·mil)/(100in^2·day·atm). This $\overline{P}_o$ value is applicable to other PP/PVDC/PP films as long as the ratio of individual thicknesses remains the same (i.e. 3:0.5:3). For multilayer films, the permeance P' (see Section 4.4.2) is frequently reported instead of $\overline{P}_o$. P' is equivalent to $\overline{P}/L$ and is not normalized for thickness. The P' value in this example is 3.45 cm^3/(100in^2·day·atm).

Example 4.2 Relative humidity inside a multilayer film

In Figure 4.6, a multilayer film consisting of PP/EVOH/PP is exposed to 100 and 40% RH. What is the relative humidity at the middle of the EVOH layer (i.e., RH$_1$) at steady state? Assume that the water vapor permeability of PP and EVOH are 0.7 and 7.8 (g·mil)/(100 in^2·day) per 90% RH difference, respectively.

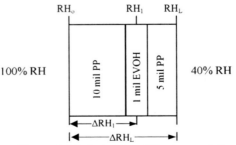

Figure 4.6 Relative humidity in a multilayer

Similar to pressure drop across a water pipe or voltage drop across a resistor, RH in the multilayer film also decreases in the flow direction due to the resistances to permeation of the individual layers. Accordingly, the RH in the middle of the EVOH layer may be expressed by:

$$RH_1 = RH_o - \Delta RH_1 \qquad (4.22)$$

where the decrease in ΔRH_1 is due to the combined resistance of the 10-mil PP and half of the 1-mil EVOH. Since the difference in relative humidity ΔRH is a driving force for permeation, the relationship in Equation 4.16 may be written as:

$$\Delta RH = \text{permeation rate} \times \text{resistance} \qquad (4.23)$$

And it follows that:

$$\Delta RH_1 = \text{permeation rate} \times \text{resistance across } \Delta RH_1 \qquad (4.24)$$

$$\Delta RH_L = \text{permeation rate} \times \text{resistance across } \Delta RH_L \qquad (4.25)$$

Substituting Equations 4.24 and 4.25 into Equation 4.22:

$$RH_1 = RH_o - \frac{\text{resistance across } \Delta RH_1}{\text{resistance across } \Delta RH_L} \cdot \Delta RH_L \qquad (4.26)$$

Furthermore, applying Equation 4.17 for the resistances to permeation:

$$RH_1 = RH_o - \frac{\dfrac{L_1}{P_1} + \dfrac{L_2}{2P_2}}{\dfrac{L_1}{P_1} + \dfrac{L_2}{P_2} + \dfrac{L_3}{P_3}} \cdot \Delta RH_L \qquad (4.27)$$

where the subscripts 1, 2, 3 refer to the individual layers counting from the left in Figure 4.6. Substituting the given values:

$$RH_1 = 100 - \frac{\dfrac{10}{0.7} + \dfrac{1}{2(7.8)}}{\dfrac{10}{0.7} + \dfrac{1}{7.8} + \dfrac{5}{0.7}} \cdot (100 - 40) = 60.1$$

Hence the relative humidity at the middle of the EVOH layer is 60.1%. This approach may also be used for films with more than three layers.

4.4 Terminology and Units for Permeation

There is much confusion in the terminology and units used to describe permeation in the literature. This situation is partly due to the lack of standard terminology embraced by all authors and partly due to the incorrect use of terms by some authors not adhering to scientific principles. The terms used in the book are based on the permeation rate equation and consistent to those adopted by the ASTM standards.

The permeation rate Q is used to describe permeation of finished packages: for example, the oxygen permeation rate of a plastic bottle may be expressed in terms of $cm^3(STP)$ day^{-1} per bottle, although 'per bottle' is usually not explicitly stated. For packaging films, three other terms are used to describe permeation: transmission rate describes permeation on per area basis, permeance describes permeation on per area per pressure difference basis, and permeability describes permeation on per area per pressure difference per thickness basis.

To define values for these terms, the polymer as well as the permeant and environmental conditions must be specified. For example, it is not sufficient to speak of permeability of high density polyethylene (HDPE)—without also specifying whether it is permeability of HDPE to oxygen, carbon dioxide, or water vapor and what temperature and relative humidity are of interest.

4.4.1 Transmission Rate

Transmission rate (TR) is the quantity of permeant passing through a film per unit area per unit time at steady state:

$$TR = \frac{\text{quantity of permeant}}{A \cdot t} \quad (4.28)$$

The transmission rate of a film is usually measured using the methods described in Section 4.7. In reporting a TR value, the permeant, the polymer, as well as the test conditions including temperature, partial pressures on both sides of the film, and relative humidity must also be stated. TR is also permeation rate normalized for area:

$$TR = \frac{Q}{A} \quad (4.29)$$

The SI unit of TR is $mol/(m^2 \cdot s)$. There are many other units reported in the literature, and some of them are described below.

Since gases are usually measured by volume, gas transmission rate (GTR) is expressed by:

$$GTR = \frac{\text{volume of gas permeant (STP)}}{A \cdot t} \quad (4.30)$$

A common unit of oxygen transmission rate (O_2TR) is $cm^3(STP)/(m^2 \cdot day)$ at one atmosphere pressure difference. 1 cm^3 at standard temperature and pressure (STP) is 44.62 μmol and 1 atmosphere is 0.1013 MPa. The units for carbon dioxide, nitrogen and other gases are also expressed in the same manner.

Since water vapor is usually measured by weight, vapor transmission rate (WVTR) is expressed by:

$$\mathrm{WVTR} = \frac{\text{mass of water}}{\text{area} \cdot \text{time}} \tag{4.31}$$

A common unit for WVTR is $g/(m^2 \cdot day)$ under the test conditions, typically 100°F (38°C) and relative humidities of 90% on one side and 0% on the other side.

4.4.2 Permeance

Permeance P′ is transmission rate normalized for pressure difference:

$$\mathrm{P}' = \frac{\mathrm{TR}}{\Delta p} \tag{4.32}$$

Like transmission rate, permeance may be used for either homogeneous or heterogeneous films. The SI unit of permeance is $mol/(m^2 \cdot s \cdot Pa)$. Below are examples of other units.

Oxygen permeance: $cm^3(STP)/(m^2 \cdot day \cdot Pa)$

Water vapor permeance: $g/(m^2 \cdot day \cdot Pa)$

4.4.3 Permeability

For homogeneous films whose properties are characteristic of the bulk material, permeability $\bar{P}$ may be obtained from:

$$\bar{P} = \mathrm{P}' \cdot \mathrm{L} = \frac{\mathrm{TR}}{\Delta p} \cdot \mathrm{L} \tag{4.33}$$

The SI unit for $\bar{P}$ is $mol/(m \cdot s \cdot Pa)$. Since $\bar{P}$ is normalized for thickness, its use is appropriate only if the transmission rate decreases linearly with thickness. This linear relationship is not obeyed by multilayer or coated films, and thus $\bar{P}$ is not meaningful for these heterogeneous materials. However, the total permeability of a multilayer film may be calculated from the permeabilities of individual layers using Equation 4.21.

According to Equation 4.15 or 4.33, the unit for permeability is:

$$\bar{P} = \left[\frac{\text{quantity of permeant}}{\text{time}} \right] \left[\frac{\text{film thickness}}{\text{film area} \times \text{pressure difference}} \right] \tag{4.34}$$

Since oxygen is typically measured by volume, below are some common units for oxygen permeability:

English unit $\dfrac{cm^3(STP) \cdot mil}{day \cdot 100\ in^2 \cdot atm}$

Metric unit $\dfrac{cm^{3}(STP) \cdot cm}{cm^{2} \cdot s \cdot cm\ Hg}$

The peculiar English unit arises from the common practice that gas transmission rate is measured in cm^{3}/day, film thickness is measured in mil (equivalent to 1/1000 inch), area for a typical food package is in the order of 100 in^{2} (certainly not 1 in^{2}!), and pressure difference is measured in atmosphere.

Below are some common units for water vapor permeability:

English unit $\dfrac{g \cdot mil}{day \cdot 100\ in^{2} \cdot atm}$

$\dfrac{g \cdot mil}{day \cdot 100\ in^{2}}$ per typically 90% RH difference

Metric unit $\dfrac{g \cdot cm}{cm^{2} \cdot s \cdot cm\ Hg}$

When reporting permeability values, temperature of the testing condition must also be stated.

Example 4.3 Conversion of Units

Convert 50 $(cm^{3} \cdot mil)/(100 in^{2} \cdot day \cdot atm)$ into $(cm^{3} \cdot m)/(cm^{2} \cdot s \cdot cm \cdot cm\ Hg)$.

$$50\ \frac{cm^{3} \cdot mil}{100\ in^{2} \cdot day \cdot atm} \left(\frac{0.001\ in}{mil}\right)\left(\frac{2.54\ cm}{in}\right)\left(\frac{in}{2.54\ cm}\right)^{2}\left(\frac{day}{(24)(3600)\ s}\right)\left(\frac{atm}{76\ cm\ Hg}\right)$$

$$= 3.00 \times 10^{-11}\ \frac{cm^{3}\ cm}{cm^{2}\ s\ cm\ Hg}$$

While this appears to be a trivial example, conversion from one unit to another is frequently necessary in this global economy, since practitioners from different countries tend to prefer different units and this situation is unlikely to change in the foreseeable future. The units in this book are commonly used in the United States and Europe.

4.5 Permeability of Food Packaging Polymers

Table 4.1 shows typical ranges of permeability values for some common food packaging polymers. It can be seen that polymers offer a wide choice of barrier properties, ranging from low oxygen barrier materials such as polyethylene (PE) to high oxygen barrier materials such as ethylene vinyl alcohol (EVOH). It is noted that polar polymer such as EVOH deteriorates in gas permeability under high relative humidity conditions (see also Section 4.6).

Table 4.1 Permeability values of polymers

Polymer	O_2 Permeability[1]	CO_2 Permeability[1]	Water vapor permeability[2]
Polyethylene (PE)			
Low density	300 - 600	1200 - 3000	1 - 2
High density	100 - 250	350 - 600	0.3 - 0.6
Polypropylene (PP)			
Unoriented	150 - 250	500 - 800	0.6 - 0.7
Oriented	100 - 160	300 - 540	0.2 - 0.5
Polystyrene (PS)	250 - 350	900 - 1050	7 - 10
Polyethylene terephthalate (PET)	3 - 6	15 - 25	1 - 2
Polyvinyl chloride (PVC)			
Unplasticized	5 - 15	20 - 50	2 - 5
Plasticized[3]	50 - 1500	200 - 8000	15 - 40
Polyvinylidene chloride (PVDC)	0.1 - 2	0.2 - 0.5	0.02 - 0.6
Ethylene vinyl alcohol (EVOH)			1.5 - 8
0% RH	0.007 - 0.1	0.01 - 0.5	—
100% RH	0.2 - 3	4 - 10	—
Ionomer	300 - 450	—	1.5 - 2
Nylon 6	2 - 3	10 - 12	10 - 20
PC	180 - 300	—	10 - 15

[1]Unit in $(cm^3 \cdot mil)/(100in^2 \cdot day \cdot atm)$ at 25°C.
[2]Unit in $(g \cdot mil)/(100in^2 \cdot day)$ at 38°C, 90% RH.
[3]Values depend greatly on plasticizer content.

Since the barrier protection against water vapor and oxygen is most important to food, it is useful to compare the oxygen permeability and water permeability together for selecting packaging material. Figure 4.7 presents the location of polymer material in two dimension matrix of oxygen and water vapor permeabilities. For designing the packaging system for a certain food, its gas and moisture permeabilities should be closely related to the food's sensitivity to oxygen and moisture.

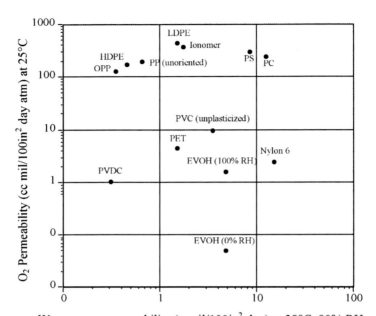

Water vapor permeability (g mil/100in² day) at 38°C, 90% RH

Figure 4.7 Oxygen permeability versus water vapor permeability

Example 4.4 Gas concentration inside a plastic container
A cylindrical HDPE container (r = 3 in, h = 6 in, L = 20 mil) is initially flushed with nitrogen. After flushing, the residual oxygen level is 1.0% and the container is stored in air. Assume the O_2 permeability of HDPE is 150 (cm³ mil)/(100 in² atm day). What is the oxygen concentration inside the container as a function of time?

A mass balance may be used to equate the rate of oxygen accumulation inside the container to the rate of oxygen permeation through the container walls:

$$\frac{dcV}{dt} = \frac{\overline{P}A}{L}(p_o - p_i) \tag{4.35}$$

where c is oxygen concentration inside the container at any time (cm³ O_2 per cm³ container volume), p_o and p_i are partial oxygen pressures outside and inside the container, respectively. Since air contains 21% oxygen, $p_o = 0.21\ p_a$ where $p_a = 1$ atm; p_i is related to c by the relationship $p_i = cp_a$. Hence Equation 4.35 becomes:

$$\frac{dcV}{dt} = \frac{\overline{P}A}{L}(0.21 - c)p_a \tag{4.36}$$

Separating the variables in the equation yields:

$$\frac{dc}{0.21-c} = \frac{\overline{P}Ap_a}{VL}dt \tag{4.37}$$

Integrating and imposing the upper and lower limits:

$$\int_{c_o}^{c}\frac{dc}{0.21-c} = \frac{\overline{P}Ap_a}{VL}\int_0^t dt \tag{4.38}$$

where c_o is initial oxygen concentration. The resulting equation is:

$$c = 0.21 - (0.21 - c_o)\exp\left[-\frac{\overline{P}Ap_a}{VL}t\right] \tag{4.39}$$

It is instructive for the reader to verify the above equation satisfies the conditions that $c=c_o$ at $t=0$ and $c=0.21$ at $t=\infty$. This is an unsteady state problem since the oxygen concentration *inside the container* changes with time. Nevertheless, the oxygen concentration profile *within the container walls* is assumed to have reached 'quasi-steady state' since $(\overline{P}A / L)(p_o - p_i)$ is used to formulate the mass balance in Equation 4.35. This quasi-steady state assumes that the effects caused by the changes in oxygen concentration within the container walls are negligible compared to those inside the container. This assumption greatly simplifies the otherwise very complicated mathematical equation.

For the cylindrical container in this problem, $c_o=0.01$ and:

$$A = 2\pi r h + 2\pi r^2 = 2\pi(3\times6) + 2\pi(3)^2 = 169.6\ in^2$$
$$V = \pi r^2 h = \pi(3)^2(6) = 169.6\ in^3 = 2780\ cm^3$$

Substituting these values into Equation 4.39:

$$c = 0.21 - (0.21 - 0.01)\exp\left[-\frac{\left(\frac{150}{100}\right)(169.6)}{(2780)(20)}t\right]$$

$$c = 0.21 - 0.20\exp[-0.00458t] \tag{4.40}$$

where t should be in day. Since different units are used in this example (an unfortunate situation frequently encountered in real life), extra care is needed to ensure the units are used correctly. The unit of $\overline{P}$ in this example requires that A and V should have the units of in^3 and cm^3, respectively.

Based on calculations using Equation 4.40, $c = 0.172, 0.203, 0.209$ (or 17.2%, 20.3%, 20.9%) at $t = 1, 2$, and 3 years, respectively. This kind of problem will also be handled in Chapter 16 for shelf life estimation of packaged foods.

4.6 Factors Governing Permeation

The major factors governing permeability fall into three categories: nature of polymer matrix, nature of permeant, and ambient conditions. Understanding these factors is helpful for selecting proper materials for barrier protection of food packages.

4.6.1 Nature of Polymer

Generally, the gas and vapor barrier of polymers is improved with increasing polarity of certain types, regularity of molecular structure, and close chain-to-chain packing in the polymer matrix [2]. As shown in Table 4.2, polar functional groups such as OH, Cl, and CN decrease O_2 and CO_2 permeabilities under dry conditions due to strong polymer interactions. High polarity results in high cohesive energy between the polymer chains and consequently lower diffusion and permeability [3].

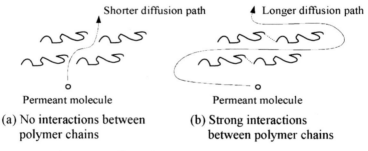

Shorter diffusion path   Longer diffusion path

Permeant molecule   Permeant molecule

(a) No interactions between   (b) Strong interactions
    polymer chains   between polymer chains

Figure 4.8 Effect of interactions on diffusion paths

Figure 4.8 explains the effects of polymer interactions on the diffusion path taken by the permeant. In situation (a), such as for polyethylene, there is no or very weak interaction between the nonpolar polymer chains, and hence the permeant molecule may diffuse through the polymer matrix via a shorter and more direct path. In situation (b), such as for EVOH, there are strong intermolecular interactions such as hydrogen bonding between the polymer chains. These strong interactions block the passage of the permeant molecule, requiring it to diffuse in the polymer matrix via a much longer path. The longer diffusion path also causes the permeability to decrease.

Permeability also decreases with increasing crystallinity and molecular orientation in the polymer matrix, since regions of crystallinity and molecular orientation are obstacles to the passage of the permeant. Polymers with regular molecular structure and close chain-to-chain packing tend to have higher degrees of crystallinity and are more easily oriented in molecular structure (see

Section 2.5). Higher crystallinity also lowers the solubility of permeant in the polymer matrix and thus the permeability. Inclusion of crystallites or inorganic platelets may also increase crystallinity and lower permeation [4].

Permeability generally increases with addition of additives, fillers, and plasticizers in the polymer matrix. Inert fillers such as $CaCO_3$ and ceramic powder have been used to increase the gas permeability of polyolefin films for fresh produce packaging applications.

4.6.2 Nature of Permeant

Permeability also depends on molecular size of the permeant and its chemical affinity to the polymer matrix. Larger permeant molecules generally have lower diffusivity and higher solubility compared to smaller molecules, although solubility also depends on the chemical similarity between the polymer and the permeant [5].

It is interesting to note in Table 4.1 that the values of CO_2 permeability are typically 3-7 times the values of O_2 permeability. How could the larger CO_2 molecule permeate faster than the smaller O_2 molecule? The answer may be found by recalling Equation 4.14 that permeability is the product of diffusion coefficient and solubility coefficient. While it is true that CO_2 diffuses slower than O_2, the solubility of CO_2 is much higher than O_2—the combined effect enables CO_2 to permeate faster than O_2. Although CO_2 molecules diffuse slower, there are more of them diffusing through the polymer matrix.

Table 4.2 Gas or water vapor permeabilities of selected polymers

Polymer	Permeability for gas (cm^3 mil 100in^{-2} day^{-1} atm^{-1}) and water vapor (g mil 100in^{-2} day^{-1} atm^{-1})				Functional group
	O_2 (dry condition)	O_2 (wet condition: 80% RH)	CO_2 (dry condition)	H_2O (38°C, 90% RH)	
PVOH	0.02	7.00	0.06	10.00	-OH
EVOH	0.05	7.00	0.23	10.00	-OH
PVDC	0.08	0.08	0.30	0.05	-Cl
PAN	0.03	0.03	0.12	0.50	-CN
PET	5.00	5.00	20.00	1.30	-COO-
Nylon 6,6	3.00	15.00	5.00	24.00	-CONH-
PP	110.00	110.00	240.00	0.30	-CH$_3$

Abbreviation: PVOH=polyvinyl alcohol; PAN=polyacrylonitrile; refer to Table 4.1 for others. From Ref. [2] with kind permission of American Chemical Society.

Although the permeabilities of most inert gases are independent of permeant concentration according to Equation 4.15, organic vapors such as aroma, flavors, and solvents show strong dependence on the concentration because of their strong interactions with the polymer.

4.6.3 Ambient Environment

The most important environmental factors are temperature and relative humidity. Temperature generally affects gas permeability by following Arrhenius equation:

$$\overline{P} = \overline{P}_o \exp\left(- \frac{E_a}{RT} \right) \tag{4.41}$$

where $\overline{P}_o$ is pre-exponential factor, E_a is activation energy, R is the gas constant, and T is absolute temperature.

As rough estimate, a $10°C$ increase in temperature increases the permeability roughly by a factor of 2. However, the temperature dependence of permeability drastically changes at glass transition temperature, i.e. the activation energy changes around that temperature. Many packaging plastics such as PET, PE, and PP show the change of activation energy in a narrow temperature range between 0 and $-12°C$.

When water under high humidity is absorbed onto the polymer and interacts with polar group to swell the polymer structure, the gas permeability is greatly increased. Moisture acts as a plasticizer in the polymer structure. This kind of behavior is observed for PVOH, EVOH, and nylon 6,6, but not for polar polymers such as PVDC and PAN (Table 4.2). Moisture interacts differently with types of polymers: it is absorbed little onto nonpolar polymers (PP) and certain types of polar ones (PVDC, PAN), and thus does not affect the permeability. The interaction of the polymer with moisture can be seen by high water vapor permeability for PVOH, EVOH, and nylon 6,6 (Table 4.2). Higher polarity of EVOH resulting with the lower ethylene content increases the water vapor permeability, which means increased sensitivity to moisture.

4.7 Measurements for Permeation Properties

Measuring permeation properties including permeation rate, transmission rate, permeance, and permeability is important for quality control or assurance and research and development. The following sections introduce the basic concepts of permeation testing. Specific procedures may be found in ASTM Standards such as D1434, D3985, F372, and F1307.

4.7.1 Basic Design Principles

Although many methods are available for measuring permeation properties of polymer films and sheets, only a few of them are used by packaging laboratories. The designs of most popular methods are based on the permeation rate equation:

$$Q = \overline{P}A \frac{p_A - p_B}{L} = \frac{\overline{P}A}{L}\Delta p \qquad (4.15)$$

For example, a sensitive detector or a precision balance is used to measure the permeation rate Q, a permeation cell is used to accurately define the exposed film surface area A, a device is used to control the pressure difference Δp (i.e. p_A - p_B) of permeant, and a precision micrometer is used to measure average film thickness L. Transmission rate may then be calculated from Equation 4.28, permeance from Equation 4.32, and permeability from Equation 4.33.

The choice of detector determines which permeant can be measured; for example, thermal conductivity detectors are used for most permeant gases, flame ionization detectors for volatile organic vapors, coulometric sensors for oxygen, and infrared detectors for water vapor.

4.7.2 Isostatic Method

The heart of this method is a permeability cell consisting of two cylindrical stainless steel compartments separated by a sample film (Figure 4.9a). Streams of permeant gas and an inert carrier gas (such as nitrogen or helium) are sweeping continuously through the compartments. It is called an isostatic method because the same total pressure is achieved on both sides of the film by balancing the two gas flows so that the film is not stressed. Yet a partial pressure difference Δp is also maintained, thereby providing the constant driving force for a portion of the permeant gas to move across the film.

To measure transmission rate, a detector is connected to the outlet of the lower compartment. Data generated from this method are permeation rate as a function of time (Figure 4.9b). There is an initial period of time for conditioning the film during which the permeation rate increases with time. The permeation rate reaches the steady state at a later time when it is no longer changing. Since the film area A and pressure difference Δp are also known, the permeance can be calculated using Equation 4.32. If the film is homogeneous, its permeability can also be calculated using Equation 4.33.

If the diffusion coefficient D is independent of permeant concentration, it can be estimated using the half-time method:

$$D = \frac{L^2}{7.205\, t_{1/2}} \qquad (4.42)$$

where $t_{1/2}$ is time required to reach one half of steady state permeation rate Q and L is film thickness. This method requires that the permeant must be totally outgassed from the film before testing begins and time zero is defined. Furthermore, the solubility coefficient can be estimated from $S = \overline{P}/D$ assuming the validity of Henry's Law.

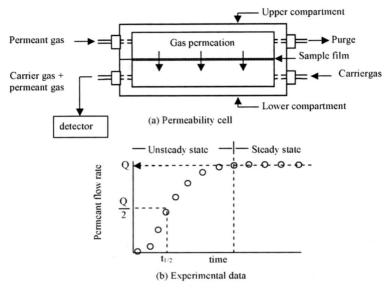

(a) Permeability cell

(b) Experimental data

Figure 4.9 Isostatic method for measuring permeability

4.7.3 Quasi-Isostatic Method

This method is similar to the isostatic method except that the flow of carrier gas in lower compartment of the permeability cell is zero (Figure 4.10a). This is achieved by first purging the lower compartment with a carrier gas and then closing the end ports to leave behind a stagnant volume of carrier gas. In the upper compartment, a stream of permeant gas creates the driving force for the permeant molecules to permeate toward and accumulate in the lower compartment.

At predetermined time intervals, small amounts of gas are taken from the compartment through the sampling port and immediately injected to a gas chromatograph for analysis. To compensate for the loss, the same amount of carrier gas is injected back to the compartment. The experiment is stopped when the permeant concentration in the lower chamber reaches about 10% of that in the upper chamber because the calculation for this method assumes zero permeant concentration in the lower chamber.

Data generated from this method are the amount of permeant accumulated in the lower compartment as a function of time (Figure 4.10b). The steady state permeation rate is approximated by the slope of the line. The permeance and permeability can then calculated using Equations 4.32 and 4.33, respectively.

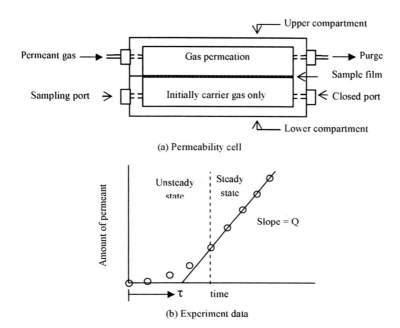

(a) Permeability cell

(b) Experiment data

Figure 4.10 Quasi-isostatic method for measuring permeability

The diffusion coefficient may be obtained from the equation:

$$D = \frac{L^2}{6\tau} \tag{4.43}$$

where τ is lag time estimated from the intersection by the steady state straight line and the x-axis (Figure 4.10b). This method also requires that the permeant must be totally outgassed from the film before time zero is defined. The solubility may be estimated from $\overline{P} = D\,S$.

4.7.4 Measuring Permeation Rate of Finished Packages

The permeation rates of finished packages instead of films can also be measured. Figure 4.11 shows a method in which O_2 (test gas) and N_2 (carrier gas) is sweeping the outside and the inside of a plastic bottle, respectively, thereby creating an O_2 pressure difference of approximately 1 atmosphere across the bottle walls. The detector is used to measure the amount of O_2 permeated through the package. The enclosure may be as a chamber or a plastic bag to ensure that the O_2 concentration surrounding the outer walls of the bottle is approximately 100%.

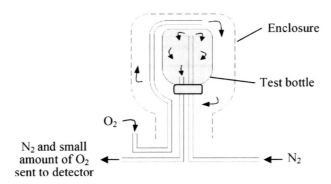

Figure 4.11 Measuring O_2 permeation of a plastic container

This method is similar to the isostatic method in Section 4.7.2 because it also involves the continuous flows of a test gas and a carrier gas. Data generated from this method are similar to those in Figure 4.9, and the O_2 permeation rate of the bottle in cm^3(STP)/day may be obtained once steady state is reached. Usually, permeability is not calculated since the bottle does not have a uniform thickness.

4.7.5 Another Method for Estimating Permeation of Plastic Containers

A method for estimating the O_2, CO_2 and N_2 permeation of a plastic container is shown in Figure 4.12. The lid has three silicon ports: the vacuum pump port and the injection port are used to flush CO_2 throughout the container, and once the container is completely filled with CO_2, the sampling port is used to periodically withdraw headspace gas samples for O_2, CO_2, and N_2 analysis. This method is similar to the quasi-isostatic method in Section 4.7.3 because it also involves the periodic withdrawal of gas samples for analysis.

As permeation occurs across the container, its headspace composition responds by changing with time. This dynamic behavior may be described by the mass balance equation:

$$\frac{dn_{i,in}}{dt} = \frac{\overline{P}_i A(p_{i,out} - p_{i,in})}{L} \qquad (4.44)$$

where the left side represents the rate of change in the O_2, CO_2 and N_2 headspace concentrations; the right side represents the permeation rates of O_2, CO_2 and N_2 through the container body and lid. The symbol n is mole number; the subscript i refers to O_2, CO_2 or N_2; the subscript 'in' and 'out' refer to inside and outside of the container, respectively. Note that Equation 4.44 assumes that the concentration profile within the container walls is at quasi-steady state (also see discussion in Example 4.4 of Section 4.5). This can be achieved by flushing the container with CO_2 through injection and vacuum ports.

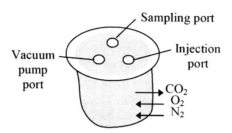

Figure 4.12 Estimating gas permeabilities of a
hermetic plastic container

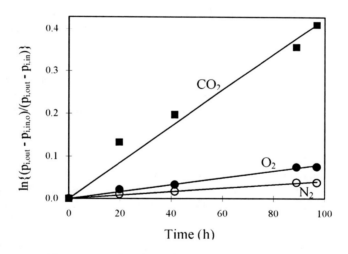

Figure 4.13 Plots for estimating package permeation

If the Ideal Gas Law $pV = nRT$ applies, then Equation 4.44 becomes:

$$\frac{dp_{i,in}}{dt} = \frac{RT}{V}\left[\frac{\bar{P}_i A (p_{i,out} - p_{i,in})}{L}\right] \tag{4.45}$$

where V is volume of the container. Furthermore, if the gas concentrations outside the container $p_{i,out}$ remain constant, the above equation upon integration becomes:

$$\ln\left(\frac{p_{i,out} - p_{i,in,o}}{p_{i,out} - p_{i,in}}\right) = \frac{RT}{V}\left[\frac{\overline{P}_i A}{L}\right]t \tag{4.46}$$

where $p_{i,in,o}$ and $p_{i,in}$ are partial pressures of O_2, CO_2, or N_2 inside the container at time zero and at any time, respectively.

According to Equation 4.46, a plot of $\ln[(p_{i,out} - p_{i,in,o})/(p_{i,out} - p_{i,in})]$ versus t should yield a straight line and from its slope $\overline{P}_i A / L$ can be obtained. We call the term $\overline{P}_i A / L$ 'package permeability' (note that this is different from the 'permeability' used throughout this chapter), because it is a measure of permeation for the entire package. The inverse of this term is equivalent to the package resistance defined in Equation 4.17.

Figure 4.13 shows sample plots for a hermetic plastic container (V=1095 mL) initially flushed with 100% CO_2 with adequate time so that container walls were fully conditioned before measurements began. The outside partial pressure $p_{i,out}$ for O_2, CO_2 and N_2 were 0.21, 0.00 and 0.79 atm, respectively. From the slopes, the calculated 'package permeabilities' $\overline{P}_i A / L$ for O_2, CO_2 and N_2 are 0.88, 4.65, 0.46 mL h^{-1} atm^{-1}, respectively.

Permeabilities may also be calculated using average values for A and L, with the understanding that these are 'estimated values' since there are some variations in thickness over the area. For A=0.074 m^2 and L=2.1 mm, the estimated permeabilities for O_2, CO_2 and N_2 are 25, 132 and 13 cm^3 mm h^{-1} m^{-2} atm^{-1}, respectively.

4.7.6 Gravimetric Method

The water vapor transmission rate (WVTR) may be measured gravimetrically using the cup method. Figure 4.14 shows an aluminum cup sealed by a film sample. A desiccant such as calcium chloride or calcium sulfate is used to maintain a low water vapor pressure inside the cup. The cup is stored in a temperature controlled cabinet at relative humidity typically around 90%; 100% RH is not used to avoid condensation of water vapor. The weight of the cup is monitored as a function of time (Figure 4.14). The WVTR is obtained from the slope of the line after steady state condition has been achieved. The water vapor permeability may also be calculated using the

$$\overline{P}_w = \frac{WVTR \cdot L}{A\,(0.9\,P_s)} \tag{4.47}$$

where L is film thickness, A is surface area of film, and P_s is saturated water vapor pressure. $0.9P_s$ represents the water vapor pressure difference between 90% and 0%.

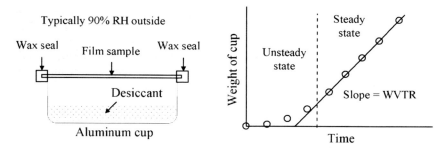

Figure 4.14 Gravimetric method for measuring WVTR

WVTR data in the literature are typically measured at 38°C, 0% RH inside (RH_i) and 90% RH outside (RH_o). This accelerated condition is chosen to shorten the testing time. In Table 4.3, this accelerated condition is compared to two more real-life conditions: Case 1 may present a package containing a moist food ($a_w = 0.9$) stored at low temperature (7°C) in a dry warehouse (45% RH), while Case 2 may represent a package containing a dry food ($a_w = 0.1$) stored at room temperature (20°C) in a humid warehouse (75% RH). The last column in the table shows the water vapor pressure driving force, P_s (RH_o - RH_i), of the accelerated condition is much higher than those of the real-life conditions. When time is not a limiting factor, conducting experiments under real-life conditions is often preferred.

Table 4.3 Comparison between testing conditions and real life conditions

	T (°C)	P_s (mm Hg)	RH_i (%)	RH_o (%)	P_s (RH_o -RH_i) (mm Hg)
Accelerated condition	38	49.7	0	90	44.7
Case 1 condition	7	7.51	90	45	3.38
Case 2 condition	20	17.5	10	75	11.4

4.8 Gas Transport through Leaks

Besides permeation, leak is also an important mechanism of gas transport. Pinholes and channel leaks are two types of leakers occasionally found in defective packages. Pinholes may be found on package walls; for example, on very thin aluminum foils of less than 1 mil thick. Channel leaks are minute channels which may found in defective seal areas caused by improper sealing conditions. Since channel leaks usually have larger depths than pinholes, their leak rates may be much slower.

Figure 4.15 illustrates that the mechanism of gas transport through a pinhole involves diffusion of gas molecules through a column of stagnant air inside the pinhole. Unlike permeation, the steps of adsorption and desorption are not involved. Also, diffusion through leak occurs in air, while diffusion in permeation occurs in solid. The diffusion velocity in gases, liquids, and solids are generally in the range of 0.00001, 0.5, and 10 cm/min, respectively—hence the transport of gas usually occurs faster through leak than permeation. Both channel leaks and pinholes should be prevented since they can significantly compromise the gas barrier of the package.

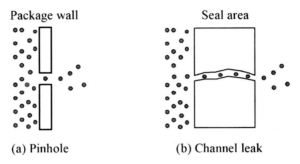

Figure 4.15 Pinhole and channel leak

Discussion Questions and Problems

1.   What is the change in permeation rate if permeability is decreased by 40%, surface area is increased by 20%, and thickness is decreased by 50%?

2.   A cylindrical HDPE container is flushed with nitrogen. After flushing, the oxygen level in the container is 2.5%, and the container is then stored in air at 25°C. The O_2 permeability of HDPE at 25°C is 150 $(cm^3$ mil)/(100 in^2 atm day).

  a.   Derive an equation to describe the oxygen concentration (%, v/v) inside the package as a function of storage time.

  b.   Plot the oxygen concentration as a function of storage time up to 12 months, for package thicknesses = 10, 35, and 50 mil. Assume package radius = 3 in and package height = 6 in.

  c.   Plot the oxygen concentration as a function of storage time up to 12 months, for package radii = 2, 3, and 4 in. Assume package height = 6 in and package thickness = 35 mil.

  d.   Estimate the oxygen concentration after 3 years.

3.   A gravimetric method was used to measure the water vapor permeation rates of two kinds of thermoformed blisters: one made of a low cost material X and the other made of PVC. Five samples were used from each material. Each sample was a square pack of material X or PVC containing 25 blisters. The blisters were first filled with calcium chloride and then heat

sealed with aluminum plastic lids. Unfortunately the lid used for material X was rather thin, and thus there was some concern about the presence of pinholes. The lid used for PVC was a thicker material with no pinhole problem. The prepared samples were stored inside a controlled environment at 37°C and 84% RH, and their weights were periodically recorded as a function of time.

Samples for material X

t (h)	X1	X2	X3	X4	X5
0.0	6.7798	6.7686	6.7360	7.0306	6.5682
22.9	6.7934	6.7786	6.7502	7.0386	6.5764
47.8	6.8411	6.8152	6.7959	7.0660	6.6048
71.5	6.8833	6.8473	6.8370	7.0886	6.6295
95.4	6.9255	6.8793	6.8779	7.1108	6.6541
114.5	6.9612	6.9059	6.9122	7.1295	6.6749
159.9	7.0579	6.9712	7.0035	7.1796	6.7313

Samples for PVC

t (h)	PVC1	PVC2	PVC3	PVC4	PVC5
0	8.3142	8.4380	7.6815	7.7038	7.7920
22.9	8.3214	8.4469	7.6904	7.7199	7.7995
47.8	8.3430	8.4702	7.7129	7.7723	7.8226
71.5	8.3622	8.4910	7.7323	7.8180	7.8432
95.4	8.3811	8.5114	7.7519	7.8640	7.8639
114.5	8.3978	8.5285	7.7686	7.9022	7.8807
159.9	8.4400	8.5748	7.8125	8.0042	7.9279

a. Calculate the means and the standard deviations of water vapor permeation rate in g/h for the material X and PVC blisters. Explain if any unreasonable data may be excluded in the calculations.

b. The standard deviation for material X will be found to be rather large, after some unreasonable data are excluded. Explain the possible cause.

c. Assume the inner and the outside RH's of the blisters are changed to 40 and 80%, respectively. Estimate the amount of moisture each blister of X and each blister of PVC will absorb per year.

4. The oxygen permeability of a film was determined using the quasi-isostatic method. The film thickness is 2.5 mil, surface area of test film is 50 cm², volume of lower cell chamber is 50 cm³, $\Delta P = 0.6$ atm, and test temperature is 25°C. The O_2 concentrations (v/v %) in the cell chamber are

Run time (h)	% O_2
0	--
1.0	--

1.5	0.001
2.0	0.060
2.5	0.120
3.0	0.228
3.5	0.396
4.0	0.782
4.5	1.256
5.0	1.720
5.5	2.256
6.0	2.712

    a. Estimate the permeability of the film in $(cm^3\ O_2\ mil)/(m^2\ day\ atm)$.

    b. Estimate the diffusion coefficient in cm^2/s using the lag time.

5.   A plastic sheet is constructed of three layers as shown in the diagram. One side is exposed to 85% RH and the other side to 40% RH. The water permeabilities of PP and EVOH can be obtained from Example 4.2. Calculate the steady-state relative humidity at the mid-point of the EVOH layer, which is shown as RH in the diagram.

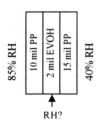

RH?

Bibliography

1. DH Chung, SE Papadakis, KL Yam. A model to evaluate transport of gas and vapors through leaks. Packag Technol Sci 16:77-86, 2003.
2. GW Halek. Relationship between polymer structure and performance in food packaging applications. In: JH Hotchkiss, ed. Food and Packaging Interactions. Washington DC: American Chemical Society, 1988, pp. 195-202.
3. B Pascat. Study of some factors affecting permeability. In: M Mathlouthi, ed. Food Packaging and Preservation. London: Elsevier Applied Science, 1986, pp. 7-23.
4. WE Brown. Plastic in Food Packaging. New York: Marcel Dekker, 1992, pp. 292-396.
5. RJ Hernandez, SEM Selke, JD Culter. Plastics Packaging. Munchi, Germany: Hanser Publishers, 2000, pp. 313-352.

Chapter 5

Migration and Food-Package Interactions

5.1   Introduction 109
5.2   Phenomenal Description of Migration Process 110
    5.2.1 Phenomenal Description of Migration Process ················· 111
    5.2.2 Kinetic and Thermodynamic Approach ···························· 114
5.3   Migration Issues in Food Packaging 116
    5.3.1 Chemicals from Plastics ·· 117
    5.3.2 Recycled Plastics ··· 117
    5.3.3 Microwave Susceptor ··· 120
5.4   Flavor Scalping and Sorption 121
5.5   Migration Testing 122
    5.5.1 General Principles ··· 123
    5.5.2 Food Simulants ··· 124
    5.5.3 Analytical Techniques ·· 124
5.6   Predictive Migration Models 125
    5.6.1 Models for Simplified Systems ··· 125
    5.6.2 Estimation of Diffusion Coefficient and Partition Coefficient ········ 128
    5.6.3 Modeling for Worst Case Scenario ··································· 130
5.7   Regulatory Considerations 132
    5.7.1 Commons and Differences between Regulations ··············· 133
    5.7.2 Sensory Tainting ·· 136
Discussion Questions and Problems 137
Bibliography 138

5.1 Introduction

While the package protects the food from the external environment, it may also interact, either negatively or positively, with the food. Typically these interactions occur at the food-package interface and may be classified as physical, chemical, or biological in nature. For example, the movement of chemical compounds across the food-package interface represents a physical interaction, the corrosion of a metal container caused by contact with a food product represents a chemical interaction, and food spoilage due to microbial contamination by the package represents a biological interaction.

This chapter is focused on physical interactions relating to movements of harmful chemical compounds from the package to the food, or in the opposite direction. Those movements may occur at macroscopic or submicroscopic scale. Macroscopic interactions involve the movement of relatively large, macroscopic fragments by processes such as soiling and rough contamination. Submicroscopic interactions involve the movement of submicroscopic compounds by specific means of molecular diffusion.

Migration, the main topic in this chapter, may be defined as a submicroscopic food-package interaction involving the movement of chemical compounds in the direction from the package to the food, with the emphasis that the movement is controlled by molecular diffusion [1]. Migration from plastics is of particular concern, since it is related to the important matter of consumer protection and regulatory compliance. In most developed countries, some regulations on migration are being enforced. While those regulations differ from country to country, the guiding principles are similar. A brief review of the regulatory aspects will be provided later in this chapter.

Scalping is a popular term used to describe the migration of chemical compounds from food to package. To underline the direction of mass transfer, scalping is sometimes known as reverse migration. Flavor scalping is a common problem which involves the movement of flavor compounds from the food to the package (especially in situations where the food contact layer of the package is made of plastics), resulting in the loss of desirable flavors in the food. Undesirable migration or scalping should be minimized and possibly avoided.

The negative impact of migration has become an important aspect for analyzing the safety of packaged foods. While empirical data are being accumulated in the literature and predictive mathematical models are being developed, due to the complexity of food-package systems, further research is still needed to expand the current knowledge and tools and apply them to make more informative and wiser decisions.

Besides migration and scalping, there are other important food-package interactions. Food contamination may cause corrosion on surfaces of metal (discussed in Chapter 8), glass, and ceramic. In some cases, migration of active compounds is intentional and desirable; for example, antimicrobials or antioxidants can be incorporate into the package in such a manner to allow these active compounds to release slowly from the package to enhance safety and quality of the food. This type of active packaging technology will be further discussed in Chapter 15.

5.2 Phenomenal Description of Migration Process

The plastics used as packaging materials are heterogeneous media, because they are always formulated with various amounts of functional additives. The additives are generally quite small molecules (low molecular mass) in comparison with the hosting medium (very high molecular mass); also the chemical nature of the additives (esters, organic salts, amines, etc.) is quite different from that of the inert medium in which they are dispersed. Therefore high mobility of these components is expected in plastics. Moreover, residuals are also present in plastics as a consequence of the process of production or converting (monomers, catalysts, solvents, adhesives, etc.), and also these molecules, for the same reasons stated above, may have a quite high mobility in plastics. Finally, it is not rare to detect in the plastics, neo-forming, or decomposition products (e.g., acetaldehyde and oligomers in PET) which are produced from the reactions that may occur during the processing or ageing of plastic materials. Therefore, three classes of substances may be defined as potential migrants in plastics intended to be in contact with food: additives, residuals, and neo-forming molecules.

5.2.1 Phenomenal Description of Migration Process

The wide spectrum of different modalities by which the interaction phenomenon may occur, led to some scheme of migration phenomenon classification and definitions, which is useful. From modification of Katan's [2] classification, migration can be categorized into three types described below: nonmigrating, volatile, and leaching systems.

▪ Nonmigrating System

In nonmigrating system, it is assumed that there is negligible migration for high molecular weight polymer, some inorganic residues, or pigments. The assumption of zero migration does not impose any concerns on this system of plastic packaging, but it may be changed to migratory system depending on nondetectable limits in analytical techniques or regulatory specifications.

▪ Volatile System

In volatile system migration does not require contact between the packaging and the food even though it may be affected by the contact. Typically this system applies to volatile migrant as shown in Figure 5.1. This type applies to dry solid foods with poor direct contact with the package. As shown in Figure 5.1, only food particle A is contacting the package, while other particles such as particle B do not have direct contact with the food. Migration to the food is limited to volatile compounds which have relatively high vapor pressures at room temperature. The migration phenomenon is usually controlled by the diffusion in packaging material, not by characteristics of food phase. Volatile compounds in the package wall may also be evaporated into outside environment without causing migration toward food.

The steps for migration from package in Figure 5.1 are as follows:

① Diffusion of migrant in package wall toward the food-package interface. The diffusional process will be described in detail later in this chapter.

② Desorption of migrant at food-package interface. For the migrant to reach food particle B (which has no direct contact with the package), the migrant must first be desorbed from the package into the headspace and then be absorbed by the food. This requires the migrants be volatile; nonvolatile compounds will not migrate to food to a large extent due to poor package/food contact.

③ Adsorption onto food. Volatile compounds from the headspace will be adsorbed by the food. At equilibrium, the partitioning of the volatile compounds will depend on the food composition and temperature.

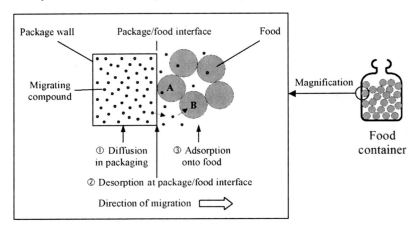

Figure 5.1 Migration in the case of poor food-package contact

- **Leaching System**

Leaching system requires the contact between the packaging and food (mostly liquid or other extractant). The type of migration phenomena includes the most common cases of migration from plastic packaging to liquid foods or moist solid foods with good direct contact with the package (Figure 5.2). Package walls from plastic packages such as plastic containers and plastic pouches in a magnified view are of particular interest here. In a simpler type of the leaching system, the migrant has a high diffusion coefficient in the plastic and is readily dissolved into the contacting food phase. In a dual interactive mode, the migrant is of low diffusion in the plastic at the initial contact with food but attains high diffusion coefficient after liquid food component has been adsorbed and diffused into the film: dual-mode interaction of swelling of plastic material subsequently accelerates diffusion of migrant.

Migration of migrant from package to food in Figure 5.2 involves three steps:

① Diffusion of migrant in package wall toward food-package interface. Frequently this is the slowest or rate-determining step of migration. In a dual interactive mode, concurrent penetration of solvent or food ingredient may change the package wall structure to swollen state increasing the mobility of migrant.

② Dissolution of migrant at food-package interface. The dissolution depends on the affinity of migrant to the package phase and food phase.

③ Dispersion or diffusion of migrant into food. The migrant will then disperse into liquid foods or diffuse into solid foods.

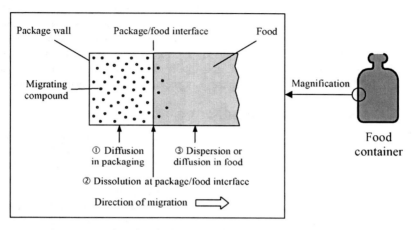

Figure 5.2 Migration in the case of good food-package contact

The physical diffusion theory can be used for describing and predicting the diffusional migration. For the swollen migration of dual interactive mode, however, any complete mathematical treatments have not been established so far and semi-empirical approach is often used for quantitative description of the migration process.

The diffusional mobility of potential migrants in the package, typically along the thickness of the package walls, mainly depends on the temperature, their molecular mass, and their chemical affinity with the plastic in which they are present. Actually, the motion of the potential migrants almost always comprises a very slow diffusion: several days are normally needed for the migrants to go through a few micrometers of plastic thickness. This motion of diffusion may be described by means of Fick's First Law, which defines the diffusion rate in proportion to the driving force of concentration gradient (see also Equation 4.1 in Section 4.2.2):

$$J = -D_p \frac{dC}{dx} \tag{5.1}$$

where J is diffusion flux (mg m^{-2} s^{-1}), D_p is diffusion coefficient of migrant in packaging material (also called diffusivity, m^2 s^{-1}), C is migrant concentration (mg m^{-3}), and x is the distance in the package layer. As for Equation 5.1, the diffusion coefficient determines the rate of migrant movement in the packaging material.

The molecules diffused out from the packaging layer may be transferred to the package headspace in volatile phase (volatile system) or to the contacting food in soluble phase (leaching system). The potential for migration is strictly related to the characteristics of the molecules, of the contact phase (i.e., the food or the beverage inside the package), and of contact condition. For instance, the volatility of the potential migrant may affect direction and degree of migration, the ability of the contact phase to dissolve promptly the migrated substances may influence the rate of migration, and finally the agitation or the turbulence induced by transportation or other causes can accelerate and promote migration.

5.2.2 Kinetic and Thermodynamic Approach

Migration of low molecular weight compounds from packaging materials toward packaged foods mostly involves diffusion process in packaging and dispersion in food with their partition working across the interface. The most general and frequently met cases are liquid foods contained in plastic containers. Figure 5.3 illustrates the progress of migration in terms of migrant concentrations in the packaging and well-mixed food solution: with start of migration, the concentration gradient due to diffusion is set up inside the package film according to Equation 5.1 while the concentration of the migrant dispersed relatively homogeneously in the food starts to increase; the process continues until the equilibrium between phases is attained with disappearance of concentration gradient in the film layer. The migration profile in Figure 5.3 is the result of the sequential progress of migrant transfer from thin plastic packaging material to the liquid food consisting of diffusion in the plastic and dissolution in the interface (Figure 5.2). In this kind of system, diffusion coefficient in liquid food is usually several orders higher than that in the packaging material, and thus the diffusion in the packaging layer limits the rate of the whole transfer process. This condition usually allows one to assume the negligible gradient of migrant concentration through the liquid phase; there may also be some mixing in liquid part by convective movement. Thus equilibrated dissolution of the migrant is assumed at the interface between plastic boundary and food layer: through the whole process of migration, at the interface, the relationship between concentration of migrant in packaging and that in food is dictated by a partition coefficient ($K_{p/f} = C_{p,\infty}/C_{f,\infty}$), which can be obtained at equilibrium condition (Figure 5.3). Please note that $C_{p,\infty}$ and $C_{f,\infty}$ are given in unit of mg m^{-3} for packaging material and food at infinite contact time, respectively.

The rate limiting diffusion process of the plastic packaging material is described for the transient state in one dimension by Fick's Second Law (see Section 4.3.1 for its derivation):

$$\frac{\partial C}{\partial t} = D_p \frac{\partial^2 C}{\partial x^2} \qquad (5.2)$$

where C is migrant concentration as function of time, t and location, x.

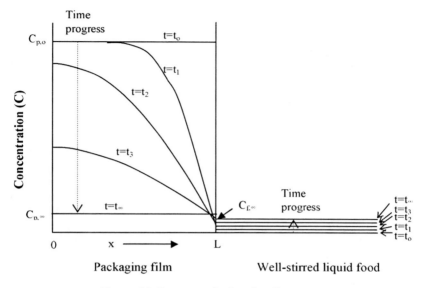

Figure 5.3 Progress of migration from $t=t_o$ to $t=t_\infty$

An analytical solution of Equation 5.2 can be obtained for specific initial and boundary conditions: the most typical one is given by following Equation 5.3 for single side contact of infinite slab with initial conditions of homogeneously constant migrant concentration in plastic layer and zero concentration in the food [3-5].

$$\frac{M_{f,t}}{M_{f,\infty}} = 1 - \sum_{n=1}^{\infty} \frac{2\alpha(1+\alpha)}{1+\alpha+\alpha^2 q_n^2} \exp(\frac{-D_p q_n^2 t}{l^2}) \qquad (5.3)$$

where t is time (s); $M_{f,t}$ is the migration (mg) at time t; $M_{f,\infty}$ is the total amount of migrant in food at equilibration (mg); l is thickness of the package layer (m); α is mass ratio of equilibrated migrant in food to that in packaging material defined by $\frac{M_{f,\infty}}{M_{p,\infty}}$ or $\frac{V_f}{K_{p/f} \cdot V_p}$; $M_{p,\infty}$ is amount of migrant remained in packaging material at equilibration (mg); V_f and V_p are volumes of liquid food and packaging material (m³), respectively; and q_n is the positive roots of the transcendent equation of $\tan q_n = -\alpha q_n$. For $\alpha \ll 1$, $q_n \approx n\pi/(1+\alpha)$, and for other α values $q_n \approx \left[n - \frac{\alpha}{2(1+\alpha)} \right] \pi$.

Even though the computation of Equation 5.3 can be attained by adopting about 50 terms in summation for very short times, summation of about 5 terms suffices to obtain relatively accurate results in most cases with substantial migration.

The other assumption for Equation 5.3 is no loss of migrant on external surface, which gives final fractional release ratio to initially contained migrant by following equation of mass balance equilibrated between package and food:

$$\frac{M_{f,\infty}}{M_{p,o}} = \frac{M_{f,\infty}}{M_{f,\infty} + M_{p,\infty}} = \frac{\alpha}{1+\alpha} \tag{5.4}$$

The migration amount given by Equation 5.3 is regulated both by the diffusion coefficient of the potential migrant in the plastic and its partition coefficient into contacting food. The diffusion coefficient of the migrant in the polymer determining the diffusion rate (Equation 5.1) depends on the size and shape of the migrant molecules and on the size and number of the holes in the plastics [6]. Higher diffusion coefficient leads to faster approach to the equilibrium (Figure 5.4). The partition coefficient influenced by polarity and solubility of the migrant determines the final level of migration at the equilibrium as shown in Figure 5.4.

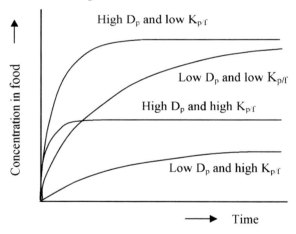

Figure 5.4 Effect of diffusion and partition coefficients on the migration process

5.3 Migration Issues in Food Packaging

A large variety of materials are currently used for food packaging, and different times and conditions of storage are normally employed for the packaged foods. The kind of food contact material as well as the time and temperature of contact can deeply affect the possible migration. Any substance that migrates from the packaging into the food is of concern if it could be harmful to the consumer. Even if the migrating substance is not potentially harmful, it could have an adverse effect on the flavor and acceptability of the food, which should be avoided. Therefore, preventing the packaged food from possible contamination from the packaging material used for protection function is the subject of constant research and regulation.

5.3.1 Chemicals from Plastics

Plastics are, by far, the materials most prone to migration. This mostly depends on their heterogeneous nature, i.e., they are formulated with several different additives according to the performance required. This is also related to the manufacturing processes which can lead to the presence of residuals and also to the production of neo-forming substances. Moreover, their synthetic origin and relatively short history of use (plastics are somehow 'recent materials' about 60 years old) account for the great concern on the migration from plastics from the consumers and the media.

Probably the first important and debated case of migration from plastics had been that of VCM (vinyl chloride monomer) from polyvinyl chloride (PVC) small bottles of liqueur offered from airlines companies in the early '60s. VCM has been ascertained as a potent carcinogen substance. At that time PVC could retain up to 1000 ppm of residual monomer, whereas today the residual is normally less than 1 ppm.

A major concern in the past has been in connection with plasticizers which are used to improve the flexibility of some packaging materials. They are used in a range of plastics but particularly in PVC films. Since it was recognized that in many PVC/food contact situations these additives would migrate, the plastics industry has moved to reformulate various grades of 'cling' films to reduce the likelihood of plasticizer migration. There is no evidence that concentrations of plasticizer found thus far in food constitute a risk to human health, but unnecessary exposure to such contaminants must be avoided. This is the background to the more complex regulations now being developed internationally.

Much more recently, a greatly debated case of migration concerns a class of neo-forming substances like the primary aromatic amines (PAA). These possibly dangerous substances have been occasionally recovered in moist food (mozzarella cheese) packed in plastic laminate structures. Different PAAs can be produced from the reaction between isocyanate and food moisture that diffuses into the laminate. Isocyanate and polyols are the starting substances for polyurethane adhesive, largely used in laminate manufacturing. If not enough time is awaited for completing the cross linking of polyurethane adhesive in the laminate manufacturing, the isocyanate residue within the structure may react with water leading to the synthesis of PAA which can migrate into foods.

5.3.2 Recycled Plastics

Starting from the '90s, pressures to utilize post consumer recycled (PCR) plastics in food packaging have increased enormously. The use of PCR plastics as a food packaging materials is a potentially very attractive application from an economical point of view but also quite problematic one.

PCR plastics, in fact, may contain chemical contaminants, originating from their previous use (or misuse), from the contact with other wastes during collecti ng procedures, or, finally, due to the techniques used for recycling. Since these c hemicals can migrate into food, the composition of recycled high density polyethylene (HDPE) and polypropylene (PP) collected from mixed solid waste has been studied in order to identify the most common low molecular weight contaminants [7]. Several identified compounds were typically fragrance and flavor constituents; in addition, alcohols, esters, and ketones were also found. Most of them are constituents of the packaged products, detergents, or beverages, thus with a low toxicity risk, but the possibility that more dangerous substances may have contaminated the material by contact with other wastes cannot be excluded.

The use of recycled plastics would not present a hazard of potential dangerous migration, as long as the PCR plastics are separated from the food by an effective barrier made from a material meeting all the existing specifications for food and designed to protect the food from the contaminants possibly found in the recycled material. This 'effective barrier', however cannot be assumed as a total, absolute barrier like that provided by a glass or metal layer when a plastic material is involved; therefore the concept of 'functional barrier (FB)' progressively arose. Actually, the expression 'functional barriers' is not new at all, being already used in biology and medicine to describe a mechanism that maintains different the compositions of two domains for specific constituents; not an absolute barrier but a system to stop selectively the transfer at a predefined level or for a specific function.

In food packaging the FBs must ensure the migration of authorized substanc es below the limit of specific migration (SML) and/or reduce the migration of nonauthorized substances to a 'nondetectable' or 'acceptable' level. Ten ppb has been proposed in European Union as the threshold for acceptable level.

To demonstrate that a standard plastic polymer of a given thickness acts as an effective barrier to migration of contaminants, the approach generally followed is to incorporate model contaminants (called 'surrogates') into the resin and to perform extraction studies with food-simulating solvents. Table 5.1 shows a list of the surrogate contaminants mostly used in experimental investigations and suggested by ILSI (International Life Science Institute), on the basis of results from specific research projects of European Union and FDA recommendations [8]. They have been selected in order to cover a broad range of different chemical structures and properties.

Table 5.1 List of model contaminants used for evaluating the efficiency of functional barriers

Surrogate contaminant	Main features
Thrichloro-ethane	Polar, volatile
Benzo phenone	Polar, not volatile
Toluene	Nonpolar, volatile
Chloro benzene	Nonpolar, volatile
Phenyl cyclohexane	Nonpolar, not volatile
Methyl palmitate	Miming organo-metallic

Presently, a large number of practical and theoretical studies have been carried out in this field, focusing on strategic information for a possible PCR plastics usages like the lifetime of a FB in specific application, its minimum thickness for a given level of migration [9], the influence of sample's shape or dimensions [10], and the temperature effects on migration [11].

The assembling of an FB to the PCR plastic can be performed in different ways, mostly by lamination (with adhesives), by coextrusion or coinjection. These two last technologies apply high temperature which makes the contamina nts diffuse through the package at a fast rate, and then the time of protection of t he functional barrier may be significantly reduced. Also this aspect has been ext ensively investigated to show that low density polyethylene (LDPE), high density polyethylene (HDPE), and polyethylene vinyl acetate (EVA) are not suit able polymers for FB design, leading to very high diffusion during processing, w hereas polyethylene terephthalate (PET), ethylene vinyl alcohol (EVOH), polyamide (PA), polyvinyl chloride (PVC), even though require higher processi ng temperatures, are better materials giving lower diffusion rates [12, 13]. Also in these kinds of researches the usefulness of mathematical modeling has been demonstrated coupling of the heat transfer with the contaminant diffusion [14]. Figure 5.5 shows the results obtained in the diffusion modeling of a fast diffusing substance during the coextrusion of a tri-layer PET polymer film of to tal thickness of 0.015 cm with the same thickness of each layer in which the exte rnal layers were assumed as FB and the inner as PCR; after the cooling of the ex truded structure the contaminant is already present in the middle of the functiona l barrier.

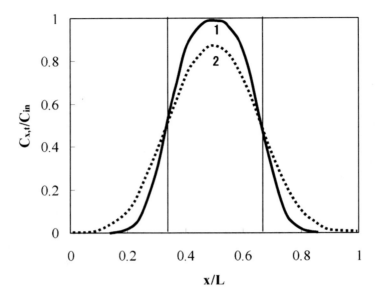

Figure 5.5. Profiles of concentration of the contaminant in the tri-layer film at the end of the cooling period of the coextrusion process, at 300°C (1) and at 330°C (2). The contaminant concentration ($C_{x,t}$) as function of location (x) and time (t) was presented as ratio of the initially contained one (C_{in}) in middle layer. From Ref. [14] with kind permission from Elsevier.

5.3.3 Microwave Susceptor

The microwave susceptors are multilayer structures used as packaging material, as part of a package (popcorn packs) or as a plate or a baking-pan, in order to add crispness and browning to microwaved foods (see Sections 14.4.3 and 14.5.1). They act to convert microwave energy into heat and provide a localized heating with very high temperature. This effect is generally obtained by means of the very thin layer of aluminum of a metallized plastic film. The metallized film may be laminated to paper, board, or other materials, but in the most applications the true food contact material is a PET or oriented polypropylene (OPP) film.

From a migration point of view, these materials are very critical packaging applications. In fact the temperature at the plastic/food interface can rise up to more than 200°C in less than 1 minute and much higher temperature have been recorded into the laminated structure [15, 16]. The effects of such high temperature on possible diffusion phenomena are significant and have been documented several times. The plastic film may deteriorate, leading to the formation of cracks, food components can rapidly migrate into the plastic and the underlying layers, and migration of adhesive compounds from the plastic film to food has been observed [17].

As these problems are known and under investigation since early '90s, there have been several test methods proposed for testing the reliability of microwave susceptors as far as migration is concerned. These methods comprise compositional analyses and migration tests. The migration tests use specially manufactured cells and liquid or solid simulants (olive oil, polymeric absorbent) to collect the migrating substances and to permit an accurate and quantitative analysis of them.

5.4 Flavor Scalping and Sorption

Scalping of flavor occurs commonly by the ability of plastic polymers to dissolve many low molecular weight substances present in foods and beverages. The phenomenon for both volatile and nonvolatile substances may take place in different ways: a sorption mostly limited to the surface of the material (called *adsorption*) or the one that penetrates through the thickness of the material to different extents and by different mechanisms (these cases are generally named *absorption*). Actually, this last absorption often follows true diffusion phenomena and the mass transport may be characterized by determining the coefficients of solubility and diffusion [18], but when they involve substances with a high level of interaction with the polymer, as frequently happens, a time dependent diffusion takes place and the classical Fickian diffusion for the modeling may no longer be valid.

Table 5.2 The effects of sorption phenomena at the interface of food/packaging

Effects on the food	Effects on the package
Loss of vitamins (mainly the oil soluble)	Increase of permeability (mainly gas and aroma permeabilities)
Loss/change of the taste	Decrease of mechanical properties
Loss/change of the flavor	Stress cracking occurrences
Change of the color	Loss of adhesion in laminated structures
Reduction of natural antioxidant protection	
	Decrease of seal strength
Shelf life reduction (as a consequence of previous effects)	Corrosion of laminated foil (for the sorption of acid substances)

These effects of the scalping are generally important and undesirable, because it reduces or alters the flavor of a packed food to such extent that it might be necessary to provide an extra amount of aroma additives to guarantee long shelf lives. The effects of sorption phenomena, however, exceed the sensory aspects and concern both the package and the product; they are summarized in Table 5.2.

The most numerous studies, nevertheless, concern the flavor scalping by LDPE and, particularly, the sorption of limonene and other aroma compounds from fruit juices [19-21]. Actually, LDPE, being the sealing layer of almost all the composite structures, is very frequently used for contact with flavored beverages. Moreover, LDPE has been proved to be very prone to sorption of such kinds of molecules.

The variables affecting the sorption phenomena are the same as those that affect any other mass transport phenomenon (e.g., permeation and migration). Table 5.3 summarizes the variables as the polymer characteristics, the sorbate, and the environment.

Recent results [22] also showed that aroma scalping, strongly depends on the sorbate solvent used. Testing 95% ethanol, sunflower oil, n–heptane, and iso-octane as solvents of d-limonene, n-decane, ethyl caproate, phenylethanol, n-hexanol, and hexanal, it has been demonstrated that polar components were more sorbed into polymers if their solvent was nonpolar and vice versa. To reduce the negative effects of scalping, therefore, the suitable polarity and morphology of polymers may be chosen for specific potential sorbate and its solvent, but a different strategy has also been tested, trying to change the structure of the polymer. Organosilane films have been electron beam grafted to polyolefin film [23], demonstrating very good anti-scalping properties against limonene.

Table 5.3 The variables of polymer, sorbate, and environment which mostly affect sorption

Polymer variables	Sorbate variables	Environment variables
Morphology	Concentration	Temperature
Polarity	Polarity	Relative humidity
Chemical structure	Presence of a comigrant	Geometry

Finally, also some useful effects of scalping have been noted and investigated: the sorption phenomenon was exploited for removal of unwanted constituents. The sorption into polyethylene of polycyclic aromatic hydrocarbons from water was studied as a means for purification of liquid smoke flavor containing possible toxic substances [24].

5.5 Migration Testing

Basically, two different approaches are possible for studying migration, i.e., for investigating rate and/or amount of a specific diffusion phenomenon from package to food: migration testing and predictive migration. Migration testing is a sort of simulation of real migration and a quantitative determination of the substance (or the substances) which is migrated under well defined conditions. The predictive migration is an estimation of migration on the basis of a mathematical model. In other words, referring to the diffusion laws and, possibly, with the aid of specific software, it is very often possible today to predict the rate and level of migration accurately. This possible approach will be described in the next Section 5.6 with some examples. However, even in adopting a predictive approach for migration studies, it is frequently necessary to perform some experimental determinations, for instance to determine the diffusion coefficient or to evaluate the temperature effect on the migration rate.

5.5.1 General Principles

In whatever migration test, it is always possible to have a conceptual distinction of the experimental design in two different steps, which are successive and strictly correlated but characterized by different complexity and problem. The first one concerns the exposure of the package (or its representative sample), to the contact with food or, more often, with one effective simulant; the second one is the quantification and the identification of each substance migrated.

In designing the first step of the migration test, it is required to establish and put under control, the temperature and the modalities of the contact as far as the media and the geometry of the contact is concerned. The chemical nature of the simulant, as well as the surface area and the ratio of 'simulant volume/migrating surface' will affect greatly the final result. Very likely these test conditions are set up differently, according to the purpose desired. The test conditions should generally be the realistic reproduction of the possible interaction, more often, with some acceleration of diffusion, providing a prudential over-estimate of the possible interaction, the 'worst case scenario'. To manage this first step of the migration test, it is almost always necessary to use dedicated devices (migration cells), as this part is completely tailored to the specific purpose of testing migration.

The second step of any whatever migration test deals with the quali-quantitative determination of the migrated substances. The difficulties and specificities of this part of migration tests are the same of whatever analytical task in whatever scientific field. The only peculiarity is that the substances to be there are always in very very low amounts, measurable as ppm level in the best case and as ppt very often.

Another useful consideration is the conformity with the regulations introduced in almost all the developed countries. The migration test may be oriented to the estimate of the total amount migrated, without any identification of the migrated substances, and this purpose coincides with what is called 'overall migration' (OM). Overall migration is somehow a test of inertness and stability of the package or the packaging materials but nothing to deal with the danger of food/packaging interaction. Otherwise, the test may be aimed to find and determine a specific molecule which needs to be monitored and controlled for safety or just for sensorial concerns. In relation to this case many official regulations define the concept of 'specific migration' (SM). It is worth it to underline that an OM test is generally performed by means of simulation test, whereas a specific migration can be estimated by a predictive model.

5.5.2 Food Simulants

Chemical composition and physical structure of food are so complex and diversified that the migration test is rarely carried out on real foods; much more often, in order to have more consistent and more accurate results, food simulant (FS) are used. Food simulants are liquid or solid substances, having a simple and known composition, able to emulate the extraction capacity and solubility of foodstuffs, thus making the migration tests easier and more reliable.

Several different FSs have been proposed and are in use today (Table 5.4), and some of them are now official simulants in the migration methodologies prescribed by mandatory regulations. Generally, in order to make easier the determination of the migrated substances, it is common to select a volatile simulant which can be easily removed after the contact. Also in selecting the simulant, however, the prudential approach of the 'worst case' is recognized and the products chosen are normally much more extractable media than the corresponding food or beverage.

Table 5.4 The food simulants (FS) used in migration testing

Category of FS	Simulant
Aqueous	Distilled water
Acidic	3% (m/v) acetic acid in water
Alcoholic	Ethanol in water (8-50% v/v)
Fatty	Olive oil, sunflower oil, corn oil, synthetic mixture of triglycerides
Solid	Adsorbent polymer (polyphenylene oxide, Tenax®), charcoal, milk chocolate)
Solvent	Heptane, iso-octane, ethanol, ethanol-water

5.5.3 Analytical Techniques

When the aim is to evaluate the amount of global migration, i.e., an estimate of OM, the most classical analytical approach is a gravimetric test. In other words, after the evaporation (for volatile simulant) or the removal of the simulant (for oil), the dry residue or the sample material after the contact is weighed to determine, by difference, the amount of solid substances migrated. An alternative analytical approach that can be applied for OM test, is the measurement of optical density of the liquid simulant in the near UV region, which has been demonstrated to represent in reliable way the traditional method [25]. As another index of overall migration, organic extractable matters in distilled water are measured by titration with $KMnO_4$ solution. Finally, a nonconventional way for overall migration test, is represented by sensorial analysis. An olfactory or gustative test performed on liquid or solid simulant, in fact, may reveal the degree of interaction with high sensitivity.

On the other hand, when the aim is the determination of a specific substance (SM), the most appropriate analytical technique for the particular molecule is used. Modern and advanced analytical techniques like FT-IR GC-MS, LC-MS, are currently used, often in association with analytical steps for the enrichment of the molecule to determine; e.g., solid phase microextraction (SPME), supercritical fluid extraction (SFE), and head space sorptive extraction (HSSE).

5.6 Predictive Migration Models

Since migration experiments are time-consuming and expensive, predictive models have been introduced to facilitate migration studies. The event of migration is described mathematically in order to predict the degree and rate of migrated amount in a certain condition. Mathematical models may use a set of experiments on a food/package system and then provide predictions for other systems consisting of different constructions. An adequate mathematical model can save the time-consuming and expensive experimental procedures of measurements by giving reasonable predictions of migration values based on scientific principles [26]. However, mathematical models have limitations in prediction accuracy and should be used in consistency with the assumptions on which the models are based. The predictions may not describe the phenomena sufficiently in case that the interactions between packaging and food not considered in the modeling occur. Validation process including experimental confirmation is required to ensure the adequate prediction. In this chapter below, simper mathematical formulas with practical usefulness are to be introduced for ready understanding and analysis of the migration phenomena.

5.6.1 Models for Simplified Systems

Simplified solution forms of Fick's Second Law are applied for easy analysis of migration. As a way to deal with the most typical and common case, limited volume package layer contacting infinite volume food ($V_f = \infty$) is often assumed ($\alpha = \dfrac{M_{f,\infty}}{M_{p,\infty}} = \dfrac{V_f}{K_{p/f} \cdot V_p} = \infty$) and the expression of Equation 5.3 converges to Equation 5.5 by the relationship of $q_n = (1 - 1/2)\pi$.

$$\frac{M_{f,t}}{M_{f,\infty}} = 1 - \sum_{n=1}^{\infty} \frac{8}{(2n-1)^2 \pi^2} \exp(\frac{-D_p(2n-1)^2 \pi^2 t}{4l^2}) \qquad (5.5)$$

For long contact times, first term approximation of Equation 5.5 can be used without significant errors:

$$\frac{M_{f,t}}{M_{f,\infty}} = 1 - \frac{8}{\pi^2} \exp(\frac{-\pi^2 D_p t}{4l^2}) \qquad (5.6)$$

For short times with $(M_{f,t}/M_{f,\infty}) \leq 0.6$, Equation 5.7 of simplified form can be used with reasonable accuracy:

$$\frac{M_{f,t}}{M_{f,\infty}} = \frac{2}{l}(\frac{D_p t}{\pi})^{0.5} \tag{5.7}$$

It is readily noted that migration is proportional to square root of time in Equation 5.7: the linearity of migration vs. square root of time is widely used for determining diffusion coefficient of migrant from the slope of the linear plot.

Another simple approach may be attained with large α in Equation 5.4 (small $K_{p,f}$ or large volume of food): $M_{p,o}$ may be close to $M_{f,\infty}$, or complete migration may be assumed. In this simplified condition Equation 5.7 leads to Equation 5.8.

$$M_{f,t} = M_{p,o}\frac{2}{l}(\frac{D_p t}{\pi})^{0.5} = C_{p,o}V_p\frac{2}{l}(\frac{D_p t}{\pi})^{0.5} \cong 1.128 C_{p,o}A(D_p t)^{0.5} \tag{5.8}$$

where $C_{p,o}$ is initial concentration of migrant in packaging material (mg m^{-3}); A is contact area (m^2).

In many cases, migration of small molecules from polymer matrix into aqueous foods or food simulant solvent is described by the swelling-controlled model, which consists of solvent penetration into the polymer matrix and diffusion transport in the matrix [27]. The relative degree of time-dependent contribution by these two processes may influence the release pattern of the migrant from the packaging layer, and other interactions between food and packaging layer may play to affect the diffusion process of migrant and simulant. The swelling-controlled diffusion of dual interactive mode is difficult to be analyzed mathematically in completeness due to its complexity as mentioned above [27, 28]. Therefore, even in these complex cases, simplified approach using equations described above is used with apparent diffusion coefficient. More complex package structure such as multilayer film can only be analyzed by numerical solution of diffusion equations [4]. Some examples of simple migration prediction are given below. Solutions in MS Excel® can be found in the website given in the preface.

Example 5.1 Simplified estimation of migration progress
A typical value of diffusion coefficient of styrene monomer in polystyrene is 5.5 x 10^{-18} m^2 s^{-1} at room temperature. Typical content of styrene monomer in polystyrene is 300 mg kg^{-1}. Estimate the progress of migration for the styrene monomer into fatty liquid foods stored for 12 months. It may be assumed that styrene monomer is readily soluble in fatty food. The package dimension is rectangular shape of 12 x 9 x 2.4 cm. The wall thickness of the packaging tray is 1 mm. The densities of polystyrene and fatty food are 1,080 and 980 kg m^{-3}, respectively. Assume that top cover of the package is not made of polystyrene and does not contact the food.

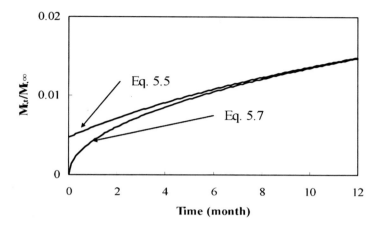

Figure 5.6 $M_{f,t}/M_{f,\infty}$ vs. time in example 5.1

A much larger volume of food than that of the package layer, and small partition coefficient due to readily soluble property of styrene monomer in food, make it possible to assume α as infinite. Equations 5.5 and 5.7 were used to estimate the migration as a function of time (Figure 5.6). Because of low and slow migration, 50 terms in summation of Equation 5.5 were used for required accuracy. After 12 months, about 1.5% of initially contained styrene was estimated to migrate into food ($M_{p,o} \cong M_{f,\infty}$ due to very large α).

The calculation using Equation 5.7 is given:

$$\frac{M_{f,t}}{M_{p,o}} = \frac{2}{l}(\frac{D_p t}{\pi})^{0.5} = \frac{2}{0.001}(\frac{5.5 \times 10^{-18} \times 12 \times 30 \times 24 \times 3600}{\pi})^{0.5} = 0.015$$

Weight of package layer is calculated as product of surface area ($0.12 \times 0.09 + 0.12 \times 0.024 \times 2 + 0.09 \times 0.024 \times 2 = 0.02088$ m^2), package layer thickness (0.001 m) and its density (1,080 kg m^{-3}). That is 0.0226 kg. Therefore the amount of initially contained styrene monomer would be 6.77 mg (0.0226 kg $\times$ 300 mg kg^{-1}). The 1.5% of the 6.77 mg in the package layer corresponds to 0.101 mg, which would be dispersed into the food of 0.254 kg ($0.12 \times 0.09 \times 0.024$ m$^3 \times 980$ kg m^{-3} = 0.254 kg). The concentration of styrene in the food is thus obtained as 0.40 mg kg^{-1} (0.101 mg/0.254 kg).

Example 5.2 Estimating diffusion coefficient from migration test
According to Lickly et al. [29], migration of styrene from expanded foam polystyrene plate (thickness: 0.26 cm, density: 0.0416 g cm^{-3}) to cooking oil at 48.9°C showed the progress of 0.05, 0.10, and 0.15 µg cm^{-2} for 1, 4, and 10 days,

respectively. Determine the diffusivity of styrene in the polymer. Initially contained styrene level was 220 μg g^{-1}.

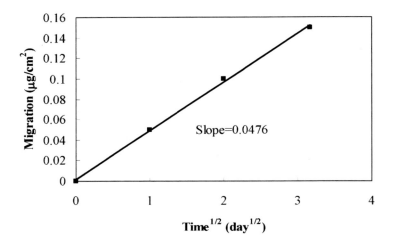

Figure 5.7 $M_{f,t}/A$ vs. $\sqrt{t}$ in example 5.2

Equation 5.7 can be converted into the following form:

$$\frac{M_{f,t}}{A} = 1.128 C_{p,o} (D_p t)^{0.5}$$

When $M_{f,t}/A$ was plotted against $t^{1/2}$, the slope would be $1.128\, C_{p,o} D_p^{1/2}$. From the Figure 5.7, the slope of 0.0476 in μg cm^{-2} day$^{-1/2}$ was attained.

$$1.128\, C_{p,o} D_p^{1/2} = 0.0476 \text{ μg cm}^{-2} \text{ day}^{-1/2}$$

If $C_{p,o}$ is provided in unit of μg cm^{-3} from multiplying 220 μg g^{-1} of initial concentration in polystyrene by its density (0.0416 g cm^{-3}),

$$1.128 \times (220 \times 0.0416 \text{ μg cm}^{-3}) \times D_p^{1/2} = 0.0476 \text{ μg cm}^{-2} \text{ day}^{-1/2}$$

$$D_p = 2.13 \times 10^{-5} \text{ cm}^2 \text{ day}^{-1} \text{ or } 2.46 \times 10^{-14} \text{ m}^2 \text{ s}^{-1}$$

5.6.2 Estimation of Diffusion Coefficient and Partition Coefficient

As shown above, the migration process can be estimated based on the correct knowledge on the diffusion coefficient and partition coefficient. The diffusion coefficient of a compound in packaging material may be obtained by measurement or literatures. However, it is often difficult to find the information exactly suitable for the conditions wanted. Therefore some literature data may be used for interpolation to get the diffusion coefficient at certain temperature by Arrhenius equation:

$$D_p = D_{po} \exp(\frac{-E_a}{RT}) \tag{5.9}$$

where D_{po} is pre-exponential factor ($m^2 s^{-1}$), E_a is activation energy ($J mol^{-1}$), R is gas constant ($8.314 J K^{-1} mol^{-1}$), and T is absolute temperature (K).

Table 5.5 presents diffusion coefficients with temperature dependence parameters for some low molecular compounds in plastic materials. More comprehensive data can be found in related handbooks or literatures [30, 31].

Table 5.5 Diffusion coefficient values of some additives in plastics contacting food

Migrant	Plastic	Food or food simulant	D_p at 23°C ($m^2 s^{-1}$)	D_{po} ($m^2 s^{-1}$)	E_a (kJ mol^{-1})	Ref.
BHA	HDPE	Air	3.56×10^{-14}	1.177×10^{-2}	65.3	[32]
BHT	LDPE		1.10×10^{-13}	1.413	74.5	[30]
BHT	PP		1.16×10^{-13}	4.276×10^3	108.8	[30]
DOP	LDPE		4.70×10^{-17}	4.571×10^{14}	175.6	[30]
Irganox 1010	PP	Corn oil	8.28×10^{-18}	2.075×10^5	127.0	[33]
Irganox 1010	PP	Distilled water	5.72×10^{-10}	1.66	53.6	[33]
Styrene	GPPS	Cooking oil	8.73×10^{-19}	28.266	110.6	[34]
Styrene	GPPS	8% ethanol	3.19×10^{-19}	240.726	118.4	[34]

Abbreviation: BHA = 2-teryiary-butyl-4-methoxy phenol; BHT = butylated hydroxytoluene; DOP = di-octyl-phthalate; Irganox 1010 = tetrakis (3-(3,5-ditert-butyl-4-hydroxy-phenyl)propionyloxymethyl)-methane; HDPE = high density polyethylene; LDPE = low density polyethylene; PP = polypropylene; GPPS = general purpose polystyrene.

For the condition where a very large value of α cannot be assumed, more accurate estimation of migration using by Equation 5.3 needs the information on partition coefficient, $K_{p/f}$. It is also required for estimating the equilibrated migration level (Equation 5.4). For migrants with high solubility in the foodstuff or food simulant, $K_{p/f}$ value of 1 can be used for safer estimation. For migrants not soluble in food and simulant, $K_{p/f}$ value of 1000 may be used. For some solute/plastic/food combinations of different polarities, closer ranges of partition coefficient can be found in Baner [35]: for example, migrants of middle polarity (aldehydes, esters, ketons, etc.) from nonpolar plastic to very polar food phase such as water have partition coefficient of 10-40.

5.6.3 Modeling for Worst Case Scenario

Safety aspect of any migrant in packaging material is determined by its maximum possible concentration in food to which it migrates from packaging material in certain period of time. This value of maximum possible concentration is obtained from toxicological examination of the compound and food consumption data. The most conservative way of analysis is based on the complete migration of the compound initially contained in the packaging material [26]. In this case the maximum initial concentration in the package layer ($C_{p,o}$ in mg m^{-3}) allowing the migrant concentration in the food lower than specific limit ($C_{f,\infty}$ in the case of complete migration assumption) is obtained by mass balance relationships:

$$M_{p,o} = C_{p,o}V_p = C_{f,\infty}V_f \qquad (5.10)$$

$$C_{p,o} = \frac{C_{f,\infty}V_f}{V_p} \qquad (5.11)$$

When the migration is considered for regulatory purpose, the level of migration for a certain time period at a given temperature is required to be estimated with a sufficient safety margin. In this case diffusion coefficient to avoid underestimation is required. Because there is scatter of measured value in diffusion coefficient, the formula for estimation needs to take into consideration variation in the experimental data. The formula for diffusion coefficient with safety margin (D_p^*) is thus given as function of molecular weight, M (dalton) and temperature, T (K) by Brandsch et al. [5] as follows:

$$D_p^* = D_o \cdot \exp(A_p - 0.1351M^{2/3} + 0.003M - \frac{10454}{T}) \qquad (5.12)$$

where $D_o=1$ m^2 s^{-1}, and A_p being 11.5 for LDPE and depending on temperature for HDPE and PP as

$$A_p = A_o - \frac{\tau}{T} \qquad (5.13)$$

with respective A_o of 14.5 and 13.1, and the same τ of 1577.

Using this value of diffusion coefficient, maximum initial concentration of a migrant in packaging material, satisfying lower migration level than that of specific migration limit of $C_{f,t}$ in food within a certain time period, can be obtained in the following equation:

$$C_{p,o} = \frac{C_{f,t}V_f}{V_p(\dfrac{\alpha}{1+\alpha})(\dfrac{M_{f,t}}{M_{f,\infty}})} \qquad (5.14)$$

The $\frac{M_{f,t}}{M_{f,\infty}}$ in Equation 5.14 can be determined by Equations 5.3, or 5.5-5.7.

The maximum initial concentration requirement of the packaging material ($C_{p,o}$ in mg m^{-3}) in Equations 5.11 and 5.14 can be converted into more familiar unit of mg kg^{-1} by dividing its density (kg m^{-3}). Some useful examples of migration prediction using worst case scenario are described below (Examples 5.3 and 5.4).

The procedures described above and some more advanced mathematical treatments were incorporated into computer programs, which are used for testing compliance with regulation. Some examples of software commercially available are MigraPass® (formerly MigraTest© Rite) developed by FABES, Germany, AKTS-SML® by AKTS, Switzerland, and Migrarisk® by INRA-University of Reims, France.

Example 5.3 An estimation of worst case migration
Edible oil is packaged in a plastic bottle with a diameter of 10 cm and a height of 18 cm. The wall of bottle has a thickness of 2 mm and contains an antioxidant of 1,500 mg kg^{-1}. The density of the plastic package material is 980 kg m^{-3}. If the antioxidant is readily soluble in edible oil, what would be the maximum concentration of the antioxidant in the edible oil? The edible oil is known to have the density of 920 kg m^{-3}.

We assume that the package is cylindrical and the whole surface except the cover is in contact with the edible oil.

$$V_f = \text{area x height} = \pi \times 0.05^2 \times 0.18 = 1.41 \times 10^{-3} \text{ m}^3$$

$$V_p = \text{surface area x thickness} = (2\pi \times 0.05^2 + \pi \times 0.10 \times 0.18) \times 0.002$$
$$= 1.44 \times 10^{-4} \text{ m}^3$$

$$C_{p,o} = 1,500 \text{ mg kg}^{-1} \times 980 \text{ kg m}^{-3} = 1.47 \times 10^6 \text{ mg m}^{-3}$$

From the relationship of Equation 5.10 or 5.11

$$C_{f,\infty} = \frac{C_{p,o}V_p}{V_f} = \frac{1.47\times10^6 \times 1.44\times10^{-4}}{1.41\times10^{-3}} = 1.50 \times 10^5 \text{ mg m}^{-3}$$

If this value is divided by the oil density of 920 kg m^{-3}, the concentration based on the oil mass would be 163 mg kg^{-1}.

Example 5.4 Estimation on safe level of additive inclusion
As with an antioxidant, octadecyl-3-(3,5-di-tert-butyl-hydroxy-phenyl) propionate (Irganox 1076, molecular weight = 531 daltons) would like to be included in HDPE sheet 0.2 cm thick with a density of 948 kg m^{-3}. In Europe, specific migration limit of this compound in food is set as 6 mg kg^{-1}. The migration experiment with single side exposure to olive oil was conducted for

10 days at 40°C. The contact area was adjusted to be 6 dm² per kg of olive oil. Density of olive oil is 919 kg m⁻³. What would be the safe maximum initial concentration of the antioxidant in the polymer?

The specific migration limit of 6 mg kg⁻¹ corresponds to 6 x 919 mg m⁻³ ($C_{f,t}$) based on the olive oil volume. The volume of packaging material based on 1 kg of olive oil in the test condition is given as the product of thickness (0.2 cm) and surface area (6 x 1 dm²), which is 0.002 x 6 x 0.01 m³.

For safer conservative estimation, partition coefficient of $K_{p/f}$ is assumed as 1, and thus α is calculated from the test condition of 1 kg oil as:

$$\alpha = \frac{V_f}{K_{p/f} \cdot V_p} = \frac{(1/919)m^3}{1 \times (0.002 \times 6 \times 10^{-2})m^3} = 9.07$$

Diffusion coefficient may be estimated conservatively by using Equation 5.12:

$$D_p^* = D_o \cdot \exp(A_p - 0.1351M^{2/3} + 0.003M - \frac{10454}{T})$$

$$= 1 \times \exp(14.5 - \frac{1577}{313} - 0.1351 \times 531^{2/3} + 0.003 \times 531 - \frac{10454}{313})$$

$$= 2.808 \times 10^{-14} \ m^2s^{-1}$$

Fractional migration calculated by Equation 5.3 is given as:

$$\frac{M_{f,t}}{M_{f,\infty}} = 0.18$$

Therefore maximum initial concentration is obtained by Equation 5.14:

$$C_{p,o} = \frac{C_{f,t}V_f}{V_p(\frac{\alpha}{1+\alpha})(\frac{M_{f,t}}{M_{f,\infty}})} = \frac{6 \times 919 \times (1/919)}{0.002 \times 6 \times 0.01 \times (\frac{9.07}{1+9.07})(0.18)} = 3.08 \ x \ 10^5 \ mg \ m^{-3}$$

If this value is divided by HDPE density of 948 kg m⁻³, the concentration based on the package mass would be 325 mg kg⁻¹.

5.7 Regulatory Considerations

The reliability of a food package is a quite complex concept, dealing with different aspects like marketability, physical protection of the content inside, and best performance for extending shelf life, but certainly a major part of reliability features is related to safety of the materials in contact with food.

All packaging materials must be safe from the possible migration of undesirable packaging constituents into the food. Therefore, the matters of inertness of food contact materials (FCM) and packaging reliability are in the domain of law in all the developed countries, where nowadays exist very huge detailed and generally severe regulations on this topic. It is not wrong to say that food packages are the consumer goods, not intended for direct human consumption, most regulated, inspected, and controlled in the world, especially if they are made of plastic materials.

A full account of the regulations established in the different areas of the world for food packages, their historical development and the social or political background for them, is well beyond the scope of this chapter and, moreover, regulations tend to change from time to time, making it almost impossible to provide complete and updated information.

The modern view on food packaging regulation has started with the passage of the Food Additives Amendment of 1958 in the United States, later on Italy enacted legislations on plastics packaging in 1963, based on the American example, and soon German authorities prepared industry recommendations specific to major polymers or applications [36]. Subsequently France, the Netherlands, and Belgium issued similar laws, and rapidly the harmonized European Union legislation became, together with US FDA Acts, the main inspiring sources for regulations and for settling possible divergences in other countries [37].

5.7.1 Commons and Differences between Regulations

Even if food packaging reliability and consequently its regulation, became a global subject due to the market globalization, it is also true that the regulatory systems employed in different countries are hugely different in their details and forces, each having its own history and its specific sets of exemption [36, 38].

Some of the common and distinctive elements in the different regulations are shortly reviewed here. Positive and negative lists are formats of regulations widely employed in different legislations. A positive list contains all the approved components and ingredients for food contact, together with possible concentration limits and exclusions. Establishing a positive list is very costly, but it ensures the maximum level of safety. A negative list covers the materials and the substances which are prohibited for their toxicity or dangerousness; it is generally short and ensures the safety less than a positive list. Regulations in Europe, USA, Japan, Korea, and South America, are all using hybrid systems, considering both formats.

The concept of overall migration limit (OML), a limit to the leaking/migration from whatever package, is accepted widely but not worldwide, and the absolute levels are sometimes different with countries. In the Mercosur countries (Argentina, Brazil, Paraguay, Uruguay, Bolivia, and Chile) the limit is 50 ppm, while in Europe it is 60 ppm. The limit of 60 ppm in Europe indicates the maximum amount that can migrate into the food mass (60 mg kg^{-1}) but it may be also expressed as 10 mg dm^{-2} (or 8 mg dm^{-2} in South America) i.e., the mass migrated from the package's unit surface. The two different ways of expressing the OMLs are considered equivalent, as Figure 5.8 shows, because a cube with each of the 6 faces of 1 dm^2 in area, has an internal volume of 1 dm^3, thus 1 liter that can be conventionally assumed as 1 kg: if each face of the cube releases 10 mg of substances (10 mg dm^{-2}) into the internal volume, the total concentration will be 60 mg per kg.

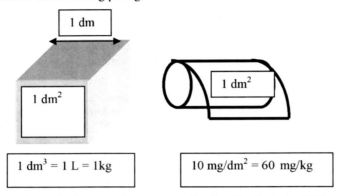

Figure 5.8 The equivalence of the two overall migration limits in EU legislation

A number of additives and adjuvants are also regulated in Europe and Mercosur countries with a specific migration limit (SML): i.e., the migrated amount of a specific substance should not exceed that value. Additionally, some of these substances have been assigned limits on the residual concentration in the finished article and this restriction is referred as QM, or maximum quantity permitted. In Table 5.6, as an example, a selection of these restrictions applied in the European Union is presented.

In USA, Japan, Korea, and other countries, the SML concept is not accepted on overall basis and the materials are mostly regulated on the basis of their own specifications and standards. In these countries the legal requirements of a food contact material, derive from a definition of food additive that includes also the substances which are reasonably expected to migrate, the so-called 'indirect food additives'. These indirect additives are all subject to FDA regulation in USA (Food Additives Amendment, FDA 1958) and to local legislation in other countries.

In USA, the only substances that are not under the FDA regulation concerning the food additives are those 'prior-sanctioned' (the substances with a specific approval prior to the enactment in 1958), the GRAS substances (Generally Recognized As Safe), and finally the substances that are not reasonably expected to become a component of food, i.e., that are not expected to migrate into food or beverage. This last category, however, is not well defined as analytical chemistry has recently made great advances and the limit of detection for several substances is now much lower than before. To manage this aspect, in 1969 a standard was introduced (the so-called "Ramsey proposal", however, was never formally adopted by FDA), which permitted the use, without a prior promulgation of an additive regulation, of indirect additives migrating in amounts not greater than 50 ppb. A more recent FDA proposal (October 1993) is the 'threshold of regulation': here the focus is shifted from 'migration values' to 'exposure values', i.e., the migration amount related to the daily intake. The level of 0.5 ppb in the diet of a substance for which no toxicological data are available, is defined as a trivial exposure that does not require any formal regulation; it is like saying that such low concentration (0.5 ppb) may be always considered as safe.

Table 5.6 A selection of the restrictions (SML and QM) applied in EU for authorized plastic monomers

Monomer	Restriction (ppm)
Acetic acid vinyl ester	SML = 12
Acrylamide	SML = not detectable (detection limit of the method indicated 10 ppb)
Butadiene	QM = 1 or SML = not detectable (detection limit of the method indicated 20 ppb)
Caprolactame	SML = 15 (with its sodium salt)
Carbonyl chloride	QM = 1
Diethylene glycol	SML = 30 (alone or with monoethylene glycol)
1.3 dihydroxybenzene	SML = 2.4
1.4 dihydroxybenzene	SML = 0.6
Ethylene oxide	QM = 1
Hexamethylenediamine	SML = 2.4
Maleic acid	SML = 30
1-octene	SML = 15
Terephtalic acid	SML = 7.5
Vinylidene chloride	QM = 5 or SML = not detectable (detection limit of the method indicated 50 ppb)
Isocyanates	QM = 1 (expressed as CNO)

The food simulants are also, in some cases, distinctive elements between the different regulations. It is a very hard task to find a solvent or a simulant which really mimics the food behavior in contact with the packaging material. The criterion normally adopted, however, is the prudential one of the worst case; i.e., the selection of a simulant which certainly leads to a greater migration than the real food. However, this prudential approach might be even too protective, preventing the use of a good package by an overestimate of migration risk. This is particular true for the simulant generally used for fatty foods, olive oil, which is able to extract much more potentially migrating substances, than any solid fat or solid fatty foods. Due to this possibility, in some legislation (e.g., EU and

Mercosur) the so-called 'reduction factors' have been introduced. These are values (2, 3, 4, and 5) for which the result of a migration test must be divided prior to compare the value with an OML or SML. The reduction factors are established as a function of food and simulant used: for instance in Europe the wrapping material for a bakery product coated by chocolate must be tested with olive oil (simulant for fatty contact) but the result of the migration test will be divided by five; the package of a processed cheese has to be tested with three different stimulants: distilled water, acetic acid, and olive oil, but the result obtained with the last simulant will be divided by three prior to declare its reliability to the migration test.

5.7.2 Sensory Tainting

In EU and Mercosur regulations, moreover, a special attention is reserved for possible tainting phenomena. The principle underlying those regulations, expressed with different words, is that any material or article intended to come into contact either directly or indirectly with foodstuffs, must be stable enough not to transfer substances which could bring about deterioration in the organoleptic properties of the food, thus a tainting phenomenon. According to a definition provided by the International Standard Organization (ISO), taint is a taste or an odor foreign to the product. An important difference is recognized between *taints* and *off-flavors*, the formers (the taints) being unwanted odors or flavors imparted to food through external sources, whereas off-flavors are consequences of internal deteriorative changes occurring in foods or beverages [39]. It is therefore clear that food-packaging interactions may lead to sensory tainting whereas wrong choices of inadequate or under-performing packages may lead to off-flavor development during the shelf life.

After the principle has been stated, however, no procedures or methods have been officially adopted by the legislation for testing packages and materials against the risk of tainting. Nevertheless, the release of packaging components altering the food's flavor, taste, or odor is a quite common phenomenon from flexible packaging and plastic materials in particular.

The compounds identified in flexible packaging materials and able to taint the packed foods are from a variety of sources; they include inks used for printing, adhesives and solvents used in composite structure manufacturing, and coating and varnishes applied on the contact surface, but they also comprise residual monomers (e.g., styrene), plastic additives, and decomposition or reaction products (e.g., acetaldehyde in PET, or oxidation products in polyolefins). Even with different sources and chemical natures all the tainting compounds are characterized by a very very low threshold of sensorial perception. Very often these substances are present in packages and migrate in foods at extremely low concentration, in the range of ppm or ppb and even ppt (10^{-12}), they are not subject to specific limits of migration, and mostly their analytical detection is more problematic than a sensorial perception. Nowadays, a lot of data have been collected about threshold of sensorial detection (particularly odorous perception) of several possible packaging contaminants, both in air and in food media, showing the potential tainting at very very low level. These substances are frequently indicated as VOC (volatile organic compounds), and in Table 5.7 a selection of their odor thresholds, from different sources, is presented.

Table 5.7 The sensory threshold detection for some possible volatile organic compounds (VOC), expressed in ppm

Odor threshold (ppm)	VOC/media
10^4	Propylene/air; ethylene/air
10^3	Octane/air;
10^2	Ethanol/air; acrylic acid/air; butanol/air; hexane/butter cream; toluene/butter cream
10^1	Toluene/air; methyl ethyl ketone/air; ethyl acetate/air; 2 butanone/butter cream
10^0	Acetic acid/air; vinyl acetate/air; styrene/butter cream; pentanal/butter cream; pentachloroanisole/egg
10^{-1}	Styrene/air; methylmetacrilate/air; amyl alcohol/air; 2.3.6 trichloroanisole/whisky
10^{-2}	Pyridine/air; chlorophenol/air; eugenol/air; amyl acetate/air
10^{-3}	Butylacrylate/air; ethylmercaptan/air; 2.3.4.6 tetrachloroanisole/dried fruit
10^{-4}	Amylmercaptane/air; trans 2 nonenal/air; ethyl acrylate
10^{-5}	1-nonen 3 one/air; 2.3.6 trichloroanisole/tea
10^{-6}	Vanillin/air; 2.3.6 trichloroanisole/beer

Discussion Questions and Problems

1. Explain the difference between specific migration (SM) and overall migration (OM).

2. Is the migration rate ($dM_{f,t}/dt$) constant during the contact time of a food with a packaging material?

3. Rationalize the concept of 'functional barrier'.

4. Susceptor materials may represent a special case of migration. Explain why.

5. Migration and scalping are not always negative and unwanted phenomena. Explain the reason with some examples.

6. According to Lickley et al. [29], migration of styrene from foamed polystyrene cup (thickness: 0.089 cm, density: 0.114 g cm^{-3}) to cooking oil at 48.9°C showed the progress of 0.51, 0.80, and 0.99 μg cm^{-2} for 1, 4, and

10 days, respectively. Determine the diffusion coefficient of styrene in the polymer. Initially contained styrene level was 378 $\mu g\ g^{-1}$.

7. An additive initially remained at 1000 mg kg^{-1} in a plastic package containing 1000 mL of aqueous food. The package wall is 0.02 cm thick and has a surface area of 0.06 m^2 contacting food. Density of both food and plastic may be assumed as 1000 kg m^{-3}. The additive has a diffusion coefficient of 1.5 x 10^{-14} m^2 s^{-1} and has partition coefficient value of 1 between plastic and food at 40°C. Estimate the migration level of the additive in the food after 10 days at this temperature.

8. A migration experiment for polypropylene sheet of 1 mm thickness was conducted using a sealed single side cell exposing only one surface of the sample to the fatty food simulant olive oil. The additive of concern having molecular weight of 326 daltons was initially distributed uniformly in polypropylene at concentration of 1500 ppm. The ratio of 6 dm^2 surface area to 1 kg olive oil was applied. What would be the conservative 'worst case' prediction of migration for 2 hours at 121°C? Densities of olive oil and polypropylene are 919 and 905 kg m^{-3}, respectively. The additive is readily soluble in olive oil.

Bibliography

1. LL Katan. Introduction. In: LL Katan, ed. Migration from Food Contact Materials. London: Blackie Academic & Professional, 1996, pp. 1-10.
2. LL Katan. Health safety. In: JH Briston, ed. Plastic Films. Essex, UK: Longman Scientific & Technical, 1989, pp. 128-189.
3. J Crank. The Mathematics of Diffusion. Oxford: Clarendon Press, 1975, pp. 44-68.
4. O Piringer. Transport equations and their solutions. In: O Piringer, AL Baner, ed. Plastic Packaging Materials for Foods. Weinhem: Wiley-VCH, 2000, pp. 183-219.
5. J Brandsch, P Mercea, M Ruter, V Tosa, O Piringer. Migration modeling as a tool for quality assurance of food packaging. Food Addit Contam 19(supplement):29-41, 2002.
6. E Helmroth, R Rijk, M Dekker, W Jongen. Predictive modelling of migration from packaging materials into food products for regulatory purposes. Trends Food Sci Technol 13:102-109, 2002.
7. W Camacho, S Karlsson. The quality of recycled resins of high density polyethylene (HDPE) and polypropylene (PP) separated from mixed solid waste. Polym Degrad Stab 71:123-134, 2001.
8. ILSI. Guidelines for recycling of plastics for food contact use. Bruxell, International Life Sciences Institute, 1998.
9. PY Pennarun, P Dole, A Feigenbaum. Overestimated diffusion coefficients for the prediction of worst case migration from PET: application to recycled

PET and to functional barriers assessment. Packag Technol Sci 17:307-320, 2004.

10. D Scholler, JM Vergnaud, J Bouquant, H Vergallen, A Feigenbaum. Safety and quality of plastic food contact materials. Optimization of extraction time and extraction yield, based on arithmetic rules derived from mathematical description of diffusion. Application to control strategies. Packag Technol Sci 16:209-220, 2003.

11. J-K Han, SE Selke, TW Downes, BR Harte. Application of a computer model to evaluate the ability of plastics to act as functional barriers. Packag Technol Sci 16:107-118, 2003.

12. PY Pennarun, P Dole, A Feigenbaum. Functional barriers in PET recycled bottles. Part I. Determination of diffusion coefficients in bioriented PET with and without contact with food simulants. J Appl Polym Sci 92:2845-2858, 2004.

13. PY Pennarun, Y Ngono, P Dole, A Feigenbaum. Functional barriers in PET recycled bottles. Part II. Diffusion of pollutants during processing. J Appl Polymer Sci 92:2859-2870, 2004.

14. AL Perou, S Laoubi, JM Vergnaud. Contaminant transfer during the coextrusion of tri-layer polymer films with a recycled layer. Effect of this transfer on the time of protection of the food. Adv Colloid Interface Sci 81:19-33, 1999.

15. KD Woods. Food-package interaction safety: European views. In: SL Risch, JH Hotchkiss, ed. Food Packaging Interactions II. Washington DC: American Chemical Society, 1991, pp. 111-117.

16. L Castle. Migration testing food contact plastics for high temperature applications. In: DH Watson, MN Meah, ed. Chemical Migration from Food Packaging. New York: Ellis Horwood, 1994, pp. 45-76.

17. TH Begley, JE Biles, HC Hollifield. Migration of an epoxy adhesive compound into a food simulating liquid and food from microwave susceptor packaging. J Agri Food Chem 39:1944-1945, 1991.

18. P Hernandez-Munoz, R Gavara, RJ Hernandez. Evaluation of solubility and diffusion coefficients in polymer film-vapor systems by sorption experiments. J Membrane Sci 154:195-204, 1999.

19. T Nielsen, M Jagerstad. Flavour scalping by food packaging. Trends Food Sci Technol 5:353-356, 1994.

20. SC Fayoux, AM Seuvre, AJ Voilley. Aroma transfer in and through plastic packagings: Orange juice and d-limonene. A review. Part I: Orange juice aroma sorption. Packag Technol Sci 10:69-82, 1997.

21. A Askar. Flavor changes during processing and storage of fruit juices-II. Interaction with packaging materials. Fruit Processing 11:432-439, 1999.

22. P Hernandez-Munoz, R Catala, R Gavara. Food aroma partition between packaging materials and fatty food simulants. Food Addit Contam 18:673-682, 2001.

23. IJ Rangwalla, SV Nablo. Electron grafted barrier coatings for packaging film modification. Radiat Phys Chem 42:41-45, 1993.

24. P Simko, P Simon, V Khunova. Removal of polycyclic aromatic hydrocarbons from water by migration into polyethylene. Food Chem 64:157-161, 1999.
25. MS Choudhry, F Lox, A Buekens, P Decroly. Evaluation of migrational behaviour of plastic food-contact materials: a comparison of methods. Packag Technol Sci 11:275-283, 1998.
26. PC Chatwin. Mathematical modelling. In: LL Katan, ed. Migration from Food Contact Material. London: Blackie Academic & Professional, 1996, pp. 26-50.
27. SS Chang, CM Guttman, IC Sanchez, LE Smith. Theoretical and computational aspects of migration of package components to food. In: JH Hotchkiss, ed. Food and Packaging Interactions. Washington DC: American Chemical Society, 1988, pp. 106-117.
28. V Gnanasekharan, JD Floros. Migration and sorption phenomena in packaged foods. Crit Rev Food Sci 37:519-559, 1997.
29. TD Lickly, KM Lehr, GC Welsh. Migration of styrene from polystyrene foam food-contact articles. Food Chem Toxic 33:475-481, 1995.
30. P Mercea. Appendix I. In: O Piringer, AL Baner, ed. Plastic Packaging Materials for Foods. Weinhem: Wiley-VCH, 2000, pp. 470-530.
31. W Limm, HC Hollifield. Modelling of additive diffusion in polyolefins. Food Addit Contam 13:949-967, 1996.
32. JK Han, J Miltz, BR Harte, JR Giacin. Loss of 2-tertiary-butyl-4-methoxy phenol (BHA) from high-density polyethylene film. Polym Eng Sci 27:934-928, 1987.
33. W Limm, HC Hollifield. Effects of temperature and mixing on polymer adjuvant migration to corn oil and water. Food Addit Contam 12:609-624, 1995.
34. PG Murphy, DA MacDonald, TD Lickly. Styrene migration from general-purpose and high-impact polystyrene into food-simulating solvents. Food Chem Toxic 30:225-232, 1992.
35. AL Baner. Partition coefficients. In: O Piringer, AL Baner, ed. Plastic Packaging Materials for Foods. Weinhem: Wiley-VCH, 2000, pp. 79-123.
36. JH Heckman. Food packaging regulation in the United States and the European Union. Regul Toxicol Pharm 42:96-123, 2005.
37. M Padula, A Ariosti. Legislacion MERCOSUR sobre la aptitude sanitaria de los envases para alimentos. In: R Catala, R Gavara, ed. Migration de Components y Residuos e Envases en Contacto con Alimentos. Burjassot (Valencia), Spain: Instituto de Agroquimica y Tecnologia de Aliemntos CSIC, 2002, pp. 45-64.
38. LL Katan, L Rossi, JH Heckman, L Borodinsky, H Ishiwata. Regulations. In: LL Katan, ed. Migration from Food Contact Material. London: Blackie Academic & Professional, 1996, pp. 277-290.
39. D Kilcast. Organoleptic assessment. In: LL Katan, ed. Migration from Food Contact Material. London: Blackie Academic & Professional, 1996, pp. 51-76.

Chapter 6

Food Packaging Polymers

6.1 Basic Concepts of Polymers 142
    6.1.1   Chemical Structure ···142
    6.1.2   Molecular Shape ···143
    6.1.3   Thermoplastics versus Thermosets ···144
    6.1.4   Homopolymers versus Copolymers ···144
    6.1.5   Polymer Blends ··145
6.2 Polymerization Reactions 145
    6.2.1   Addition Polymerization ··145
    6.2.2   Condensation Polymerization ···146
6.3 Plastics versus Polymers 147
    6.3.1   Production of Polymers and Plastic Resins ·······································147
    6.3.2   Advantages and Disadvantages of Plastics ·······································148
6.4 Composition/Processing/Morphology/Properties Relationships 148
6.5 Characteristics of Packaging Polymers 149
    6.5.1   Molecular Weight ···149
    6.5.2   Chain Entanglement ··151
    6.5.3   Summation of Intermolecular Forces between Polymer Chains ·····151
    6.5.4   Polymer Morphology ···152
6.6 Food Packaging Polymers 154
    6.6.1   Polyethylene (PE) ···155
    6.6.2   Polypropylene (PP) ···157
    6.6.3   Polystyrene (PS) ···157
    6.6.4   Polyvinyl Chloride (PVC) ··158
    6.6.5   Polyethylene Terephthalate (PET) ··158
    6.6.6   Polyvinylidene Chloride (PVDC) ··159
    6.6.7   Ethylene Vinyl Alcohol Copolymer (EVOH) ····································160
    6.6.8   Ionomer ···160
    6.6.9   Ethylene Vinyl Acetate (EVA) Copolymer ·······································161
    6.6.10 Polyamides (Nylons) ··162
    6.6.11 Polycarbonate (PC) ··162
    6.6.12 Edible Coatings and Films ··163
    6.6.13 Recycling Symbols ···163
6.7 Polymer Processing 163
    6.7.1   Extrusion ··164
    6.7.2   Coextrusion ··164
    6.7.3   Cast Film Extrusion ··165
    6.7.4   Blown Film Extrusion ··166
    6.7.5   Injection Molding ···168
    6.7.6   Blow Molding ···168

6.7.7   Extrusion Coating ··· 171
6.7.8   Extrusion Lamination ·· 172
6.7.9   Adhesive Lamination ··· 173
6.7.10   Thermoforming ··· 174
6.7.11   Vacuum Metallization ··· 174
Discussion Questions and Problems ··· 175
Bibliography ··· 176

6.1 Basic Concepts of Polymers

During the past two decades, packaging polymers have become the material of choice for many food packaging applications which were once dominated by paper, metals, or glass. Polymers are a vast subject whose details require several books to fill, and hence this chapter is limited to an introduction of those basic concepts and applications most relevant to food packaging. More information can be found in the Macrogalleria® website of polymer science and general references [1-6] devoted solely to plastic packaging.

Polymers encompass a wide range of natural and synthetic materials, only a small portion of which is suitable for packaging. Paper and paperboard are biopolymers of cellulose; foods are composed of biopolymers such as protein and starch. Packaging polymers are confined mostly to low cost synthetic polymers which can be formed into various package shapes and have properties suitable for protecting foods during distribution.

6.1.1 Chemical Structure

The word polymer is derived from the Greek "poly" meaning "many" and the Greek "mer" meaning "part". A polymer is a large molecule (known also as macromolecule) composed of thousands of relatively small repeat units, known as monomers, joining together by the polymerization reaction. Polymer is a term which may refer to the polymer molecules or to the material made of these polymer molecules.

Figure 6.1 shows the chemical structure of a polyethylene (PE) molecule, in which the repeat units (C_2H_4) are covalently bonded to form a long chain. The atoms or groups at the two ends of the chain depend on the type of polymerization and are normally not specified. For convenience, the long chain may be expressed by a condensed formula, where the index 'n' is called degree of polymerization and is the number of monomer units in the molecular chain.

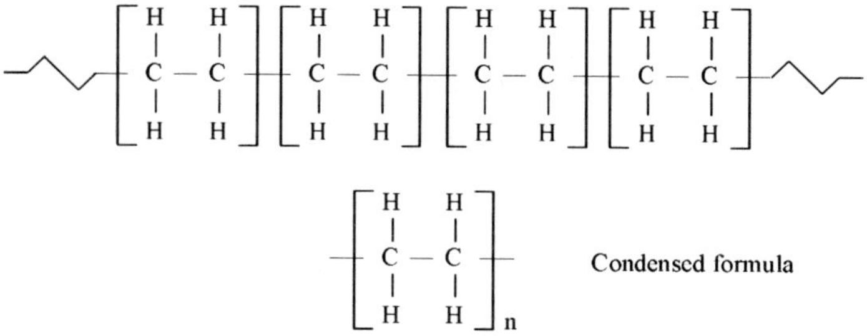

$$\sim\!\!\!\wedge\!\!\!\!\left[\begin{array}{cc} H & H \\ | & | \\ C - C \\ | & | \\ H & H \end{array}\right]\!\!\!\!\left[\begin{array}{cc} H & H \\ | & | \\ C - C \\ | & | \\ H & H \end{array}\right]\!\!\!\!\left[\begin{array}{cc} H & H \\ | & | \\ C - C \\ | & | \\ H & H \end{array}\right]\!\!\!\!\left[\begin{array}{cc} H & H \\ | & | \\ C - C \\ | & | \\ H & H \end{array}\right]\!\!\!\wedge\!\!\!\sim$$

$$\left[\begin{array}{cc} H & H \\ | & | \\ C - C \\ | & | \\ H & H \end{array}\right]_n \qquad \text{Condensed formula}$$

Figure 6.1 Chemical structure of polyethylene

Carbon and hydrogen are most commonly found in polymers, but oxygen, chlorine, nitrogen, and other atoms are also found in some polymers.

6.1.2 Molecular Shape

A polymer molecule may have one of the basic shapes shown in Figure 6.2.

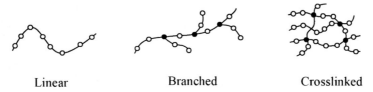

Linear   Branched   Crosslinked

Figure 6.2 Basic shapes of polymer molecules

A linear polymer is composed of macromolecules with no or very few branch points (dark circles in Figure 6.2) and side chains. This simple shape allows the polymer chains to arrange into compact structures, thereby increasing the density, degree of crystallinity, strength, and gas barrier properties of the polymer. Examples of linear polymers are high density polyethylene (HDPE) and polyvinyl chloride (PVC).

A branched polymer is composed of macromolecules characterized by short or long side chains attached along the main chain. These side chains hinder packing and folding of the main chain, thereby decreasing the density and crystallinity of the polymer. Linear low density polyethylene (LLDPE) is an example of branched polymer with short chains, while low density polyethylene (LDPE) is one with long chains.

A crosslinked polymer is a three-dimensional network of interconnected polymer chains. This crosslinked network restricts the mobility of polymer chains, making the polymer stronger and more resistant to heat and chemicals.

6.1.3 Thermoplastics versus Thermosets

Packaging polymers may be classified into thermoplastics and thermosets, depending on whether or not crosslinks are present.

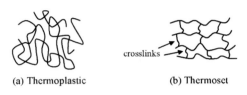

(a) Thermoplastic　　　　　　　　(b) Thermoset

Figure 6.3 Thermoplastics versus thermosets

As shown in Figure 6.3(a), thermoplastics are a group of uncrosslinked polymers, which have either amorphous or semi-crystalline structures. They can be easily shaped by heat and pressure into various packaging products such as films, bottles, and trays. They can withstand several cycles of heating to become polymer melt and cooling to become solid, without significant losses in properties. This unique characteristic allows reprocessing of industrial thermoplastic scraps and recycling of post-consumer thermoplastic plastics (such as PET and HDPE bottles), thereby reducing the amount of these materials to the waste stream. Most food packaging polymers are thermoplastics.

Thermosets, also called thermosetting plastics, are crosslinked polymers as shown in Figure 6.3(b). They are hard and stiff materials formed, usually at high temperature, by crosslinking polymer chains with covalent bonds into three-dimensional networks. Examples of thermosets are phenolics and epoxies. As its name implies, once a thermoset is crosslinked or set into permanent shape, it cannot be remolded or recycled. Heating will not soften it into melt, and severe heating can only irreversibly destroy its covalent crosslinks, causing it to become charred. The applications of thermosets in food packaging are limited mostly to coating and adhesives.

Crosslinks make a polymer stronger, more rigid, and more resistant to heat and chemicals. Crosslinking may be induced using a crosslinking agent (such as organic peroxide), UV irradiation, or electron beam. Although crosslinking is applied mostly to thermosets, it is occasionally also applied to thermoplastics which may be crosslinked.

6.1.4 Homopolymers versus Copolymers

A homopolymer is synthesized using a single type of monomer; for example, polyethylene homopolymer is produced from ethylene monomer. A copolymer is synthesized using more than one species of monomer; for example, ethylene vinyl acetate copolymer is produced from a mixture of ethylene and vinyl acetate monomers. Some polymers are available commercially in both homopolymer and copolymer; for example, propylene homopolymer may be produced using propylene monomers, while propylene copolymer may also be produced using a mixture of propylene and ethylene monomers.

6.1.5 Polymer Blends

As another way to obtain a polymer material with desirable properties, two or more polymers (homopolymers or copolymers) are often blended with help of compatibilizer. Miscibility and adhesion between the phases are important for successful blending.

6.2 Polymerization Reactions

Polymerization is a chemical reaction in which hundreds or even thousands of monomers are covalently linked together to form a polymer. Traditionally, polymerization reactions may be divided into additional polymerization and condensation polymerization, depending on whether byproduct molecules are formed. There are also more sophisticated classifications of polymerization reactions which are beyond the scope of this chapter.

6.2.1 Addition Polymerization

Addition polymerization, known also as chain-growth polymerization, is a reaction in which monomers of the same species are chemically linked together into a long-chain without the formation of byproduct. As shown in Figure 6.4, this process typically involves an initiator ($I\cdot$) attacking an unsaturated monomer ($H_2C=CHR$), such as ethylene or styrene, to create a reactive immediate ($IH_2C-CHR\cdot$), followed by repeated addition of monomer units to form a growing chain:

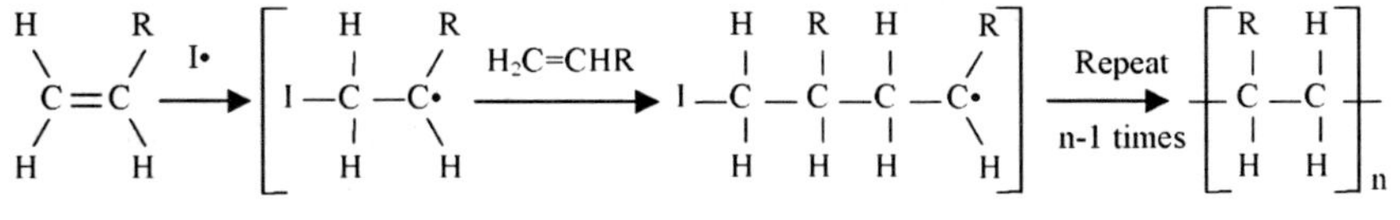

Figure 6.4 Mechanism of additional polymerization

The initiator $I\cdot$ may be a free radical, a transition metal catalyst, a cation, or an anion. Accordingly, the polymerization process may be further divided into free radical polymerization, anion polymerization, cation polymerization, and catalytic polymerization.

As an example, the production of polyethylene, a free radical polymerization using ethylene as monomer and peroxide as initiator, is described below. This reaction is characterized by three sequential steps: chain initiation, chain propagation, and chain termination.

Chain initiation is the first stage that involves the dissociation of the peroxide into two radicals under heating, followed by the reaction between a radical and a monomer to form a larger radical species:

$$R - O - O - R \xrightarrow{\text{Heat}} 2R - O\bullet$$

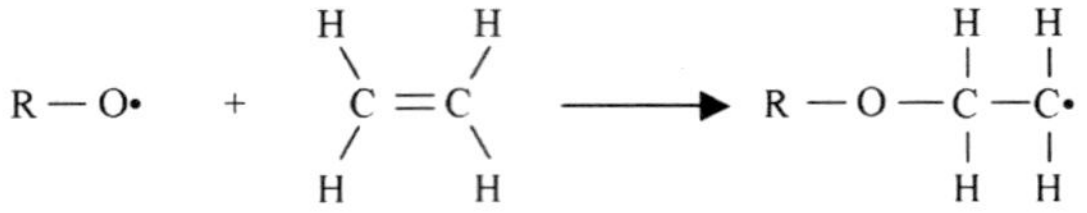

Chain propagation is the rapid second stage that involves repeatedly adding monomers, one at a time, onto the growing chain:

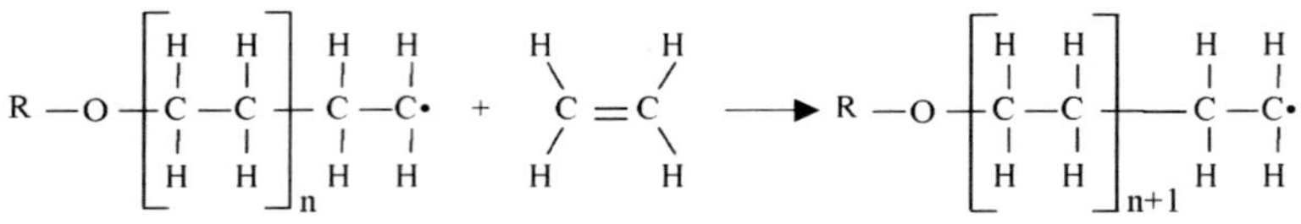

Chain termination is the final step which occurs when two radicals react to form a stable molecule:

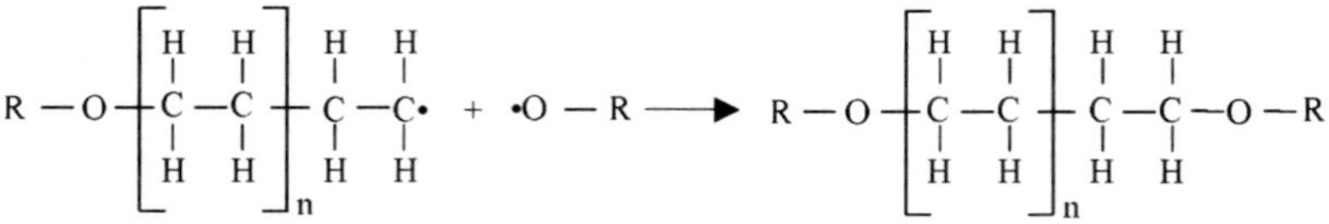

6.2.2 Condensation Polymerization

Condensation polymerization, known also as step-growth polymerization, is a process in which molecules from two different monomer species are linked together to form a polymer, with the important characteristic that small byproduct molecules are also formed. Typical condensation polymers are polyamides, polyesters, and certain polyurethanes.

For example, polyethylene terephthalate (PET) is formed by the following condensation polymerization reaction, in which water molecules are formed as a byproduct (Figure 6.5).

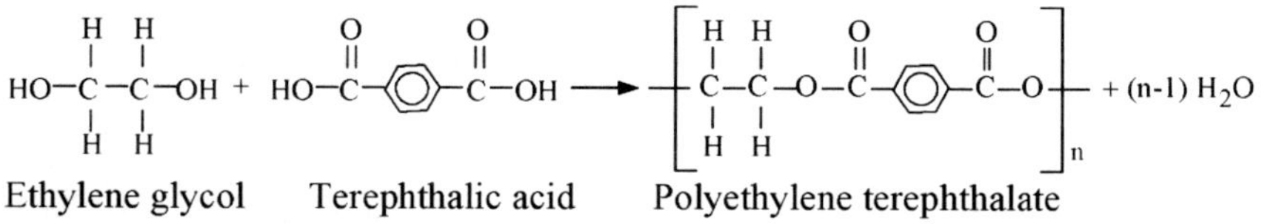

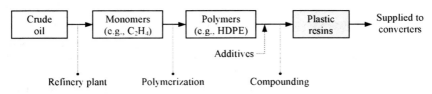

Figure 6.5 Reaction of condensation polymerization

6.3 Plastics versus Polymers

In the packaging industry, the terms plastics and polymers are frequently interchangeably used, but they are not identical. While packaging plastics are composed predominately of polymers (typically > 99%), they also contain small amounts of additives (typically < 1%) such as antioxidants, plasticizers, and fillers to provide enhanced stability or functionality.

The polymers used in packaging plastics have molecular weights typically between 50,000 and 200,000, an optimum range suitable for shaping the polymers into bags, containers, or other forms which have the necessary mechanical, gas barrier, and other properties for packaging applications. Packaging plastics (both the polymers and additives) must also have regulatory approvals for food contact applications.

6.3.1 Production of Polymers and Plastic Resins

Crude oil, usually obtained from land or sea drilling, is the raw material of plastics. Figure 6.6 shows the steps for producing plastic resins by chemical companies. First, crude oil is broken down into various petrochemical products in a refinery plant, some of which such as ethylene, propylene, or styrene are used as monomer feedstock. Then, the monomer molecules are joined together to form polymer molecules via the polymerization reaction, followed by a process known as compounding in which the polymer and small amounts of additives are blended together in an extruder and then chopped into small pellets. These pellets, known as plastic resin or resin, are sold to converters to make films, trays, and other package forms.

```
┌────────┐   ┌────────────┐   ┌────────────┐   ┌─────────┐
│ Crude  │──▶│  Monomers  │──▶│  Polymers  │──▶│ Plastic │──▶ Supplied to
│  oil   │   │ (e.g., C₂H₄)│   │ (e.g., HDPE)│   │ resins  │    converters
└────────┘   └────────────┘   └────────────┘   └─────────┘

                                  Additives

   Refinery plant      Polymerization        Compounding
```

Figure 6.6 Production of plastic resins

A plastic resin may be available in several different grades, each grade tailored for a specific application such as blow molding, injection molding, and extrusion coating. Grades of the same resin may differ in properties such as melt flow index, melt strength, density, molecular weight, and additives.

6.3.2 Advantages and Disadvantages of Plastics

There are three major advantages of using packaging plastics: variety, versatility, and efficiency. Variety because plastics offer a wide range of shapes and forms (such as films, pouches, bottles, cups, and trays), as well as a myriad choices of structures including single-layer or multilayer films, coated or uncoated films, and flexible or semirigid containers. Versatility because plastics are microwavable, heat sealable, thermoformable, and nonbreakable, and their properties can be tailored for numerous food packaging applications. Efficiency because plastic packages are amenable to high-speed manufacture and due to their light weight, higher product volume can be delivered per weight of material used.

Virtually all plastics are permeable to gases and vapors to some extents as discussed in Chapter 4. This is usually a major disadvantage since many foods are moisture or oxygen sensitive, and thus packing those foods in plastics will result in shorter shelf lives than packing them in glass or metals. On the contrary, the permeable nature of plastics is advantageous for packing respiring products such as fresh produce where gas exchange across the package is desirable.

Another major disadvantage is the tendency of plastics to interact with food, leading to undesirable migration or flavor scalping (see Chapter 5). Some small molecule additives incorporated into plastics in the processing, may move into the food interacting with the packaging layer. Conversely plastics may adsorb some active compounds dissolved in the contained food. The progress and degree of migration and flavor scalping are influenced mostly by intermolecular forces among plastics, food, and water which possess many similar bonds and functional groups.

6.4 Composition/Processing/Morphology/Properties Relationships

The plastic resins (for simplicity, the term polymers will be used in the discussion hereafter) supplied by chemical companies are usually purchased by converters for extruding into films or sheets or blow-molding into bottles. Figure 6.7 shows the major factors influencing the properties of these products.

The solid line from ❶ to ❹ shows that polymer composition is a primary factor controlling the properties. Composition is determined by the chemical structure and molecular weight of the polymer. For polymer blends, it is also determined by the weight ratio of individual polymer components. Additives, albeit present in small amounts in plastics can also greatly influence properties.

The solid line from ❷ to ❹ shows that polymer processing is another primary factor controlling the properties. Processing is determined by the type of converting process such as cast film process or blown film process. In addition, operating conditions such as barrel temperature, residence time, and screw configurations are also important.

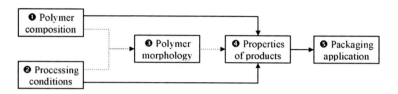

Figure 6.7 Composition/processing/morphology/property relationships

The dashed lines from ❶ and ❷ to ❸ show that composition and processing may affect the polymer morphology (crystallinity, molecular orientation, etc.). For example, rapid cooling leads to more crystalline product, slow cooling leads to more amorphous product, and stretching leads to more orientation of polymer molecules. As shown by the dashed line from ❸ to ❹, morphology is a secondary factor that can significantly influence the barrier, mechanical, optical, and other properties of the product.

From the practical point of view, the application ❺ is usually the most important consideration. In package design, the paths in Figure 6.7 may be considered in the reverse order: the application ❺ determines the required properties ❹, which in turn determines the selections of composition ❶ and processing ❷. Manipulating morphology ❸ is sometimes also useful to achieve certain desired properties.

6.5 Characteristics of Packaging Polymers

Packaging polymers are macromolecules with characteristics quite different from those of smaller molecules. Below are some characteristics responsible for the mechanical, barrier, and other properties of packaging polymers.

6.5.1 Molecular Weight

An important characteristic of polymers is their molecular sizes relative to the much smaller organic or inorganic molecules. The molecular weight of a polymer, a measure of its molecular chain length, can significantly affect its physical properties. As molecular weight increases, tensile and impact strengths increase sharply before leveling off, while melt viscosity increases slowly and then sharply (Figure 6.8).

When selecting molecular weight for a specific application, the strength and melt viscosity should be compromised. The molecular weight should be high enough to acquire adequate strength, but low enough to avoid high melt viscosity which makes the polymer difficult to process, such as flowing into a mold or through an extruder die. Typically, the practical molecular weight range for packaging polymers is between 50,000 and 200,000 as mentioned above.

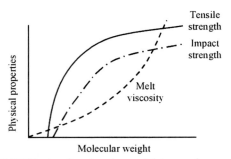

Figure 6.8 Effects of molecular weight on polymer properties

Unlike small molecules such as oxygen or water, a polymer does not have a single molecular weight. This is because polymer molecules of different sizes are formed during the polymerization reaction. The relative fraction of different sizes may be characterized using a plot of frequency versus molecular weight, known as molecular weight distribution (MWD), as shown in Figure 6.9.

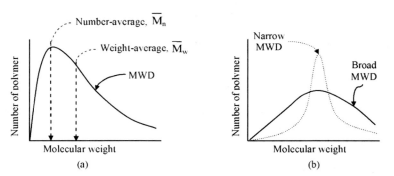

Figure 6.9 Molecular weight distribution

Two average molecular weights are often obtained from the MWD. The number-average molecular weight $\overline{M}_n$ is defined as

$$\overline{M}_n = \frac{\sum_i N_i M_i}{\sum_i N_i} \tag{6.1}$$

where N_i and M_i are the number and molecular weight of polymer molecule i, respectively. The weight-average molecular weight $\overline{M}_w$ is defined as

$$\overline{M}_w = \frac{\sum_i w_i M_i}{\sum_i w_i} = \frac{\sum_i N_i M_i^2}{\sum_i N_i M_i} \qquad (6.2)$$

where w_i is weight of polymer molecule i.

The polydispersity index (PDI) is defined as

$$PDI = \overline{M}_w / \overline{M}_n \qquad (6.3)$$

which is a measure of the spread of the molecular weight distribution. The value of PDI is always ≥ 1. PDI = 1 means that all the polymer molecules have the same molecular weight and thus $\overline{M}_n = \overline{M}_w$. However, PDI = 1 is difficult if not impossible to achieve; most commercial packaging polymers have PDI greater than 2. A higher PDI is associated with a broader MWD (Figure 6.9(b)), meaning more variation in molecular weights among the polymer molecules.

A broader MWD tends to decrease the tensile and impact strengths of a polymer but increase its ease of processability, due to the smaller molecules in the distribution acting as a plasticizer or processing aid.

6.5.2 Chain Entanglement

Another unique characteristic of packaging polymers is chain entanglement, the ability of polymer chains to become entangled with one another. The degree of chain entanglement increases with length of the polymer chain or molecular weight of the polymer. Polymers consisting of short chains are weak and brittle but become strong and ductile above some critical length. A minimum of 1000 to 2000 repeating units, for example, is necessary for polyethylene to acquire the strength adequate for packaging foods.

Chain entanglement is important for packaging polymers. For example, during thermoforming or blow molding of a thermoplastic, the polymer is in molten state with its molecular chains resembling cooked spaghetti (or more vividly, wriggling long worms) tangled up on a plate. Chain entanglement holds the polymer chains together but also allows them to slide pass one other, thereby providing both the necessary strength and pliability for the polymer to be molded into a film, bottle, or other shapes. After cooling, sliding between polymer chains is more difficult and chain entanglement makes the polymer strong and resilient.

6.5.3 Summation of Intermolecular Forces between Polymer Chains

As explained in Chapter 2, intermolecular forces are responsible for holding neighboring molecules together. Depending on the chemical structure of molecules, these forces range from weak dispersion forces to strong dipole-dipole forces (especially strong are the forces from hydrogen bonds). While intermolecular forces exist between all molecules, their influences are more pronounced for polymers. Summation of intermolecular forces refers to the total force between neighboring molecules—this total force is particularly significant for polymer molecules due to their long chains.

The effects of summation of intermolecular forces may be observed in the physical states of large and small molecules: for example, polyethylene of solid versus ethane of gas. Except for the difference in size, the chemical structures are the same and weak nonpolar intermolecular forces operate on both compounds. Although weak nonpolar forces are operating in both compounds, the summation of intermolecular forces is much greater for the long polyethylene chains of solid than for ethane gas. Yet polyethylene molecules are held more strongly together due to greater summation of intermolecular forces.

Summation of intermolecular forces and chain entanglement can significantly influence the mechanical, thermal, rheological, and other properties of thermoplastics.

6.5.4 Polymer Morphology

Polymer morphology may be described as the arrangement of polymer molecules characterized by two or more distinguishable regions which can be studied using techniques such as scanning electron microscopy, X-ray diffraction, and differential scanning calorimetry (DSC). As mentioned in Section 6.4, the development of polymer morphology depends on composition and processing conditions.

Thermoplastics have either amorphous or semi-crystalline structures as shown in Figure 6.10. Amorphous thermoplastics consist of only amorphous regions with randomly arranged polymer chains. Semi-crystalline thermoplastics consist not only of amorphous regions but also of crystalline regions where polymer chains are arranged in orderly fashion (see also Section 2.4.2).

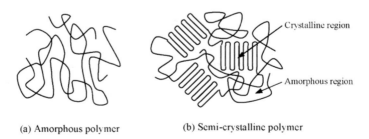

(a) Amorphous polymer   (b) Semi-crystalline polymer

Figure 6.10 Amorphous and semi-crystalline polymers

When sufficient time is allowed to cool a molten crystallizable polymer, the polymer chains will align themselves into the orderly pattern of crystalline region. Crystallinity may be defined as the presence of three-dimensional order on the level of atomic dimensions. Films and containers made from crystallizable polymers are typically semi-crystalline, since pure crystalline morphology is difficult to obtain. Semi-crystalline morphology has a desirable property that combines the strength of crystalline region with the flexibility of amorphous region. The amount of crystallinity relative to pure crystalline is known as degree of crystallinity, which may be measured using methods such as DSC and X-ray diffraction.

Many packaging polymers are crystallizable, polyethylene being an example. HDPE (high density polyethylene) has high crystallinity typically ranging 65-90%, due to its linear polymer chains readily aligning with each other. LDPE (low density polyethylene), however, has lower crystallinity typically ranging 45-70%, due to its chain branching (see Section 6.6.1) hindering molecular alignment. The degree of crystallinity of a polymer film and containers also depends on cooling rate and any additive used in the production. Polymers with bulky pendant groups, such as polystyrene, are usually considered noncrystallizable since their polymer chains cannot align with each other easily.

As shown in Table 6.1, amorphous and semi-crystalline polymers have distinctly different properties which determine their applications (see also Sections 2.4.2 and 3.2.5).

Table 6.1 Comparison between amorphous and semi-crystalline polymers

Amorphous polymers	Semi-crystalline polymers
Clear in appearance	Cloudy to opaque, since crystalline region hinders transmission of light
No melting temperature, T_m	Specific melting temperature, T_m
Relatively weak and flexible	Relatively strong and brittle, but lower impact resistance
Poor to moderate gas barrier	Moderate to good gas barrier
Moderate chemical resistance	Good chemical resistance
Cooling rate and orientation have weak to moderate effects on properties	Cooling rate and orientation have strong effects on properties

As shown in Figure 6.11, molecular orientation involves stretching an unoriented thermoplastic below T_m (melting temperature) but above T_g (glass transition temperature) either uniaxially (in machine direction only) or biaxially (in both machine and transverse directions), to induce alignment of polymer chains and then retain the acquired polymer morphology thereafter, with the purpose of achieving better optical, mechanical, barrier, or other properties. The oriented films have a tendency to shrink with heating and are used for shrink packaging (see Section 11.7.3). When dimensional stability is required, the oriented film is further heated to relax or set the structure.

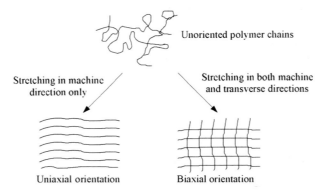

Figure 6.11 Stretching polymer chains to induce molecular orientation

Stretching of thermoplastics to induce molecular orientation typically occurs in processes such as blown film extrusion, injection stretch blow molding, and cast film extrusion (when a tenter frame is used). Molecular orientation is achieved with proper control of temperature and rate of stretching to develop and retain the desirable molecular orientation.

6.6 Food Packaging Polymers

Packaging polymers are typically low cost materials suitable for high volume production. They have relatively limited stress and low temperature resistance compared to engineering polymers. Most food packaging polymers are thermoplastics used for bags, pouches, tubs, and trays. Thermosets are used at much lower volume to limited applications such as caps and trays. Table 6.2 presents typical properties of thermoplastics widely used for food packaging applications. It needs to be mentioned that plastic material properties may vary greatly depending on composition and processing conditions: the information in Table 6.2 serves as only a preliminary and rough guide to the properties for typical applications.

Table 6.2 Typical properties of plastic packaging materials

Material	Density (kg m^{-3})	Mechanical property	Moisture barrier	Gas barrier	Resistance to grease and oil	Trans-parency	Use temp. (°C)
PE							
LDPE	910-925	Tough, flexible	High	Very low	Good	Poor-fair	-50 to 80
LLDPE	910-940	Tough, extensible	High	Very low	Good	Poor-fair	-30 to 100
HDPE	945-967	Tough, flexible	Very high	Very low	Good	Poor	-40 to 120
PP	900-915	Moderately stiff, strong	High	Low	Good	Fair	-40 to 120
PS							
General	1040-1089	Stiff, strong, brittle	Low	Low	Fair-good	Very good	-20 to 90
Impact	1030-1100	Tough, strong	Low	Low	Fair-good	Poor	-20 to 90
PVC							
Unplastic.	1350-1450	Stiff, strong	High	Moderate	Good	Good	-2 to 80
Plasticized	1160-1350	Soft, extensible	Moderate	Moderate	Good	Good	-2 to 80
PET	1380-1410	Stiff, strong	Moderate	Moderate	Good	Very good	-60 to 200
PVDC	1600-1700	Stiff, strong	Very high	Very high	Good	Good	-20 to 130
EVOH	1140-1210	Stiff, strong	Low	Very high*	Very good	Good	-20 to 150
Ionomer	940-960	Tough, extensible	Moderate	Low	Good	Very good	-40 to 65
EVA	920-940	Tough, extensible	Moderate	Low	Good	Very good	-75 to 65
Nylon	1130-1160	Strong, tough	High	High*	Good	Fair-good	-2 to 120

*Decreases with high humidity.

6.6.1 Polyethylene (PE)

$$\left[\begin{array}{cc} \underset{|}{\overset{|}{H}} & \underset{|}{\overset{|}{H}} \\ C & C \\ \underset{|}{\overset{|}{H}} & \underset{|}{\overset{|}{H}} \end{array}\right]_n$$

PE is the most frequently used polymer in food packaging applications because of its low cost, easy processing, and good mechanical properties. PE also has the simplest chemical composition of all polymers, being essentially a straight-chain hydrocarbon. The major classifications of polyethylene are high density polyethylene (HDPE), low density polyethylene (LDPE), and linear low density polyethylene (LLDPE) (Figure 6.12). These classifications differ in density, chain branching, and crystallinity. Within each classification are numerous grades that have different melt viscosities, additives, molecular weights, and molecular weight distributions.

LDPE	HDPE	LLDPE
Branched structure	Linear structure	Linear structure
Many side-chain branches	Few side-chain branches	Many side-chain branches
Long side-chain branches	Short side-chain branches	Very short side-chain branches
Density (0.910 - 0.925 g/cc)	Density (0.940 - 0.965 g/cc)	Density (0.900 - 0.939 g/cc)
Lower crystallinity	High crystallinity	Highly crystallinity
More transparent	Less transparent	Ziegler-Natta polymerization
Free radical polymerization	Ziegler-Natta polymerization	

Figure 6.12 Differences between HDPE, LDPE, and LLDPE

HDPE is a linear polymer with relatively few side-chain branches, and hence its molecules can fold and pack into an opaque, highly crystalline structure. Compared to LDPE, HPDE has a higher melting point (typically 135°C versus 110°C), greater tensile strength and hardness, and better chemical resistance. HDPE is used mostly to make blow-molded bottles for the packaging of products such as milk or water. It is also used for food containers, bags, films, extrusion coating, bottle closures, etc.

LDPE is a homopolymer with many side-chain branches. It is used mostly as film for packaging fresh produce and baked goods, as an adhesive in multilayer structures, and as waterproof and greaseproof coatings for paperboard packaging materials. The packaging film made from LDPE is soft, flexible, and stretchable. The film also has good clarity and heat sealability, and is used for extrusion coating.

LLDPE is a copolymer having 1-10% of alkene comonomers, which result in linear polymer with many short side-chain branches. It has the similar clarity and heat sealability of LDPE, as well as the strength and toughness of HDPE. With its superior properties, LLDPE has been replacing LDPE in many food packaging applications. The main uses of LLDPE are stretch/cling wrap and heat-sealant coating. In recent years, there have been many development efforts focused on using single-site or metallocene catalyst to produce LLPDE films with superior heat seal and hot tack properties, clarity, abuse resistance, and strength. Metallocene polymer can be better controlled in its MWD incorporating new comonomers.

6.6.2 Polypropylene (PP)

$$\left[\begin{array}{cc} H & CH_3 \\ | & | \\ C & - C \\ | & | \\ H & H \end{array}\right]_n$$

PP is a linear, crystalline polymer that has the lowest density (0.9) among all major plastics. Compared to PE, PP has higher tensile strength, stiffness, and hardness. It also has a higher melting temperature (165°C), and hence it is more suitable for hot filling and retorting applications.

A major application for PP is in packaging films. Commercial PP films are available in oriented and unoriented forms. Orientation can be achieved by stretching the film either uniaxially or biaxially during the film forming process. Oriented PP film (OPP) has improved strength, stiffness, and gas barrier; however, it is not heat sealable. Unoriented PP film has excellent clarity, good dimensional stability, and good heat-seal strength. Beside films, other major applications for PP are in containers and closures.

6.6.3 Polystyrene (PS)

$$\left[\begin{array}{cc} H & \bigcirc \\ | & | \\ C & - C \\ | & | \\ H & H \end{array}\right]_n$$

PS is an amorphous polymer that has excellent clarity. Solid PS is a clear, low gas barrier, hard, and low impact strength material. It has relatively low melting point (88°C) and can be readily thermoformed or injection molded into items such as food containers, cups, closures, and dishware. Because of its excellent clarity, PS film is often used as windows in paperboard boxes to display products (such as baked goods) that do not require a good gas barrier.

Expanded polystyrene (EPS) of various bulk densities are manufactured by adding foaming agents in the extrusion process. EPS is used to make cups, bowls, plates, meat trays, clamshell containers, and egg cartons, as well as lightweight protective or thermal insulation packaging materials for shipping. In oriental countries, there have been concerns and controversies on using ESP cups for packaging instant noodles: because the hot water is poured onto the noodle in the cup or the cup with water is heated in microwave ovens. High temperature conditions may allow the migration of styrene dimers and trimers, which are suspected as possible endocrine disorder materials.

High impact polystyrene (HIPS) is an immiscible blend of polystyrene, polybutadiene, and grafted polystyrene-polybutadiene copolymer. As shown in Figure 6.13, it consists of two different phases of polystyrene and polybutadiene bound together by polystyrene-polybutadiene copolymer. Compared to PS, HIPS is more resilient and impact resistant due to the presence of polybutadiene.

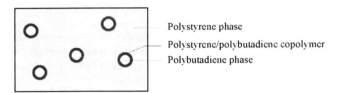

Polystyrene phase

Polystyrene/polybutadiene copolymer

Polybutadiene phase

Figure 6.13 Morphology of HIPS

6.6.4 Polyvinyl Chloride (PVC)

$$\left[\begin{array}{cc} H & Cl \\ | & | \\ C & -C \\ | & | \\ H & H \end{array} \right]_n$$

PVC is a clear, amorphous polymer that is used mostly for films and containers. Most often, plasticizers (organic liquids of low volatility) are added to the polymer to yield widely varying properties, depending on the type and amount of plasticizers used. Plasticized PVC films are limp, tacky, and stretchable, and the films are commonly used for packaging fresh meat and fresh produce. There have been safety concerns and controversies about possible migration of plasticizers (diethylhexyl adipate, diethylhexyl phthalate, etc) in plasticized PVC 'cling' film (see also Section 5.3.1). Unplasticized PVC sheets are rigid, and the sheets are often thermoformed to produce inserts for snacks such as chocolate and biscuits.

PVC bottles have better clarity, oil resistance, and barrier properties than those of HDPE. However, the use of PVC bottles in food packaging is relatively small due to poor thermal processing stability and environmental concerns with chlorine containing plastics.

6.6.5 Polyethylene Terephthalate (PET)

$$\left[\begin{array}{ccccc} H & H & & O & & O \\ | & | & & || & & || \\ C & -C & -O- & C & -\langle\bigcirc\rangle- & C & -O \\ | & | & & & & \\ H & H & & & & \end{array} \right]_n$$

PET (also abbreviated PETE or PETP) is the major polyester for food packaging. The amorphous form of PET (APET) is used mostly as injection blow-molded bottles for carbonated soft drinks, water, edible oil, and juices. PET bottles are stronger, clearer, and have better gas barrier than HDPE bottles, although they are more expensive.

PETG, glycol-modified polyethylene terephthalate, is a copolyester that contains 6% cyclohexane dimethanol as comonomer. It is a clear amorphous thermoplastic and used in thermoforming applications. PETG sheet has high stiffness, hardness, and toughness as well as good impact strength.

APET is also used to produce films that have high strength, high melting point (267°C), high scuff resistance, good clarity, good printing characteristics, and excellent dimensional stability. Because of their poor heat sealability, PET films are often laminated or extrusion coated with a heat sealable layer such as PVDC and LDPE.

The crystallized form of PET (CPET) can withstand temperatures up to 220°C without deformation. Thus CPET food trays are used in microwave/conventional dual-ovens.

6.6.6 Polyvinylidene Chloride (PVDC)

$$\left[\begin{array}{c} H \quad Cl \\ | \quad | \\ C - C \\ | \quad | \\ H \quad Cl \end{array}\right]_x \left[\begin{array}{c} H \quad Cl \\ | \quad | \\ C - C \\ | \quad | \\ H \quad H \end{array}\right]_y$$

PVDC, known under the trade name Saran, is a copolymer of vinylidene chloride (85-90%) and vinyl chloride. The most notable advantages of PVDC are its excellent oxygen and moisture barriers. PVDC films have good clarity and grease/oil resistance.

PVDC is used in films, containers, and coatings. Monolayer PVDC films are used in household wraps. PVDC is often coextruded or laminated with other lower cost polymers (such as OPP and PET) to form multilayer films or sheets. The multilayer films are used to package foods that require a good oxygen barrier, and the multilayer sheets are often thermoformed into semi-rigid containers. PVDC copolymer film coextruded with polyolefins is widely used as shrinkable films for tightening meat and cheese products. PVDC is also used in the form of latex for coating paper, film, cellophane to achieve better oxygen and moisture barrier, grease resistance, and heat sealability.

PVDC is more costly than most other commonly used food packaging polymers. Similar to PVC, PVDC has environmental concerns associated with its use.

6.6.7 Ethylene Vinyl Alcohol Copolymer (EVOH)

EVOH is a crystalline copolymer of ethylene (usually between 27 and 48%) and vinyl alcohol, and its properties are highly dependent on the relative concentrations of these comonomers. EVOH is best known for its exceptional high oxygen barrier, even compared to PVDC. The oxygen barrier increases with decreasing ethylene concentration and decreasing relative humidity. EVOH also has good resistance to hydrocarbons and organic solvents. EVOH is the most expensive of all commonly used food packaging polymers.

EVOH is mostly used as an oxygen barrier layer in multilayer structures, either laminated or coextruded. The multilayer structures containing EVOH are used as films or containers for packaging oxygen sensitive foods. An example of a multilayer structure is PP/adhesive/EVOH/adhesive/PP, where the PP layers provide strength and moisture barrier, and the EVOH layer provides oxygen barrier. This multilayer structure is suitable for hot filling and retorting applications.

6.6.8 Ionomer

An ionomer is a polymer composing of a small but significant portion of ionic units. Surlyn is the trade name of a family of packaging ionomer resins produced by DuPont from ethylene methacrylic acid copolymers. The methacrylic acid groups typically constitute less than 15% of the molecule, with some or all of these groups neutralized by Na^+ or Zn^{-2} ions.

In this chemical structure there are nonpolar ethylene groups, polar methacrylic acid groups, and carboxylate ionic pairs acting together to provide the ionomer with a combination of properties.

Some of the properties that make Surlyn$^{®}$ excellent for packaging applications are its sealing performance, formability, clarity, oil/grease resistance, and high hot draw strength. Good hot draw strength allows faster packaging line speeds and reduces packaging failures. Another well known application of Surlyn$^{®}$ includes it use in the outer covering of golf balls.

An ionomer is an 'ionically crosslinked' polymer, with the crosslinks largely responsible for the unique properties of the ionomer. These ionic crosslinks, unlike covalent crosslinks, are reversible because when an ionomer is heated, the forces of the ionic crosslinks are greatly diminished and the polymer chains become free to move.

6.6.9 Ethylene Vinyl Acetate (EVA) Copolymer

Ethylene vinyl acetate (EVA) copolymers are long chains of ethylene hydrocarbons with acetate groups randomly distributed along the chains.

EVA copolymers are used as heat seal layers for snacks and cheese, bags for ice, and frozen food. They are used as lidding sealant and hot melt.

Due to their high versatility in hot-melt formulations, EVA copolymers are leading polymers for hot-melt manufacturing. EVA based hot-melts are able to fulfill various requirements in applications such as packaging, bookbinding, or label sticking. EVA resins are used in the manufacture of packaging film, heavy duty bags, extrusion coating, wire and cable jacketing, hot melt adhesives, and cross-linked foam. Adhesives such as hot melt packaging adhesives and glue sticks contain EVA copolymers. EVA copolymer may be used as an extrudable adhesive resin or sealing layer for lidding applications for adhering to PE, PP, PET, and PS.

6.6.10 Polyamides (Nylons)

$$\left[\ -\overset{\displaystyle H}{\underset{\displaystyle |}{N}}-(CH_2)_6-\overset{\displaystyle H}{\underset{\displaystyle |}{N}}-\overset{\displaystyle O}{\underset{\displaystyle \|}{C}}-(CH_2)_4-\overset{\displaystyle O}{\underset{\displaystyle \|}{C}}-\ \right]_n$$

Nylon 6,6

$$\left[\ -\overset{\displaystyle H}{\underset{\displaystyle |}{N}}-\overset{\displaystyle O}{\underset{\displaystyle \|}{C}}-(CH_2)_5-\ \right]_n$$

Nylon 6

Nylon is a generic name for a family of polyamide thermoplastics characterized by amide groups [-CO-NH-] in the backbone chain. Nylon 6,6 and Nylon 6 are the two most common nylons used for food packaging films which have good gas barrier, puncture resistance, and heat resistance properties. Nylon 6 is synthesized using the monomer caprolactam ($C_6H_{11}NO$) with 6 carbon atoms, and the "6" in its name is associated with its carbon number. Similarly Nylon 6,6 derived the "6,6" because it is synthesized using adipic acid ($C_6H_{10}O_4$) and hexamethylene diamine ($C_6H_{16}N_2$), each of these monomers has 6 carbon atoms.

6.6.11 Polycarbonate (PC)

$$\left[\ -\!\!\!\bigcirc\!\!\!-\overset{\displaystyle CH_3}{\underset{\displaystyle CH_3}{\overset{\displaystyle |}{\underset{\displaystyle |}{C}}}}-\!\!\!\bigcirc\!\!\!-O-\overset{\displaystyle O}{\underset{\displaystyle \|}{C}}-O-\ \right]_n$$

Polycarbonate (PC) is an amorphous thermoplastic which can be injection-molded, blow-molded, and thermoformed. PC is a good replacement of glass due to its clarity and toughness. It can withstand temperatures above 200°C and was once used as dual-ovenable meal trays in the 1980s but was later replaced by lower cost materials. Presently, the major application of polycarbonate in food packaging is 5-gal (19-L) reusable water bottles, due to its toughness and clarity. Since this polymer is much lighter than glass, it provides easier handling and fuel saving during distribution. There is a concern, even negligible for general adults, about possible migration of bisphenol A, a suspected endocrine disruptor, which is a monomer for PC.

6.6.12 Edible Coatings and Films

Biopolymers of proteins, polysaccharides, and/or lipids can be formed into edible coatings or films with the addition of plasticizers such as glycerin, polyethylene glycol, sorbitol, and sucrose. The edible coatings and films can be produced as coating layers on food surfaces and separate self-supporting films. The most common method of film formation is the controlled drying of the biopolymers solutions or emulsions, which are previously applied onto food surfaces or cast on smooth surfaces. Lipid films are often formed by molten casting which includes steps of heating and cooling.

The functions of edible coating or films include retardation of migration of moisture, fat component and solute, reduction of moisture and gas transfers, improvement of structural integrity, retention of volatile compounds, and conveyance of food additives. Typical use examples of edible coatings and films are wax coatings on fresh fruits and vegetables, collagen films for sausage casings, hydoxymethyl cellulose films of soluble pouches for dried food ingredients, zein coatings on candies, sugar coatings on drug pills, and gelatin films for soft capsules [7]. More detailed and extensive information on edible coatings and films can be found in literature [8].

6.6.13 Recycling Symbols

For easy identification, most plastic containers are marked with recycling symbols, each consisting of a triangle formed by three circular arrows with a specific number to indicate the plastic resin from which the container is made (Figure 6.14).

Figure 6.14 Recycling symbols for plastic containers

The "V" in Symbol 3 stands for PVC. The "others" in Symbol 7 stands for plastic resins other than those of Symbols 1-6; an example for Symbol 7 is a multilayer structure consisting of PP/adhesive/EVOH/adhesive/PP.

6.7 Polymer Processing

Various processes are available for converting thermoplastics into films, bags, bottles, trays, and other packaging forms. As mentioned in Section 6.4, processing can also significantly affect crystallinity and molecular orientation and hence the properties of finished products. In the following sections, schematics are provided to illustrate the basic principles of some common polymer converting processes. Photographs of equipment may be found on the Internet using a search engine such as Google® or Yahoo®.

6.7.1 Extrusion

Extrusion involves melting a thermoplastic resin with an extruder, which is the first step for extrusion coating, cast film extrusion, blown film extrusion, and other polymer processes.

Figure 6.15 shows a single-screw extruder which consists of an electrically heated metal barrel, a hopper for feeding the resin usually in pellet or powder form, a motor for rotating a screw, and a die where the polymer melt (known as extrudate) exits. The heat from the barrel is responsible for softening the surfaces of the resin pellets making them sticky to each other, while the rotating screw is responsible for melting the resin pellets and forcing the polymer melt through the die. The melting of polymer is mainly due to the mechanical energy input from the motor which rotates the screw, compresses, and shears the resin pellets, finally dissipating as frictional heat to melt the resin pellets.

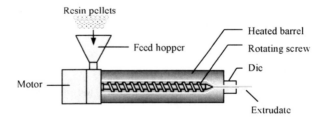

Figure 6.15 Single-screw extruder

A twin-screw extruder has the same configuration as a single-screw extruder except it has two screws instead of one screw. A corotating twin-screw extruder is one whose screws are rotating in same direction, while a counter-rotating twin-screw extruder is one whose screws are rotating in opposite directions.

6.7.2 Coextrusion

Coextrusion involves the use of two or more extruders. Figure 6.16 shows the coextrusion process in which the extrudates from two extruders are merged and welded together in a feed block (a metal block with specially designed flow channels) to form a multilayer structure. Coextrusion may also be used as the first step in extrusion coating, cast film extrusion, blown film extrusion, and other polymer processes.

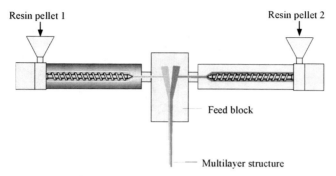

Figure 6.16 Coextrusion process

Coextrusion is used to manufacture multilayer structures for applications where the benefit/cost ratio is favorable. Below are some coextruded structures:

PVDC/adhesive/nylon

PP/adhesive/EVOH/adhesive/PP

PP/adhesive/EVOH/adhesive/regrind PP/PP

PET/adhesive/regrind PP/adhesive/PVDC/regrind PP/PP

As shown above, a thin layer of adhesive resin, known also as tie layer, is used to improve the adhesion between two different polymer layers and avoid the problem of delamination. Some common polymers used for tie layers are maleic anhydride modified PE, PP, and EVA. A thin layer of barrier polymer such as EVOH or PVDC is sometimes used to provide gas barrier protection, since coextruded structures are often used in food applications where good gas barrier is required. Coextruded sheets may be thermoformed into food trays or containers. Layers of regrind resins from industrial scraps are sometimes used to save cost and divert these materials from entering the waste stream.

6.7.3 Cast Film Extrusion

Cast film process involves extruding a polymer melt through a slit die and drawing the melt around two or more chill rolls to form a film (Figure 6.17). The chill rolls are typically chrome plated and water cooled. Coextruded cast films may be produced using multiple extruders.

Due to rapidly cooling by the chilled rolls, cast films typically have low degree of crystallinity and transparent appearance. Beside good optical properties, cast film extrusion has the advantages of high production rate, good control of film thickness and uniformity, and little or no additive required for processing.

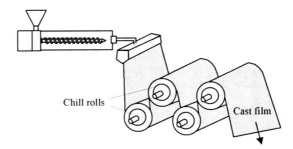

Figure 6.17 Cast film extrusion

Depending on the rotating speeds of the chill rolls, cast film extrusion typically results in no or some molecular orientation in the machine direction. However, molecular orientation in the transverse direction may be induced by attaching a tenter frame to the extrusion die as shown in Figure 6.18. The film is fed into the tenter frame consisting of belts or chains fitted with clips. These clips grip the film and stretch it in both the machine and transverse directions. The temperature of the heated tentering is carefully controlled. Upon leaving the tenter frame, the film is cooled by passing over a series of cooling rollers before being wound. Biaxially oriented PP film (BOPP) is often produced by this process to achieve increased stiffness, enhanced clarity, and improved barrier properties.

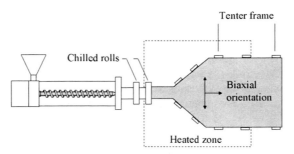

Figure 6.18 Cast film extrusion with tenter frame

6.7.4 Blown Film Extrusion

As shown in Figure 6.19, blown film extrusion, known also as tubular film extrusion, involves extruding a thermoplastic through an annular slit die, followed by introducing air via a hole in the center of the die to blow up the polymer melt into a large tube of film. The tube of film then continues upwards, continually cooling, until it passes through nip rolls where the tube is flattened to create a 'lay flat' tube of film, which is then taken back down the extrusion tower via more rollers.

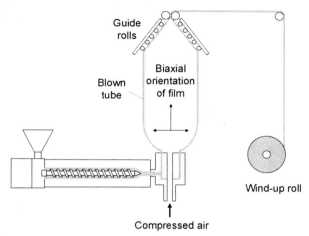

Figure 6.19 Extrusion blown film process

The air within the tube stretches the film to obtain the desired thickness. The tube of film may be made into bags by sealing across the width of film and cutting or perforating to make individual bags. Alternatively, the tube of film may be cut into two flat films.

In blown film extrusion, molecular orientation of the polymer is achieved in both machine and transverse directions, yielding a film with biaxial properties. Typically the expansion ratio between die and blown tube film is 1.5 to 4 times the die diameter. The stretching is controlled by adjusting the air pressure inside the bubble and the haul off speed. Precision temperature control is also important.

HDPE, LDPE, and LLDPE are the most common polymers used in this process, but a wide range of other polymers can also be used. Multilayer structures may be obtained by using multiple extruders. Since good melt strength is required to blow film without any direct support, polymers such as PET have limitations or difficulties when used in this process. The food packaging applications of extrusion blown films are similar to those of extrusion cast films.

6.7.5 Injection Molding

As shown in Figure 6.20, injection molding involves melting a thermoplastic by extrusion, injecting the polymer melt into a mold, cooling the part, and finally ejecting the part.

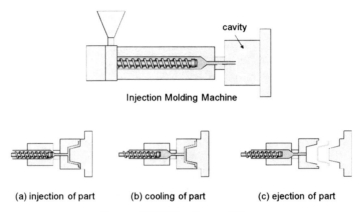

Figure 6.20 Injection molding

Most polymers can be injection molded so long as they can flow and fill the mold cavity easily. The commonly used polymers are PET, PS, PP, HDPE, LDPE, Nylon, and PVC. Injection molding may be used to manufacture a wide variety of parts such as bottle caps, food trays, containers, and preforms for blow molding.

6.7.6 Blow Molding

Blow molding is a process of blowing up a hot thermoplastic tube (parison or preform, as described later) with compressed air to conform to the shape of a chilled mold, and releasing the finished product from the mold. There are three common types of blow molding: extrusion blow molding, injection blow molding, and injection stretch blow molding.

- **Extrusion Blow Molding**

Extrusion blow molding, as shown in Figure 6.21, begins with extruding a polymer melt into a hollow tube known as a parison. The chilled mold is then closed, followed by blowing air through a blow pin to inflate the parison to conform to the shape of the mold cavity. After cooling the plastic, the mold is opened and the part is ejected. It is a low pressure process with typical air pressures of 170 to 1000 kPa.

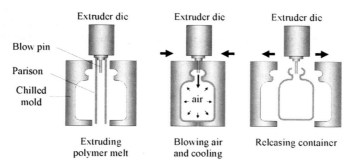

Figure 6.21 Extrusion blow molding of plastic container

Extrusion blow molding requires polymers of relatively high melt strength to support the weight of freely suspended parison for several seconds as it emerges from the die. Low melt viscosity is also desirable for the details of the part to be fully formed with a relatively low blowing pressure. Basic polymers for extrusion blow molding are HDPE, PP, and PVC, and these polymers are sometimes coextruded with EVOH or nylon to provide better gas barrier.

Extrusion blow molding is typically used for larger bottles such as the one-gallon milk bottles and water bottles. Compared to injection based blow molding, extrusion blow molding has less expensive tooling costs and fewer limitations in bottle shape allowing long and narrow, flat and wide, doubled-walled, and other odd shapes. It is still the only low cost process for manufacturing bottles with integrated handles. However, the dimensions of extrusion blow molded bottles are less precise especially in the neck area.

- **Injection Blow Molding**

As shown in Figure 6.22, injection blow molding is a two-step process for making plastic containers. In Stage 1, a molten plastic is injection molded into a thick hollow tube known as a preform. While synonymous, the terms 'preform' and 'parison' are often associated with injection based process and extrusion based process, respectively. The neck-ring in the perform mold is used to provide the preform with a precision neck. In Stage 2, the preform is first softened by conditioning it at an elevated temperature or reheating it with infrared heater, and then it is blow molded into a bottle. This stage is similar to the extrusion blow molding process except that a preform is used.

Injection blow molding requires less degree of melt strength than extrusion blow molding. Common polymers for injection blow molding are PS, LDPE, LLDPE, HDPE, PP, PVC, and PET. While PS achieves some degree of molecular orientation from this process, other polymers unfortunately do not.

The use of preform allows the manufacture of bottles with more precise detail in the neck and finish (threaded) area than extrusion blow molding. The process is flash or scrap free and the molded bottles do not require trimming. Compared to extrusion blow molding, the tooling cost of this process is higher, and this process is typically limited to the production of relatively small bottles.

Stage 1. Injection molding

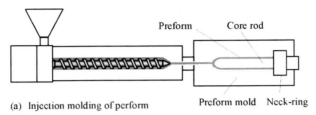

(a) Injection molding of perform

Stage 2. Blow molding

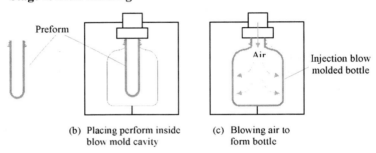

(b) Placing perform inside   (c) Blowing air to
      blow mold cavity   form bottle

Figure 6.22 Injection blow molding process

- **Injection Stretch Blow Molding**

As shown in Figure 6.23, injection stretch blow molding is an extension of injection blow molding with two modifications: (1) the perform is significantly shorter than the bottle, and (2) a stretch rod is used to stretch the preform in the axial direction. This process is used for the production of high quality bottles.

While all blow-molding processes involve blowing air to stretch the parison or preform in some fashion, injection stretch blow molding is designed to achieve and retain biaxial orientation to significantly improve gas barrier properties, impact strength, transparency, surface gross, and stiffness. Biaxial orientation is achieved by elongating the preform with the stretch rod and blowing air to stretch the perform in direction perpendicular to the axis of the preform, while precisely controlling temperature warm enough to allow rapid inflation and molecular orientation but cool enough to retard relaxation of its molecular structure once oriented.

Stage 1. Injection molding

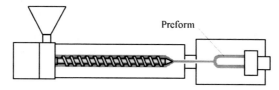

(a) Injection molding of perform

Stage 2. Blow molding

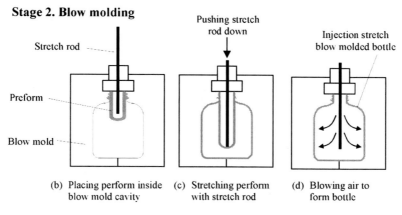

(b) Placing perform inside blow mold cavity
(c) Stretching perform with stretch rod
(d) Blowing air to form bottle

Figure 6.23 Stretch blow molding process

There are two common techniques for injection stretch molding. The first technique is a continuous process of injection molding preforms, conditioning them to proper temperature, and blowing them into bottles. This technique is typically used for low production of specialty bottles. The second technique is a two-stage processes of injection molding preforms, storing them for a short period of time (typically for 1 to 4 days), and blowing them into bottles using a reheat-blow machine. This technique is commonly used for high volume production.

PET bottles for carbonated soft drinks are the most common food packaging application of this process. The combination of stretching by the rod and blowing air at high pressure (about 4 MPa) induces biaxial molecular orientation, thereby making the bottles a better barrier to carbon dioxide and stronger to withstand the internal pressure.

6.7.7 Extrusion Coating

As shown in Figure 6.24, extrusion coating involves extruding of a thin layer of molten plastic onto a wide variety of substrates including plastics, paper, and metal foil. The extrudate can be drawn to desired thickness in the melt state and coated on the substrate. The polymer is then cooled and solidified on a set of chill rolls. For coextrusion coating, the die is fed by multiple extruders.

While LDPE is commonly used, most packaging polymers can be extrusion coated. A requirement is good adhesion between the coating polymer and the substrate. A barrier extrusion coating of PVDC is sometimes used to improve oxygen and moisture barrier.

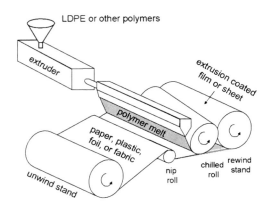

Figure 6.24 Extrusion coating

Some applications of extrusion coating are coated paper cartons for milk and juices, coated aluminum foil to provide heat seal ability, and plastic structures which cannot be coextruded due to processing or equipment constraints.

6.7.8 Extrusion Lamination

As shown in Figure 6.25, extrusion lamination involves combining two different substrates with a molten polymer. This process enables the combination of a wide variety of substrates and different thicknesses with a high level of quality.

Unlike coextrusion which is limited to combining polymers only, extrusion lamination can use substrates such as paper board, aluminum foil, OPP film, oriented PET film, and metallized film. Adhesives used as tie layers for extrusion lamination are typically ethylene acrylic ester copolymers, but other polymers such as LDPE and EVA may also be used so long as they provide sufficient adhesion to both substrates. Extrusion laminated structures are used quite commonly for food packaging applications requiring high moisture and gas barrier, and excellent print quality. Examples include retort pouches, snack packages, and meat and cheese packages.

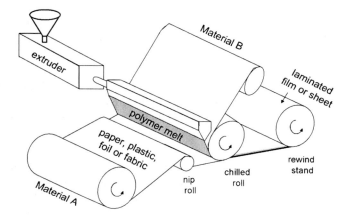

Figure 6.25 Extrusion lamination process

6.7.9 Adhesive Lamination

Adhesive lamination is similar to extrusion lamination except that liquid adhesives instead of polymer melts are used. Figure 6.26 shows a process, known as *dry bond lamination*, in which a liquid adhesive is coated onto a substrate, dried with heat and airflow, and then laminated to a second substrate via a heated compression nip.

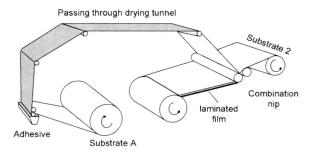

Figure 6.26 Adhesive lamination process

Most flexible substrates can be used in this lamination process. Common adhesives used for dry bond lamination are reactive solids (such as two-part polyurethanes, one-part moisture curing polyurethanes, and UV/electron beam curable acrylates), high solids solvent based (e.g., silicone), and waterborne adhesives (e.g., acrylic emulsions).

Wet bond lamination is different from dry bond lamination in that the webs are joined before the drying step while the adhesive is still wet. One of webs should be pervious to water or solvent to facilitate drying.

6.7.10 Thermoforming

Thermoforming is a process of clamping a thermoplastic sheet, softening it with heat, and applying vacuum or pressure to stretch the sheet into or over a temperature controlled mold. Infrared radiation is usually used to heat the thermoplastic sheets. Figure 6.27 shows the method using vacuum to form a container.

Sometimes, air pressure or a plug is used to assist vacuum thermoforming, and a male mold is used instead of a female mold. A wide range of thermoplastics including PP, LDPE, LLDPE, HDPE, PET, PS, and nylon may be thermoformed. The sheets may be monolayer, multilayer, coated, or laminated structures. Food packaging is the largest application for thermoformed containers, trays, cups, and tubs.

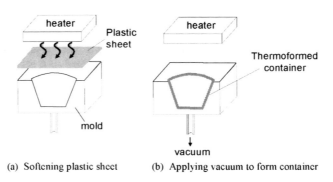

(a) Softening plastic sheet   (b) Applying vacuum to form container

Figure 6.27 Vacuum thermoforming

6.7.11 Vacuum Metallization

Very thin metal layer mostly of aluminum is coated onto plastic substrate in evaporated vapor under the high vacuum condition. Normal thickness of coating ranges from 8 to 50 nm. In a chamber evacuated to less than 3 kPa, the plastic film is unwound from a roll and passed past a chill roll on which the vaporized aluminum from heated crucible (about 1700°C) is condensed (Figure 6. 28). The thickness of the metal coating is controlled by the speed of film movement and metal temperature. The coated film web is wound on a take-up roll in the vacuum chamber. The metallized plastic films have excellent barrier to moisture, gas, and light. Particularly, metallized film is resistant to flexing stress in maintaining gas and moisture barriers. The metallized plastic films carry decorative opaque appearance, and may be selectively demetallized on specific surface to give a local transparency. They can be used for concentrated microwave heating in susceptor. The other usage of metallized film in microwavable food packaging needs a cautious approach to avoid undesired arcing (see Chapter 14).

The thickness of metallized coating is often represented by optical density (see Section 3.3.2). Most metallized films are predominantly based on PET, PP, and nylon.

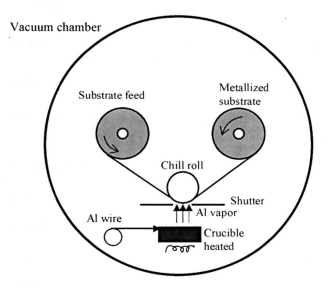

Figure 6.28 Vacuum metallization process of plastic film

Discussion Questions and Problems

1. Determine the degree of polymerization (i.e., number of repeat units) for a polyethylene chain with molecular weight of 80,000.

2. Compare low density polyethylene (LDPE) to high density polyethylene (HDPE) in terms of (a) chemical structure, (b) crystallinity, (c) barrier properties, (d) mechanical properties, and (e) transparency.

3. Draw the chemical structures for LLDPE, PET, and EVOH. Rank the oxygen permeabilities of these polymers from high to low and explain the ranking based on the chemical structures.

4. Draw the chemical structures for PVDC and EVOH copolymers. Explain why these polymers have very low oxygen permeability compared to polyethylene

5. Describe some possible processing or converting methods to improve gas barrier properties of plastic films. Discuss each method with its merits and disadvantages.

6. Compare PVDC and EVOH as a barrier layer in retortable semirigid plastic trays for a meat soup.

7. List the plastic packaging materials which can be used as household wrap. Discuss each material with good points and weak points.

8. What kinds of properties are required or desired for plastics to be used as heat sealant?

9. What properties of plastic packaging materials are desirable for recycling?

Bibliography

1. RJ Hernandez, SEM Selke, JD Culter. Plastics Packaging. Munich, Germany: Hanser Publishers, 2000, pp. 1-311.
2. O-G Piringer, AL Barner. Plastic Packaging Materials for Food. Weinheim, Germany: Wiley-VCH, 2000, pp. 1-77.
3. WE Brown. Plastics in Food Packaging. New York: Marcel Dekker, 1992, pp. 1-291.
4. WA Jenkins, JP Harrington. Packaging Foods with Plastics. Lancaster, PA: Technomic Publishing, 1991, pp. 1-99.
5. EM Abdel-Bary. Handbook of Plastic Films. Shawbery, UK: Rapra Technology, 2003, pp. 5-262.
6. KR Osborn, WA Jenkins. Plastic Films. Lancaster, PA: Technomic Publishing, 1992, pp. 1-130.
7. JH Han, A Gennadios. Edible films and coatings: a review. In: JH Han, ed. Innovations in Food Packaging. Amsterdam: Elsevier Academic Press, 2005, pp. 239-262.
8. JM Krochta, EA Baldwin, MO Nisperos-Carriedo. Edible Coatings and Films to Improve Food Quality. Lancaster, PA: Technomic Publishing, 1994, pp. 1-335.

Chapter 7

Glass Packaging

7.1   Introduction 177
7.2   Chemical Structure 178
7.3   Glass Properties 180
    7.3.1 Mechanical Property ··· 180
    7.3.2 Thermal Property ··· 182
    7.3.3 Electromagnetic Property ·· 183
    7.3.4 Chemical Inertness ·· 184
7.4   Glass Containers Manufacturing 185
    7.4.1 Glass Making ·· 185
    7.4.2 Container Manufacturing ··· 188
    7.4.3 Post Blowing Operations ·· 189
7.5   Glass Container Strengthening Treatments 192
7.6   Use of Glass Containers in Food Packaging 193
7.7   Other Ceramic Containers 194
Discussion Questions and Problems 195
Bibliography 195

7.1 Introduction

Glass and glass containers are unique and deserve a special place in the presentation of the different designs and package forms for foods. Glass is certainly, among the materials nowadays used in contact with foods and beverages, the oldest one with its history longer than 5000 years. Likely, the first glass objects with containing functions, were obtained by sculpting blocks of *obsidiana* which is a natural material coming from volcano's magma and similar in appearance and composition to synthetic glass. However, the human production of molded glass is ancient and probably goes to Phoenicians (from 2500 to 333 B.C.) many, many centuries ago [1]. The fundamental step in glass manufacturing evolution, however, has been the discovery of a technique for blowing molten glass and consequently producing hollow objects with great liberty of shape. This happened, approximately 100 years before Christ, and, starting from that moment, the technique for transforming glass into useful objects evolved slowly, particularly in most areas of ancient Roman Empire, with interest both for artistic and trivial applications; regarding these last applications, mechanization of the production (around the 19th century) reduced the industrial costs for the glass containers to be acceptable for food packaging uses.

The word *glass* applies more to a physical state of the matter than to a chemical composition; in fact the composition of glasses can be quite different, according, for instance, to their color, to thermal or mechanical resistance (see Table 7.1); glasses can be made from phosphates, aluminates, borates, or inorganic halides (a ionic compound containing a halogen), and the name *glass* is attributed even to special forms of organic material [2]. Among several definitions of glass, the oldest perhaps belongs to Eraclito (a Greek philosopher of the 5th century B.C.) who stated that glass was *an artificial stone, produced by fire*. More accurate scientific definition was proposed by ASTM in 1999 [3], according to which, *glass is an amorphous, inorganic product of fusion that has been cooled to a rigid condition without crystallizing*; glass is also indicated as a super-cooled liquid. Glass is also a subclass of ceramic materials i.e., inorganic, nonmetallic materials. Some ceramic materials (china, porcelain, ceramic, etc.) are also used for packaging functions; to them, however, just a few notes will be reserved in this chapter. Therefore this chapter deals, prevalently, with what everybody knows as *glass* (mainly a silicate glass, often named as soda-lime glass). All the different definitions of glass tend to emphasize the physical state of this material which deserves to be described and well understood.

Table 7.1 Typical glass compositions (%)

Constituent	Flint	Amber	Green	Pyrex®	Lead glass
SiO_2	73.20	72.60	72.10	67.5	60
Na_2O	11.90	12.80	2.90	13.6	1.0
K_2O	0.46	1.01	0.87	1.8	14.9
CaO	11.20	11.10	9.80	9.4	-
MgO	1.70	0.23	1.74	-	-
BaO	0.02	-	-	-	-
PbO	-	-	-	-	24.0
Al_2O_3	1.17	1.81	1.93	5.0	0.08
$Fe_2O_3 + TiO_2$	0.08	0.34	0.37	0.15	0.02
Cr_2O_3	-	0.002	0.17	-	-
SO_3	0.18	0.08	0.09	-	-

7.2 Chemical Structure

Whatever is the chemical composition of common glasses, almost three quarters of the matter are represented by silicon (Table 7.1), which is the second most abundant element on earth (it constitutes about 28% of earth's crust being the second only to oxygen). Silicon does not occur free in nature, but it is found as silica of silicon dioxide (SiO_2 in quartz, sand, cristobalite, and many other ores), or as silicate (feldspar, kaolinite, etc.) where the silicon dioxide is joined to other oxides, mainly aluminum oxide. The silica typically shows polymorphism being able to crystallize differently, leading to the different ores available in nature for glass making [4]:

$$\alpha \text{ quartz} \quad \overset{579°C}{\Longleftrightarrow} \quad \beta \text{ quartz} \quad \overset{857°C}{\Longleftrightarrow} \quad \gamma \text{ tridymite} \quad \overset{1470°C}{\Longleftrightarrow} \quad \beta \text{ cristobalite}$$

The four valences of silicon lead, in all the crystalline forms, to a simple structural unit in which every silicon is located at the center of a three-dimensional tetrahedra, having oxygen at the four corners (Figure 7.1a); the tetrahedra can be arranged continuously, each oxygen being connected to two silicon, leading to a well ordered crystalline organization of the matter (Figure 7.1b). This crystalline structure is present in all the silica-containing ores which are used for manufacturing the glass. However, the ores are neither transparent, nor chemically inert as glass is, and above all they cannot be directly formed into useful shapes and sizes.

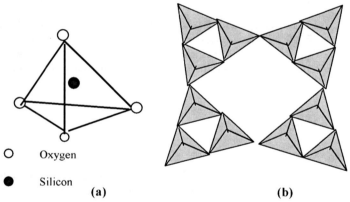

○   Oxygen

●   Silicon

(a)   **(b)**

Figure 7.1 Schematic representation of silica structural unit and of its crystalline organization

The process of glass making transforms the inorganic ingredients into the matter we call glass, mainly through physical transformation above 1450-1500°C not through a chemical change; any crystalline structure is lost in the process and the tetrahedra rearrange in a-periodic, messy, i.e., amorphous structure (Figure 7.2a). The randomized structure is made possible by the capacity of the 'Si-O-Si' bond to establish angles variable from 120 and 180° with a maximum probability at about 144° [5-7]; this structure also leads to the presence of several empty spaces between the silica tetrahedra, which are filled (Figure 7.2b) by alkaline earth elements (mainly sodium, calcium, and magnesium) coming from minor ingredients of the glass making mixture of raw materials. In fact, they make the mass melt and rearrange at lower temperature, giving it a much lower viscosity. Basically, all the physical and chemical properties of glass containers derive from this exclusive amorphous structure which is retained in a solid state by means of an adequate quick cooling of the molten mixture.

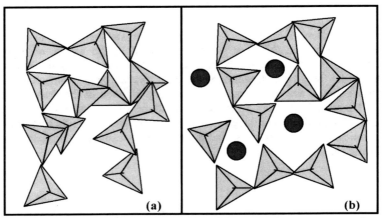

Figure 7.2 Schematic representation of the amorphous structure of silica glass (a) and silicate glass (b). In silicate glass, sodium (•) interrupts the network continuity and changes melting temperature and viscosity.

7.3 Glass Properties

7.3.1 Mechanical Property

Technically, glass is considered a hard material (Table 7.2), but at the same time glass containers are known as fragile objects, easily susceptible to mechanical failures. This is also the most critical factor of glass bottles and jars in the competition with other materials in food packaging. Scientific investigations on the strength of glass are dated to Griffith's pioneering studies of the 1920s, which established basic concepts leading to the modern theories of brittle fracture and to the main technological remedies developed [7]. Despite the high strength of the covalent bond between silicon and oxygen (see Table 2.4 in Section 2.2.3), an initiated fracture quickly propagates in glass objects, with little resistance in a large thickness and surface, leading to what is known as brittle fracture: the breakage is characterized to come about at widely varying stress values much lower than the theoretical figures (coming from the strength of chemical bond).

The reasons for this lie in two facts: first, the continuous and rigid structure of silicate glass (the inter-connected tetrahedra) does not permit any plastic flow of the material, and thus there is no stress absorption causing the fracture to propagate with destructive effects; second, the glass objects always have practically sharp cracks, flaws, or other superficial or internal defects even sometimes invisible, and at their tip a huge stress amplification occurs. For this last reason, the finished glass packages are tested in practice. Generally on randomized sampling, the finished glass objects are tested for internal pressure resistance, vertical load strength, and resistance to impact.

Table 7.2 The main physical properties of silicate glass

Property	Typical value
Mass thermal capacity (kJ kg^{-1} °C^{-1})	0.84
Thermal expansion α (°C^{-1} x 10^{-6})	6
Thermal conductivity (W m^{-1} °C^{-1} at 30 °C)	0.96
Glass transition temperature (°C)	≈ 550
Relative refraction index (20°C, 589.3 nm)	1.50 – 1.90
Specular transmittance, 550 nm (180 μm)	92.1
Mass per unit volume (g cm^{-3})	2.40 - 2.80
Static coefficient of friction (glass/glass)	0.9 – 1.0
Modulus of elasticity (MPa)	70 000
Breaking point (MPa)	70

Differential pressure stresses across the glass container walls develop from carbonated beverages or vacuum-packed food-products; high pressures are also produced by heat treatments such as pasteurization or sterilization of already packed food or beverages. The internal pressure of various carbonated beverages may vary, at room temperature, from 0.15 to 0.5 MPa, but the temperature of a pasteurization step may double these values [8]. The stresses generated in a closed glass container are considered circumferential and longitudinal, the latter being about half of the former in the cylindrical container. In this part, the most relevant stress (σ, circumferential) is related to the package diameter (d), the wall's thickness (ℓ), and the internal pressure (P) as follows:

$$\sigma = \frac{P \cdot d}{2\ell} \tag{7.1}$$

Equation 7.1 is a simplified relationship, and more accurate equations are required to take into account rapid changes of curvature and thickness in the different parts of the containers (Figure 7.3), being well beyond the scope of this chapter [9]. However, the resistance of a glass container is generally proportional to its wall thickness, and very heavy glass bottles, like those for sparkling wines or Champagnes, can resist internal pressure of up to 7 MPa (about 70 atm), a value ten times higher than the normal pressure of these products.

Also vertical loads are quite commonly met in real life of glass packages both during stacked storage and during closure application. The pressures applied in these cases are also quite high and can achieve 7 MPa, generating dangerous tensile components on the shoulder and on the bottom (Figure 7.3). An appropriate design can improve the resistance of glass containers, i.e., reducing the difference between the neck and body diameter, but usually the strength increases with glass weight and thickness [9].

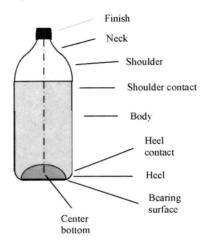

Figure 7.3 Names of the typical parts of a glass bottle

7.3.2 Thermal Property

As already stated, glass is an amorphous material, thus it does not show a real melting temperature (the commonly presented melting temperature is that of its ingredients), but it progressively softens achieving a liquid-like state. At the same time, the cooling process that fixes the amorphous structure into a solidified state is not a sudden change of phase, but occurs over a temperature range. In the soda-lime glass, the higher the concentration of alkali, the more nonbridging oxygen sites are created; this reduces the tetrahedral connectivity and thus decreases the glass transition temperature [5, 6]. A lower glass transition temperature means that the glass is less expensive to produce as it melts at lower temperature and will have also a lower temperature of deformation.

Silica has the lowest coefficient of thermal expansion among any known substances (Table 7.2), and this property is maintained in glass, leading to potential failures when a glass container, with a metal closure, is submitted to thermal treatment. The main thermal characteristic of a glass container is its ability to endure a thermal shock, i.e., a sudden change of its temperature. Glass is a rather poor thermal conductor, but the changes of temperature in a glass object deeply alter the stress equilibrium in it: when it is quickly cooled, on the surface from which the heat is removed first (generally the outer one), tensile stresses are established, which are compensated from compressive stresses on the opposite surface (generally the inner one); on the contrary if the object is suddenly heated, on the surface where its temperature increases first, compressive forces are set up with tensile forces on the opposite side. When the temperature equilibrium is reached these stresses disappear. Glass is more sensitive to tensile stresses than to compressive ones and thus shows high vulnerability to sudden cooling. Therefore, the resistance to thermal shock for many glass containers is determined in standardized tests where preheated objects (e.g., at 63°C for 5 minutes) are immersed in a cold water bath (e.g., at 21°C). The thermal strength of glass objects is influenced not only by its thickness (thinner glass is more resistant) but also by its chemical composition (the presence of boron and aluminum oxide, like in Pyrex$^{®}$ or Corning$^{®}$ glasses, increases the resistance) and possible surface treatments, which will be mentioned later on. Finally, concerning the thermal properties, it is important to emphasize that glass may be indefinitely melted and molded without any loss of its original properties.

Table 7.3 Coloring ingredients of glass mixture

Color	Constituents
Colorless	CeO_2, TiO_2, Fe_2O_3
Blue	Co_3O_4, Cu_2O + CuO
Purple	Mn_2O_3, NiO
Green	Cr_2O_3, Fe_2O_3 + Cr_2O_3 + CuO, V_2O_3
Brown	MnO, MnO + Fe_2O_3, TiO_2 + Fe_2O_3, MnO + CeO_2
Amber	Na_2S, S, carbon
Yellow	CdS, CeO_2 + TiO_2
Orange	CdS + Se
Red	CdS + Se, Au, Cu, UO_3, Sb_2S_3
Black	Co_3O_4 (+ Mn, Ni, Fe, Cu, Cr oxides)

7.3.3 Electromagnetic Property

The largely appreciated transparency of glass objects is determined by the amorphous structure of glass and by the chemical nature of its ingredients. The pure silica has a UV cut-off around 150 nm, but the presence of alkaline oxides in the glass composition makes the UV barrier effective at higher wavelengths (see Figure 3.4 in Section 3.3.3). In the visible region the glass transparency may be modified by selecting the appropriate ingredients. The most common ingredients for coloring are listed in Table 7.3, however it must be underlined that frequently these ingredients alter physical properties, such as density and mechanical or thermal resistance. Glass is perfectly transparent to microwave, leading to low energy dissipation, i.e., the negligible reduction in energy as a microwave travels through the glass wall.

7.3.4 Chemical Inertness

The most generally recognized advantage of glass containers is their inertness to food and beverage contact. No chemical reactions and no relevant leaching phenomena are expected when the material in contact with food and beverage is glass. Eraclito again (5th century B.C.), wrote as: *objects for drinking made of this material are better than those made of gold or silver.* The consumers' perception is exactly still the same even today. Generally an amorphous structure is much less reactive, compared to crystalline one, and silica has little solubility at neutrality or acidic pH values (except for concentrate HF and hot H_3PO_4 solution); however, glass can also lead to some interactions phenomena with contacting aqueous phases.

The behavior of glass is completely different in an acidic or alkaline environment. In an acidic media, an ion exchange takes place to a limited extent: the hydrated surface leads to a leaching of positive ions (principally sodium) exchanged with hydrogen ions in the solution. This kind of phenomenon (acidic corrosion) may be relevant only for glasses rich in sodium (common silicate glasses) and for lead glasses which can allow some heavy metal to leach out. The ion solubility, however, is limited to the first time contact with an acidic solution and to a superficial layer of the material. Much more serious is the alkaline attack to the glass structure; at higher pH values, the silica network is progressively dissolved together with all the remaining constituents, generally leading to a kind of surface etching [6, 7]. The alkaline corrosion depends on the pH value and the temperature, and is not likely to occur in real food/beverage contact.

From the above discussion, the most important form of interaction is the contact of glass with water at high temperature. Distilled water is generally a weakly acidic media and performs an exchange of small ions; the sodium and calcium ions are released, progressively elevating the pH and leading to the alkaline type interaction which has much more serious consequences [6, 7]. Therefore, admitting the glass containers filled with distilled water to autoclaving temperature (121°C) for 1 hour appears to be an aggressive test activating the two possible mechanisms of glass deterioration, and these conditions are used in Europe as testing procedures for glass objects. As Table 7.4 shows, the only significant substance extracted in 3% acetic acid solution is sodium oxide, whereas the matter predominantly leached in water is silica.

Table 7.4 Substances released (ppm) from flint glass bottle, after 1 hour at 121°C, in different media

Constituent	Distilled water	Acetic acid (3%)
SiO_2	26	0.2
Na_2O	6.5	3.8
K_2O	0.09	0.025
CaO	4.1	0.52
MgO	0.35	0.02
Al_2O_3	0.55	0.03

Another problem concerning glass as food contact material is the possible presence of glass fragments. The emotional impact of a chip or sliver found by the consumer in his bottled beverage is awful even if the event is rare, and unfortunately it is not always easy and possible to discriminate and detect the minute pieces of glass that a mechanical shock at closing operation might produce. During last decades a lot of preparatory controls and countermeasures were addressed to limit this kind of possible physical contamination [10].

7.4 Glass Containers Manufacturing

The glass container plants have a singular specificity: in a single plant raw materials enter to be transformed into completely finished containers, bottles, or jars that can be used immediately in the filling lines of food industries frequently even with prelabeled appearance. The same is not true for many other food packages, where several intermediate steps of production and converting are always undergone, from the raw materials (ores, wood, oil, etc.) to the ultimate package (can, board, plastic film, etc.). This does not mean, however, that glass container manufacturing is a trivial process; on the contrary the new technologies introduced for strengthening the light glass containers added complexity and imposed high standard onto the plant, innovating old manufacturing practices.

7.4.1 Glass Making

The different ores (generally transported to the plant by trucks or rails, in bulk and in large amounts) have each a specific function in glass making (Table 7.5). The main ingredient, silica, is responsible for the fundamental structure of final material, but also boron (B_2O_3) and partially aluminum (Al_2O_3) can take part in the amorphous network and their presence is generally responsible for improved performance in terms of thermal and mechanical resistance (see Table 7.1). Recycled glass takes a fundamental function in glass manufacture providing significant energy saving. In fact, it is already an amorphous structure and does not require as much energy as the other raw material; even if its cost (due to the possible operations for color selection and purification from foreign materials) may be higher than that of sand and other ores, the usage of cullet or recycled glass is always welcome because of it's willingness to melt and mix with other ingredients. The availability and supply for cullet have been increased more and more by the policies of waste collection and recycling. Even before pressures from environmental concerns, glass makers have always used a large amount of cullet collected in house. As already stated, in fact, glass is probably the only material (at least among those in the food packaging area) that is endlessly recyclable: melting and new molding, never reduce the original performance of the material. Better than switching off the plant, even without adequate orders of final containers, it is appropriate to run the factory by using the recycle of overproduction: restart of the stopped plant is too costly.

Table 7.5 Raw materials and functions in glass making

Ingredient	Function
Silica (sand), boric anhydride, recycled glass (cullet)	Network formers
Soda ash (Na_2Co_3), K_2CO_3	Melting
Limestone ($CaCO_3$), Mg and Ba carbonate	Stabilizer
Sulfite, sulfate, nitrate	Refining
Metal oxides	Decoloring
Metal oxides, carbon black, sulfur	Coloring
Recycled glass (cullet)	Energy saving

In order to make the mass less viscous and melt at lower temperature, at least one tenth of the mixture is constituted of sodium carbonate and about the same amount is represented by calcium carbonate (soda-lime glass). These two ingredients lead to a chemical transformation during glass making, as they decompose to oxides, releasing a huge amount of carbon dioxide, almost 200 times the volume of glass produced [11]. Both sodium and calcium oxides are not directly linked to the silica network, and rather interrupt the silica continuum. However, their presence is essential for speeding up the process of container manufacturing and making it more feasible from an economic point of view. Because sodium silicate is quite soluble and may tend to bloom on surface impairing transparency, calcium carbonate, which is less soluble, is also added as a stabilizer of the mixture.

The large quantity of carbon dioxide released in the glass making, together with lesser amount of sulfur dioxide and water vapor, is not only an environmental problem but also a technological one; the small gas bubbles remaining in the final containers place a serious risk of possible mechanical breakage. A specific category of ingredients is therefore used to aid in melting and removing gas from the molten glass mass, and is represented by sulfate, nitrate, or sulfite of alkaline ions.

A large variety of simple substance and metal oxides (see also Table 7.3) are finally used to get the right wanted color or the right level of colorlessness. Transparency as well as a specific color tone of a bottle or a container is an important tool for food/beverage brand identification, and thus is a critical point in the aspect of marketing; it is therefore necessary to understand the essential necessity of such ingredients, even at low level, frequently less than 0.05%.

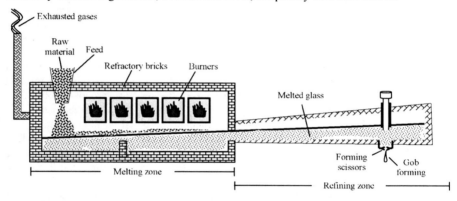

Figure 7.4 Schematic representation of a furnace (side port) for glass making

All together, the exactly weighed ingredients are almost continuously charged in the furnace (Figure 7.4), which is a large covered and heated basin where they are melted and transformed into glass. Typically, the surface of a furnace may range from 50 to 65 m^2 and its depth from 1 to 2 m. The entire furnace is made of refractory bricks able to resist against the high temperature required, which is reached by electric boosting, fossil-fuel firing, or mixed arrangements. The way the fuel is introduced and the exhausted gases eliminated (by several ports on the sides or two ports on the end), leads to the names of the principal types of furnaces: side-port and end-port. From the point of charging the raw materials to the exit of finished glass gobs, an appropriate gradient of temperature is maintained. The ingredients start to melt around 1250°C, the mass is refined at about 1500°C (for better removal of the gases), and then the molten glass is cooled to about 1150°C, which is a better temperature for forming hollow objects. Typical furnaces can produce 3.5–8.0 tons of glass each hour and thus must be fed with a little larger amount of raw materials continuously, because the glass level must be generally held constant within 1 mm [11]. This obviously takes place 24 hours each day, 7 days a week, 365 days a year, because it would be nonsense to switch off the furnace for the weekend and ignite it on Monday. Fortunately, glass can be removed and charged again in the furnace without loss of continuity and without any degradation of the material; actually, some furnaces work continuously for several years and are stopped only when maintenance is required for the brick walls.

7.4.2 Container Manufacturing

Glass containers are made by several different types of forming machines; whatever machine, however, has to perform the same two fundamental functions, which are shaping the drops of molten glass (the cylindrical gob exiting from the furnace and having the same weight of future container) into hollow objects and removing enough heat from the forming container in the amorphous state to prevent it from deforming under its own weight. There are two types of forming machines that are predominant throughout the world: *blow and blow* (B&B, Figure 7.5) and *press and blow* (P&B). In both techniques, two different molds are used in two subsequent steps. In the first mold the gob is transformed into a preform (named *parison*) which has just a sketch of hollowness with the top and inside surfaces of the finish already formed. In the second step, always by blow air (pressure around 160-250 kPa), the final shape of the container is reached (Figures 7.6 and 7.7), pushing the glass against the metal surface of mold. The difference thus concerns exclusively the way the parison is made in the first mold (blank mold), i.e., by pressing a plunger in the gob (P&B) or blowing pressured air into it (B&B). It is not a trivial difference, however, and for a long period of time the B&B process was the only right one for producing bottles or narrow neck containers, whereas the P&B process is used for jars or wide mouth containers. Nowadays, things are changing, and it is now possible to produce narrow neck glass packages with the press and blow process (*narrow neck press & blow*), which produces containers with less glass, thus lighter containers [12]. Producing the parison by a plunger allows its walls to have the same thickness at any position and therefore gives less thermal inhomogeneities in the structure.

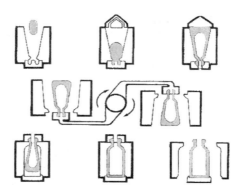

Figure 7.5 Schematic representation of blow and blow process

The time required for producing a glass container, from the time that the gob falls in the blank mold to the time the formed package exits, is about 10-12 seconds for both processes, which corresponds to a relatively slower production rate compared to other food packaging containers.

Figure 7.6 Parisons and final shapes of bottles and jars

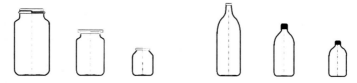

Figure 7.7 Typical shapes of glass containers with wide mouths and narrow necks

7.4.3 Post Blowing Operations

Because the temperature of the falling gobs is about 1100-1250°C and at the exit of the mold the container's temperature is much less than 500°C, a dramatic rapid cooling takes place during short time of about 10 seconds and this introduces strong stresses in the formed container, making it quite fragile. The thermally nonequilibrated contact of the parison and formed container with the mold's walls and tools of the machine is moreover likely to bring about superficial defects which might be critical for containers integrity. Some operations are therefore performed in order to strengthen the final containers, after the bottle or the jar exits the mold.

- *Hot end* - Practically all the glass containers for food/beverage products are at first submitted to the so called *hot-end* treatment, which is an outer surface coating, able to fill up the microcracks, the flaws and all the surface defects that can act as stress amplifiers; moreover, the hot-end treatment facilitates the subsequent *cold-end* treatment, changing the surface properties of the container [13]. The sheathing is performed mostly by spraying on the hot bottle or jar, in a coating hood maintained at 500-600°C, a solution/suspension of tin or titanium compounds, which can be inorganic ($SnCl_4$, $TiCl_4$) as well as organic (dimethyl tin dichloride [$(CH_3)_2SnCl_2$], tin alkyl chloride, organo titanates). Tin tetrachloride is the most widely used coating agent. The operation instantaneously leads to pyrolysis due to high temperatures and the presence of water vapor, coating the surface with a very thin metal oxide layer as with:

$$SnCl_4 + 2\ H_2O \rightarrow SnO_2 + 4\ HCl$$

HCl gas released is removed from the hood where the treatment has been done. A coating level of 3 $\mu g\ cm^{-2}$ is regarded as a minimum amount for effective protection from abrasion [14]. The friction coefficient of the glass surface is increased by the hot-end treatment while the transparency of the container is affected little. The increased friction coefficient of the containers may increase the chance of the surface damage during their handling and can be overcome by the successive cold-end treatment.

- *Annealing* - After the hot-end step, the glass containers are addressed to the annealing process, which is the thermal treatment required for removing the tensile and compressive stresses concentrated in the glass because of the rapid cooling during the container forming. To do this, along a path in a covered tunnel, which may be as long as 25-35 m for the containers to pass through about 30-45 minutes, their temperature is raised to 540-550°C (just above softening temperature), held constant for a couple of minutes, then cooled slowly well below the softening point and finally cooled quickly to 35-40°C. The annealing tunnel physically divides the glass container plant between the so-called hot-end and cold-end operations.

- *Cold end* - Before the end of the annealing process, when the containers reached the best temperature for the *cold-end* treatment (around 120°C), their outer surface is sprayed with a water solution containing stearates, waxes, silicones, polyethylene, or other organic substances. The aim is to anchor an organic layer on the surface modified by the hot-end treatment, reducing the coefficient of friction of the containers and thus indirectly making them more resistant. Due to the growing automation in the filling plants and the high speed that is experienced there (in brewer plant even 100,000 bottles are filled each hour), a high coefficient of friction in the containers may be a critical factor. The possible contact damages of commercial glass containers can be related to the stress generated on a surface as a function of coefficient of friction, normal load, and the radius of the contacting object [15]. According to Sanyal and Mukerji [13, 14], a monomolecular layer of organic substances with long hydrocarbon chains, may produce a considerable reduction of friction.

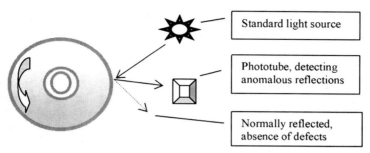

Standard light source

Phototube, detecting anomalous reflections

Normally reflected, absence of defects

Figure 7.8 An example of optical device for glass container quality control. If a scratch is present on the surface of the containers rotating at high speed, the detector catches an anomalous reflection.

- *Quality control* - Some special quality controls are performed by sampling the containers in a random way and measuring their attributes, e.g., the thickness of hot-end coating or the internal pressure resistance, but many feature measurements are realized on all the containers produced during the glass container inspection, which follows the annealing and cold-end treatments. Microprocessor-based devices permit to control both the geometrical features of the containers and the possible defects in the glass. Particles of unmolten refractory material (stones), bubbles of gas trapped in the matter, blemishes, and other defects are detected by means of different optical devices (Figure 7.8); the thickness of the finish, of the walls and other dimensional tolerances are checked continuously; in some plants, even an impact simulator that squeezes each container between rollers is used in the inspection. These inspections on 100% of the production, performed for less than 1 hour after the exit out of the mold, effectively permit a quick intervention in the hot end part of the plant, guaranteeing the best possible level of such a delicate production. One specific feature of glass containers is the constancy of their volumetric capacity. As rigid containers of constant capacity, bottles and jars can easily guarantee the nominal content of the food or beverage inside, when filled at the same level. As a first approximation the glass bottle or jar capacity strictly depends on the gob amount and the volume capacity of mold: container capacity = mold capacity – gob volume, where gob volume can be obtained by dividing the mass by its density. For assuring the highest steadiness in the production, the amount of dropping glass should be carefully controlled, and the molds and tools for blowing parisons need to be maintained adequately.

At the end of this description, it should be evident that the price of the glass containers are not due to the cost of raw materials (the ores are cheap and largely available almost everywhere) but mainly to the energy costs of the process; the complexity of glass making is not due to special converting operations but to the demand for maintaining temperatures, levels, and composition constant during a long time and throughout different parts of the furnace and finally to the sophisticated quality control activity performed.

7.5 Glass Container Strengthening Treatments

Besides the conventional operations of hot-end treatments, cold-end treatments, and annealing, several other methods to increase glass strength, leading to further possible weight reduction, have been extensively investigated even if they are not widely used. Some of them will be described, regardless of their importance or diffusion in the market. Actually, the need for the lightweighting of single use glass containers is still quite strong, despite the great progress that has been made in the '70s in reducing the weight of one way bottle to less than 50%.

- *Chemical toughening* – A chemical ion exchange is realized on the surface of the glass (outside and possibly inside) by salt solution spray or by a molten salt bath in a heated lehr, substituting sodium with lithium or potassium ions. It produces a surface layer of some tens of μm, resulting in quite high compressive strength that reinforces the container. Lithium reduces the thermal expansion coefficient of the glass surface, and its surface contracts less than the interior when the glass cools, leading to a compression layer on the surface. The role of potassium ions in chemical strengthening is related to their atomic radius much bigger than the sodium radius: the metallic radiuses of sodium and potassium atoms account for 180 and 220 pm (pm=10^{-12} m), respectively [16]. Unfortunately the compressive layer (15-20 μm) of chemical treatment on soda-lime glass is much too thin and tends to be easily removed [17].

- *Thermal toughening* – This is another way of creating a compressive strength on the internal and external surfaces of a glass object. The evaporation of volatile organic compounds or blasts of cool air are used to reduce the surface temperature much faster than the interior, generating compressive layer on the surfaces balanced by equivalent tensile layer in the interior. The strength involved is weaker than in the chemical toughening but the thickness of modified layer is much bigger. Thermal toughening of glass containers has never become a commercial success beyond the test level because the process is complex and gives good results only if the glass is thick enough (an already strong glass). Another problem related to this process is the high number of glass fragments that results when the glass breaks: for a glass food/beverage container, in fact, it seems more dangerous to have many little pieces than a few large fragments.

- *Pre-labelling* – More than strengthening the container, the aim of this technique is to reduce the adverse effects of possible impact by covering a large part of the bottle body with a wrap that is functional and decorative at the same time. The plastic covering may be in form of shrink band of plastic films such as polyvinyl chloride or polyolefin. The plastic bands are mostly films uniaxially oriented for desired shrinking property. Another type of plastic label is a body sleeve made of foam or homogeneous plastic. These protective wrappings or sleeves have a somehow cushioning effect, reduce the risk of abrasion during handling, reduce the noise on filling lines, and also retain the fragments if the glass should break.

- *Acid polishing* – The etching phenomenon (see Section 2.5.3) is exploited for chemically dissolving a layer of glass at the surface, removing the faults that might have been introduced during container making. It is rather difficult to use this technique commercially because dangerous chemicals (hydrofluoric acid) are used and the treatment is not permanent with the glass surface being flawed or scratched again.

- *Crystallization* – It was found that a few nucleating agents (titanate, zirconate, and fluoride particles) can be melted in a homogeneous way into the glass and subsequently be precipitated for use as starting nuclei to crystallize bulk material leaving a small amount of residual glass. These materials called glass-ceramics (for many reasons not suitable for containers) have a structure that prevents fracture propagation and account for better mechanical reliability. Several unusual properties are typical of these crystallized glasses: *sapphire* is harder than any glass and *mica* is extremely soft; certain crystalline families have unexpected dielectric or magnetic properties, and some are semiconducting or show piezoelectric properties [18].

7.6 Use of Glass Containers in Food Packaging

Glass containers are used for a wide variety of foods covering the states of liquid, solid, and semi-solid. They are usually delivered from glass manufacturers to food processors in bulk palletized form shrink-wrapped. Depalletization, cleaning, filling, capping, in-bottle heat treatment, labeling, and distribution are the usual sequences followed. Conveying is an essential operation linking all these steps. Through all the operations, the line should be designed to prevent damage and contamination to the glasses. Bunching of the containers in the conveyor line should be avoided by proper means when the line speed slows down at the event of accidental breakdown or stoppage of a machine: clustering of the containers in the line should be detected for the timely feed back into the line. In any operation, the glass containers should not be exposed to drastic temperature change. When the containers are in heating phase, the limit of allowed temperature differential is considered as 60°C. In the cooling cycle stricter temperature differential of 40°C is regarded as the maximum limit.

Cleaning empty bottles before filling operation may be accomplished by air blowing, warm water rinse, washing with detergent solution, or combination of the above. The returned containers for reuse are subjected to washing with caustic soda solution. In case of use for hot-filled products, the container temperature needs to be increased in the cleaning step or a separate heating tunnel to a certain level in order to prevent thermal shock. Filling of liquid foods into glass containers is usually based on level control and operated under the force of gravity, vacuum, pressure, or their combination. Closures are applied for hermetic sealing of the containers usually in an integrated system of the filler. For details of closures and filling operation, refer to Chapter 11. Heat processing of filled glass bottles such as pasteurization and sterilization should be controlled in temperature program avoiding abrupt thermal shock. The temperature after cooling phase is desired to be around 40°C to allow the evaporation of moisture on the surface of closure and bottle. As regards designing and operation of thermal processing of the packaged foods, refer to Chapter 12. The labels in paper or laminated film are attached onto sealed bottles by proper adhesive, and then the labeled bottles move to secondary packaging such as plastic wrap or corrugated box for distribution in the market.

7.7 Other Ceramic Containers

As stated before, *ceramic* is a term that may be generally assigned to whatever inorganic, nonmetallic material, but is also the common word to indicate a broad class of materials obtained by cooking a knead of clay and water; selecting the appropriate clays, the water proportion, the molding, and the firing techniques, several different and useful objects from bricks to porcelain have been produced by men during a millenary history of progress and development. Ceramic jars have been extensively used as food and beverage containers in ancient past from Greeks and Romans, and actually permitted a huge development of trade and exploration [19]. In the modern world, ceramic materials are widely used for tableware, cookware, and storage vessels. The ceramic containers or dishes may be classified into porous earthenware and nonporous porcelain [20, 21]. The nonvitreous earthenware with porosity may be glazed for properties of impermeableness and brilliance. The porcelain is usually produced in glazed form vitrified from the selected and pure raw materials. Earthenware potteries of different glazing treatments have been used as containers for food storage and ripening, and are said to play a fundamental role for attaining natural ripening in traditional oriental fermented foods [22]. One of the most common ways of manufacturing these ceramic containers involves drying of the wet shaped body to a required moisture content and preliminary cooking at 800–1000°C, after which a glazing or enamel is applied and the cooking is completed by a long treatment (about 12 hours) at 1300-1500°C. Some less expensive earthenware potteries may be fired only once after optional glazing treatment. The earthenware with microporous structure may be tailored with controlled glazing to offer unique high gas and moisture permeation properties [22]. The CO_2/O_2 permeability ratios close to 1 may offer a new potential for fresh produce packaging. Evaporation from moisture percolated from porous and permeable earthenware wall is reasoned to have the effect of cooling the contained wet food in hot climates, giving the thermal stability [20].

Earthenware and porcelain are strong and completely opaque to the light and particularly have a quite high thermal capacity that makes them useful as thermal insulators. The glazes used for ceramic coatings may contain harmful heavy metals. One typical example of glaze composition is respective molar ratio of 2.17, 1.16, 0.37, 0.19, and 0.12 in SiO_2, B_2O_3, Al_2O_3, Na_2O, and K_2O [21]. Lead compound of PbO has been used as an ingredient for lowering the melting temperature of the glaze, but is not permitted for food containers due to safety concerns [20, 21]. Both European legislation and FDA regulations contain provisions on migration levels for lead and cadmium of ceramic tableware and cookware.

Discussion Questions and Problems

1. What are the advantages and disadvantages in using glass containers for sterilized liquid food?

2. List the advantages and disadvantages of glass packaging in food marketing and distribution.

3. Discuss the interrelationship between food quality preservation and optical properties of glass containers, using beer as an example.

4. List the main ingredients in glass making and explain their major functions.

5. Referring to its chemical structure and manufacturing process, explain the fragile behavior of glass bottles.

6. Present glass surface treatment methods to augment the mechanical strength of glass containers and reduce their damage in handling.

7. Discuss the microwave heating of foods packaged in glass containers, in aspect of the thermal and electromagnetic properties.

8. Construct functions/environments table for a glass bottle package of orange juice by referring to Section 1.7.3.

Bibliography

1. H Tait. Five Thousand Years of Glass. 2nd ed. London: The British Museum Press, 1999, pp. 214-229.
2. HF Mark. Encyclopedia of Polymer Science and Engineering. 2nd ed. New York: Wiley, 1985, pp. 472-475.
3. ASTM. Standard terminology of glass and glass products, in Annual Book of ASTM Standards C162. 1999, American Society for Testing and Materials: Philadelphia, PA.
4. T Demuth, Y Jeanvoine, J Hafner, JG Angyan. Polymorphism in silica studied in the local density and generalized-gradient approximations. J Phys Condens Matter 11:3833-3874, 1999.

5. YM Chiang, D Birnie III, WD Kingery. Physical Ceramics. New York: John Wiley & Sons, 1997, pp. 80-93.
6. AK Varshneya. Fundamentals of Inorganic Glasses. San Diego, CA: Academic Press, 1994, pp. 61-408.
7. H Rawson. Properties and Applications of Glass. Amsterdam: Elsevier, 1980, pp. 98-291.
8. AL Griff. Carbonated beverage packaging. In: M Bakker, ed. The Wiley Encyclopaedia of Packaging Technology. New York: John Wiley & Sons, 1986, pp. 121-124.
9. DL Hambley. Glass container design. In: M Bakker, ed. The Wiley Encyclopedia of Packaging Technology. New York: John Wiley & Sons, 1986, pp. 370-374.
10. JS Gecan, SM Cichowicz, PM Brickey. Analytical techniques for glass contamination of food: a guide for administrators and analysts. J Food Protect 53(10):895-899, 1990.
11. NT Huff. Glass container manufacturing. In: M Bakker, ed. The Wiley Encyclopaedia of Packaging Technology. New York: John Wiley & Sons, 1986, pp. 377-389.
12. SBM Jaime, SA Ortiz, TB Dantas, CF Damasceno. A comparison of the performance of lightweight glass containers manufactured by the P&B and B&B processes. Packag Technol Sci 15(4):225-230, 2002.
13. AS Sanyal, J Mukerji. Moessbauer study of hot end coated tin on soda-lime glass. Phys Chem Glasses 26(5):135-136, 1985.
14. AS Sanyal, J Mukerji. Chemical vapour deposition of hot end coatings on glass from stannic chloride. Glass Technol 23(6):271-276, 1982.
15. RD Southwick, JS Wasylyk, GL Smay, JB Kepple, EC Smith, BO Augustsson. The mechanical properties of films for the protection of glass surfaces. Thin Solid Films 77(1/3):41-50, 1981.
16. PW Atkins. General Chemistry. New York: Scientific American Books, 1989, pp. 256-264.
17. PWL Graham. Lightweighting, strengthening, and coatings. Glass Technol 25(1):7-12, 1984.
18. GH Beall, JF MacDowell. Fine ceramics from glass. Chemtech 11:673-679, 1988.
19. D Twede. The packaging technology and science of ancient transport amphoras. Packag Technol Sci 15:181-195, 2002.
20. M Chandler. Ceramics in the Modern World. Garden City, NY: Doubleday & Company, 1967, pp. 41-87.
21. J Hlavac. The Technology of Glass and Ceramics. Amsterdam: Elsevier Scientific, 1983, pp. 244-416.
22. GH Seo, SK Chung, DS An, DS Lee. Permeabilities of Korean earthenware containers and their potential for packaging fresh produce. Food Sci Biochnol 14(1):82-88, 2005.

Chapter 8

Metal Packaging

8.1   Introduction 197
8.2   Aluminum..198
        8.2.1 Aluminum Foil ···201
8.3   Coated Steels...203
        8.3.1 Tinplate···204
        8.3.2 Tin Free Steels ···207
        8.3.3 Polymer Coated Steels···208
8.4   Stainless Steel..209
8.5   Metal Corrosion..210
        8.5.1 Basic Corrosion Theory···210
        8.5.2 Corrosion of Tinplate ···215
        8.5.3 Corrosion of TFS ···218
        8.5.4 Corrosion of Aluminum ··218
        8.5.5 Corrosion of Stainless Steel···218
        8.5.6 Microbiologically Induced Corrosion···219
8.6   Metal Container Manufacturing 220
        8.6.1 Three-piece Can ···221
        8.6.2 Two-piece Can - D&I···224
        8.6.3 Two-piece Can - DRD··225
        8.6.4 Can Ends···226
        8.6.5 Double Seaming ···228
        8.6.6 Kegs and Drums ···229
        8.6.7 Trays···230
        8.6.8 Collapsible Tubes and Aerosol Containers··230
8.7   Protective Lacquers 232
        8.7.1 Types of Coatings ···232
        8.7.2 Methods of Coating Application ···236
8.8   Recent Developments and Trends 237
Discussion Questions and Problems 238
Bibliography 239

8.1 Introduction

In chemistry, the word *metal* has a precise and clear meaning, and *metallic elements* are easily distinguished from the other elements by several unique characteristics such as their compact chemical structures (almost no diffusion phenomenon can take place in normal condition through a metal), their malleability (it is possible to obtain any shape of a metallic solid), and their high thermal and electric conductivities. Many metal properties are useful for packaging applications.

Three quarters of known elements are metals, but in food packaging just a few metallic materials (metal alloys) are used: aluminum and steels (iron alloys) mostly coated by tin, chromium oxides, and/or varnishes in special cases. Stainless steel is widely used for food contact purpose as containers, vessels, kitchen utensils, and pipelines in homes and food factories. Occasionally, copper and cast iron may also come in contact with food (e.g., dairy boilers and grill tools) and recently titanium also generates some interest for food packaging due to its peculiar mechanical properties and high inertia. Other minor metals in metal packaging include nickel, which is applied as a sandwich layer between steel and tin in low-tincoating steel (LTS), lead used for soldering, and copper for welding in side seam of a three-piece can.

In this chapter, however, only the main metallic materials (aluminum, stainless steel, tinplate, and tin free steel) and the most widely used metallic containers will be discussed.

8.2 Aluminum

The name of the lightest metal, *aluminum*, has a long history, starting from the word *alum* of the ancient Greeks and Romans who used an impure hydrated aluminum oxide ore in medicine as an astringent and as a mordant in colorants. Many centuries later (around 1800), the word *aluminum* or *aluminium* came to common use, in conformity with the ending *ium* of many chemical elements. At that time pure metal (Al) was not discovered yet. The process of obtaining pure aluminum metal by electrolysis of alumina (Al_2O_3) was developed only at the end of 19th century.

Aluminum is the most abundant metal in the earth's crust (8.1%), where it is found as aluminum ores and the third element after silicon and oxygen. The main ore is mined as bauxite, which contains aluminum oxide, together with oxides of silicon, iron, and other metals. Despite the element's wide availability, it is difficult and expensive to get aluminum in a pure form as reduced metal. As already stated, the production of pure aluminum is an electrolytic process applied to aluminum oxide (Al_2O_3) which is obtained from bauxite. The process requires high amount of energy and produces carbon dioxide. The production of 1 kg of aluminum, which takes 4 kg of bauxite, requires 7-10 folds the energy to produce the same mass of tinplate or steel. The reason is that aluminum has a high negative value in the electrochemical series (see Table 8.7 in later part of this chapter), thus a high thermodynamic tendency to ionize; the consequence is that aluminum reacts with oxygen and other elements spontaneously, whereas the counter reaction is not thermodynamically favored at all and its reduction to pure metal requires a great energy. The high energy requirement for aluminum production makes aluminum the most expensive material used in food packaging.

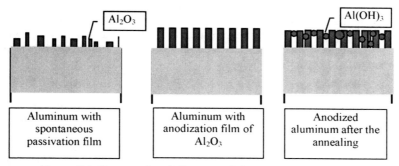

Figure 8.1 Crosssectional diagram of aluminum before and after the anodization treatment

Aluminum for packaging is usually produced in tempers indicated as 'O' and 'H'. The O temper refers to the softest temper of the product and produced after annealing and recrystallization. H temper applies to products that have their strength increased by strain-hardening with or without supplementary thermal treatments.

Table 8.1 The main physical properties of aluminum

Property	Typical values
Mass thermal capacity (kJ kg^{-1} $°C^{-1}$)	0.90
Thermal expansion α ($°C^{-1}$ x 10^{-6})	24
Thermal conductivity (W m^{-1} $°C^{-1}$ at 25 $°C$)	237.00
Melting temperature ($°C$)	670
Boiling temperature ($°C$)	2518.8
Mass per unit volume (g cm^{-3})	2.70 - 3.20
Static coefficient of friction	1.9
Modulus of elasticity (MPa)	70 000
Tensile strength (MPa)	70 – 210

The spontaneous reaction of the metal with the air oxygen, leads to the formation of a thin passivated film (1-5 nm) that gives just a slight protection to the corrosion phenomena. Because this oxide film is thin and also not homogeneous, it is therefore a good industrial practice to proceed with a chemical or electrochemical passivation (anodization) which increases the aluminum oxide thickness to 50-200 nm and makes it more regular (Figure 8.1). Finally, a heat treatment reduces greatly (but does not eliminate completely) the porosity of the oxide film, resulting in a very thin ceramic layer onto a metallic surface. However, the oxide layer on the surface of aluminum is not a complete protection to the metal, because it is removable both at low (< 4.0) and at high (> 8.0) pH values and is a porous coating permeable to many ions. In order to improve corrosion resistance and surface properties of final aluminum packages, other different types of chemical treatment are also employed, and the aluminum surface is protected by a lacquer adhesion or a polymer coating particularly as concerned with food packaging.

Table 8.2 Composition (%), strength, and typical use of some aluminum alloys for packaging

Alloy code	Main use	Mn	Mg	Si[1]	Fe[1]	Cu[1]	Cr[1]	Tensile strength (MPa)
1050	Foils and flexible tubes	0.05[1]	-	0.25	0.4	0.05	-	140
3004	Body stock	1.0-1.5	0.8-1.3	0.3	0.7	0.25	-	220-270
5182	End stock	0.2-0.5	4.0-5.0	0.2	0.35	0.06	0.1	380
5052		0.1[1]	2.2-2.8	0.45	0.45	0.1	0.15-0.35	
5042	Tab stock	0.2-0.5	3.0-4.0	0.2	0.35	0.15	0.1	380
5082		0.15	4.0-5.0	0.2	0.35	0.15	0.15	
8079	Foils for lamination	-	-	0.05-0.3	0.7-1.3	0.05	-	-

[1]Maximum amount allowed.

Aluminum has been used for food packaging applications quite recently and it may be considered the newest material in food packaging. Actually, its main industrial use (about 90%) still remains in the other sectors like building, automotive, and furniture; however its properties are much appreciated in food packaging for flexible, rigid, and semi rigid applications (Table 8.1). Different types of alloys, in which the amount of aluminum is always high, are used [1-3]. Their mechanical and chemical properties can differ according to their composition and, particularly, to the amount of two alloying metals: manganese and magnesium. The former increases the corrosion resistance slightly, whereas magnesium increases the strength of material but reduces corrosion resistance against acids and alkalis. Pure aluminum types (1000 series alloy) are used for foils, deep-drawn containers, and impact-extruded cans or tubes. The 3000 series alloys, containing manganese of 0.3-1.5%, are used for manufacturing bodies of two-piece cans. The 5000 series alloys, containing magnesium of 0.5-5%, are used for shallow drawn parts such as can ends and ring-pull tabs. 8000 series alloys containing iron are used for foil of laminate film structure. Table 8.2 presents chemical composition and end use of typical aluminum alloys.

Beside high mechanical resistance of all the aluminum alloys, this material shows an extraordinary malleability that make it possible to reduce the thickness of aluminum foil to 3-6 μm. It can be cast in any forms, it can be rolled, roll-formed, stamped, drawn, spun, forged, and extruded into a variety of shapes, making quite easy the manufacturing of whatever aluminum packaging. Due to the high purity of aluminum alloys, their recycling is easy and convenient: the production of aluminum by recycling uses only 5% of the energy used in primary production.

Also some thermal properties of aluminum deserve a special mention because they can be useful for food packaging applications. At high temperatures (200-250°C) aluminum alloys tend to lose some of their strength; however, at negative temperatures, their strength increases while retaining their ductility, making aluminum an extremely useful low-temperature material that even at ultra freezing temperatures does not become fragile. Moreover, at reduced pressure it is possible to evaporate pure aluminum and let it condense onto a flexible surface (plastic film, paper sheet), as a very thin layer of just some hundreds of nm (see Section 6.7.11). The main drawbacks of aluminum, besides its high price, are difficulties of welding and the poor chemical resistance.

A small percentage of the population seems to have allergy to aluminum, showing contact dermatitis, but, in general, this metal is considered safe and in fact its oxide (alumina) finds many pharmaceutical applications and the hypothesis that it might be involved in Alzheimer disease is nowadays completely abandoned [4].

8.2.1 Aluminum Foil

Aluminum foil is the sheet of aluminum alloy that was reduce-rolled to thickness in the broad range of 4-150 μm. It started to be produced in 1910s and has been used for wrapping candies and chewing gums. Home cooking foil came to market in 1920s, heat-sealable foil was developed in 1930s, and then semi-rigid pressed trays were introduced in late 1940s.

The commercial production of aluminum foil is done by rolling through heated rolls or cold rolling after continuous casting. The importance and diffusion of the latter technique are growing because it is less expensive from an energetic point of view. After rolling, a continuous aluminum oxide (Al_2O_3) layer is formed by the reaction between oxygen and the freshly created aluminum surface. During the subsequent annealing step (approximately 300°C), the alumina layer grows thicker due to fast oxygen diffusion through the oxide and the metal reactivity at that temperature. This growth occurs at the interface between oxide and metal. The top oxide layer becomes more compact and resistant with a loss of water during annealing.

The foil can be produced from a variety of alloys such as 1000, 2000, 3000, 5000, and 8000 series (see Table 8.2). Usually types of 1100, 1145, and 1235 are used for flexible packaging, and 3003 type is for thick foils with stiff property. The foil can be processed further into a variety of films and containers. It can be press-formed into semi-rigid containers, laminated onto the other film structure, printed, and/or coated. In some cases, embossing treatment is given for desired shape.

Usually, for a final foil thickness below 60 µm, two layers of foil are wound together before the last rolling step. Rolling oil is sprayed between the two layers as a release agent, and the 'twin foil' is rolled to the desired final thickness. After the final rolling step, the two foil webs are separated, slit, and wound to the desired length and width. Always, the production process gives a unique feature to aluminum foil that presents one side quite shining and the other definitely mat, due to the different surface states: during this final rolling, the surface contacting the roller becomes shiny while the other surface facing the opposite foil has dull finish.

Aluminum foil has excellent barrier properties to water, gas, and aroma. Its barrier is absolute when the thickness is above 25 µm. As it becomes thinner, the barrier deteriorates because of pinholes formation. For example, foil thick 8.9 µm has water transmission rate of 0.3 mL m^{-2} day^{-1} at 38°C and 100% relative humidity. It is lightweight and resistant to most foods except those with high acid and/or salt. It can withstand high and cold temperature conditions. Aluminum foil has dead-fold property, which is desirable for most packaging applications, but needs cautious handling to prevent pinhole formation.

Typical current usages of aluminum foil include cooking foil, laminated film structure, and semi-rigid trays. For flexible packaging material, the foil is laminated onto plastic film or paper to provide barrier against light, oxygen, and moisture. The laminated films are used for retortable pouch, bottle closure liner, or composite can. A relative new application concerns the flexible capsules covering sparkling wines and champagne corks. These are made by a poly-laminate, consisting of a polyethylene film between two aluminum foils; this unusual configuration permits to have the required properties of flexibility and resistance for rolling on the bottle neck as a tight, elegant, and hermetic capsule.

As rigid containers, thick foil is formed into dishes or cups by pressing to be used for packaging convenience foods, cakes, or frozen foods. The aluminum trays have been extensively used for chilled foods in catering, institutional cooking, and take-away hot meals. Aluminum trays were not used for heating purpose in microwave oven formerly because of concerns on arcing, but recent technological developments solved this problem for the wisely-designed trays to be applied for improved temperature uniformity. Aluminum surface area may be used in the design for microwave shield to improve heating uniformity (see also Chapter 14).

8.3 Coated Steels

Steel is a generic term indicating a large family of iron alloys, with a low content of carbon (between 0.2 and 2%), widely used as structural materials because they are hard, strong, durable, and easy to shape (Table 8.3). On a world-wide basis, steels are the most used structural materials just after cement.

Carbon is the binding agent in the alloy, locking the iron atoms into a rigid lattice, thus its amount largely controls the properties of the resulting steel. The higher the carbon content of the steel, the higher its tensile strength but also the higher its brittleness. Steels are produced from molten iron ores, coal, and limestone in a blast furnace to obtain what is called 'pig iron'. As pig iron is hard and brittle, steelmakers refine the material by purification steps and adding other elements to strengthen it. Therefore, steels often contain other constituents such as manganese, chromium, nickel, molybdenum, copper, tungsten, cobalt, and silicon, depending on the desired properties. According to the chemical compositions, standard steels can be classified into three groups: *carbon steels*, *alloy steels*, and *stainless steels* (see Section 8.4); in the first category which is the most important for food packaging applications, the alloying elements do not exceed these limits: 1% carbon, 0.6% copper, 1.65% manganese, 0.4% phosphorus, 0.6% silicon, and 0.05% sulfur.

Table 8.3 The main physical properties of carbon steels

Property	Typical values
Mass thermal capacity (kJ kg^{-1} $°C^{-1}$)	0.45-2.08
Thermal expansion α ($°C^{-1}$ x 10^{-6})	11-16.6
Thermal conductivity (W m^{-1} $°C^{-1}$ at 25 °C)	24.3-65.2
Melting temperature (°C)	1426–1538
Mass per unit volume (g cm^{-3})	7.85
Modulus of elasticity (GPa)	190-210
Tensile strength (MPa)	276-1882

Steels are generally recognized as cheap and strong materials, but their chemical inertness is never sufficient and particularly several problems may occur in food contact. It is a common practice, therefore, to cover the surface of steel objects to increase the chemical resistance of the material and to avoid unwanted interactions with foods or any other products in contact with it. The coating may be inorganic (generally other metals or metal oxides) or also organic (varnishes and resins). Even if the sheathing is generally lower than 0.5% of the total weight, it may represent more than 10% of the final price of the steel plate. By far, the most important and diffused coated steel is tinplate which had a huge importance in food preservation, making feasible the development of an industry of canned foods. Nowadays, tinplate and the other coated steels are largely used not only for can production but also for closures, barrels, pails, and drums.

8.3.1 Tinplate

Tinplate is the oldest metallic material used as food packaging material (introduced about the middle of eighteenth century), but its fine current structure is greatly different from the early one used by the pioneers of preserved foods.

The word *tinplate* nowadays refers to the low carbon steel (called *blackplate*), having a thickness of 0.13-0.38 mm and coated with tin for a usual thickness of 0.2-2.5 μm, but the coating load may be different on each side [1-3]. The final structure of tinplate is quite complex (Figure 8.2) deriving from a metalworking process that includes mechanical, thermal, and chemical steps on the blackplate. Blackplate is the name assigned to the low carbon steel after the tempering process, due to the presence of iron oxides on its surface. Next to iron, manganese is the main alloy component of the steels used for tinplate production (Table 8.4). Composition of the base steel influences its formability and thus may determine its application and usage. The most commonly used one is type MR generally of low metalloids and residual elements. Type L of high purity is used for requirements of good corrosion resistance [5]. Type D is for severe drawing applications.

Table 8.4 Chemical composition of base steel for tinplate manufacture

Element	Composition (%)		
	Type D	Type L	Type MR
Carbon	0.12	0.13	0.13
Manganese	0.60	0.60	0.60
Phosphorus	0.020	0.015	0.020
Sulfur	0.05	0.05	0.05
Silicon	0.020	0.010	0.010
Copper	0.20	0.06	0.20
Nickel	-	0.04	-
Chromium	-	0.06	-
Molybdenum	-	0.05	-
Other residual elements; each	-	0.02	-

The slabs of refined low carbon steel coming out from the furnace are hot-rolled to a strip with thickness of 1.91-2.54 mm. After a step of chemical removal of the iron oxides, the strip is submitted to subsequent steps of thickness reduction (up to 10-folds the initial value) by cold-rolling. In case of *single reduced plate*, the thickness reached is close to the final wanted gauge, whereas the *double-reduced* steels are not rolled heavily in this stage and are subject to another cold roll after the annealing process. The cold-rolled product is brittle due to its crystalline structure, and should be subjected to annealing. Annealing is carried out in batch or continuous processes in an inert gas at temperature of 580-700°C, leading to different levels of malleability, thus to different possible food packaging applications. In case of double-reduced product, the annealed plate passes through a second cold roll to reduce the thickness by 35%. The double reduced plate is stiff at thin gauge and thus is used for making lighter weight cans. Finally both single and double-reduced plates pass thorough temper roll which reduces the thickness by 0.5-2% and determines surface roughness [1, 3, 6].

Temper refers also to the state or condition of the plate as well as to its hardness or toughness. Table 8.5 shows tempers of tinplate in Rockwell hardness scale varying with its application. The Rockwell hardness scale is measured by indenting the test material with a diamond cone or hardened steel ball indenter. For tinplate, a variant of the test method, 30T is usually used.

Table 8.5 Tempers of tinplate

Temper	Rockwell hardness 30T scale	Applications
T-1	52 max.	Nozzles, spouts, and closures; deep drawn parts
T-2	50-56	Shallow-drawn and specialized can parts, closures
T-3	54-60	Can ends and bodies, large diameter closures, crown caps
T-4	58-64	Stiff can ends and bodies for noncorrosive products, crown caps
T-4-CA[1]	58-64	Fair stiff applications, closures, can ends, and bodies
T-5-CA[1]	62-68	Increased stiff applications requiring buckling resistance, can ends, and bodies
DR-8[2]	73	Small diameter round can bodies and ends
DR-9[2]	76	Large diameter round can bodies and ends
DR-9M[2]	77	Beer and carbonated beverage can ends

[1]Continuously annealed; [2]double-reduced.

The tin coating of finished steel is made by an electrolytic process that since the '50s has replaced the old dipping process, in which the plates were dipped in molten tin for the coating. The electrolytic tinning occurs in an acidic bath of tin sulfate, where the steel is covered by a thin layer of metallic tin (about 0.4 μm); this step is followed by a thermal treatment (260-270°C) and a rapid quenching that leads to the formation of an iron-tin alloy (about 0.13 μm thick). The next step is the chemical passivation that consists of passing the tin-covered plate in a solution of sodium dichromate where tin and chromium oxides are produced on the surface (about 0.01 μm thick). This step also stabilizes the coating, adding a thin film of metallic chromium/hydrated chromium oxide on tin oxide layer. An oily lubricant is finally applied (about 0.005 μm thick) to improve slip characteristics of the plate, enhance the protection against surface scratch, and also increase resistance to environmental corrosion, avoiding the contact with atmospheric moisture. Lubricants mostly used are acetyl tributyl citrate and dioctyl sebacate. Usual thickness of single reduced tinplate is in the range of 0.16 and 0.38 mm, and that of double-reduced one is between 0.13 and 0.29 mm [2]. Tin coating load is generally expressed as basis weight (grammage) of the sheating, equally or differentially on the two surfaces: E 5.6/5.6 means the equal tin coating of 5.6 g m^{-2} on both sides and D 5.6/2.8 means the differential tin coating of 5.6 and 2.8 g m^{-2} on each side, with the inner side, destined to food contact, being protected with the higher amount of coating [1].

The thickness reduction of both the blackplate and the tin oxide film drove steadily the tinplate industry in the last half century. In about 60 years the weight of the blackplate has been reduced by 35% and the passivation film by 80% [6]. This trend, completely justified by economic reasons, brought about consequences and effects on the manufacturing of tinplate containers. The use of double reduced plate (DR), for instance, induced a depth increase of expansion rings and, consequently, required a higher flexibility of lacquer coatings applied for better protection of certain foods. On the other hand, the introduction of LTS (low-tincoating steels) increased the attention to possible scratches on the surface and to the phenomena of corrosion; the reduced amount of tin, in fact, reduces the electrochemical protection and the residual oxygen in the headspace of the cans may become a critical factor to cause corrosion.

Recently Japanese steel makers introduced low-tincoating steel (LTS) with a nickel layer deposited between tin and steel [7]. It can minimize the tin coating while providing weldability on the side seam. Particularly concaved tin coating shape is obtained on the surface in a melting process after electrolytic tinning. This LTS plate is excellent in welding property and adhesion with plastic lacquer. Figure 8.2 compares a cross-sectional structure of typical LTS plate with that of tinplate.

Mostly, tin coating layer protects the steel from the corrosion, a phenomenon that may have many different causes and different behaviors, like metal dissolution, pinholing, and others, but the tin layer itself may also be subjected to corrosion phenomena that can be really unpleasant to the food inside the tinplate cans; all these aspects are better described later, in Section 8.5. Because the protective role of the tin layer is limited and at times inadequate, a number of synthetic lacquer coatings are applied on both the inside and the outside of the tinplate to provide further resistance to corrosion from food products and environmental conditions, respectively [3]. This issue will be detailed later in Section 8.7, presenting the protective lacquering; it is already evident from present discussion that in several cases the primary packaging material, i.e., the food contact material, is a synthetic resin, and the metal is just the structural component of the package.

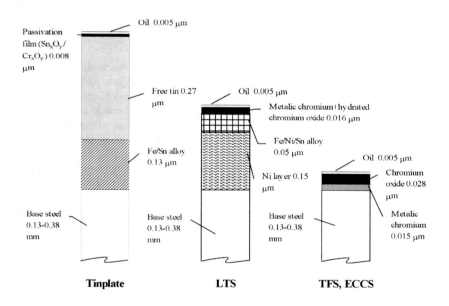

Figure 8.2 Cross-sectional diagram of typical tinplate, LTS, and TFS

8.3.2 Tin Free Steels

Tin is an expensive metal because its availability is limited and the major mines are located in countries (Bolivia, Malaysia, Indonesia, Thailand) far from the main user markets. As already stated, tinplate manufacturers progressively have reduced the load of tin onto the steel plate, but obviously this trend cannot continue indefinitely. This situation encouraged the development of steel plates protected with a different passivation film but leading to the same performance in can making and food protection. These plates have been generally called *tin free steels* (TFS), and after several trials with aluminum, chromium, and phosphate the cold-rolled plate coated electrolytically with chromium and chromium oxide (ECCS, electrolytic chromium coated steel) has been chosen as the best alternative to tinplate (Figure 8.2).

The surface film layer has thickness equal on both sides of the plate of about 0.04 μm; thus, very thinner than the tin oxides layer, however it is compact and hard. In more details, the grammage of total chromium on the plate ranges from 0.090 to 0.140 g m ^{2}: that of metal chromium ranges from 0.080 to 0.120 and the oxide basis weight is about 0.01-0.02 g m ^{2}. Also ECCS plate receives, at the end of manufacture, an oil coating of di-octyl sebacate or butyl stearate for protective purposes. The TFS plate has good functionality in heat resistance, lacquer coating adhesion, and printing, but it is less resistant than tinplate to corrosion in acidic environment, whereas it is more resistant at neutral or alkaline pH. Thus TFS plates are always used in forms coated with organic lacquer.

The principal uses of ECCS in food packaging are, at this moment, for food can ends, crown caps, vacuum closures for glass preserve jars, and deep-drawn cans. The typical systems of welding used for tinplate cans, however, are not feasible with traditional ECCS, because the chromium hydroxide has extremely low electrical conductivity and local overheating would occur where the welding current passes, leading to a defect named surface flash that does not allow stable welding. Therefore, the side seam of a TFS can is usually carried out by cementing or adhesive bonding, or the chromium layer is removed if welding has to be applied. However, recently a Japanese company developed and started to produce new TFS products with thinner chromium oxide film that can be welded without grinding the surface to remove the metallic coating layer. The elimination of the grinding process allows the manufacture of cleaner cans.

8.3.3 Polymer Coated Steels

As mentioned before, plastic lacquer coating is applied onto tinplate and TFS plate to offer additional protection. Also in nonfood industrial applications (for the production of pails and drums to be used for chemicals and paints, for instance) varnishes or synthetic resins have been directly applied on the blackplates for the protection from corrosion. More recently, these processes have been developed by introduction of the polymer laminated steel process to meet the stringent demands put on today's food packaging materials.

The steel substrates (whether tinplate or ECCS) are combined with thermo-plastic polymers, merging the advantages of steel with those of plastic polymers. Besides an excellent appearance, the benefits of polymer coated steels are abrasion and corrosion resistance as well as high moisture barrier. In a process under license from Carnaud/Metalbox, the steel substrate is preheated prior to entering the film application stage where polyethylene terephthalate (PET) is applied on the external surface of the steel and polypropylene on the internal surface. Different polymers and thicknesses can be applied simultaneously to opposing surfaces. The manufacturing process under license from Toyo Kohan allows simultaneous coating of both surfaces with different coating materials of various thicknesses. Different coating methods using thermal lamination and solvent free lamination are used in this technology. Lacquer coating or film lamination is usually made on the plate excluding the area for side seam if the plate is intended for three-piece cans with welded side seam. Recently TFS laminated with PET has been used for manufacturing two-piece and three-piece cans. Particularly two-piece cans, formed by stretch drawing of the laminated TFS, do not have any further lacquer coating step and therefore can reduce the emission of organic solvent [8].

8.4 Stainless Steel

Stainless steels are iron alloys with better corrosion resistance than any other steels because they contain large amounts of chromium. An iron alloy containing a certain minimum amount of chromium (about 11%) is able to reach spontaneously a state of auto-passivation. It means that chromium reacts with oxygen producing a molecular layer of chromium oxide protecting the alloy from corrosion. Stainless steels have equal composition both on the surface and in the mass, and do not need any further protective coating.

They can be divided into three basic groups according to their crystalline structure, named *austenitic, ferritic,* and *martensitic*. The austenitic type typically contains 18% chromium and 8% nickel, is widely known as 18-8 and is the most important in food applications. It is nonmagnetic in annealed condition and can only be hardened by cold working. The ferritic steels contain a small amount of nickel and either 17% chromium or 12% chromium with other elements such as aluminum or titanium. Always magnetic, also this grade can be hardened only by cold working. The martensitic usually contain 12% chromium and no nickel: it is magnetic and can be hardened by heat treatment.

Table 8.6 The main physical properties of stainless steels

Property	Typical values
Mass thermal capacity (kJ kg^{-1} $°C^{-1}$)	0.420-0.500
Thermal expansion α ($°C^{-1}$ x 10^{-6})	9.0-20.7
Thermal conductivity (W m^{-1} $°C^{-1}$ at 25 °C)	11.2-36.7
Melting temperature (°C)	1371-1454
Mass per unit volume (g cm^{-3})	7.75-8.1
Modulus of elasticity (GPa)	190-210
Tensile strength (MPa)	640-2000

The AISI (American Iron and Steel Institute) classified many different stainless steels, and this classification is recognized world wide. Many of them are approved for food contact. Stainless steels are expensive alloys and for this reason their use as food packaging material is practically limited to only one case of returnable container for beverages (kegs, see Section 8.6.6), but they are the leading materials for processing plants, large storage containers, kitchen tools, and many other objects intended for food contact. The most important characteristics of stainless steels are their high resilience, good thermal conductivity, and ease of welding and casting, but the most appreciated and peculiar feature is certainly their hygienic properties. Studies performed on different materials artificially contaminated and subjected to standard cleaning processes, have clearly and repeatedly demonstrated the higher bacterial removing capacity and the lower bacterial retention of the stainless steel surfaces [9, 10]. This phenomenon, probably related to the difficulty of biofilm formation onto the glossy surface of this material (see also Sections 2.5.2 and 8.5.6), is also enhanced by the mechanical, thermal, and chemical resistance permitting to use heavy/drastic cleaning cycles, i.e., using high temperatures and concentrated detergents.

8.5 Metal Corrosion

Metallic packaging materials provide excellent properties of mechanical strength, ductility, and barrier to food packaging applications. However, the chemical nature of metals may cause corrosion through interaction between the metal surface and the contained food. Corrosion or rust may also come from the interaction with the outside environment. Corrosion reaction is an intrinsic property of all the metallic materials, even though its rate and degree vary widely with material. Corrosion takes place on the metal surface, and therefore its rate can be controlled or reduced by altering the surface conditions. One example is lacquer coating on the surface. The corrosion may weaken the hygienic quality of packaged foods and hermetic integrity of the package. Therefore the corrosion control of metal packaging is of great concern to the package manufacturers, food processors, and consumers.

8.5.1 Basic Corrosion Theory

Reaction of metal in aqueous solution or wet conditions is an electrochemical one involving the transfer of electrons. Electrolytic solution (contained food) is the medium transferring the electric current created by electron transfer. In this process, the metal surface acts as an electrode whose electron transfer equals that in the electrolytic medium.

When the metal surface corrodes, the metal oxidizes to its cationic form dissolving into the aqueous solution and leaving the generated electrons on the surface. The electrode with oxidation reaction is called anode. Oxidation of metal, M, in aqueous medium, producing ions, M^{z-} and leaving z electrons at anode, can be described as:

$$M \rightarrow M^{z+} + z\,e^- \qquad (8.1)$$

Accumulation of electrons in the metal would establish a negative charge on the metal and the electrolytic medium would be reduced receiving the remaining electrons. The electrode with reduction reaction is called cathode. At any moment, cathodic current is equal to anodic current.

The thermodynamic tendency of an electrode to oxidize or reduce may be expressed by standard reduction potential (E^o) with reference to standard hydrogen electrode assigned to zero value of the potential (Table 8.7). The higher positive standard reduction potential means the higher tendency to be reduced under the conditions where all the species in the reaction are of unit activity. That is, the standard electrode potential has the relationship of Equation 8.2 with standard Gibbs free energy, whose negative value represents the tendency of spontaneous reactivity under standard conditions.

$$\Delta G^o = -zFE^o \qquad (8.2)$$

where z is the electron mole numbers per reaction mole, and F is Faraday constant (96,500 coulomb/mol). Metals with higher positive value of E^o have the tendency to exist as a reduced form and therefore stable in contact with aqueous medium: gold (Au) at the most upper location in Table 8.7 is the most stable, while aluminum (Al) at the lowest location is the most prone to oxidize and corrode in its pure bare state.

Table 8.7 Standard reduction potential for some selected electrode

Electrode	Standard reduction potential (V)	
$Au^{2+} + 2\,e^- \rightarrow Au$	+1.50	
$1/2\,O_2 + 2\,H^+ + 2\,e^- \rightarrow H_2O$	+1.23	+0.81 at pH=7
$Ag^+ + e^- \rightarrow Ag$	+0.80	
$Fe^{3+} + e^- \rightarrow Fe^{2+}$	+0.77	
$1/2\,O_2 + H_2O + 2\,e^- \rightarrow 2\,OH^-$	+0.40	+0.81 at pH=7
$Cu^{2+} + 2\,e^- \rightarrow Cu$	+0.34	
$2\,H^+ + 2\,e^- \rightarrow H_2$	+0.00 (reference)	
$Pb^{2+} + 2\,e^- \rightarrow Pb$	-0.13	
$Sn^{2+} + 2\,e^- \rightarrow Sn$	-0.14	
$Ni^{2+} + 2\,e^- \rightarrow Ni$	-0.25	
$Fe^{2+} + 2\,e^- \rightarrow Fe$	-0.44	
$Cr^{3+} + 3\,e^- \rightarrow Cr$	-0.74	
$Al^{3+} + 3\,e^- \rightarrow Al$	-1.66	

The standard reduction potentials in Table 8.7 are given for half reactions with reference to hydrogen electrode, but the electrode reactions should be coupled for the oxidation and reduction reactions to occur. The electrode with higher E^o would work as a cathode to have reduction reaction as forward direction in Table 8.7, while that with lower E^o would work as an anode to have oxidation as reverse direction of the table. The resultant E^o would be the difference between two electrodes as described by:

$$E^o_{overall} = E^o_{cathode} - E^o_{anode} \qquad (8.3)$$

As for the tin in acidic aerated water at standard state, tin would be an anode with oxidation and the other electrode would be the cathode with reduction of dissolved oxygen as follows:

	Reaction	E^o (V)	
At anode	$Sn \rightarrow Sn^{2-} + 2\ e^-$	+ 0.14	(8.4)
At cathode	$1/2\ O_2 + 2\ H^+ + 2\ e^- \rightarrow H_2O$	+ 1.23	(8.5)
Overall	$Sn + 1/2\ O_2 + 2\ H^+ \rightarrow Sn^{2+} + H_2O$	+ 1.37	(8.6)

The tin would be dissolved into the solution under the standard state where the activities of all the species are 1. Activity of the pure solid or liquid may be assumed as 1, that of gas component is commonly partial pressure in bar unit and that of solute in dilute solution is molar concentration (mol L^{-1}).

However, electric potential in real food solution, E would be different from standard electric potential, and can be described for the case of tin in acidic aerated water at 25°C by Nernst equation:

$$E = E^o - \frac{RT}{zF} \ln \frac{a_{Sn^{2+}}}{(a_{O_2})^{1/2} \cdot (a_{H^+})^2} = 1.37 - 0.0296 \ln \frac{a_{Sn^{2+}}}{(a_{O_2})^{1/2} \cdot (a_{H^+})^2} \qquad (8.7)$$

where T is absolute temperature (K), R is gas constant (8.314 J K^{-1} mol^{-1}), $a_{Sn^{2-}}$ represents the activity of Sn^{2+} ion, and so on. The reaction proceeds forward when the potential E is positive, and becomes more spontaneous with higher positive value of E. If the activity (concentration) of Sn^{2+} decreases, partial pressure of oxygen increases, and activity of H^+ increases, the E in Equation 8.7 becomes more positive to make the reaction more spontaneous. It is obvious that lower pH and high oxygen concentration help the tin corrosion.

If Sn^{2-} ion is depleted by the reaction with other components, E in Equation 8.7 moves to more positive direction, which increases the spontaneity of the reaction. The chemical compound that removes the metal ion to favor the spontaneity of the corrosion reaction is called *depolarizer* and plays an important role in the internal corrosion of food can.

Internal corrosion of the can containing food involves oxidation of tin and/or iron, reduction of H^+ and oxygen, and other reactions, whose combination and electric potential differ with environmental conditions. Oxide film formed on metal surface makes it hard to explain the corrosion by the simple electrochemical cell reaction. Chromium (Cr) and aluminum (Al) having more negative standard reduction potential (E°) than tin (Sn) and iron (Fe) may seem simply to be more likely to be oxidized, but do not respond so. Those metals form oxide films, and thus become more resistant to corrosion. Other reactions not listed in Table 8.7 may occur through interaction of the metal with various ions and solutes. For that reason it is difficult to understand and predict the resultant corrosion phenomena. Since corrosion phenomena depend on several factors, each one being function of several variables, the attempt to manage these problems with a statistic/probabilistic approach has been under investigation for many years [11, 12]. This topic is beyond the scope of this book, but it is easy to find more pertinent references.

Oxide metal surface film is stable in electrolyte solution within defined pH ranges and is resistant against a corrosive environment. The oxide film stops the anodic oxidation reaction and suppresses the cathodic reaction by providing impervious barrier layer. The process of applying or favoring the insoluble oxide film is called passivation treatment and it has been already described for aluminum, tinplate, and TFS production.

Thermodynamic stability of metal may be explained as a function of electrode potential and pH by the Pourbaix diagram. It represents the relative stabilities of the solid phase and soluble ions that are produced by reaction between a metal and an aqueous environment [13]. Simplified Pourbaix diagrams for iron-water, tin-water, aluminum-water, and nickel-water are shown in Figure 8.3. Each region in the diagram shows the form of stable species in aqueous medium, while the region between two dotted slope lines represents where water is stable. The domain in which soluble ion is the stable species represents the condition labile for corrosion. The domain in which the metal is stable species accounts for the condition of immunity of the metal. That of metal oxide or insoluble solid as stable species is for passivity. The diagram for tin has no zone of stability for the metal in common with that of water, but the stannic oxide film is stable at pH between about 3 and 10. Outside this range within stable domain of water, corrosion of tin may likely to occur. At negative potentials, the tin metal is stable and thus corrosion is impossible: this region is called immunity domain.

Even though thermodynamic treatment of interaction between metal and environment provides a basis for corrosion tendency, it does not give or predict the rate of corrosion, which is important in metal food packaging [14]. The corrosion rate is related to nonequilibrium state at which the system tends to move toward equilibrium. The corrosion rate of anodic area on a metal surface is proportional to the corrosion current which is determined by electric potential change called electrode polarization. Therefore corrosion rate per unit metal surface can be expressed as current density in A m^{-2}. The corrosion rate on ideal theoretical basis can also be described as the product of metal's equivalent weight (g eq^{-1}) and corrosion current (A) divided by Faraday's constant (96,500 coulomb eq^{-1}). The electrode polarization is explained by ion concentration change on the metal surface controlled by diffusion in the electrolyte, activation energy of the reaction, and/or potential drop through a surface film of metal-reaction product [13-15]. Corrosion rate of the metal can be obtained from the estimation of polarization, which is beyond the scope of this book. Interested readers are suggested to consult the related references.

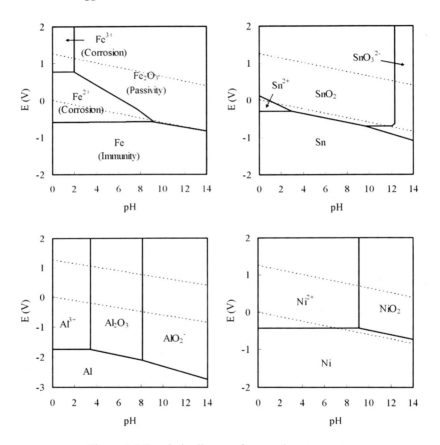

Figure 8.3 Pourbaix diagram for metal-water system

Corrosion resistance of metal containers can be evaluated by several tests [16, 17]. Food product or simulant is packaged by the test container and stored under controlled conditions with observations on the internal surface: any signs of change or failure are measured or monitored such as internal vacuum, headspace, product appearance and composition, lacquer condition, detinning, pitting, and corrosion [5]; the amounts of metals dissolved out of the metal surface may also be measured. Storage period may extend maximally to one and two years at 38 and 25°C, respectively. High temperature of 38 or 50°C may be used for accelerated test. As an electrochemical test, an electrode is installed onto can body, small electric potential is applied between can body and electrode for a certain period of time, and electric current is measured and compared with the visible state of corroded internal surface. Food is filled into the container with the electrode working as cathode and can body as anode. For the lacquered cans, the coating may be intentionally damaged and exposed to the filled content for the surface conditions to be examined usually after 5 days at 50°C. Another simple method immerses the sample metal plate into a standard solution, whose weight is measured after storage.

8.5.2 Corrosion of Tinplate

Tin coating is more prone to corrode than iron plate when their standard reduction potentials were compared in Table 8.7. However, the potential reverses in most cans seamed in vacuum condition. Tin acts as a sacrificing coating protecting the iron [15]. Complex-ion formation between food constituent and Sn^{2+} ion apparently causes this reversal. The activity of Sn^{2+} is thus lowered greatly and tin becomes anodic. Basically tin is generally resistant to atmospheric corrosion, dilute mineral acids in the absence of air, and many organic acids, but it is corroded by strong mineral acids.

Even though most cans are used as lacquered state for protecting the steel plate, some food products are packed in plain tinplate cans [1]. Typical products normally packed in plain tinplate cans are white peaches, pineapple, citrus juices, mushrooms, and tomato products. High tin coating of 11.2 g m^{-2} is employed for plain tinplate cans. The corrosion of plain tinplate in the presence of oxygen and acid can be described by Equations 8.4 to 8.6. Oxygen will be consumed after a few weeks of packing, and cathodic reactions will occur at reduced rate with lower current, resulting in controlled and limited detinning. Other common cathodic reactions are reduction of nitrates, pigments, and sulfurous compounds. Normal corrosion process of internal tinplate can surface is slow and can expose some part of base plate after 1.5 to 2 years at 25°C [18].

If the tin coating is not enough to protect the underlying steel and/or excess oxygen is inside the container to spend all the tin, iron dissolution may take place producing hydrogen from acid as here described:

	Reaction	E° (V)	
At anode	$Fe \rightarrow Fe^{2+} + 2\,e^-$	+ 0.44	(8.8)
At cathode	$2\,H^+ + 2\,e^- \rightarrow H_2$	+ 0.00	(8.9)
Overall	$Fe + 2\,H^+ \rightarrow H_2 + Fe^{2+}$	+ 0.44	(8.10)

The hydrogen gas evolved into the headspace would relieve the can vacuum and cause the swollen can. This kind of rapid internal corrosion may cause complete detinning and/or pitting in as early as 3 months. Rapid detinning may be caused by fast corroding steel, abnormally corrosive product and/or too thin tin coating. Dissolution of iron may also lead to browning of the food containing phenolic compounds (such as tannin). Too low fill weight, low filling temperature, and side seam leakage could result in excessive oxygen inside the can accelerating the corrosion process: level of oxygen inside the can should be minimized by applying good vacuum and proper headspace.

Anodic dissolution of iron ultimately results in perforation and leaking of the product, which will help the external corrosion by the contaminated liquid. Usually perforation failure arises from pitting corrosion or partial detinning. Corrosion pits can be found on the body beads, on the bead crests, at side seam weld, or can wall area of headspace. Deformation of tinplate through can-making process produces microcracks in the steel sensitive to pitting.

Even lacquer coating on tinplate surface does not completely eliminate the concerns related to corrosion. Normal corrosion process in lacquered cans is a slow attack on the tin and base steel at microscopic pores in the organic coating. Anodic reaction is concentrated in small areas of tin and iron at pores or scratches in the lacquer. Unlike plain tinplate cans, tin's sacrificial protection against steel is rapidly lost, and iron is soon able to corrode. Abnormal rapid corrosion can occur at metal exposed areas in the organic coating, which may include locations of side seams, body beads, the expansion ring on the ends, and the die code where stresses are concentrated during fabrication. Symptoms of corrosion include pitting corrosion, sulfide blackening, under-lacquer corrosion, and lacquer flaking. Failures may include hydrogen swell and perforation. Pitting corrosion is the most frequently encountered one in the lacquered cans. Improved coverage of the metal surface is the best preservative measure to control the abnormal corrosion of the lacquered cans. The protection effectiveness of the lacquer coating depends on its resistance to ion transfer which may occur even in the absence of pores, scratches, or blisters. The resistance becomes higher with thicker coating. Thickness of 8-12 μm is required for aggressive products such as tomato paste while 4-6 μm is sufficient for nonaggressive products [16].

Nitrates, sulfites, pigments, organic acids, chloride, and trace metals are known to promote detinning speed [1, 5, 16, 18-20]. Nitrates act as depolarizer below pH 5.4 and may accelerate the corrosion. Nitrates are originated from plants, water supplies, and fertilized soils. During storage of canned foods, the nitrates are reduced eventually to ammonia and act as corrosion accelerator. The corrosion rate generally increases with acid content of the product, but it depends also on the type of acid: acetic, lactic, and fumaric acids are more active in corrosive attack. Certain acid solutions can accelerate the corrosion process by forming soluble tin-acid complexes and acting as depolarizer. However, acid such as citrate can also form a kind of passivation layer and protect the tin somehow. Therefore it is hard to predict corrosion behavior of tinplate simply from acid content. Certain trace metals such as copper and nickel in the food are known to increase the corrosivity. Copper at a level of 0.5 to 2 ppm can promote the corrosion. Because of the potentiality of the welded side seam area to increase the copper level of the product, high quality stripe of organic coating on the side seam is suggested as a protective measure.

Sulfur compounds in canned products can lead to accelerated corrosion and unpleasant blackish staining of tinplate surfaces. Sulfur or sulfur dioxide (SO_2) may be reduced to hydrogen sulfide in soft drink and wine cans, giving an offensive smell. Even 1 ppm of sulfur dioxide in acid medium is known to accelerate the corrosion. Low molecular weight sulfhydryl compounds such as hydrogen sulfide may also be formed from breakdown of thioprotein containing amino acids such as cysteine, cystine, and methionine. The sulfide ions in the compounds can react with tin and iron to produce tin sulfide of blue-black color or iron sulfide of black deposit, respectively, which cause a visible staining problem on the metal surface. The sulfide stains usually take place during or after heat processing. Though these staining compounds are only moderately soluble and pose no toxic risk, they give an unpleasant appearance and lead to consumer complaints. Lacquers added with zinc compounds are a solution for avoiding this cosmetic problem because alternatively formed zinc sulfide is white and is not a noticeable objection (see Section 8.7.1).

Natural pigments such as antocyanins may act as depolarizer and enhance the dissolution of metal. Reduction reaction of these pigments may lead to undesirable color change in the fruits such as strawberry, raspberry, and cherry. Phosphates naturally present or artificially added in the meat products may lead to increased corrosion with formation of iron phosphate.

External surface of tinplate may corrode to form rust or localized stain. Overcooling below 40°C after heat processing and an excessive use of chlorine in cooling water may cause external corrosion of tinplate. Long-time exposure to wet environment may induce the moisture condensation on the can surface and therefore should be avoided. Temperature fluctuation during transportation and storage should also be minimized. Damaged lacquer coating and food contamination of can surface may accelerate the external corrosion. When cans with external scratch defects are stored at relative humidity of 60-95% at 21-35°C, another type of external corrosion called filiform corrosion takes place [21]. Filiform corrosion proceeds with tunneling through metal surface and subsequent formation of the spike-like filaments starting from scratch defects and extending over the surface under the coating. Filiform corrosion can also occur on empty cans and ends. It can be prevented by minimizing scratch defects and handling the cans properly through washing, drying, and storage.

High temperature storage of the canned products increases internal corrosion rate. Even though generally most chemical reactions doubles in the rate with temperature increase of 10°C, corrosion rate does not often follow this trend: it is known that corrosion rate at 37°C is 2.6 to 4.3 times of that at 20°C [22]. At high temperature there may also occur severe nonenzymatic browning in tomato products producing carbon dioxide, which contributes to swelling.

8.5.3 Corrosion of TFS

The layer of chromium/chromium oxide is stable and is always lacquered in can manufacturing, and thus TFS plate has excellent resistance against rusting and corrosion [16, 18]. Lacquer adhesion quality of TFS is excellent and lacquered TFS is also resistant to corrosion. However, TFS cans are labile to filiform corrosion which may cause perforations, if there are scratch defects on the outside. TFS is unsuitable for products with high acidity [19].

8.5.4 Corrosion of Aluminum

Although aluminum is highly electro-positive (Table 8.7), it is not liable to corrode because the aluminum oxide (Al_2O_3) is formed as passivation layer when exposed to oxygen or water (see Figure 8.3). However, aluminum oxide may dissolve in aqueous solution to generate Al^{3+} or AlO_2^- ions under acidic (pH<4) or alkaline (pH>8) conditions and/or in the presence of complexing ligands [16], for this reason the products containing citric, malic, and phosphoric acid, sulfites, oily components, or antocyanin pigments may promote the corrosion of aluminum cans.

Aluminum cans are particularly corroded by chloride ion. Chlorides in the food products can cause perforation of the aluminum can body and easy-opening end particularly at the score [17]. High temperature, low pH values, and high oxygen concentration conditions, which are quite frequent in food packaging, can accelerate the corrosion of aluminum foil [23]. To improve corrosion resistance some special alloys are selected, but particularly lacquer coating of 3-10 μm thickness is applied on the surface.

External corrosion of aluminum cans occurs mainly around stay-on-tab of easy open end and is promoted by mechanical stresses. Due to internal pressure of carbonated beverage, tensile stress is applied onto the can surface and can cause corrosion crack if there is product contamination or moisture condensation on the surface. The stress corrosion crack can be minimized by maintaining clean and dry surface of cans during the storage and transportation.

8.5.5 Corrosion of Stainless Steel

Even though stainless steel is the least corrodible material among industrial food contact and packaging materials, there is no metal that will not corrode in some environment. As previously underlined, all the metals derive their corrosion resistance from the protective oxide which can be produced on the surface. According to the nature of the oxide film, metals may be classified in two groups, named *active* and *passive*. In the former, the oxide film, more or less slowly, grows continuously and continuously sloughs off, till the metal is completely consumed. Examples of metals with active oxides are iron and copper. Passive film metals, on the contrary, form an extremely thin oxide layer (less than 300 atoms thick), which then stops growing. These films remain stable, until something breaks this equilibrium conditions. Stainless steel (together with titanium, gold, platinum, and silver) is a good example of passive film metals.

The simplest form of corrosion with steel is the general corrosion that occurs over large areas of the surface and mainly results in a chemical dissolution of the metal. Stainless steels used in food industry are subject to general corrosion only in some concentrated acids and salt solutions, and they are known to be not subject to general corrosion in water. However, electrochemical (called *galvanic*) corrosion may occur in stainless steels: this may occur when the passive film is lost in one spot and the metal becomes active in that area. Thus, the metal has both passive and active sites on the same surface. This is the mechanism for pitting and crevice formation which is the most dangerous phenomenon of stainless steel corrosion [24].

In pitting corrosion the chromium in the passive layer is dissolved leaving the iron labile to corrosion and an active site. The electric potential (E) between the passive and active layer on an austenitic stainless steel is quite high, being 0.78 V. Chlorides are dangerous to stainless steel because they form chromium chloride ($CrCl_3$), which is soluble in water. This leads to the removal of the chromium from the passive layer, leaving only the active iron oxide. As the chromium is dissolved, the electrically driven chlorides drill the metal, creating an almost spherical pit. Into the pit remains a ferric chloride ($FeCl_3$) solution which is corrosive to stainless steel: in fact this reaction is largely used in corrosion tests for stainless steel. Therefore, the rate of pitting corrosion accelerated by all these phenomena can lead to the complete piercing of the object in relatively short time. Also crevice corrosion is a type of electrochemical corrosion, leading to tight crevices and occurring mainly when stainless steel is in close contact with another metal or a tape of plastic, a rubber gasket, or any other material. Like pitting, the reaction proceeds at a faster rate in the presence of chlorides.

8.5.6 Microbiologically Induced Corrosion

The microbiologically induced corrosion (MIC) or biological corrosion, is a form of attack on metallic materials that can cause damages to metals in contact with water or liquid foods like in water supply systems, as well as in cooling circuits of processing plants, sewage treatment plants, and in several other cases. When a metal surface remains in contact with water or humid products for a long time, it may be colonized by micro-organisms such as bacteria, fungi, micro-algae able to form the so-called biofilms (see Section 2.5.2). The microorganisms make up from 5 to 25% of the volume of a biofilm and their colonies are held together by substances (mainly acidic exopolysaccharides and lipopolysaccharides) produced by the cells themselves. These substances make the biological film adhere to the metal surface and modify locally the chemical composition, the pH value, and the oxygen content. Microbial biofilms cause problems in many fields: in medicine, for instance, they cause infections in biomedical devices, and in the food industry they can impede heat flow over heat exchanger surfaces and particularly contribute to corrosion [25-27].

Biofilm contributes to MIC in several ways, the simplest one being buildup of the dissolved oxygen concentration differential across the biofilm [28]. Biofilm also leads to the accumulation of acidic metabolic products near the metal surface which accelerate the cathodic reaction. In some cases, the metabolic byproducts react with the environmental constituents to create a corrosive media. One particular metabolic product, hydrogen sulfide will also promote the anodic reaction through the formation of highly insoluble ferrous sulfide [28].

There are several families of bacteria able to cause corrosion: the most relevant to steel alloys, include those able to oxidize iron and manganese (*Gallionella, Sidercapsa, Spheaerotilus* spp.), the ones able to oxidize iron as well as sulfur and sulfide to sulfates (*Thiobacillus thiooxidans, Thobacillus ferroxidans*) and those able to reduce sulfates to sulfides (*Desulfovibrio, Desulfomaculum* spp.). These last bacteria are anaerobic and cause the worst damages, catalyzing the reduction of sulfates to sulfides in addition to increasing the electromotive force for the oxidation of iron, which eventually give rise to the highly black and insoluble iron sulfide. Although it can occur virtually anywhere, incidents of MIC tend to be more prevalent in areas with warmer climates and rarely affect food packages.

8.6 Metal Container Manufacturing

A variety of cans and metal containers are used for packaging foods. They vary in shape, dimensions, material, structure, and/or lacquering. The most diffused and used are certainly the cans. Traditionally cans in round, square, oval, or trapezoidal shapes have been used for the preserved foods as well as for beverages; recently irregular shapes and embossed surfaces have been introduced particularly for drinks and beers cans. Up to 1970, cans have been formed only from three-pieces (body and two ends), using tinplate as the only material. However, these days, two-piece cans have been developed from a variety of materials (aluminum, tinplate, TFS, and polymer coated steels) and widely used for packaging beverage and solid foods.

Usual volume size of food/beverage in metal packaging largely ranges from 110 mL of small cans to 50 L of the stainless steel kegs or 180-200 L of cylindrical and tapered drums for bulk or semi-bulk quantities. Dimension of cylindrical metal cans are usually given for overall diameter and height in inches plus sixteenth of an inch in three digits. The dimension of 211 × 300 means (2+11/16) inches in diameter and three inches in height. Necked cans may be reported also with neck diameter before the height. In stating the dimensions of oval and rectangular cans, the dimensions of the opening are stated first, followed by the height in the same way. The first two are the long and short axes of the opening. An oval can given as 402 × 304 × 612 in its size, means that the oval opening is (4+ 2/16) × (3+4/16) inches, and the height is 6+12/16 inches. There are several national and international standards for can size, volume, and resistance.

Metal containers should be able to withstand the mechanical stresses experienced in transporting, filling, processing, storage, and distribution. The container filled and sealed should have physical strength enough to undergo temperature and pressure of thermal sterilization. There occurs pressure differential of 0.05-0.2 MPa (about 0.5-2 atm) across the can wall to process in the range of 90-125°C. Mechanical strength of filled cans are measured and presented in axial strength, paneling resistance, and peaking resistance [1]. Axial strength represents the resistance to collapse from top load or compression, and is related to the strength to withstand the loads imposed during filling, seaming operation, palletization, and stacked storage. Axial strength can be measured by compression test. Panelling resistance refers to the strength to resist against collapse inward due to vacuumized condition and is measured as deformation under the application of external overpressure. Peaking resistance tells about the resistance against pressure differential mainly generated in the thermal sterilization, particularly the cooling stage. It can be measured by examining the expansion characteristics of can body and ends under the application of internal pressure. Beading on the can body can strengthen the paneling and peaking resistances, while axial strength is reduced by that. Expansion ring or bead on the end also improves the resistance to distortion caused by internal pressure. The mechanical strength of a can is a function of metal thickness, temper, can body design, and end structure.

8.6.1 Three-piece Can

A three-piece can consists of body blank and two ends. Cylindrical cans are used for most commercially available canned foods and therefore cylindrical body blank is the most common case. This section will therefore describe the manufacturing process of three-piece cylindrical cans.

Figure 8.4 describes the basic sequence of three-piece can making and its use. The large coil of metal is cut into square sheets of about 1 × 1 m on the shear press. An inside protective coating is placed on the sheets and then cured. The sheets are decorated with printing design and then an overcoat of varnish is placed on the decorated sheet and cured. The body sheets are now stacked on pallets for shipment to a fabricating plant or stage. Body sheets containing up to 35 body blanks per sheet are slit into individual body blanks. The corners are notched to remove extra thickness of metal on the flange, which is to be combined with a curl on the end. The body blank is rolled into a cylinder and seamed on side.

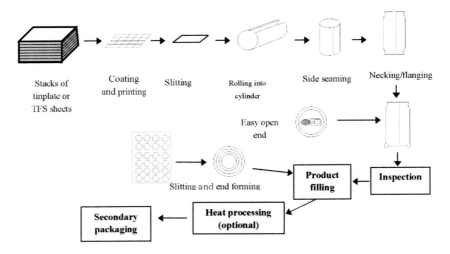

Figure 8.4 General outline of three-piece can manufacturing and use

Figure 8.5 Schematic comparison of soldered, cemented, and welded side seams. The final seam thickness corresponds, respectively, to 4, 2, and 1.3 times the initial sheet thickness.

There are three ways of side seam: soldering, cementing, and welding (Figure 8.5). Soldering is attained by passing the seam area through a bath of molten solder. Bare metal should be exposed to accept the solder, which readily penetrates into the side seam structure. The locked seam needs to be coated inside and outside. Formerly, solder consisted of lead, tin, silver, or antimony in various portions [3]. However lead-soldered cans have been inhibited by hygienic regulations in some countries or eliminated almost from the market because of concerns on lead toxicity. Today, solder consisting of tin and silver is used to have proper bonding and mechanical resistance at required food safety. These days most steel cans are welded on side seam. Overlap of curled plate is subjected to a high electric power of alternating current (about 3500A), which produces high temperature for welding (Figure 8.6). The rolls with moving copper wire electrode also apply pressure for the molten metal to flow and attain high integrity bond. New techniques involving laser use have been developed for better welding quality and higher production speed, but did not succeed in commercialization. Welding offers a stronger side seam of narrow margin with less use of metal. Minimum diameter of a welded can is limited to 52 mm due to accommodation of the welding rolls [29]. Cans soldered or welded need side striping on internal and external surfaces on the seam for protection of exposed metal area. Cemented side seam is used when soldering or welding is not desirable such as for a TFS can: the polyamide-based adhesive forms the seal on the heated overlap. It does not require further coating or striping operation due to nylon coverage on the seam. The cemented side seam does not affect the external appearance of cans that, on the contrary, is improved due to the absence of bare metal exposed.

The formed cylinder moves to the flanger, where the metal on both ends is rolled to form a flange on each end of the can. The cans are often subjected to beading treatment to attain resistance against collapse due to the pressure differential across the can wall, which occurs severely during cooling cycle of thermal processing. The beading also increases the resistance to high vacuum conditions inside the can, as mentioned above. On the other hand, beading has a negative aspect in increasing the consumption of the body blank for a given volume of the can, reducing the compression strength and producing difficulty in labeling on the surface. The creation of beads in a welded can cylinder is achieved by rolling the can wall in the interlocking tool [29, 30]. The bead tooling has another negative impact of imposing severe stress on the side wall and side seam area.

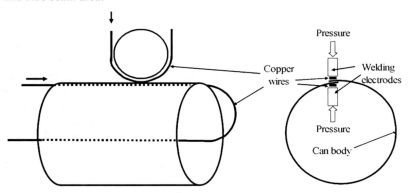

Figure 8.6 Principle of welded side seam

The formed flange is combined mechanically with one end, top or bottom, depending on customer specifications. The other end will be obviously combined with the can after the filling and in a different site but with the same operation (Figure 8.4); this operation, called double seaming, is identical to the one used in closing two-piece cans after filling and will be described later in Section 8.6.5.

Necking is sometimes applied on three-piece cans, which reduce a top diameter and will be explained later with two-piece can manufacturing.

Protective coating is finally sprayed on the interior surface of the can. The final interior coating is baked and cured through a high-temperature oven. The cans completing the manufacturing process pass through a leak detector and are shipped to the food packer usually in palletized load. Three-piece cans may afford delicate decoration while their production is relatively low with maximum of 1000 cans/min for each production line [8].

8.6.2 Two-piece Can - D&I

Two-piece cans consist of a drawn body and one end, i.e., the bottom end is not applied to the body but originates from the same part as the wall. Two-piece cans have advantages of providing wider decoration area and better hermetic integrity due to the absence of a side seam. Their production also saves metallic material by about 35% compared to a three-piece can. The bottom and necking designs of two-piece cans allow easier and more compact stacking. They may be manufactured in two methods: D&I (drawn and wall-ironed, or DWI) and DRD (drawn and redrawn). Production rate of a two-piece is higher than that of a three-piece: D&I cans are produced at rate of 1000-2000 cans/min per each production line.

A D&I can has thick bottom and thin wall, and is exclusively used for beers and carbonated beverages, whose pressure helps to support the mechanical structure. Production of D&I cans had been formerly only from aluminum, but later have been extended to steel plates with technological developments. Aluminum or steel strips in huge coils are lubricated with a thin film of oil and then fed continuously through a cupping press that stamps and draws shallow cups (Figure 8.7). Each cup is pushed and rammed through a series of tungsten carbide rings. This ironing process stretches the wall, thins to 1/3 of the original plate thickness, and raises the can height to three times of its original height. The bottom remains in its original diameter and is domed to provide strength and stability. Typical thicknesses of bottom and side wall for D&I cans are about 0.3 and 0.1 mm, respectively. Ratio of height to diameter can reach 2:1. Trimmers remove the surplus irregular edge and cut each can to a precise, specific height. The trimmed can bodies are passed through highly efficient washers and then dried in preparation for coating inside and outside. The clean cans are coated externally with a clear or pigment base coat that forms a good surface for the printing inks. The cans pass through a hot air oven to dry the lacquer. The sophisticated design is printed with a varnish by printer/decorator. A coat of varnish is also applied to the base of each can by a rim-coater. The cans pass through a second oven which dries the inks and varnish. The inside of each can is sprayed with lacquer and baked again in an oven. The cans are passed through a necker/flanger. Necking to decrease the top diameter reduces the material use for the top end. In case of typical can size of 211 diameter (2+11/16 inches), the necking reduces the end diameter to 206 diameter (2+6/16 inches). Diameter reduction is achieved by a rotating disk forcing against top side wall of the rotating can. The necking process consists of several serial tooling steps. The necking also has the effect of an increase in metal thickness and compressed stress on the coating. Here the diameter of the wall is reduced or 'necked-in'. The top of the can is flanged outwards to accept the end once the can has been filled. The sequences of D&I manufacturing shown in Figure 8.7 are achieved in a short time in a tool pack.

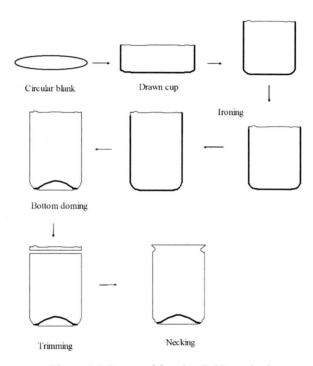

Figure 8.7 Stages of forming D&I can body

8.6.3 Two-piece Can - DRD

Shallow cans have been produced directly from a single drawing operation and used for fish products such as smoked oysters, tuna, or anchovies. These single drawn cans have the height to diameter ratio less than 1:1. Blanks for drawn cans may be printed before drawing, taking into account, with an appropriate design, the subsequent deformation of the metal [31]. Manufacturing DRD cans involves a series of drawing to attain higher height. DRD can manufacturing is the same as the D&I method at initial stage of cupping, but differs in that subsequent redrawing reduces the diameter of the can with increase of its height (Figure 8.8). The thickness of the bottom and side wall does not change in the process of drawing and redrawing. DRD cans are produced from lacquer-coated tinplate, TFS, and aluminum sheets. In the aspect of material use, DRD cans are inferior to D&I cans because the former needs a thicker gauge for structural integrity. Typical wall thickness of the DRD cans is about 0.2 mm excluding the lacquer coating layer. Maximum ratio of height to diameter is about 1.5:1. DRD cans can stand inside vacuum conditions and thermal sterilization process in their strength. Therefore they are usually used for thermally processed solid or semisolid foods rather than beverages.

The classification between the D&I and DRD cans became less distinct through recent technological developments in two-piece can manufacturing. The drawing and ironing operations are combined or closely interrelated to make the difference less clear. DTR (draw and thin redraw) method draws and stretches precoated ultra-light-gauge steel plate (0.18 mm thick, double reduced). Cup with a large diameter is first drawn with flange formed and then stretch forming is undertaken in series to reduce the diameter and increase the height while retaining the flange [2]. The sidewall of the finished can is thinner by about 15% compared to the original metal and bottom end. Stretch forming is better than the ironing process of D&I method in protecting the coating integrity.

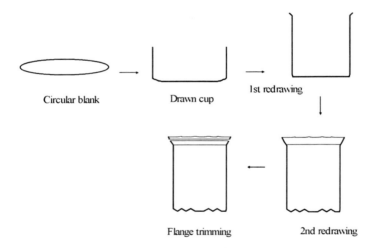

Figure 8.8 Stages of forming DRD can body

8.6.4 Can Ends

Until recent years, fixed can ends of round, oval/irregular, or rectangular shape were made from tinplate or TFS, and used for most canned food products. They need can-opener for opening the cans. However, easy-opening ends were introduced for beverage cans in 1960s and then extended to most food cans, providing convenience and competence to the canned foods and beverages. They were initially made of aluminum, but are also currently made mainly of tinplate. Several elaborate designs were developed for improving convenience and environmental friendliness. Even though easy-opening ends are widely used for canned foods, at least the bottom part of the can consists of fixed end.

Fixed ends are formed through a press that stamps out circular panel from lacquered metal sheet and curls the edges at the same time. A precise bead of compound sealant is applied around the inside of the curl. The sealing compound is based on natural or synthetic rubber, which is applied as dispersion solution in water or solvent. Beaded circles called expansion rings are made on the end surface to resist against pressurized internal expansion experienced during thermal processing. The expansion rings help to prevent wrinkling of the double seam in the process of sterilization, particularly in the cooling cycle. Figure 8.9 shows the typical structure of a fixed end.

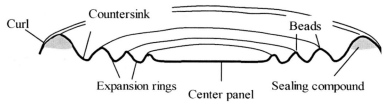

Figure 8.9 Cross section of a fixed can end

Easy open ends are classified into pour-aperture types for beverages and full-aperture types for thermally processed foods. Recently ends with stay-on-tabs sometimes called ecology ends have been developed for pour aperture types. The end with the stay-on-tab eliminated the problem of litter by the detached tab. The tab is designed to be pushed by hand so that the scored line may be torn apart while tab still stays on the rivet. The easy-open covers are made from lacquered tinplate, TFS, or aluminum alloy. A blank is cut from a coil and drawn into a suitable profile. A seaming compound is applied under the curl and dried. The rivet is formed through a multi-stage press with subsequent scoring on the panel and attachment of the tab. Figure 8.10 shows the major stages of fabricating easy opening ends. The scoring depth should be controlled to be taken out easily, but also be retained safely in place without leaking before opening. The aluminum easy open end has a scoring depth corresponding to 30-60% of the total thickness. Because scoring and rivet formation may damage the coating layer, the inside and outside of the end are sometimes spray-coated to repair the possible metal exposure.

The design of easy-open ends differs depending on the normal pressure inside the can, i.e., vacuumed or pressurized conditions. Thickness, material, strength, and shape of the end also differ with product and process conditions. For example, alloys of 5182 and 5082 are used for pressurized containers, while that of 5052 is used for vacuumed cans. A variety of designs are available for easy-open ends with technological developments. Some beads are also formed on the ends for improved performance. An easy-open end design with a raised conical part has been developed to open inward by pushing the scored part. It has the advantage of not leaving the protruded edge outside the end. Another type of easy-opening end available is laminated foil or metallized plastic lid, which is adhesive-bonded to the end with opening and is peelable. The foil disc is stuck onto the flat surface of the rim of opening by adhesives or heat-sealable lacquers [30]. The opening may be a small pour aperture or a full one.

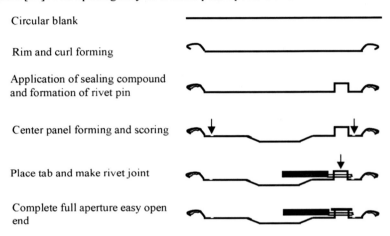

Circular blank

Rim and curl forming

Application of sealing compound and formation of rivet pin

Center panel forming and scoring

Place tab and make rivet joint

Complete full aperture easy open end

Figure 8.10 Stages of fabricating an easy-opening end

8.6.5 Double Seaming

With the only exception of peelable lid, all the different types of can ends are attached to a can body by double seaming. The hermetic airtight joint between body and ends is maintained by overlapped interlocking helped by a sealing compound. It is formed by interlocking the can body flange and end curl, folding and pressing firmly together five thicknesses of the two plates (Figure 8.11).

The operating cycle of the double seamer consists of two steps: initial step of feeding and rolling up the end curl and body in the concave profile of a seam roll; second step of further compression and flattening against the side of the container body in another seam roll. The chuck, base plate (lifter), and seam rolls are required for forming the double seam. Chuck fits into the recessed area or countersink of the can cover, centers the cover on the can body, and acts as an anvil against which the double seam is formed. Base plate raises and supports the can with proper pressure. The first and second operation seam rolls form the seam interlock with the proper dimension. The profile contour of the second operation seam roll is more shallow allowing final compression of the seam. Different change parts are required when changing can sizes and/or styles. Because double seam plays a role of securing the hermetic seal and safety of the canned product, it is controlled strictly in both canmaker and food packer. Sealing compound acts as a flexible gasket material filling the crevice between metal surfaces. The quality of the seam is checked by cutting the part by jeweler's saw and examining its cross section with magnification by optical seam projector. There are allowable limits on the measured lengths and thicknesses. The double seamer machineries may have from 4 to 12 seaming heads and run at the speed from 100 to 1000 cans/min.

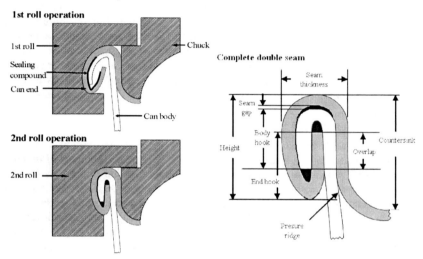

Figure 8.11 Double seaming stages and cross section of a complete seam.

Adapted from Ref. [2, 32].

8.6.6 Kegs and Drums

Beside the widely diffused two- and three-piece cans, some other metallic containers are used by food industry mainly for bulk quantities. They are known as kegs, drums, barrels, and pails. A few details are reserved here on kind of metal packaging.

Kegs are stainless steel refillable containers, ranging from 10 to 50 liters used for beverage distribution mainly to catering and restaurant. The body is manufactured by soldering along the circumference (by plasma process) two drawn cups and then attaching two large rings for handling and sustaining. An opening is threaded on the top end for withdrawal. The stainless steel interior without any protective coating can be cleaned and sterilized easily; the outside can be painted or coated by polyurethane foam for better protection from possible shocks. Kegs with pouring-out devices are largely used to hold and deliver wines, beers, and drinks, available in bars, cafeterias, and fast food restaurants.

Drums are essentially cylindrical or tapered big containers, ranging from 180 to 200 liters. Thanks to their shape the tapered empty drums can be nested into each other with a considerable saving of storing and transport expenses. They are generally manufactured using cold rolled steel and the body is electrically welded and reinforced with one or more beads. They are generally protected inside by synthetic lacquers and outside they are just painted. They are offered in two different ways of closing: tight head and open head. The first has the top and bottom ends attached to the body by double seaming. The top bottom offers a threaded opening used for filling and withdrawing liquids. The open head style, usually used for solid products, has the body double seamed to the bottom end, and the top head fixed to the body by a mobile ring is completely removable. Sometimes these open head drums are used as secondary packages for big flexible sacks (plastic laminated to aluminum foil) which are aseptically filled with liquid or semi solid foods.

8.6.7 Trays

Recently two-piece metal tray containers are made from tinplate, TFS, or aluminum sheet by the simple drawing or DRD process. Steel or aluminum sheets 0.2-0.3 mm thick are formed into rigid trays, or thin aluminum foils of 0.05-0.20 mm thickness are shaped into semi-rigid trays [33, 34]. The plates or sheets are coated with protective lacquer and then formed. Hermetic sealing of rigid trays is attained by double seaming like normal cans. Semi-rigid trays of aluminum foil are usually heat-sealed with the help of a polypropylene layer laminated onto the foil. The internal polypropylene may be applied alone or on top of an initial lacquer layer. These metal trays can withstand the thermal processing at 110-125°C, and are used for heat-preserved convenience foods. The rectangular-shape can reduces the sterilization time by as much as 60% compared to cylindrical one, saving production cost and improving high retention of nutrients in canned food. The large tray often serves the purpose of shipping containers, heating containers, and serving containers. The volume of trays may usually be in range of 200 mL to 3 L.

8.6.8 Collapsible Tubes and Aerosol Containers

Collapsible aluminum tubes are used for some viscous products or paste foods. Their use is limited to shelf-stable product of concentrated form, but is growing with the growth of prepared convenience foods. Typical usages include mayonnaise, mustard, cheese spread, and tomato puree, which are dispensed in small amounts.

Manufacturing of the aluminum tube follows the process called impact extrusion, in which a slug located in die is struck with great force by a punch (Figure 8.12). Under the high impact pressure, the aluminum (a special alloy with more than 99.5% of pure metal and high deformability) is drawn into a long cylindrical shape. The shoulder and opening of the tube are formed as part of the process. The formed collapsible tube is annealed at 480-650°C to remove the stiffness of the wall that was caused by the impact extrusion. The tube is trimmed to the desired length, coated internally by an epoxy type lacquer and decorated externally by roller. The thread is turned into the closure and sent for filling products. The tube is filled through the bottom, and the end is flattened, folded several times, and crimped to provide sealing and a barrier to the outside. The crimped sealing may sometimes be assisted with heat-set sealant.

The aluminum tube has an absolute barrier to moisture, oxygen, and light. And its dead-fold characteristic does not allow the air to come into the tube during the consumption cycle of frequent partial dispensing. This property is effective to preserve the quality of the product that is sensitive to oxidative deterioration. Some tube packages are sterilized for storage and distribution at an ambient temperature. Nowadays, also for foods, after toothpastes and cosmetics, a big competition is coming to the metal tube packaging from laminated plastic tubes.

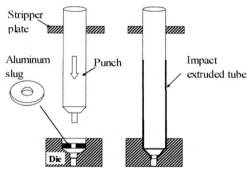

Figure 8.12 Concept of impact extrusion process of collapsible aluminum tube

A quite marginal metal container, in food applications, is represented by the *aerosol containers*, a kind of pressurized container consisting of a combination of concentrate liquid or viscous product and a propellent gas, in a pressure resistant container, equipped with a dispenser fitted with a valve. Therefore, these containers have typical dispenser devices, because the propellent (carbon dioxide, nitrogen, or nitrous oxide), under pressure in the container, makes convenient pouring out of the product possible. Food applications, may include cheese spreads, cake icings, whipped cream, etc. The aerosol containers are manufactured as three-piece tinplate cans as well as two-piece ones of aluminum or tinplate. Always the top end is dome shaped, for better pressure resistance, and provided with an attachment for dip tube, the valve, and the actuator for dispensing the product. Always the bottom end is curved to allow the dip tube to reach and use all the content. In the US, the country with the major use, just a few hundred million of aerosol containers are manufactured each year. Some concerns about the disposal of used aerosol containers, however, exist even for these cans that contained only food products because they must be treated as hazardous waste, as the residual pressure in them can cause the container to explode if exposed to heat or punctured.

8.7 Protective Lacquers

Cans and almost all the metal packages intended for food contact, are mostly coated on the inside and outside of the body and ends in order to protect the metal surfaces and the foods. Coatings also help to facilitate can manufacturing process by providing lubrication and abrasion resistance in drawing and forming. Coatings on the metal surface are called lacquer or enamel. Clear external coating is usually referred to as varnish. Preliminary coating to promote adhesion of subsequent coating is called coating primer or size. Internal lacquer protects the metal from contained foods and prevents the contamination of foods from metallic ions of the container. External lacquer may serve as a basis for printing and provides barrier layer against external lust and abrasion. External coating may be white pigmented for further decoration or clear form for varnish.

Protective internal coating should not transfer any odor or flavor or unwanted substance to the contents and acts as stabilized barrier between the food and metal surface. The coating should be able to stand the physical stresses in the fabrication procedure of the container and have chemical resistance. The coating should also be flexible and be adhered strongly and evenly onto the metal surface without any break. Its integrity should be able to undergo processing conditions of packaged foods such as thermal sterilization. The coating grammage ranges from 3 to 9 g m^{-2} and the thickness from 4 to 12 μm.

8.7.1 Types of Coatings

Selection of coating type depends on the fabrication method of container, and the nature of food to be packaged [29, 31, 35-37]. The conditions of applying the lacquer, its required barrier property, and metal surface for coating, need to be considered. Processing conditions of the canned food and required shelf life should also be taken into account. There are various resin types for coating, whose properties may determine their usage. Many proprietary and confidential formulations are being used for specific applications in practice. Typical coating types shall be described below.

Oleoresinous type is the oldest coating of natural origin. Natural gums (rosin) and other resins are blended with drying oils of tung and linseed, heated for polymerization, and dissolved in solvent. It has a yellowish color and a relatively cheap cost in application. It is still widely used for fruits and vegetables. However, it is not resistant against heat and oily products. It is not used on drawn cans due to limited flexibility. The coating is cured by passing through baking oven of 200-210°C for 10 minutes. The oleoresinous type lacquer added with zinc oxide (ZnO) is called *C enamel* and used for sulfur-containing foods, such as corn and meat products, which cause blackish sulfide staining, while regular oleoresinous type lacquer is often called *R enamel*. The C enamel was initially used for corn products from which the letter 'C' was adopted. Before sulfur compounds such as hydrogen sulfide (H_2S) originating from sulfur-containing foods reach the tin surface, the zinc oxide in the lacquer absorbs the sulfur and forms zinc sulfide (ZnS) preventing the blackish staining.

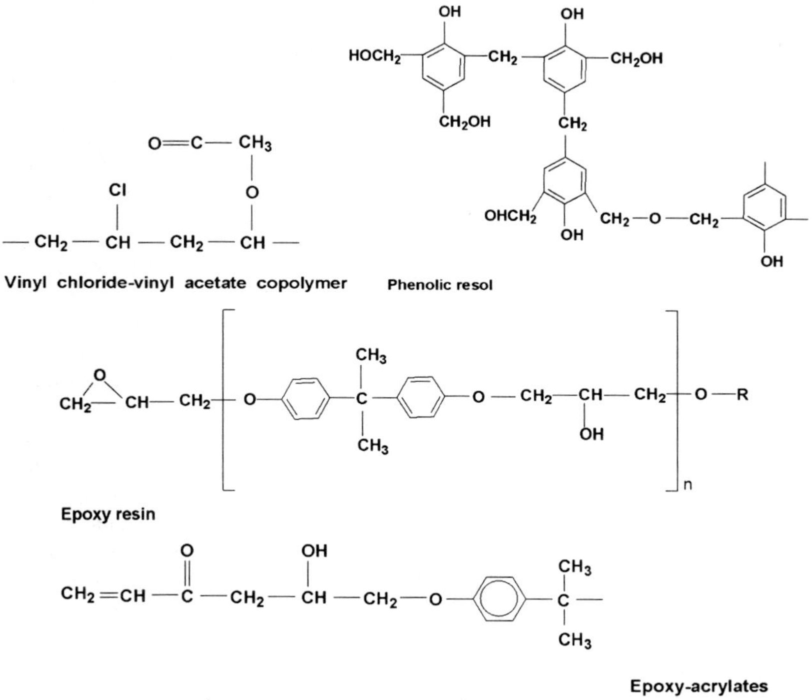

Vinyl chloride-vinyl acetate copolymer   **Phenolic resol**

Epoxy resin

Epoxy-acrylates

Figure 8.13 Chemical structures of compounds used for metal container coatings

Vinyl type lacquers range from solution of low solid content to dispersion of high solid. Solution vinyl coatings are made by dissolving vinyl chloride and vinyl acetate copolymer in solvent (Figure 8.13). The lacquer film has no taste and odor, is resistant against acid, alkali, oil and detergent, and is also flexible and tough with good adhesion. However, its heat stability is limited and the resin may be modified with other resins for improved heat resistance. Usually the vinyl lacquer is used for cans of dry foods, beers, and beverages, but not for retorted foods. They are often applied as top coats over an epoxy-based coating. Due to high flexibility the vinyl lacquer is often used for drawn cans and closures. Vinyl dispersions of high molecular weight and high solid content are called organosols or plastisols, and consist mainly of polyvinyl chloride (PVC). Organosols have properties of flexibility, flavor absence, and adhesion with better heat resistance than solution vinyl coatings, and thus used widely for processed food cans. Typical areas of use include internal coating for DRD cans and the ends. They are relatively resistant to sulfide staining, but are prone to absorb food colorants.

Phenolic lacquers are obtained from polymerization reaction between phenols and formaldehydes at high temperature of 190-195°C (Figure 8.13). Resols of one-stage phenolics or novolacs of two-stage phenolics, both in extremely complex molecular structure, are obtained depending on the ratio of reactants and catalysts. They are resistant to oil, acid, alkali, heated steam, and sulfide staining. However, they are basically brittle and easily peel off from the coated surface. Their poor flexibility limits their usage to three-piece can bodies and ends, where coatings can be applied flat. They are used for acid fruits, fishes, and meat products. They are often used in combination as epoxy-phenolic coatings.

Epoxy resins are produced from condensation polymerization between epichlorohydrin and bisphenol A (Figure 8.13). They are widely used for protective and decorative coatings usually in combination with phenolics, esters, polyamides, and amino compounds. Epoxy-phenolic lacquers are one of the most widely used coatings in metal cans. They display good adhesion and flexibility of epoxy group together with high chemical resistance of phenolic group. The relative degree of flexibility and chemical resistance can be adjusted by the ratio of epoxy and phenolic contents. They are extensively used for acidic and non-acid foods/beverages such as fruits, vegetables, meats, fishes, juices, beers, and soft drinks. They are sometimes applied in a form pigmented with aluminum or zinc carbonate to hide sulfide staining. One use of the epoxy-phenolics is base lacquer for meat-release lacquer combined with wax, which does not allow sticking of meat onto the internal surface of can. Epoxy resins blended with amino resins such as urea formaldehyde produce nearly colorless coatings of high chemical resistance, which can be used for external decoration and internal protection of beverage cans. Recently there has been a rise in attention about the risk of migration of bisphenol A (BFA) from epoxy type coating. Bisphenol A is suspected to be an endocrine disruptor which reduces male fertility [38, 39], and the monitoring of the migration of BFA and bisphenol A diglycidyl ether (BADGE) in canned foods started in Europe [40-42]. The can manufacturing industry is trying to reduce its level in the coating, even though level of contamination and health hazard risk are low.

Acrylics are derived from polymerization of acrylic acid (CH_2=CHCOOH) and its derivatives. They provide good color retention and heat stability. They are used as internal and external coatings. High solid content stripe on three-piece welded cans is a possible application. They are often used as epoxy-acrylates for spray-coating beverage cans (Figure 8.13). The epoxy-acrylates are applied in water-based spray coating, which can reduce the air-pollution problem of volatile organic solvent in the drying or baking. Typical water-based spray coating contains 20% of polymer resin, 65% water, and 15% of organic co-solvent.

Alkyd resins are a type of polyester produced by esterification between a polyhydric alcohol, such as glycerol or penta-erythritol, and a polybasic acid or anhydride, such as phthalic anhydride. They are normally modified by drying oils such as linseed, tung, soybean, or dehydrated caster oil. They are extensively used for external coating as size, pigmented coating, and overprint varnish, but are not suitable for internal lacquering due to imparting odor and taste. Titanium oxide (TiO_2) is sometimes added for basic white color. There is a trend toward use of polyesters as a replacement of alkyds. Polyesters can be formulated to have good color stability, flexibility, product resistance, and abrasion endurance. Polyester film can also be laminated onto the metal surface to be drawn and ironed.

Table 8.8 Properties and applications of common lacquers for cans

Type	Properties[a]			Application and use			
	Flexi-bility	Adhe-sion	Retort-ability	Coating primer	External white pigmented coating	External lacquer	Internal lacquer
Alkyd	3	3	1	✓	✓	✓	
Acrylic	2-3	3	2		✓	✓	
Epoxy-amino	3	3	3	✓		✓	✓
Epoxy-ester	3	3	3		✓	✓	✓
Epoxy-phenolic	3-4	3	3-4				✓
Oleoresinous	1-2	3	3				✓
Organosol	4	4	3				✓
Phenolic	1	1	4				✓
Polyester	2-3	3	3		✓	✓	
Polyester-phenolic	3	3	3				✓
Vinyl (solution)	5	3	1	✓	✓	✓	✓

[a]1=poor; 5=excellent. From Ref. [30] with permission.

Any single lacquer does not often confer the required protective and processing functions. Therefore several combinations are frequently used for attaining the desired function. Different resins may be blended, or multiple coats of different formulations are applied on the metal surface. Degree of coating thickness may also differ with product characteristics, tin coating in case of tinplate, processing conditions, and desired shelf life. Recently, lacquers of high solid content and water-based solution have been explored to reduce the emission of volatile organic solvent into atmosphere. Solid content of organic solvent-based lacquers ranges from 35 to 65%: typical solid content of epoxy phenolic lacquer is 35%, while the organosols have 50-65% [34]. Any lacquer on the food-contact surface must comply with the applicable hygienic standards and regulations. Table 8.8 lists the properties and applications of major lacquer types.

8.7.2 Methods of Coating Application

Coatings, whether protective or decorative, are coated in one form of polymer dissolved in solvent, powder of thermoset and thermoplastic resins, and laminated film [3]. Solution form of polymer is applied by roller or spray coating, while fine powder type is spray-coated by help of electrostatic charge. Laminated films such as polypropylene and polyethylene terephthalate are adhered by help of adhesive.

Roller coating is the most widely used method and is suitable where direct contact between roller and surface is physically possible. It is used for coating sheet or coil type surface, and also for external coating of cylindrical cans. Spray coating is used when direct contact is impossible or difficult. Main area of spray coating is the internal coating of a two-piece can.

Because coatings are usually coated as solution or dispersed state in solvent, the coated layer should be heated and dried for evaporation of solvent, oxidation, and polymerization reactions. This procedure called *curing* or *stoving* is attained by passing the coated plate through the convection-heated oven usually of 210°C for about 15 minutes. Coating thickness after curing ranges in 5-10 μm. Use of organic solvents as lacquer carrier raised the concerns of pollution from emission of volatile organic compounds (VOC) and task of solvent recovery. The concerns led to the development of water-based lacquer system such as epoxy-acrylates. However, water-based coating system requires high energy consumption for evaporating the water. Recently resins to be cured for short time at lower temperature is available by application of ultra-violet (UV) light radiation. Radiation of UV light facilitates polymerization reaction with help of photo-initiators, and gives advantages of low energy consumption and space saving: example resin systems include acrylates, urethanes, polyesters, cycle-aliphatic epoxides, and vinyl ethers for polymerization. UV curable systems are mainly used for external decoration due to hygienic concerns.

Coating in powder is another method recently developed. Resin in fine powder is coated by induced electrostatic attraction and is applied for the place such as side seam which needs thick coating. Curing is accomplished by infra-red light heating or high-frequency induction heating. Hot air curing cannot be used due to flying of powder. Powder coating has advantages of low emission of VOC and low energy consumption, which are valid also for UV curable system.

Electrophoretic deposition of organic coating can be attained by applying electric current onto electrically conductive metal surfaces. Direct extrusion coating is also another new technology of coating. Table 8.9 summarizes methods of coating metal cans.

Table 8.9 Method and technology of applying coating onto metal cans

Technology	Method of application			
	Roller coating	Spraying	Immersion	Other
High-solids solvent based	✓	✓		
Waterborne	✓	✓		
Powder		✓	✓	✓
Electrophoretic		✓	✓	✓
UV and electron-beam curable	✓	✓		
Lamination				✓
Direct polymer extrusion				✓

From Ref.[30] with permission.

8.8 Recent Developments and Trends

High price of tin and aluminum pushed can manufacturing toward reduction and saving of their use. LTS cans using low amounts of tin are expected to grow in production and usage. According to some forecasting [43], it will also soon be possible to produce, by electrolytic deposition of iron, very thin sheets with the same flexibility as aluminum foil but with higher mechanical resistance; after a coating with corrosion preventing oxide or lamination to plastic films, and these new materials may meet the same uses as aluminum foil. Figure 8.14 shows trends of light-weighting for aluminum D&I can. The thickness of a tinplate can body decreased from 200 to 175 μm through 1984 to 1997 [7]. High aluminum prices also help to increase its recycle ratio. Polyester film is laminated onto TFS or aluminum sheets, and used for two-piece can manufacturing by DTR process. As a future prospect, aluminum D&I cans and polyester-laminated cans are expected to be main stream products [8]. Cost of manufacturing D&I cans was analyzed to be 84% of that of welded three-piece cans [30].

Environmental concerns pushed the increase of steel can recycle as shown also in Figure 8.14. Environmental considerations have also driven the can manufacturing process toward a more environment-friendly direction: reduction of VOC emission, lower energy consumption, stay-on-tab on the easy-opening end, and use of safer coating materials.

With technology development, a variety of can shapes became possible. So called 'shaped can' having attractive nontypical cylinder shape appeared in the market. Cans shaped with a bottle image are an example. With technology developments, stretching locally up to about 10% in radial direction became possible. Those with longitudinal beads called fluted cans were also developed. Surface design became diverse including embossing or delicate decoration. A variety of shaped cans offer differentiation of canned products and convenience features, while increasing production cost by 10-25%. Recently, three-piece bottle cans with recloseable screw caps have been used for beverage packaging. Two-piece bottle cans with recloseable screw caps are also available.

Functions of easy opening and injury-prevention have also been improved. Necking-in of the cylindrical container made it possible to stack the cans easily.

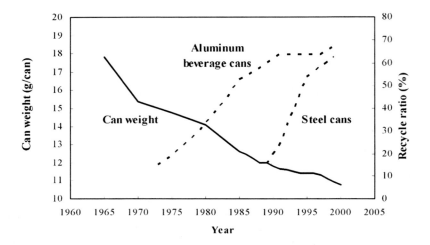

Figure 8.14 Trend in lightweighting (solid line) of aluminum D&I can and metal can recycle ratio (dotted lines) in USA. Reconstructed from Refs. [7, 30].

Even with all these developments, metal containers are being increasingly replaced with plastic and glass packaging. There have been a great deal of strives made in reducing cost, increasing productivity, improving the quality, and enhancing the recycle for improving competence. We will see the better shape of metal packaging competing with other packaging types in future.

Discussion Questions and Problems

1. List and explain all the possible methods to improve the corrosion resistance of tinplate cans used for acidic and salty foods.

2. Compare the can coatings for protecting the contained product against the migration from metal.

3. Describe the structure of double seam in the metal cans. Give the specifications for the satisfactory double seam of a can size you choose.

4. How can you reason the cause of swelling, if you see a canned product of tomato paste in the store or your storage? How can you confirm experimentally about your reasoning?

5. Why is it possible to find pure gold in nature (nuggets) and not possible to find pure aluminum?

6. How can you reason the superior properties of resistance to corrosion provided by stainless steel?

7. Compare the plain tinplate can and the lacquered can for packaging orange juice. It is noted that orange juice has low pH and high amounts of ascorbic acid and carotenoids labile to oxidation.

8. How can the tin coating layer protect the iron layer of a tinplate can in normal conditions of food packaging?

9. Discuss safety aspects of using easy-opening end for aluminum cans of carbonated soft drinks.

10. How have been the historical efforts to reduce the material usage in can fabrication?

11. Construct functions/environments table for aluminum can packages of carbonated soft drinks with consulting Section 1.7.3.

Bibliography

1. TA Turner. Packaging of heat preserved foods in metal containers. In: JAG Rees, J Bettison, ed. Processing and Packaging of Heat Preserved Foods. Glasgow: Blackie, 1991, pp. 92-137.
2. H Matsubayashi. Metal containers. In: T Kadoya, ed. Food Packaging. Boston: Academic Press, 1990, pp. 85-104.
3. JF Hanlon, RJ Kelsey, HE Forcinio. Handbook of Package Engineering. 3rd ed. Boca Raton, FL: CRC Press, 1998, pp. 329-360.
4. VB Gupta, S Anitha, ML Hegde, L Zecca, RM Garruto, R Ravid, SK Shankar, R Stein, P Shanmugavelu, KS Jagannatha Rao. Aluminium in Alzheimer's disease: are we still at a crossroad? Cell Mol Life Sci 62(2):143-158, 2005.
5. JC Charbonneau. Container Corrosion. Washington, DC: National Food Processors Association, 2002, pp. 22-26.
6. D Twede, R Goddard. Packaging Materials. Surrey, UK: Pira International, 1998, pp. 55-65.
7. H Kuguminao. Metal materials (1), steel for 3-piece can (in Japanese). Food & Containers 40(8):438-446, 1999.

8. ToyoSeikan. Current status and prospect of metal containers. Packpia 43(8):48-56, 1999.
9. GM Ridenour, EH Armbruster. Bacterial cleanability of various types of eating surfaces. Am J Public Health 43(2):138-149, 1953.
10. JT Holah, RH Thorpe. Cleanability in relation to bacterial retention on unused and abraded domestic sink materials. J Appl Bacteriol 69(4):599-608, 1990.
11. DG Harlow, RP Wei. A probability model for the growth of corrosion pits in aluminum alloys induced by constituent particles. Eng Fract Mech 59(3):305-325, 1998.
12. DG Harlow, RP Wei. A critical comparison between mechanistically based probability and statistically based modeling for materials aging. Materials Sci Eng A 323(1/2):278-284, 2002.
13. D Talbot, J Talbot. Corrosion Science and Technology. Boca Raton, FL: CRC Press, 1998, pp. 71-153.
14. HH Uhlig, RW Revie. Corrosion and Corrosion Control. 3rd ed. New York: John Wiley & Sons, 1985, pp. 35-59.
15. MG Fontana. Corrosion Engineering. 3rd ed. New York: McGraw Hill, 1987, pp. 244-245, 445-481.
16. GL Robertson. Food Packaging. New York: Marcel Dekker, 1992, pp. 173-231.
17. H Tsuchiya. The characteristics of aluminum can and its corrosion problems. Zairyo to Kankyo 51:293-298, 2002.
18. JE Charbonneau. Recent case histories of food product-metal container interactions using scanning electron microscopy-X-ray microanalysis. Scanning 19:512-518, 1997.
19. C Mannheim. Interaction between metal cans and food products. In: JI Gary, BR Harte, J Miltz, ed. Food Product-Package Compatibility. Lancaster: Technomic Publishing, 1986, pp. 105-133.
20. K Galic, M Pavic, N Cikovic. The effect of inhibitors on the corrosion of tin in sodium chloride solution. Corros Sci 36:785-795, 1994.
21. JE Charbonneau. Application of scanning electron microscopy and X-ray microanalysis to investigate corrosion problems in plain and enamelded three piece welded food cans. Food Struct 10:171-184, 1991.
22. Anonymous. Discussion on the selection of plain and lacquered metal cans (in Japanese). Food & Containers 43(3):177-179, 2002.
23. L Piergiovanni, P Fava, S Ciappellano, G Testolin. Modeling acidic corrosion of aluminum foil in contact with foods. Packag Technol Sci 3(4):195-201, 1990.
24. JC Tverberg. Stainless steel in brewery. Technical Quarterly-Masters Brewers Association of the Americas 38(2):67-82, 2001.
25. CG Kumar, SK Anand. Significance of microbial biofilms in food industry: a review. Int J Food Microbiol 42:9-27, 1998.

26. SH Flint, PJ Bremer, JD Brooks, F Martley. Biofilms in dairy manufacturing plant - description, current concerns and methods of control. Biofouling 11:81-97, 1997.
27. BJ Webster, SE Werner, DB Wells, PJ Bremer. Microbiologically influenced corrosion of copper in potable water systems - pH effects. Corrosion 56(9):942-950, 2000.
28. GG Geesey. What is biocorrosion? In: HC Flemming, GG Geesey, ed. Biofouling and Biocorrosion in Industrial Water Systems. Heidelberg: Springer-Verlag, 1991, pp. 155-164.
29. TA Turner. Canmaking: The Technology of Metal Protection and Decoration. London: Blackie Academic & Professional, 1998, pp. 1-131.
30. TA Turner. Canmaking for Can Fillers. Sheffield: Sheffield Academic Press, 2001, pp. 44-391.
31. W Soroka. Fundamentals of Packaging Technology. Hernon, Virginia: Institute of Packaging Professionals, 1995, pp. 147-171.
32. JD Malin. Metal containers and closures. In: SJ Palling, ed. Developments in Food Packaging. London: Applied Science, 1980, pp. 1-26.
33. P Sirbat, J Petitdemange. Metal and plastic trays suitable for sterilization. In: G Bureau, JL Multon, ed. Food Packaging Technology. New York: VCH Publishers, 1996, pp. 221-230.
34. A Lopez. A Complete Course in Canning. Book 2. 12th ed. Baltimore: The Canning Trade, 1987, pp. 60-133.
35. TP Murphy, JP Amberg-Muller. Metals. In: LL Katan, ed. Migration from Food Contact Materials. London: Blackie Academic & Professional, 1996, pp. 111-144.
36. RH Good. Recent advances in metal can interior coatings. In: JH Hotchkiss, ed. Food and Packaging Interactions. Washington DC: American Chemical Society, 1988, pp. 203-216.
37. S Kojima. Trend on the internal coatings of cans. Japan Adhesion Society J 36:33-38, 2000.
38. JB Colerangle, D Roy. Profound effects of the weak environmental estrogen-like chemical bisphenol A on the growth of the mammary gland of noble rats. J Steroid Biochem Molecular Biol 60(1-2):153-160, 1997.
39. P Perez, R Pulgar, F Olea-Serrano, M Villalobos, A Rivas, M Metzler, V Pedraza. The estrogenic of bisphenol A related diphenylalkanes with various substitutes at the central carbon and the hydroxyl group. Environ Health Persp 106(3):167-174, 1998.
40. W Summerfield, A Goodson, I Cooper. Survey of bisphenol A diglycidyl ather (BADGE) in canned foods. Food Addit Contam 15(7):818-830, 1998.
41. C Simoneau, A Theobald, P Hannaert, P Roncari, T Rudolph, E Anklam. Monitoring of bisphenol A diglycidyl ether (BADGE) in canned fish in oil. Food Addit Contam 16(5):189-195, 1999.
42. C Simoneau, A Theobald, D Wiltschko, E Anklam. Estimation of intake of bisphenol A diglycidyl ether (BADGE) from canned fish consumption in Europe and migration survey. Food Addit Contam 16(11):457-463, 1999.

43. PJ Louis. Food packaging in the next millennium. Packag Technol Sci 12(1):1-7, 1999.

Chapter 9

Cellulosic Packaging

9.1   Introduction 243
9.2   Cellulose Fiber-Morphology 245
9.3   Cellulose Fiber-Chemistry 247
      9.3.1   Lignin 247
      9.3.2   Hemicelluloses 247
      9.3.3   Cellulose 249
9.4   Paper and Paperboard Production 251
      9.4.1   Pulping Technology 251
      9.4.2   Papermaking 253
      9.4.3   Converting and Special Paper Products 257
9.5   Paper Bags and Wrappings 259
9.6   Corrugated Board and Boxes 261
9.7   Folding Cartons and Set-up Boxes 264
9.8   Composite Cans and Fiber Drums 266
9.9   Cartons for Liquids 267
9.10 Molded Cellulose 270
9.11 Cellophane 270
9.12 Other Quasi Cellulosic Materials 272
      9.12.1 Wood 272
      9.12.2 Cork 273
      9.12.3 Nonwovens 273
Discussion Questions and Problems 274
Bibliography 274

9.1 Introduction

Cellulosic packaging is a rather broad category including bags, wrappings, corrugated boards and boxes, folding cartons, composite cans, and fiber drum which are used in primary, secondary, and tertiary packaging. The only common feature of these different packaging items (Table 9.1) is the presence of cellulose fibers at least in the raw materials used. Thus the main source of the required raw materials is the plant world. The cellulosic packages, uniqueness among the packaging materials, refer to a real renewable resource naturally biodegradable more or less depending on ingredients and processing. Obviously due to the large and universal availability of plant materials cellulosic packaging in primitive forms such as leaves for wrapping or reeds for containers is certainly the earliest of man's inventions for food protection and containment. Some natural cellulosic materials are still used as food contact materials; for instance, wooden boxes for cheeses, fish, or fruits, oak barrels for wine ageing and textiles for cheese making and ripening. This chapter deals with the most common and advanced forms of cellulosic packaging.

Table 9.1 Common varieties of cellulosic packaging

Function	Types
Wrapping	Leaves, straw, textiles, tissues, papers, vegetable parchment, cellophane, etc.
Primary packaging	Reeds, baskets, chord nets, chests, paper pouches, fiber sacks, casings, boxes, folding cartons, blisters, molded pulp, etc.
Secondary packaging	Wood chest, corrugated and solid fiberboard boxes, pre-lined boxes, folding cartons, etc.

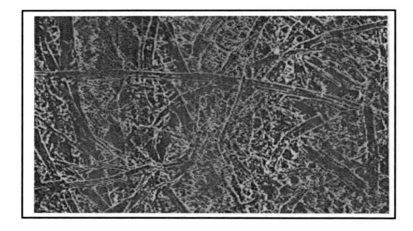

Figure 9.1 Plane network matrix of cellulose fibers in a Kraft paper

A great number of different cellulosic packaging (e.g., paper, solid board, corrugated boxes, prelined cartons) are produced starting from the same basic papermaking process, which, even if very advanced now, basically relies on the ancient manufacturing technique developed by the Chinese minister Ts'ai Lun (104 A.D.). The significant commercial use of paper and its derivatives for packaging purposes was realized only in the last century, but nowadays in the developed countries the usage of cellulosic raw materials in packaging applications almost amounts to that for printing.

The huge variety of the modern cellulosic packaging comes from several factors like the fiber source, the processing technique used for separating the fibers and arranging them in a plane network matrix (Figure 9.1), and the coating, laminating, or more generally converting processes that can greatly transform the original cellulose based material into different forms.

Table 9.2 The main sources for cellulose fibers

Source	Types
Trees	Spruce, pine, birch, eucalyptus, fir, poplar, jasmine,…
Seed hairs	Cotton, cannabis, jute, flax,…
Grasses and leaves	Papyrus, cannabis, flax, jute, straw, bagasse, bamboo, esparto, sisal, banana,…
Recycled materials	Textiles, rags, paper, board

The plant source of cellulose fibers is the first level of diversity and it can greatly affect the ultimate properties of the final cellulosic packaging. There are several, industrially used and economically feasible sources of fibers that are listed in Table 9.2, but by far the most important one for packaging and also printing usage is wood, both the conifer (softwood) and deciduous (hardwood) woods, and therefore the general description provided in this chapter refers to the wood fiber.

9.2 Cellulose Fiber-Morphology

Plants and trees mostly grow vertically, and since they do not have skeleton, membranes that enclose their cells must be thicker and more rigid than those of animal cells. This is achieved by an array of coaxial layers of cellulose fibrils embedded in an amorphous matrix of hemicellulose and lignin forming the cellulose fiber. The generally recognized model of wood fiber cell walls is represented in Figure 9.2, according to electron microscopic observations.

Fibers are composed of the primary wall P1 and secondary walls S1, S2, and S3. The primary wall is very thin, and it is constituted of cellulose for half its weight, the remaining being peptic substances and hemicelluloses. It is linked by a large amount of lignin to the adjacent cell wall, forming the middle lamella which has fundamental supporting functions for the living plant tissue. But lignin is the main byproduct of the papermaking process whose aim is to remove it from cellulose. The secondary walls are not different in chemical composition from the primary ones but rather so in their structure: They are more than twice the thickness of the primary wall and contain several layers of fibrils. The S2 layer (the middle secondary wall) is the biggest layer (70-90% of the fiber) in the cell wall [1]. The fibrils may be either perpendicular or parallel to the fiber and their orientation is measured by the fibril angle. Fibrils aligned with the fiber have a 'low' angle and fibrils not aligned have 'high' angles [2]. The fibril angle of the S2 layer particularly, influences the individual fiber's physical properties and therefore the properties of the final cellulosic product. Materials composed of fibers with low fibril angle (well aligned) withstand greater load and less elongation than those with high angles (fibrils perpendicular to the fiber). Fibril angles may be different within the same fiber source depending on early (spring) or late (summer) wood and position of the fibers in the wood.

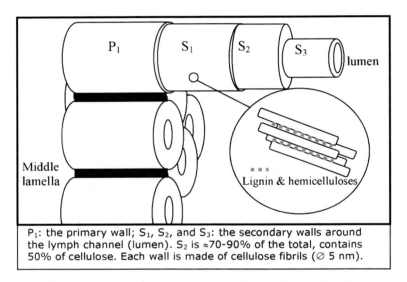

P$_1$: the primary wall; S$_1$, S$_2$, and S$_3$: the secondary walls around the lymph channel (lumen). S$_2$ is ≈70-90% of the total, contains 50% of cellulose. Each wall is made of cellulose fibrils (∅ 5 nm).

Figure 9.2 Schematics of the wood cellulose fiber cell wall layers

At the center of cell wall structure is the lumen, the thin channel throughout which the lymph flows. Even this empty part of the raw material is related to the ultimate physical properties of the material. During the cooking phase of the papermaking process, the lumen may collapse or may maintain its original shape; the resulting sheet will be compact and dense in the first case, voluminous and thick in the second. Finally of a great importance for the future performance of the cellulosic materials are the length and the diameter of the fibers: the longer the fibers the more integrated the network matrix. In softwood the length of the fiber may reach 4 mm, and the diameter may reach 30-40 μm. Those for hardwood are 1-2 mm and 10-40 μm, respectively. These figures, the length particularly, are always much reduced by the recycling process.

9.3 Cellulose Fiber-Chemistry

From a chemical point of view the main components of wood fibers are cellulose, hemicellulose, and lignin. The proportion of these fundamental constituents can vary slightly according to the different parts of the trunk, the kind of tree, and the season; the general composition of hard (deciduous) and soft (conifer) woods are presented in Table 9.3. Apart from their cellulose content, different trees are different in their contents of lignin and hemicellulose and these differences can justify different papermaking techniques and final applications.

Table 9.3 Chemical composition of wood fibers from different trees

	Cellulose (%)	Hemicellulose (%)	Lignin (%)
Softwood trees (e.g., pine tree)	41-47	22-32	26-32
Hardwood trees (e.g., birch tree)	40-48	28-42	17-26

9.3.1 Lignin

Though the chemistry of lignin is not completely understood its function in the plant world is perfectly known: it forms a cementing matrix that holds together the cellulose fibrils in the fiber structure, giving them high mechanical resistance. Lignin is a highly branched alkyl-aromatic, amorphous polymer, exclusively found in wood; the polymerization degree is quite high, and probably different forms of lignin exist in different species. 5,5-biphenyl structures constitute the most important branching structures in lignin (Figure 9.3). It is relatively easy to remove the lignin of the middle lamella, but it is much more difficult to eliminate the small lignin among the fibrils. Because lignin reacts with the chemicals used in papermaking leading to dark brown soluble derivatives, it is almost always necessary to adopt bleaching processes in order to increase the whiteness of the final material.

9.3.2 Hemicelluloses

This term is commonly assigned to a category of different substances that are found in association with cellulose. Hemicelluloses are amorphous copolymers of one or more sugars like xylose, mannose, arabinose, and galactose, combined with uronic acids (Figure 9.3); the average molecular weight is as low as the degree of polymerization of about 100-200, and because of their high solubility in diluted alkali, hemicelluloses are easily separated from cellulose in the pulping process. Their content is measured in the papermaking industry to identify the characteristics of pulp produced. In fact, higher hemicellulose content increases greatly tensile, burst, and folding strengths. Hemicelluloses moreover affect the paper-water relationships, as they are much more hydrophilic than the other constituents. Moisture uptake is greater in cellulosic materials with higher hemicelluloses, as they reduce the total crystallinity: the more crystalline is the paper, the lower is its water absorption. If the manufacturing technique requires a swelling step (like for vegetable parchment, see Section 9.4.3) the pulp should contain a good amount of hemicelluloses for favoring the process; if a wrapping becomes easily wet in contact with food, it might have a too high content of hemicelluloses and lignin.

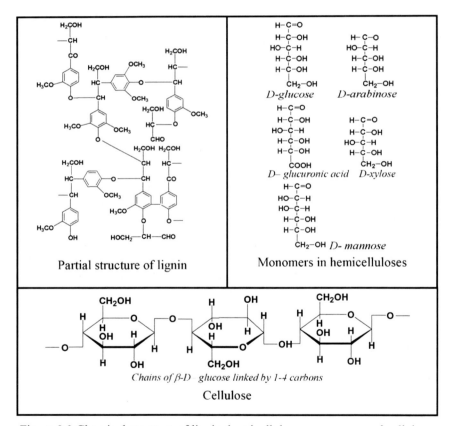

Figure 9.3 Chemical structure of lignin, hemicellulose monomers, and cellulose linear polymer

9.3.3 Cellulose

This is the principal and also, from a packaging point of view, the most important constituent of wood fibers and all other fiber sources. Cellulose is a linear, crystalline, polymer of β–D(+)–glucose, linked by 1-4 carbons (Figure 9.3) and with a high degree of polymerization (3,000-15,000). This particular configuration justifies some general properties, such as nonreducing behavior (resistance to oxidation), poor or null water and alkali solubility (easiness to separate cellulose from other constituents), inability to digest by nonruminants animals (not a form of energy storage like starch and glycogen but rather a structural component), fiber forming attitude, and absence of any thermoplastic behavior.

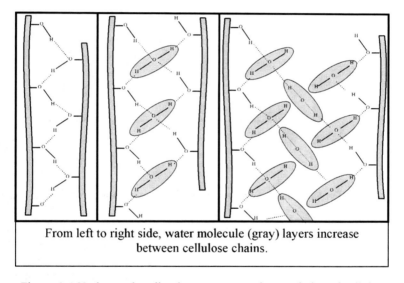

From left to right side, water molecule (gray) layers increase between cellulose chains.

Figure 9.4 Hydrogen bonding between two polymer chains of cellulose

The cellulose fibrils are bundles of cellulose linear polymer chains linked together by hydrogen bonding; these bonds are tight when they link directly two adjacent chains of cellulose (this happens in the crystalline regions of pure cellulose), and they become less tight when a monolayer of water molecules is in between; the link strength decreases as the number of water molecule layers increase (Figure 9.4). Mechanical strength of paper is partly due to inter-fiber hydrogen bonding and also partly due to inherent strength of individual fibers [3]. The hydrogen bonding between cellulose polymer chains determines the tenacity of plane fiber network of paper or paperboard, but they cannot satisfy the requirements of strength for packaging applications and therefore other nonfibrous substances are usually added to increase the bonding. Also these bonds are affected by water molecules and it is a well known phenomenon that a wet paper loses its strength. However, the relationship between relative humidity (RH) and strength properties is quite complex. The tensile strength of fibers increases with relative humidity of low level, but the fiber network reduces its strength when water under highly increased RH competes for the sites of hydrogen bonding between fiber surfaces [3]. The result is that a sheet of wet paper has a tensile strength that amounts to 5-25% (depending on the additives used) of that in dry conditions (Figure 9.5). The paper fibers gain plasticity doubling their elongation limit if RH passes from 10 to 100%. Moisture sorption under high humidity also increases tearing strength of the paper.

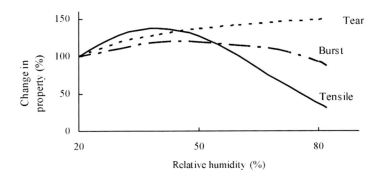

Figure 9.5 Typical effect of relative humidity on mechanical strength of paper

In wood and several other cellulose fiber sources, the cellulose polymer is tightly combined with lignin, hemicellulose, and other minor constituents, which represents a technological problem to overcome as already underlined. However, there are a few cases where almost pure cellulose fibers exist in nature like cotton seeds, whose downs (hairs) are free cellulose fibers some centimeters long, and with cannabis, flax, and jute where long fibers are associated with small amounts of crusting substances. Highly pure cellulose is used for some special cases in the packaging field; for instance in the cellophane production (see Section 9.11) and in the preparation of cellulose esters (cellulose acetate and propionate, for instance), a useful thermoplastic behavior is attained by transformation of the cellulose polymer through the reaction between the hydroxyl groups and organic acids.

9.4 Paper and Paperboard Production

9.4.1 Pulping Technology

Pulping is the operation to transform the raw material (generally wood chips or logs) into an intermediate product (pulp) rich in cellulose fibers purified from cementing substances (mainly lignin). The pulp is suitable for the subsequent process of papermaking. Different technologies for pulping are used based on fiber source available and aesthetic and functional characteristics desired for the final product [1]. Different technologies therefore produce pulps for papers or boards different in price, color, strength, etc.: they are commonly differentiated in their yield, the percentage ratio of dry pulp weight to dry wood weight (Table 9.4). The installations where the pulping is made are generally close to the forests where the conifer or the deciduous trees are grown and not always close to the papermaker company. These installations are effective, able to produce even more than 1000 tons of pulp each day.

Table 9.4 Yield and brightness of pulp from different technologies

Process	Yield (%)	Brightness*
Mechanical (groundwood)	95	50-60
Thermo-mechanical	85-90	50-60
Chemical–thermo-mechanical	85-88	50-60
Semichemical–neutral-sulfite	65-85	40-50
Chemical–sulfate	50-55	15-30
Chemical–sulfate–bleached	40-45	80
Chemical–sulfite	40-50	40-50
Chemical–sulfite–bleached	35-45	90

*Measured as reflectance of light at effective wavelength of 457 nm and expressed in comparison to the standard of magnesium oxide with 100% brightness [4].

- *Mechanical pulp* – This is the cheapest pulp, due to the high yield obtainable; no chemicals are required in mechanical pulping, and the only means for producing the *groundwood pulp* is a fast rotating grindstone, against which the wooden logs or chips are forced. The heat generated is removed by a continuous casting of water and the pulping product looks like sawing powder. Actually the operation is driven in such a way that the fibers are substantially saved and disconnected from the middle lamella which is made softer due to the frictional heat. This process is not a true purification step of cellulose in nature as the high yield is due to large quantity of lignin and other wood constituents contaminating the fibers; however the ability of forming a network matrix is attained at very low cost. Often some refining steps are included in this pulping technology. The groundwood pulps are stiff and have low mechanical resistance and are mainly used in manufacturing papers for the press sector (newspaper) and for cartons, especially as an intermediate layer. Better mechanical characteristics of the pulp are obtained if the wooden chips are preliminarily steamed at 110-150°C (*thermo-mechanical pulping*) in order to reduce the fiber breaking. With this thermal treatment the fibers show a plastic behavior becoming more malleable and less fragile.

- *Chemical pulp* – The chemical processes for pulp production are more recent than the mechanical ones and consist of cellulose refining steps resulting in low yield (Table 9.4). The first industrial process was appeared in England the mid 19th century using diluted (5%) sodium hydroxide and high temperature (170°C). The soda process was soon modified because of the low yield and the poor mechanical resistance of cellulose, leading to the most preponderant *sulfate* process known as *kraft* process for the excellent mechanical properties of the paper produced (in fact the word *kraft* means *strong* both in German and Swedish). This process uses solutions of sodium hydroxide, sodium sulfide, and small quantities of sodium sulfate. Several advantages characterize the kraft process: the length of cellulose fibers is saved during the pulping; even the most resinous kind of wood can be treated by this process and the required chemicals are mostly obtained by the operation of leach recovery. The recovery provides a good source for energy retrieval as it contains large amounts of organic combustible substances (mainly lignin and cortex). A drawback of the sulfate process, however, is that the cellulosic material obtained tends toward brown color and additional bleaching processes may be necessary. The *sulfite* process, instead, with almost the same yield, provides whiter products. This process uses sodium sulfite solution and sulfurous acid, and is generally more suitable for nonwood cellulose sources (e.g., straw or agricultural refuses). Also the classic sulfite process, like the sulfate, attacks the cellulose crusting materials and dissolves lignin in an alkaline environment. Different variants of the process are in use differing in the pH value of the cooking solution and even an acidic process (pH about 4) is sometimes used. All of them are based on the ability of sulfur dioxide as main component of the pulping liquor.

- *Combined techniques* – To achieve the fundamental aim of pulping technology, which is the liberation of the cellulose fibers as pure as possible from crusted substances of the plant source, any measures, chemical, thermal, and mechanical can be combined for improved yield and strength. Each factor having its peculiarity in saving the fiber integrity, in accelerating the process, or in establishing the cost of the technology should be carefully balanced. Therefore it is not surprising that several combined technologies have been proposed for getting cellulose pulp. The most important process in this category is the *neutral-sulfite semichemical* which is used mainly in producing paper for corrugated boards. The conditions used in this technology permit to get high yield because more lignin remains in the pulp and also because a larger quantity of hemicelluloses is retained with an improved strength of the finish product.

- *Bleaching* – Brightness is a common index of evaluation of pulp and paper, and thus the manufacturing of cellulosic materials is accompanied by the chemical bleaching treatment. The color of the cellulosic materials depends very little on the cellulose source (wood species or other plant resource) but much more on the process used, the consequent colored bodies, and residual lignin staying in them. Several different chemicals can be used to remove encrusting residual materials and/or to oxidize the chromophoric groups, and their choice is related to the pulping technology used. For mechanical pulps with the high amount of residual lignin a fair increase in brightness is possible with some hours of treatment with hydrogen peroxide at 40°C; with chemical pulps where the bleaching action is mainly related to the oxidation of colored substances, chlorinated chemicals (chlorine gas, hypochlorite, chlorine dioxide) and oxygenated substances (oxygen gas, ozone, peroxide, and hydrogen peroxide) are commonly used. This step of processing cellulosic materials is frequently questioned for the possible pollution related to the use of these chemicals and to the possible dioxin production. The bleaching process also always reduces the strength of the pulp and the tensile properties of the paper [5].

9.4.2 Papermaking

Pulps are prepared for papermaking by subjecting the fibers to mechanical action and adding a wide variety of chemical compounds. The stock is distributed uniformly as a thin web, pressed for mechanical water removal, and finally dried in paper manufacture.

- *Beating/refining* – The papermaking process always starts with a mechanical operation applied on the diluted pulp (5-7% of solid content). This operation which is called *beating* or *refining* has the goal of increasing the available surface area of fibers for water absorption and inter-fiber bonding. It determines to a large extent the physical properties of the future paper bags or folding cartons. Fundamentally it is a size reduction unit operation and its effects are on both the length and width of the fibers: fiber length may be reduced in the beater by the knives set on the roll and on the bedplate around it and fiber bundles may be detached at the beating/refining operation to enlarge the space between them. The hemicellulose amount present in the pulp has a great importance for the best pulp refining. As this component is located close to cellulose chains and swells readily much more than any other component, the beating of hemicellulose-rich pulps leads to a faster operation and produces fibers that are more flexible and able to deform plastically during the next operations of paper making. As the beating time increases, the tensile strength of the future paper sheet increases and the burst strength increases to a maximum followed by subsequent decrease, whereas the tear resistance decreases after initial sharp increase (Figure 9.6) [5-7].

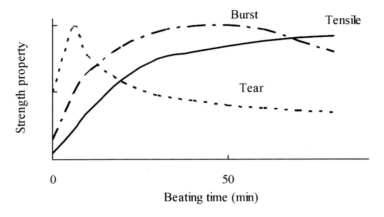

Figure 9.6 Effect of beating time on mechanical strength of paper

The beating/refining step in the papermaking process is also the step where the addition of chemical additives is done (sometimes called as internal sizing in relation to surface sizing in later step controlling moisture penetration). Several nonfibrous additives to increase performance and to reduce cost of manufacturing are commonly used in the production of cellulosic materials for food packaging and a list of them is presented in Table 9.5. Some sensorial and potential safety problems of packaged foods are related to the possible transfer of such chemicals, particularly as residuals in recycled paper and board used in folding carton manufacturing [8, 9].

Table 9.5 Possible nonfibrous chemical ingredients in paper and board making

Function	Additives
Filler	Calcium carbonate, silicates, talc, barium sulfate, caolin, celite,…
Sizing	Rosin, glues, synthetic resins, waxes, silicones, CMC, starch,…
Coloring	Titanium oxide, various colorants and pigments, whitening agents,…
Moisture proof	Formaldehyde-urea resins, melamine-formaldehyde resins, polyamides,…
Plasticizer	Gums,…
Technological aids	Antifoaming, draining agents, floating agents, chelating agents,…
Other additives	Softeners, strengtheners, flame retardant, antimicrobial,…

- *Papermachines* – Once the refining process is completed, a diluted suspension of fibers (less than 1% of solids) is added to the machine where the cellulosic sheet will be prepared. The majority of the modern continuous machines in use around the world still work on principles of the first papermachines invented about two centuries ago in Europe and are commonly indicated as *Fourdrinier* (after the name of the English owner of the first French patent on this system) and *cylinder* machines. The consolidation of the cellulosic sheet occurs by draining the diluted suspension onto a fine screen in the Fourdrinier system or on rotating wire-mesh cylinder connected to a vacuum pump able to aspire the water from the suspension in the second system. Fourdrinier machine is mainly used for paper and thinner board. The cylinder techniques are used for the production of heavy multilayer sheets of board. The fibers in cellulosic sheet on the machine has a tendency to run along the wire direction (machine direction), which is also called the grain of paper [10]. Natural tearing goes along the grain direction, while the paper curl due to expansion with moisture is across it.

The thickness of the sheet produced, or its grammage (g/m^2, see Section 3.4.1), is related to amount of fibers laid per unit surface, and it is modulated by the number of head boxes from which the diluted fiber suspension is provided in the *Fourdrinier* machine. The thickness of paper and paperboard in the USA is usually expressed in *points*, which are thousandths of inches. In the USA base weight is generally used, which is the weight in pounds of a standard ream of 500 sheets with 24 x 36 inches (total area: 3000 ft^2) for papers and 12 x 24 inches (total area: 1000 ft^2) for paperboard [4]. The basis weight and thickness at times are the main criteria used around the world to discriminate between *paper* and *paperboard*, but the borderline is not clear unfortunately with no agreement between standards or countries. According to the ISO standard board or paperboards are materials having grammage higher than 200 g m^{-2}, and in spite of *globalization* the limit is > 250-300 μm of thickness in UK or other countries [1, 7, 11, 12]. In Italy a three level classification is also used where the paper has basis weight less than 150 g m^{-2}, cards are between 150 and 400 g m^{-2}, and boards higher than 400 g m^{-2}. The system of *head boxes* also permits to give different textures on the two faces of the sheet even though many more modifications of appearance and partially the cellulosic material properties are obtained on the subsequent step of sheet drying in the papermaking process.

- *Pressing and drying* – When the sheets leave the papermachines, they still have a moisture content of 75-90%, thus they have to be submitted to a drying operation in two steps: first, a mechanical pressing by means of rotary presses with internal suction reduces the water content to about 60%, then a number of steam-heated rolls bring the paper to the final moisture content that ranges below 10%. It is worth underlining that from an energy consumption point of view this is the most expensive step in the cellulosics manufacturing process and a large amount of byproducts are used as fuel to eliminate refusals and get the thermal energy needed. The appearance of the surface can be changed through the mechanical and thermal operations of pressing and drying, leading also to interesting properties like those of *creped* papers or machine-glazed papers. The creping is made by slowing down the press rolls relative to line speed.

The water that remains in the celluosic materials is both bonded to cellulose and free in the interstices of fibers having a great importance for final products, particularly for food packaging. Paper-moisture relationships are complex and the rate and amount of water regain depend on the used raw materials as well as the process and the finishing operations applied. However in a broad range of relative humidity change (i.e., 5-90%) the moisture content might increase by 15-25% and the dimension (in the machine direction) might increase by about 1%. The moisture content of cellulosic packaging materials can affect not only dimensional stability or mechanical resistance (see Figure 9.5) but also the quality and shelf life of sensitive products. Relevant barrier properties of the package can be influenced by the moisture content. At the end of drying step surface sizing of water-soluble polymer like starch is often applied onto the paper surface to improve moisture resistance and other surface property. The surface sizing is usually followed by another drying step to remove the water picked in the process.

9.4.3 Converting and Special Paper Products

As already mentioned, a variety of paper properties can be obtained through pulping and papermaking operations. The most typical paper products for packaging produced from the papermaking process are natural kraft paper made from sulfate pulp. Light brown in color, it is the strongest type of paper and is usually used for heavy weight packaging for sugar, grains, and feed. Bleaching treatment of the kraft pulp produces the bleached kraft paper of white color. It may also be calendared for smooth surface or creped for extensibility. However, the most performing functions of food packaging cellulosic materials are obtained by several techniques of coating, impregnating, laminating, and finishing process that can be collectively called as *converting* and take place after the paper exits the paper machine (often in a different location or factory). According to the papermaking processes and the converting techniques used, a great number of different cellulosic materials are currently available and an exhaustive description of them is beyond the scope of this chapter; however some special papers which are so widely used or important for food packaging, deserve some attention.

- *Coating and impregnating* – Coating is a well known process for covering a surface with a thin layer of useful or functional substance like a moisture barrier, an oil repellent, an adhesive, or a better printing substrate. In case of cellulosic materials the coating equipments are also used for saturating the fibrous network to change not only the surface properties of the material but also the basic properties [13]. The oldest substances for coating papers and boards are paraffin waxes, and the newest are probably the fluorochemical emulsions. In 'wet waxing' process, paraffin waxes, microcrystalline waxes, hot melts, fluorochemicals, and others are applied to settle onto the surface, where the internal penetration into the fibrous network is impeded by the moisture inside and a quick chill of the coated substrate. In 'dry waxing' they are driven into the body of the substrate with aid of heating roll, filling the interstices of the paper substrate. Wet waxing is more appropriate for heat sealing. Folding cartons, cups, corrugated containers, trays, and wrapping sheets can take great advantages from these kinds of coating or impregnating systems giving the cellulosic materials much more resistance to moist and fatty foods (e.g., wrappings and packages for fast foods), hindrance to liquid penetration (e.g., folding cartons for frozen foods), good gas and odor barriers (e.g., packages for pet-foods), stiffness, and/or even just an appealing finish (e.g., satin, orange-peel effects).

- *Laminating* – Board and paper sheets assigned to food contact are frequently combined with other flexible materials to optimize their physical and chemical properties. Papers, boards, aluminum thin foils, and plastic films are put together throughout a laminating operation which also permits the combination of layers of virgin cellulose with recycled materials.

Laminates are produced on continuous machines where two or more plies of materials are combined by means of appropriate adhesives, which can be coated on the surface or provided by an extrusion step. Bonding between plies is the crucial point in any laminating operation: the strength of the bond can vary depending on the process variables and the chemicals used (solvents and glues) can also cause several problems in food packaging applications like sensorial tainting, meeting food safety regulations, and providing the required performance. The general description and discussion of the machinery used for laminating is omitted here, as the same has already been mentioned for plastics in Sections 6.7.8 and 6.7.9, but in Table 9.6 a list of possible laminated structure including paper or board is presented to give an idea of the possible choices.

Table 9.6 Possible laminated structures including sheets of paper or board

Structure	Uses
Paper/LDPE	Bags for sugar, flour, dry foods,...
LDPE/paper/aluminum foil/LDPE	Cartons for milk, water, oil, juices, liquids,...
Glazed board/recycled pulp/virgin paper	Folding cartons for several solid dry foods,...
LDPE/board/LDPE	Folding cartons for frozen foods,...
Glazed paper/aluminum foil/LDPE	Wrappings for spices, butter, condiments,...
LDPE/paper/aluminum foil/LDPE	Bags for biscuits, breakfast cereals,...

- *Greaseproof* papers are resistant to fat and oil penetration and can be used for wrapping fatty foods. The greaseproof property of the papers is obtained through a refining process (very prolonged beating) which leads to an extensive fiber breakage and the closure of many interstitial voids. Unfortunately, nowadays the greaseproof property is obtained mainly by resin saturation or by coating with synthetic materials considering the cost and process effectiveness. Greaseproof paper is also named *imitation parchment* because vegetable parchment is practically impermeable to fats.

- *Vegetable parchment* (parchment was obtained in ancient times from animal skin and has been one of the first materials used for writing and wrapping) has a surprising performance that without any chemical additive the paper is made impermeable to fats and much more resistant to moisture. An unsized paper produced from a chemical pulp is rapidly passed (10-15 seconds) in a concentrated (65%) sulfuric acid bath at low temperature (10-15°C). This way the fibers swell and partially dissolve leading to a gel which fills and glues the inter-fiber voids. After the acid treatment the paper is washed in alkaline solution and in fresh water to eliminate any chemicals, and then dried to obtain a cellulosic material having high wet strength and resistance to grease and oils. The price of the product and the environmental impact of its preparation are reducing its use in food packaging with its substitution by *wet-strength paper*.

- *Wet-strength paper* is the typical parchment substitute. To get its performance of high wet strength and grease resistance, the wet-strength paper is produced by removing the water between hydroxyl and carboxyl groups in cellulose and linking the chains directly by hydrogen bonding (see Figure 9.4). A more convenient way to reach this in wet-strength paper, in comparison with parchmentizing process, is to add several different cross-linked synthetic resins like polyamides. These chemicals are added to the pulp during the manufacturing process in order to cross-link with cellulose, increasing the natural hydrogen bonding.

- *Glassine* is a glossy, translucent, almost transparent paper, reasonably resistant to fats and slightly brittle. Glassine is produced by treating a greaseproof paper through a battery of steam-heated rollers. Both the mechanical compression and the thermal treatment change the cellulosic inter-fiber matrix inducing many more hydrogen bonds and giving a sort of transparency to the paper compatible with an amorphous behavior of cellulose. Glassine is still widely used in food contact applications, particularly for bakery and confectionery products.

9.5 Paper Bags and Wrappings

Paper bags and wrappings are the most traditional and simple forms of flexible packaging and are widely used all over the world for both consumer and industrial products, for food and nonfood items. Bags, in particular, are produced in several different types and sizes for fine powders, particulates, and solid products. They are used alone or as inner bag for folding cartons. They may be produced by a single layer of paper, two (duplex) layers, or multi-walls. Generally the paper bags are made of kraft paper but it is not uncommon to find bags of glassine and other kind of paper like greaseproof or sulfite as well as the plastic laminated or coated paper. The tubes are pasted whereas the bottom and the mouth may be sewn or pasted; the paper can be folded during the bag manufacture to introduce gusset and thus to allow easier filling operation and greater capacity. The design of the bottom is especially important leading to useful features of the bags for filling operation particularly like the SOS bag (self opening satchel) or *automatic bag*. Figure 9.7 shows schematically the different styles of the most common paper bags which are generally produced on specialized bag-making machinery and generally sold as preformed ones. The food manufacturers will fill with foods and close them. The bagging method is also presented elsewhere in this book (see Section 11.5.2).

Wrapping is the simple method of packaging in which a product or some products are enveloped in a sheet of material; this can be done manually or by an automatic machine. It may be a primary form of packaging (i.e., in direct contact with the product) or a way of assembling prepacked products (secondary packaging – multipacks). Several different materials can be used for wrapping, like aluminum foil, plastic films, and even vegetable leaves, but by far the most important and traditional ones are cellophane and papers. This kind of elementary, somehow primitive, form of packaging is still important as primary packaging for several food products like fruit, chocolate bars, butter and margarine, chewing gum, bread and baked goods, pot pies, pizzas, individual candy pieces, trays of meat, produce, and several other items. Generally each product requires a different pattern of folding and therefore a different machine. The final form of the wrapping obviously follows shape and dimension of the wrapped product. The tucking, sealing, and cut-off between packages are mostly done in similar ways as the form-fill-seal pouches and paper bags are done, and the sealing can be done by gluing, thermo-sealing, or just by tightly folding the paper. Section 11.5.1 describes some different patterns of wrapping for different shapes.

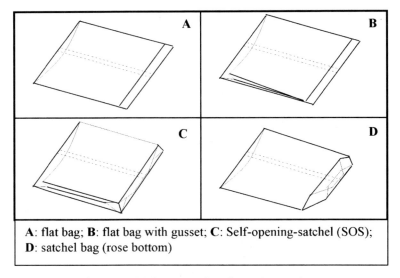

A: flat bag; **B**: flat bag with gusset; **C**: Self-opening-satchel (SOS);
D: satchel bag (rose bottom)

Figure 9.7 Main types of preformed paper bags

9.6 Corrugated Board and Boxes

Nine products out of ten are shipped in boxes made of corrugated board. Even though this cellulosic material is seldom in direct contact with food (some relevant exceptions however exist, like boxes for take-away pizzas or protective wrapping for fragile bakery products), its position in food packaging is great. In many markets/countries the food packaging sector uses about half of the total corrugated boards produced, and thanks to their exceptional mechanical resistance, transportation can be safe, and due to the standardized shapes of corrugated boxes the logistics of shipment can be economically convenient. It is surprising that such a light material mechanically resistant can be manufactured simply by overlapping at least two sheets of paper. The reason is that at least one of these sheets is corrugated producing a *flute* along the other which is the actual *wall* (Figure 9.8); actually the mechanical features of such a structure is well known and largely used in housing technologies. Flutes normally running from the top to the bottom of the box act like columns that resists the static load above. The flutes also resist against flat crushing like arches providing cushioning and insulating properties [4].

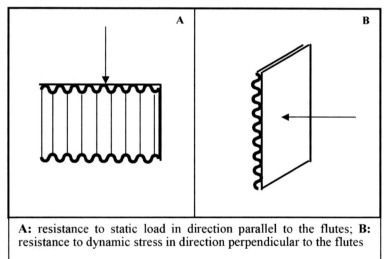

A: resistance to static load in direction parallel to the flutes; B: resistance to dynamic stress in direction perpendicular to the flutes

Figure 9.8 Mechanical resistance features of a corrugated board

The fluting is made by passing the paper sheet onto a profiled roll heated at 160°C after which the *flute* is glued on a plane sheet of the *wall, liner, or facing* by a starch based adhesive. Changing the number of flutes and walls as well as the basis weight and chemical nature of the paper used, makes several different corrugated boards available for different applications, and a typical situation is summarized in Table 9.7. The most common type is the double-faced with C flute. Recycled fibers are used increasingly for manufacturing corrugated boards.

Table 9.7 Possible different structures and characteristics of corrugated board

Types of corrugated board	Structure	Cross-sectional sketch	Main use
Single-face	1 flute/1 wall		Wrapping, cushion
Double-faced	1 flute/2 walls		Standard box
Double-wall	2 flutes/3 walls		Heavy duty box
Triple-wall	3 flutes/4 walls		Heavy duty box
Constructive material	Paper type	Grammage of the paper used	Recycled fiber use
Liners or walls	Natural kraft paper, bleached kraft paper	127 to 440 g m^{-2}	Some virgin pulp
Flutes	Semichemical unbleached paper	112, 127, 150 g m^{-2}	Mostly recycled fiber
Types of flutes[*]	Flute numbers/m	Flute height (mm)	Main usage
A	104-125	5.0	Cushioning wrap
B	150-184	3.0	Small boxes
C	120-145	4.0	Most boxes
E	275-310	1.5	Substitute for folding carton

[*]The alphabetical character (A, B, C, and E) means the type of flute.

The corrugated board may be coated with wax coatings onto the liner and/or medium alone or the complete board [4]. The purpose of the coating is to impart the resistance to moisture and oil, reduce abrasion, and protect printing. The wax coatings consist of paraffin, microcrystalline waxes, and resins (typically polyethylene). Wax-coated corrugated boxes are criticized for being nonrecyclable due to difficulty of separating the wax from the paper fibers.

Single-face corrugated board is sometimes used in the form of pads, wrappers, and inserts like cushioning material. But due to its limited shock absorption and sensitivity to moisture it is less effective compared to expanded or foamed plastics. The corrugated boards are mainly used for manufacturing shipment boxes and also in this case a large number of different shapes and dimensions are available. The sizes for corrugated boxes are always stated in terms of inside dimensions with the order of length (L) × width (W) × height (H); in order to avoid long and unclear verbal description indicating different types a reasoning job has been made by international organizations like FEFCO (European Federation of Corrugated Board Manufacturers) which assigned a numerical code to each type. The most common types (RSC, Regular Slotted Containers and OSC, Overlap Slotted Container) are shown in Figure 9.9.

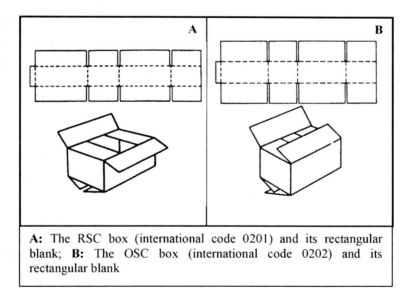

A: The RSC box (international code 0201) and its rectangular blank; **B:** The OSC box (international code 0202) and its rectangular blank

Figure 9.9 The most common styles of corrugated boxes

9.7 Folding Cartons and Set-up Boxes

Solid boards (consisting of one or more layers of the same material, with typical basis weights ranging from 300 to 500 g m^{-2}) plain, laminated, or clay-coated boards are used to manufacture folding cartons. Folded cartons of small and medium sized packages are widely used for a broad range of different products, both as a form of direct food contact and as an outer package for already wrapped or packed products. Folding cartons can be categorized into two standard styles: *tube* and *tray* styles [4]. Tube style cartons are the most common in forms of tubes with end flaps tucked or sealed. Tray styles consist of solid base and walls sometimes with an extending lid. There are a variety of carton designs with creative features whose description is beyond the scope of this book.

The manufacturing of folding cartons starts from a web or a sheet of already printed boards which is then cut and creased. After the separation from the web the blanks produced may be folded, glued on the longitudinal side, and then flattened. They then reach the machinery where they are erected from the flat condition, closed or sealed on one end, filled with the content, and then closed/sealed on the other end. The formation and closing are attained or assisted by heat sealing, interlocking between tabs and slots, gluing, and/or taping. The folded carton blanks with a manufacturer's joint in a flat condition can be erected and filled in separate cartoning machinery. When selecting a cartoning system several different factors should be considered both regarding the machinery needed and the products involved [14]. The filling operation is vertical if the product is free-flowing (like powders or small particulates) or horizontal if the product is solid and shaped (see also Chapter 11); in the latter case the operation may be a wrap-around packaging step. There is a wide variation of cartoning and filling operations depending on product type, carton design, and machines available, which cannot be described in this chapter.

Whatever the system used, the fundamental and particular step in manufacturing folding cartons is the cutting and creasing of the paperboard. Both operations take place in the same machine and at the same moment by using a die with steel cutting knives and creasing rules fitted in a plywood form (Figure 9.10). Cutting and creasing paperboard require great precision with limited tolerance for the blanks to be erected, filled, and closed by automatic devices. Carton is usually designed for grain direction to run around the carton perimeter or cross the corners for maximum carton stiffness [15]. Furthermore, in order to optimize the board consumption and the possible stiffness of the future packages, the fitting of the blanks on the paperboard sheet must be designed very carefully and remains a crucial step of the process (Figure 9.11). In some cases these packages used as a secondary package to protect a primary package have been considered as useless forms of 'over packaging' but actually the segment of folding cartons is continuously growing because of their ability to assure safety of the products, convenience to the consumers, and motivation to the selling.

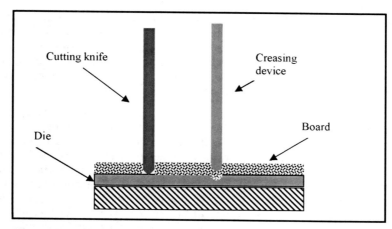

Figure 9.10 Dies for cutting and creasing in folding carton manufacture

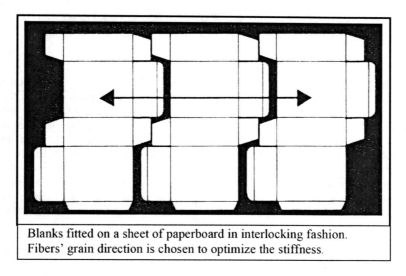

Blanks fitted on a sheet of paperboard in interlocking fashion.
Fibers' grain direction is chosen to optimize the stiffness.

Figure 9.11 Design of blanks on the paperboard sheet

Practically the folding carton materials and technology are often combined with other useful packaging forms: tray cartons are used inside flexible plastic pouches and 'carrier cartons' are typical examples of wrap-around system used for multipacks of cans, bottles, jars, and many other primary packages. There are also several variants of paper cartons or tubs. Cartons may be lined with flexible plastic bag. Paperboard tray coated with polyethylene terephthalate (PET), polypropylene (PP), or acrylic coatings of about 30 μm thickness are used for microwave heating. Corners are made round to avoid sharp localized heating (see Chapter 14 for microwave packaging). Some paper tubs or cups coated by polyethylene (PE) or wax are used for containers of ice cream and dessert, and also as serving containers for hot and cold drinks in vending machines. The PE and wax coatings provide the resistance to oil and water. The empty paper tubs or cups are moved to the service place and stored in nested stacking.

Set-up boxes are produced in a finished shape with a bottom of nonbending paperboard (0.4-1.6 mm thick). They are manufactured through the stages of cutting of the blank board, cutting out the corners of the blank, folding up of the box sides and ends frequently by help of gummed paper tape, and fitting between the base and lid. Covering paper may be glued on the surface of the board structure and/or along the corner for decorative image and strengthened design. Set-up boxes are not only of rectangular shapes but also of others like hexagon or heart forms. They are not suitable for high speed production but may offer a convenient long-term storage and distribution of delicate food items [15]. Set-up boxes are usually made from thicker paperboard and are stronger compared to folding cartons. They also have a disadvantage of occupying space even in empty state from the package manufacturers to food processors. Set-up boxes are usually used for quality candies, chocolates, and bakery products.

9.8 Composite Cans and Fiber Drums

Solid boards are also used to manufacture cylindrical packages (composite cans and fiber drums) which find useful applications for packing solid and dry products, e.g., powders and snacks. The bodies of the cans and drums are constructed by two methods with adhesive-bonding of the layers: *spiral winding* which produces a cylindrical tube by winding several webs of paper or laminated layer around a stationary mandrel, spiraling the material fed from reels in a continuous way, and *convolute* or *straight winding* that produces the tube, winding individual rectangular sheets around a rotating mandrel. The latter is stronger than spirally wound packages but more expensive with low production rate. Usual shapes of the cans and drums are cylindrical, but rectangular bodies with smooth corners are also produced.

Small size cylindrical packages (3-17 cm in diameter and 3-33 cm in length) with paperboard body of protective liner materials are called composite cans. Their bodies are made of paperboard or paperboard laminated with aluminum foil and/or plastic film to offer the strength and barrier to gas and moisture. Aluminum foil or plastic film layer in the structure can offer protection against moisture and oxygen. Composite can body usually contains a high percentage of recycled materials resulting from an environmental push. The closures are of paperboard, plastic, or double-seamed metal ends with lever lid, snap-on, or ring-pull lids. The plastic closures may be equipped with a dispenser or sifter for convenience. The metal ends may be wedged with a tear strip [4]. A flexible diaphragm may be located as a secondary functional barrier and another tamper-evident feature in dry food packaging. An optional plastic overcap allows consumers to reclose the can after initial opening. Composite cans are extensively used to package food products such as snacks, dry infant formula, nuts, ground coffee, candy, refrigerated dough, and pastries. Even though they are not as strong as metal cans, composite cans have strength enough for usual food distribution channel, with less cost and lighter weight. In case of chilled or frozen products they protect the products with better insulation against abrupt exposure to abused temperature conditions. Recently they are used with modified atmosphere packaging. A disadvantage of the composite cans compared to metallic cans has been inability to withstand wet thermal processing conditions with water and steam. However, the composite cans have been developed to be used for aseptic packaging of liquid foods and beverages thermally processed.

Lightweight fiber drums are able to contain and transport both dry and semi-liquid foods [11]. The drum capacity may range from 20 L to 280 L. The drum ends and closures are based on fiber, metal, or plastic material. The closing method may be with a lever-locking band, metal lug, tape, or plug-in. Depending on the closure, the top-rim of the drum may be reinforced by metal. Inside the drum, liners of PE, PET, or aluminum foil may be incorporated as a laminated layer or separate insert.

9.9 Cartons for Liquids

About 100 billion liters of liquid foods are packed each year, worldwide, in paper-based containers that are different in shape, dimension, and many other minor details. Different laminate structures dependent on shelf life required are used to provide relevant barrier properties. The structure with a thin aluminum foil (even less than 7 µm) gives barrier properties required for a longer shelf life, whereas the structure coated only with a heat-sealable polyethylene layer is used for short shelf life products. These laminated structures whose typical schematic layer combinations are presented in Figure 9.12, lead to manufacturing of containers in a range of volumes between 150 mL and 2 L with shapes like those depicted in Figure 9.13.

The classical tetrahedral shape represents a smart packaging idea because that format is easily achieved on automatic form-fill-seal machinery by means of two seams at right angles to each other performed on the tube obtained by the longitudinal seal on the laminate unwound from the reel (Figure 9.14). Moreover, the tetrahedral shape has a surface/volume ratio lower than any other format, leading to a significant saving in packaging material, but unfortunately that shape is not comfortable during handling and has poor visibility of the possible advertising messages on the shelf. Thus it is almost given up in the most developed countries. The cartons of tetrahedral and some parallelepiped shapes are form-fill-seal containers which derive from a reel of laminate and are manufactured in the same machine where they are filled with liquid food. This feature is extremely important for aseptic packaging operations as the web material can be easily sterilized by passing through a hydrogen peroxide bath.

Recently these cartons adopted a heat-resistant structure and extended their application to retortable foods with particles which are sterilized in retort after filling (e.g., Tetra Recart®).

LDPE/paperboard/LDPE

LDPE/paperboard/ LDPE/ aluminum/LDPE/ LDPE

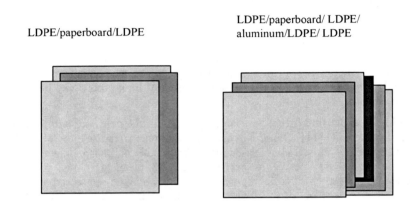

Figure 9.12 Design of laminate layers on the paperboard carton

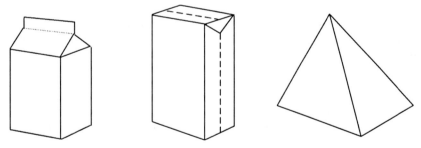

Figure 9.13 Most common shapes of cartons for liquids: gable top, brick, and tetrahedron styles from left to right

Other cartons are preformed containers and reach the filling machine in a similar way as the folding cartons do. The most common shape of such preformed containers is a parallelepiped one with a square section and a gable top, which also presents a convenient reclosing system and useful spiller. This format of carton for liquid is much older than the classical Tetra Standard® (the first tetrahedral container, 1951), as the first gable top style of preformed container made of a wax coated paper was patented in the USA in 1915 well before the 'plastic revolution'. Compared to form-fill-seal cartons, preformed cartons are thicker, generally more expensive and give higher filling rates. Preformed cartons contain some air or inert atmosphere in the head space, whereas the form-fill-seal containers may be even completely deprived of a head space; i.e., the transverse seal is done under the filling level, leaving no air in the headspace of the container which is completely full of the food.

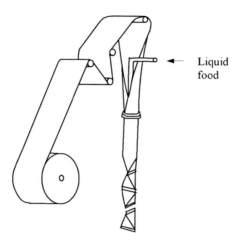

Liquid food

Figure 9.14 Concept of making tetrahedral packages (Tetra Classic®). Adapted from Ref. [16].

For several reasons this kind of package is important in food packaging because huge number of pieces are sold worldwide and also because, as a container for liquids, they come in contact with somehow perishable foods like milk, wines, juices, and sauces, which are sensitive to environmental effects. Consequently, light irradiation, oxygen permeation, and external microbial contamination are serious problems to face [9]. Moreover, the contact of the inner layer (generally of LDPE) with such liquid foods sometimes leads to unwanted phenomena of adsorption/migration that can modify the sensorial and hygienic properties of the foods [9, 17].

Finally, the appearance of these packages led to two epoch-making breakthroughs: a technological invention and a social event of the last 40 years. The technological invention is the aseptic packaging which appeared around 1961 in Europe when the first Tetra Standard® aseptic packaging machine was introduced (see Chapter 12). The aseptic packaging consequentially experienced a tremendous growth in areas of different materials and containers. The social event is the progressive shift from the returnable system to the one-way packages that started with the substitution of milk glass bottles with the cartons for liquids. The one-way carton packages then grew so quickly all around the world in different areas. This logistic revolution was great and fast in many countries for the economical advantages it brought, and attracted much attention on the themes of resource control and ecology considerations. Taking into account the comprehensive perspectives on entire life cycle of the packages the one-trip cartons have been argued to be comparable to a returnable bottle in several real circumstances [16].

9.10 Molded Cellulose

All the cellulosic packaging considered up to now are obtained by folding and gluing sheets of paper, boards, or laminates of cellulosic materials, but the network formation of cellulose fibers is also exploited in manufacturing three-dimensional packaging that are generally called *molded pulp* or *molded cellulose*. They are fabricated directly by squeezing aqueous fiber suspension onto screen mold to remove water. Probably the most well known packages of this type are the one dozen egg carton and bottle sleeve. The other items that deal with food packaging are trays for meat, fruits, vegetables, and cafeteria servings. The mold often fits shape of foods to be packed.

There are basically, two preparation methods of molded pulp items, which are different in the production speed, total cost of management, and accuracy in shape's details. The *pressure injection process* forms the article by blowing hot air pressurized to 0.4 MPa into the mold containing pulp and water mixture. The molded article is subjected to further drying with sterilization effect. The *suction process* applies vacuum to remove water from the slurry of cellulose fiber in the mold. The formed article has higher residual moisture and thus is subjected to subsequent drying. When compared to the latter process the pressure injection molding is generally slower and produces less smooth and denser products. The smoother surface of the mold from the suction process may be used for printing.

Molded pulp packages have rigid thick walls (0.2-6 mm) but are light in weight (density: $0.2-1.0$ g cm^{-3}). They can easily absorb shocks and resist against stacking stress. They are made of very low-cost raw materials and can be a good application for recycled matter. They can also be easily recycled after use. For these reasons the molded pulp packages are seen as *green packages* and favorably accepted in some niche markets. Nevertheless, this material has also interesting functional properties: besides its shock absorption ability, it is permeable to water vapor and shows a high gas transmission rates, together with a low selectivity (the similar rates of gas permeability to carbon dioxide and oxygen). These are properties especially useful for respiring products like fresh fruits and vegetables (see Sections 13.3 and 17.8).

9.11 Cellophane

Cellophane was originally Du Pont's brand name of clear, thin film obtained by a chemical conversion of very pure cellulose, which in fact comes from two French words meaning *transparent cellulose*. This product which may be considered as the archetype of all modern plastic films was invented by a Swiss textile engineer (E. Brandenberger, 1908) whose true objective was to find a way to make cloth waterproof. Serendipitously he developed a completely new material that for a long time represented the only transparent wrapping material. Twenty years later Du Pont (1927) patented a cellophane lacquer coating that made it possible to use cellophane for food packaging. The manufacturing process is quite complex: a very pure cellulose pulp (often derived from a sulfite wood pulp) is treated in sequential processes with caustic soda solution and carbon disulfide so that fibers are partially dissolved and the cellulose partially depolymerized to form a continuous web of solid regenerated cellulose in a sulfuric acid and sodium sulfate bath. The polymer recovered is mainly in a linear and shorter form of highly closed macromolecules, and it becomes the well-known clear film after some steps of purification, softening, and drying (generally to 7-8% moisture).

All the physical properties of the cellophane sheet largely depend on the degree of depolymerization, the viscosity of the slurry, the time of ageing or ripening, and the softening step. Uncoated cellophane has very limited applications in nonfood packaging, whereas the coated one covers a broad spectrum of food packaging functions. A moisture proof coating is an essential requisite because both moisture absorption and moisture loss are really detrimental to cellophane performance: an excessive water absorption reduces dramatically the gas and aroma barrier properties of regenerated cellulose, whereas the sheet drying invokes shrinkage and brittleness. The most common coatings for better performance are polyvinylidene chloride (PVDC) and nitrocellulose. In the 70-80s the packaging market of cellophane was hugely reduced by the growth of oriented polypropylene (OPP), because this plastic has similar properties of clearness and mechanical performances but costs much less. However, coated cellophane possesses several excellent properties that should not be underestimated: the package forming efficiency is very high because it retains dead-fold during twist-wrapping and transfers heat efficiently during sealing steps without any distortion; the gas and aroma barrier is comparable to those of the so called barrier polymers; the market appeal of a cellophane pouch is always high for its stiffness, clarity, natural antistatic behavior, and good printing attributes. In Table 9.8 the properties of uncoated and coated cellophanes are compared with those of an OPP film.

Despite its high price, cellophane is still broadly used for packaging candies, chocolates, cheeses, snacks, and dried foods. Its ecological image of biodegradability and using renewable resources helps its stay in food packaging.

Another derivative of cellulose used in packaging is cellulose acetate whose properties differ with acetylation degree and plasticizer type used. It has very high clarity and gloss, and thus is mainly used for carton windows or blister packages. Cellulose acetate usage in these applications has mostly been replaced by polyethylene terephthalate (PET), polypropylene, and polystyrene.

Table 9.8 Properties comparison between OPP and cellophane films of 25.4 μm thickness

Property	Uncoated cellophane	PVDC-coated cellophane	Nitrocellulose-coated cellophane	OPP (homopolymer)
Specific gravity (g cm^{-3})	1.5	1.5	1.5	0.88-0.91
Haze (%)	3.0	3.5	3.0	2-4
Elongation (%), machine direction	16	22	16	460
Elongation (%), cross direction	60	60	60	35
Tensile strength (MPa), machine direction	124	124	124	160
Tensile strength (MPa), cross direction	55	55	55	260
Heat seal range (°C)	-	165	170	162-166
O$_2$ permeability[*] 50% RH	45	7.5	30	2000
O$_2$ permeability[*] 95% RH	750	7.5	150	2000
Water vapor transmission rate[**]	1500	7.0	7.5	4-10

[*]cm^3 m^{-2} 24 h^{-1} bar^{-1} at 23 °C, [**]g m^{-2} 24 h^{-1} at 38°C and 90% RH.

9.12 Other Quasi Cellulosic Materials

9.12.1 Wood

Wood has been used in packaging or containing foods for long time mostly in relation with distribution and shipment. Wooden pallets are the most universal form of unitizing the load in distribution of the products. Wooden boxes, crates, and baskets have long been used for handling fresh fruits and vegetables. Wooden barrels are still used for storing or ripening alcohol-based liquids. Wood materials are used for supporting the package structure by their natural mechanical strength. The cost and durability of wooden containers depend on the choice of wood species: the denser and stiffer woods are generally more durable and expensive[18]. In rare cases for dimensional stability and strength at affordable cost, plywood and fiber board fabricated from wood chips are used for boxes and containers.

In Japan and Korea, wood packages specific for certain types of dry foods such as rice biscuits, teas, and dry mushrooms are used for high quality image. Sometimes bamboo baskets are used for traditional dry cakes and fruits. Many types of secondary packaging made of wood are also widely used for expensive food items such as honey, seasoned fish sauce, dry tea, and confectionery.

9.12.2 Cork

Predominant usage of cork in food packaging is in wine bottle closure and liner material in the closure. Their elastic and compressible structure with low thermal conductivity is ideal for this kind of function. When compressed onto the container mouth, it confers the barrier to the oxygen ingress and moisture loss.

Most corks are acquired from the cork oak tree in Spain and Portugal. Cork suitable for wine bottle stoppers can be obtained from trees more than 30 years old. The cork bark taken from the tree is dried for about 6 months and boiled for about 90 minutes to shrink the internal pores and kill the insects and molds. After resting for 3 weeks, it is cut and punched into final shape with direction that any knots or fissures are across the cork column to prevent leakage through them. With a diminishing supply of cork at a reasonable price, cork as a wine stopper is being widely replaced by elastic plastic cork having a microporous structure. However, natural corks are still favored because of traditional custom.

Beyond the price, one major problem of natural corks is their sensitivity to fungal contamination liable to impart the so called 'corky taste'. Several different substances can be transferred from corks and give unpleasant flavors, but the most well known is 2,4,6-trichloroanisol (TCA). TCA is produced by microorganisms that methylate the 2,4,6-trichlorophenol, a normal constituent of cork phenol. An interaction between packaging material and microorganism, followed by the subsequent migration even without contact (see Section 5.2.1), is the phenomenon responsible for this serious and common product tainting.

9.12.3 Nonwovens

Nonwoven sheets or webs are made by bonding the fibers together by mechanical, thermal, or chemical means. They are flat, porous structures that are made directly from separate fibers or from molten plastic or plastic film. They are called nonwovens as they are not made by weaving or knitting. Even though paper is a kind of nonwoven according to its process of manufacturing, it is not conventionally called nonwovens. Rather, nonwoven materials usually refer to other nonwoven fabrics or sheets as alternatives to conventional textiles. Nonwovens can provide a variety of unusual functions such as absorbency, liquid repellency, resilience, stretch, softness, strength, cushioning, filtering, bacterial barrier, and sterility, which are very useful in food packaging.

Both cellulosic and synthetic fibers may be converted into nonwovens. Bleached cotton fibers and wood pulp fibers are bonded to nonwovens for use as disposable diapers and other absorbent materials by mechanical intertwining of fibers through water jet or air stream. A nonwoven pulp fiber web has a flow pore size of 18-100 μm, low density of 60-120 kg/m^3, and total absorptive capacity of 500%, which may be useful for liquid absorption purpose [19]. The most popular polyolefin fibers for nonwovens are polyethylene and polypropylene. These polymers are either converted into a few cm long fibers which are subsequently converted into nonwoven web with help of binder or heat, or else converted into spunbonded nonwovens by extruding the polymers to form filaments which are formed into webs and bonded by thermal processes. The largest use for a nonwoven sheet in food packaging is the microporous sachet for oxygen or moisture absorbents, which are highly permeable to surrounding oxygen and water vapor. Du Pont's Tyvek® is a typical product of nonwovens for this purpose.

Discussion Questions and Problems

1. What are the benefits and advantages of paperboard packaging compared to plastic packaging?

2. How can you preserve or improve the strength properties of a carton box when it is exposed to humid weather conditions?

3. What kinds of function do the molded pulp containers offer for food packaging?

4. Survey in the fresh produce market and suggest the best type of corrugated box package for 10 kg of apples. Provide reasons for your suggestion.

5. How can you design a paper-based package for sweet candy?

6. What would be the specific properties of plastic cork required to replace natural cork in wine packaging?

7. What are useful properties and main uses of nonwoven films in food packaging?

8. Construct functions/environments table for a paper carton package of orange juice referring to Section 1.7.3.

Bibliography

1. GA Smook. Handbook for Pulp & Paper Technologists. 3rd ed. Vancouver: Angus Wilde Publications, 2002, pp. 1-356.
2. A Dwan. Paper complexity and the interpretation of conservation research. J Amer Inst Conserv 26(1):1-17, 1987.

3. JC Roberts. The Chemistry of Paper. Cambridge, UK: The Royal Society of Chemistry, 1996, pp. 56-61.
4. D Twede, SEM Selke. Cartons, Crates and Corrugated Board: Handbook of Paper and Wood Packaging Technology. Lancaster, PA: DEStech Publications, 2005, pp. 227-504.
5. GL Robertson. Food Packaging. New York: Marcel Dekker, 1992, pp. 144-172.
6. JF Hanlon, RJ Kelsey, HE Forcinio. Handbook of Package Engineering. 3rd ed. Boca Raton, FL: CRC Press, 1998, pp. 31-57.
7. FA Paine. The Packaging User's Handbook. Glasgow, UK: Blackie, 1995, pp. 36-80.
8. L Castle, CP Offen, MJ Baxter, J Gilbert. Migration studies from paper and board food packaging materials. Food Addit Contam 14(1):35-44, 1997.
9. H Brown, J Williams. Packaged food quality and shelf life. In: R Coles, D McDowell, MJ Kirwan, ed. Food Packaging Technology. Oxford, UK: Blackwell Publishing, 2003, pp. 65-95.
10. M Sumimoto. Paper and paperboard containers. In: T Kadoya, ed. Food Packaging. Boston: Academic Press, 1990, pp. 53-83.
11. MJ Kirwan. Paper and paperboard packaging. In: R Coles, D McDowell, MJ Kirwan, ed. Food Packaging Technology. Oxford, UK: Blackwell Publishing, 2003, pp. 241-281.
12. BW Attwood. Paperboard. In: M Bakker, ed. The Wiley Encyclopedia of Packaging Technology. New York: John Wiley & Sons, 1986, pp. 500-506.
13. D Satas. Coating equipment. In: M Bakker, ed. The Wiley Encyclopedia of Packaging Technology. New York: John Wiley & Sons, 1986, pp. 186-191.
14. FA Paine, HY Paine. A Handbook of Food Packaging. Glasgow, UK: Leonard Hill, 1983, pp. 113-121.
15. W Soroka. Fundamentals of Packaging Technology. Hernon, Virginia: Institute of Packaging Professionals, 1995, pp. 89-141.
16. C Nermark. Packaging of beverages in drinks cartons. In: GA Giles, ed. Handbook of Beverage Packaging. Sheffield, UK: Academic Press, 1999, pp. 139-164.
17. JO Choi, F Jitsunari, F Asakawa, HJ Park, DS Lee. Migration of surrogate contaminants in paper and paperboard into water through polyethylene coating layer. Food Addit Contam 19(2):1200-1206, 2002.
18. D Twede, R Goddard. Packaging Materials. Surrey, UK: Pira International, 1998, pp. 44-47.
19. HK Barnes, RF Cook, CH Everhart, AL McCormack, FR Radwanski, PM Rosch, AJ Trevisan. Hydraulically needled nonwoven pulp fiber web. US Patent 5137600, 1990.

276

Part Two

Packaging Technologies

Food packaging technologies may be described as a set of practical solutions, preferably science-based, for delivering high quality and safe food products to the consumer in an efficient manner. The next six chapters are devoted to some technologies important for food packaging.

Chapter 10 describes 'end-of-line operations' including printing, labeling, and sealing. Typically those operations are performed at the end of a line for producing roll stocks of a packaging material, at the end of a line for producing aluminum cans, at the end of a line for producing preformed pouches, and so on. Chapter 11 describes operations after empty containers and pouches have been made and are ready for filling of the food and sealing of the package.

Chapter 12-15 describe packaging technologies for safety, quality, or convenience. Retortable packaging and aseptic packaging are technologies for ensuring the microbial safety of food products. Vacuum/modified atmosphere packaging is for keeping the quality or extending shelf life of food. Microwaveable packaging is designed for providing convenience to the consumer. Active and intelligent packaging is a set of advanced technologies for enhancing food quality and safety.

Chapter 10

End-of-Line Operations

10.1 Introduction 279
10.2 Printing 279
    10.2.1 Image Printed ··280
    10.2.2 Conventional Printing Methods ·····································281
    10.2.3 Digital and Novel Printing Methods ······················288
10.3 Label and Labeling 291
    10.3.1 Label Types and Materials ·······································292
    10.3.2 Labeling Process ···294
10.4 Coding 296
    10.4.1 Code Imprinting···296
    10.4.2 Bar Code ··297
10.5 Sealing of Plastic Surfaces 301
    10.5.1 Heat Sealing··301
    10.5.2 Cold Seal ··307
    10.5.3 Seal Strength Test··308
10.6 Case Study: Finding Sealing Conditions for an LLDPE Film 309
Discussion Questions and Problems 311
Bibliography 312

10.1 Introduction

Even though packaging materials or packages are made or fabricated by container manufacturers, converters, or food processors, they should pass through decoration and final fastening steps for functioning properly as food packages. Most food packages are printed, labeled, and coded for transfer of information to the consumers. The format and structure of written information on the package is under legal control. The decoration also works for aesthetic and marketing purpose. Sealing of plastic bags on the side and bottom forms the package structure and also provides hermetic separation of product from the environment. All these operations usually constitute a finishing step integrated in the line of package formation, even though there are wide variations in their combination with the fabrication steps. Basic concepts and principles of the decoration and the protective sealing will be reviewed in this chapter.

10.2 Printing

Printing can be performed in sheet or web forms before container or bag fabrication. The fabricated containers may also be printed after forming or even after filling and sealing in rare cases. Substrates for printing include paper, plastic, metal, and glass surfaces. Printing on the package or a label plays a role to convey product information and facilitate marketing. The image is usually printed onto the front side of package surface, and the printed surface may be coated by transparent coating or varnish for protecting the image and also enhancing its glossy appearance. Clear plastic films are often printed on the back side with reverse image, known as reverse printing or subsurface printing, which is laminated onto a backing film or sheet layer. Metallized back film is often applied to give a reflective metallic background for the image. Printing methods depends on the package type, substrate, desired quality, speed, cost, and coordination with other converting steps.

10.2.1 Image Printed

Most artwork in package consists of text and images in multi-color. Theoretically any color can be produced by mixing three primary colors of cyan, magenta, and yellow [1]. However, combining only these three colors cannot produce perfect absorption and reflectance of relevant light wavelengths and thus has limitation in reproducing color with desired hue and brightness. For this reason, black is usually added to attain the desired contrast and shadow effect. Therefore at least four colors (cyan, yellow, magenta, and black, also known as CYMK) need to be combined in order to produce any multi-colored image. Sometimes six or more colors are used for quality printing.

Solid text and design of a single hue can be printed by lay-down of single ink from a printing plate. Half-tone image with different degree of color saturation is produced by applying fine patterns of dots and/or lines on a white background. Paper substrate or the coating of titanium oxide for metal can is often used as white background. Varying degree of dot/line size and density gives different degree of saturation and appearance. Dot size in printing is represented by dots per inch (dpi), also referred as screen. Density of printing refers to fraction of substrate surface covered by the ink.

Reproduction of a full color image in the printing starts by separating it into its component colors, typically trichromatic ones. In the conventional printing processes, four photographic negatives (three primaries and black) are prepared to develop four halftone printing plates, one for each color component. Printing all four components in an exact position or registering over each other reproduces an original image. More than four color components are often used to produce a variety of color attributes. These days, electronic and digital tools are increasingly adopted in various steps of data transfer, printing plate preparation, and final image generation. The use of digital tools in package printing is expected to increase over time.

Printing ink consists mainly of pigment, resinous vehicle, solvent, and additives. Broad range of colored pigments exists, which are mostly composed of metal oxides and organic complexes. Compounds derived from heavy metals such as lead, mercury, and cadmium has almost been eliminated from ink pigment lists due to safety concerns. The vehicle is usually dissolved in solvent, and plays a role in binding pigment particles and adhering them to the substrate. The solvent fluidizes the ink formula and wets the substrate for transferring ink. Generally, two categories of solvents are used: water based solvent and volatile organic solvents. The former is less used but poses fewer problems of possible food contamination, the latter is more effective but are sometimes responsible for sensorial damages of packed food. The drying or curing of printed ink is accomplished by evaporation, absorption, and chemical reaction. While most solvents used for dissolving ink are dried by evaporation, absorption can aid drying and stabilization of ink on an absorptive substrate such as paper. Chemical reactions such as oxidation or polymerization aided by heat or UV light can help the vehicle to dry and/or solidify. Radiation curing technologies use electron beam, ultraviolet light, or visible light to polymerize reactive and usually solvent-free inks or coating materials. UV is the radiation source most frequently used to cure these coatings, accounting for about 90% of the volume and market. A photoinitiator ingredient in the ink formula absorbs light and generates free radicals that start the chain reaction of polymerization. The risk of photoinitiatior residues, like 2-isopropyl thioxantone (ITX), in package and migration into food is of concern causing the need for identification of new and safer molecules. The processing and distribution conditions should also be considered for selecting an ink system for food packaging. A highly thermal-resistant ink system is required for retortable food packaging. A water-based solvent system is desirable to reduce environmental problem of volatile organic compound (VOC) emission during drying. The petroleum-derived organic solvents should be dried completely in the final package or their use should be minimized, as they can migrate to the food product causing off-flavor and health issues. Vegetable-oil based inks are also used for printing.

Toners used for digital printing is a pigmented powder made up of microscopic particles. They are available in dry or liquid form. Dry toners may be a single component of toner particles or dual component with toner particles and powder carrier helping its transport and binding to the substrate. Liquid toners consist of a pigment and liquid carrier, usually hydrocarbon/mineral spirit. Dry toners are fused onto the substrate by radiant heat or heat pressure. Liquid toners are absorbed by hot transfer or hot air. Liquid toners generally provide better printing quality such as bright color, small particle size, and high transparency, which is similar to conventional printing methods.

10.2.2 Conventional Printing Methods

The conventional printing processes follow three major steps: conversion of original art into image carrier, presswork of final image print, and finishing operation. Printing image carriers are processed to create two separate areas each for printing and nonprinting. The image carriers are in forms of plate, cylinder, or stencil made from metal, rubber, photopolymer, or composite materials, depending on printing method. The image areas can be created by exposing the assembled film onto photo-sensitive area, laser engraving, or digital or chemical transfer [2]. Printing area on the image carrier accepts the ink by physical or chemical means, and then transfers the image onto the substrate. The transfer of image from the image carrier to the substrate can be direct or indirect. In indirect printing sometimes called offset printing, the image is transferred first from the image carrier to an intermediate rubber-covered blanket cylinder and then finally moved to the substrate. Conventional printing methods can be classified into four types: relief (flexography or letterpress), gravure (intaglio), lithography (planography), and screen.

In relief printing, the image to be printed or the printing area containing ink is raised above the surface of the printing plate. The image is then transferred to the substrate which is pressed against the printing area. Letterpress is the oldest and simplest form of relief printing. Flatbed letterpress made of movable metal plate is directly inked and then transfers the image to the substrate. It is used mostly for printing labels and tags. Flexography relief printing is a variation of letterpress used mainly for packaging applications such as papers, plastic films, foils, corrugated boards, and labels. It uses a cylindrical rubber printing plate contacting anilox roller, which controls the amount of ink transferred (Figure 10.1). Engraved cell depth in anilox roller and doctor blade on its surface help to control the amount of ink. The printing plate is rotated for contacting the substrate which is pressed against the impression cylinder. The squeezing action during image transfer may produce a light 'halo' effect along the solid edge, which makes it difficult to attain a sharp image. Individual color station is completed by drying and followed by stations for other colors. Water-based or organic solvent-based inks are used: use of water-based inks can reduce air pollution, which is an advantage of flexography. It is widely used for printing plastic films and paper materials where moderate quality is affordable. Corrugated paperboard and flexible plastic web roll are substrates frequently printed by flexography. Printing plate for flexography has screens normally of 60-120 dpi.

Flexographic printing of a full-color image needs at least four printing stations separated by individual drying sections. The arrangement of printing stations follows two typical types: stack press and central impression press (Figure 10.2). The stack press places the printing stations vertically one above the other. The central impression press locates the printing stations around a large central impression cylinder. The stack press is used for paper and laminated films while the central impression press is for high quality webs with minimum distortion.

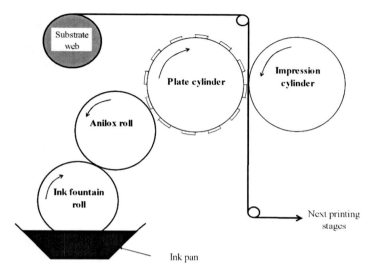

Figure 10.1 Flexographic printing

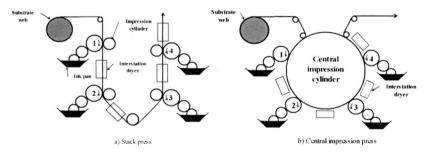

Figure 10.2 Individual color stations arranged to deal with multi-color image in flexography

Gravure printing uses a recessed surface to hold ink on the printing area (Figure 10.3). The image in the printing area is created by chemical etching or mechanical engraving into copper-plated cylinder. Electromechanical engraving is increasingly used with the aid of digital data technology. The image is composed of many small cells with different diameters and depths giving different levels of halftones. The engraved metal surface is filled with thin fluid ink and then carefully wiped out by doctor blade, leaving ink only in recessed area. The printing plate cylinder then rotates in close contact with substrate which is pressed by impression cylinder and absorbs the ink in the cells. Ink transfer from printing plate to the substrate is often assisted by opposite electrostatic charges on respective phases. Each individual printing station is followed by a dryer for evaporation of solvent and multi-color image is produced by multiple stations and dryers arranged in a row. Gravure printing is operated continuously at high speed in web process giving good quality color printing. Therefore highly volatile solvent-based ink is used for quick drying. Because gravure lays down ink as dots from tiny cells, a slightly ragged or dotted outline is formed on a solid edge. Typical gravure screen is around 150 dpi. Gravure printing is characterized as providing consistent good printing quality under long production runs but with high cost for gravure cylinder plate. There are also environmental and migratory health concerns due to the use of organic solvent-based inks. Like flexography, gravure is widely used for high-volume web-printing applications such as cartons, foils, films, labels, bags, and laminations.

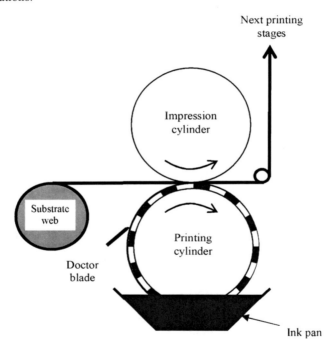

Figure 10.3 Gravure printing

As a variant of gravure printing, pad printing is used as an indirect form to print uneven and irregular surfaces. Like gravure the recesses of the image area are filled with ink and then pressed with a resilient pad. The pad subsequently is pressed against the substrate to transfer the image.

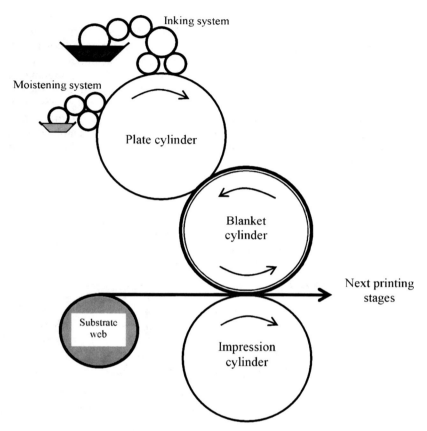

Figure 10.4 Lithographic printing

Lithography uses a planographic printing plate on which printing and nonprinting areas are on the same plane. Separation of printing and nonprinting areas is based on different wettability. Hydrophobic printing area is created chemically to accept oil-based ink and repel water. Hydrophilic nonprinting area is developed to accept water and repel the ink. In operational process, the developed printing plate made of aluminum alloy on the plate cylinder receives the ink from the ink reservoir through a series of rollers (Figure 10.4). The same cylinder also contacts a moistening system. The plate on the cylinder accepts the ink and repels water on the printing area, while the reverse is true for the nonprinting area. Then a blanket plate made of rubber on an intermediate or blanket cylinder takes the image from the printing plate (on the plate cylinder). The image is transferred to the substrate which passes between the blanket and the impression cylinder. The lithography is therefore indirect or offset printing involving intermediate blank cylinder. Lithographic presses are operated on the rotary principle to handle both sheet-fed and web-fed substrates. Multi-unit presses are used for printing multi-color images. A major advantage of lithography is a smooth and sharp image with excellent detail. It commonly attains screen of 130-150 dpi although a finer degree is possible. Lithography is the most widely used printing process for paper packaging materials such as carton stock and labels [1, 2]. It is also the most common printing for metal sheet intended for forming containers [3]. The drying stage of the printed substrate may be accelerated by heat application or by a combination of absorption and oxidation. The inks for metal application are cured in a hot oven or under a UV light.

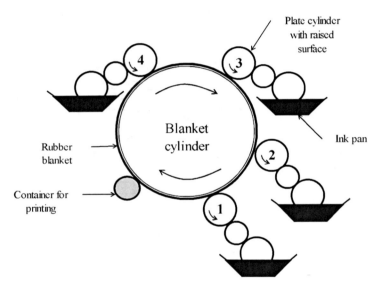

Figure 10.5 Dry offset printing

A combined variant of lithography and flexography, called dry offset, is used to print the objects such as metal cans, plastic cups, jars, and closures. A central large rubber blanket works as carrier of the image like the lithography (Figure 10.5). The blanket accepts the image from the printing plate cylinders whose surface in turn receives it from an ink fountain through a series of rollers. The entire multi-color image is assembled and printed onto a blanket cylinder from successive printing cylinders: the offset blanket is inked by each printing plate, which has the copy for one color. The printing area on the printing plates (plate cylinders) is usually raised for inking like flexography, which deletes the water dampening step of lithography and thus is called dry offset. The blanket roll then transfers the image in complete register to the contacting container in a single operation. In order to supply the pressure required for printing, containers are supported by a mandrel or inflated with air pressure. The color image on the blanket is wet until being transferred to the container. The ink in dry offset is usually dried and cured by heat or UV light, producing less waste. Dry offset printing of a can end uses a raised blanket segment to access the recessed area.

In screen printing, a viscous ink is squeezed onto the substrate through a fine screen made of silk, polyester, nylon, stainless steel mesh, or cloth stretched over a frame (Figure 10.6). A stencil carried on the screen plays a role of image carrier: blocked nonprinting area and open printing area on the stencil is formed photographically or mechanically. In the operation, the substrate material is positioned in close contact with the screen, soft ink is placed on the stencil, and then it is forced through open image area by using a rubber squeegee. Screen printing has a heavy lay down of ink, being good for line art: halftone screen is limited to about 85 dpi, giving coarse degree of reproduction. Heavy ink lay down is costly but desirable for some cases such as ceramic glass coating and light color image on dark colored plastic containers. Usages of UV curable and water-based inks improve the printing quality and environmental friendliness. Screen printing is generally slow in speed and may need extra drying time. It runs in rotary process as well as sheetfed process. It can be used for decorating items that would be difficult to print any other way, including contoured and irregular shapes. The substrates printed include paper, metal, wood, plastic, glass, cork, and fabric [2].

Pressure

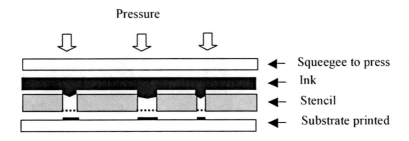

← Squeegee to press

← Ink

← Stencil

← Substrate printed

Figure 10.6 Screen printing

Any printing method has advantages and disadvantages. Table 10.1 compares the conventional printing processes. Figure 10.7 shows magnified characteristics of images printed by different methods.

Table 10.1 Summary of traditional printing processes

Attribute	Relief	Gravure	Lithography	Screen
Printing area on printing plate	Raised surface	Recessed surface	Plane hydrophobic surface	Open stencil on meshed screen
Direct or indirect	Direct	Direct/Indirect	Indirect	Direct
Speed	High	Very high	High	Slow
Screen of halftone	60-120	150	130-150	85
Visual appearance of printed image	Haloed edge	Dotted edge	Clean sharp edge	Meshed body with dotted edge
Environmental friendliness	Good	Poor	Poor	Moderate
Applications	Flexible films, corrugated board	Cartons, foils, films, labels, bags, laminations	Metal sheets, cartons, labels	Glass, metal containers, round tubes

Relief   Gravure   Lithography   Screen   Inkjet

Figure 10.7 Exaggerated characteristics of images from different printing processes

10.2.3 Digital and Novel Printing Methods

Even though the traditional printing processes have been developed and improved a lot for last decades, they have some drawbacks such as a long time to press, high cost at low production, and lack of flexibility [3]. In order to ease the problems in the conventional printings, computers are used to save time and money with better quality and flexibility. Involvement of computers in the printing process covers color separation of the artwork, manufacture of printing plates, and final image transfer to the substrate. There has been a great push for computer driven print system with development of telecommunication data transfer and processing. So the word, 'digital printing' was born. Digital printing can be defined as the process in which the printing press or print engine accepts computer-generated data and converts it into a printed image [3]. The print-on-demand concept has emerged to handle images frequently changing with consumer demand. Typical digital printing is making the print from a computer file without film or plates. Digital printing is ideal for low-quantity and price-per-piece tends not to be dependent on the volume or run-length.

Inkjet offers noncontact printing onto target substrates including irregular shape and recessed surfaces. Basic inkjet process is a sprayed projection of tiny ink drops onto a target surface (Figure 10.7), which is available in two types: continuous flow and drop-on-demand (Figure 10.8). In continuous flow type, the electroconductive liquid ink is forced from a reservoir through a minute nozzle under hydrostatic pressure. Multiple nozzles are required for multi-colored image. With vibration at nozzle orifice, the ink jet breaks up into a stream of tiny, discrete drops. The droplets are passed through an electrostatic field so that the charged ones are directed onto the precise location of the substrate with uncharged ones being collected for returning to the ink reservoir. In the nonprinting cycle all the droplets become uncharged and are deflected to a recycling reservoir. The drop-on-demand type produces the ink droplets only when is required. Earlier version using piezoelectric crystal uses a programmed pressure pulse to push the droplet through the nozzle. When electric pulse is induced onto the crystal, it expands and pushes the liquid ink for creating the droplets. Thermal inkjet, another drop-on-demand type uses rapid cycle of alternate heating and cooling as a means of creating ink droplets. The heating cycle creates a bubble of volatizing solvent (commonly water) which pushes the ink out through the nozzle. The bubble collapses with cooling to suck the new ink into the printing head. The thermal inkjet system is limited to the water-based inks while the piezoelectric inkjet is able to use UV-curable, water-, oil-, and solvent based inks. Water-based inks require specially coated media to optimize drying and dot gain. Solvent-based inks require high printhead maintenance due to evaporation in the nozzles. Oil-based and UV-curable inkjet inks require low maintenance, but the latter are growing in use for adherence to a wide range of substrates. There is a new type of inkjet ink, hot melt ink. It is jetted at a very high temperature ($125°C$) and becomes solid at room temperature. There came out recently a thermochromic inkjet ink, which changes color through heat processing [4]. Inkjet printing is widely used for simple designs and codings on a variety of food packages including metal cans, glass bottles, paper labels, and plastic bags and containers.

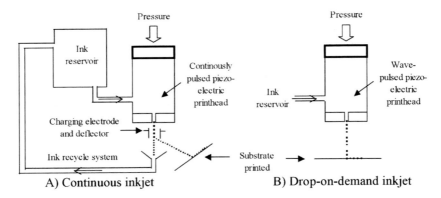

A) Continuous inkjet   B) Drop-on-demand inkjet

Figure 10.8 Two types of inkjet printing

There is a range of thermal transfer printing processes using an array of heating elements to transfer the prepared mirror image to the substrate. A typical thermal transfer printer utilizes the heated contact between the substrate and a durable polyester ribbon film coated with a dry thermal transfer ink and a heat-release coating. Thermal print head heats the ribbon to transcribe the pigment ink onto the substrate surface, where it cools and anchors to the media surface. The polyester ribbon is then peeled away with aid of heat-release coating, leaving behind a stable, passive image. A wax-coated ribbon in a different method is used to give image of fused minute wax dots on the substrate. This process is suitable for the labeling and printing of glass, plastic, and metal containers. Simpler method of direct thermal printing produces heat-pulse onto the paper substrate with heat-sensitive coating, forming characters consisting of tiny dark dots. The thermal printing is combined with a computer for printing the data transferred and has an advantage of no waste from the printing but with limited durability. Dye sublimation printing, another thermal transfer printing, uses sublimable dye inks embedded as a mirror image onto paper or film. As a next step, the printed image layer and the substrate come in intimate contact and heat is subsequently applied from the print head. The inks are vaporized to migrate from the intermediate image layer and adhere to the substrate surface. Dye sublimation printing is applied to decoration of two-piece can, but is unsuitable for retortable cans due to blurring of image caused by lateral dye diffusion.

Electrostatic printing systems such as photocopiers and lasers have mainly been used for paper printing and publishing, and have limited applications in packaging mostly of paper labels and plastic films. Recently hard substrates such as metal two-piece cans can be printed by electrophotography. The electrophotography is based on the principle that a photoconductive drum surface stores latent image temporarily in the form of electric charge pattern created by a suitable light source such as laser (Figure 10.9). The drum is usually a smooth aluminum cylinder coated with a photoconductive material such as selenium, arsenic selenide, or zinc oxide. The selectively charged drum surface is then exposed to a source of charged ink or toner. The toner is guided to be attracted electrically onto the image area with controlled depth. The image is then transferred by electrostatic attraction to the substrate that passes between the drum and an oppositely charged pressing roller called transfer roller. The printed image is finally fused into permanent one by controlled heating. The remaining ink on the drum is cleaned by wiper blade and the drum is passed through a constant electric flow to form uniform electric charge erasing the old image. In certain printing systems, the magnetic drum is magnetized to store the temporary image and the technology is called magnetography. Magnetically attracted pigments are used for the image.

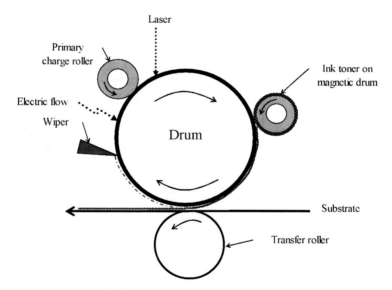

Figure 10.9 An electrophotographic printing

10.3 Label and Labeling

Food labels are affixed or printed on the packages to display its contents, nature, claims, instruction, manufacturer, and other information codes. The label material usually takes the form of the substrate strip containing written, printed, coded, or graphic matter. Labeling is the operation of applying or attaching the label to the correct position of the product package. Labeling of the packaged food products is legally mandatory in most cases for communicating the required information such as product name, quantity, ingredient, manufacturer's name, etc. There are specific regulations or guidelines for displaying the required information in an easy readable format. In USA most foods are legally required to bear nutrition labeling on their package; in Europe as well, specific directives or regulations monitor and regulate the ways foods packaged are labeled and carry nutrition and health claims. Optionally, slogans, claims, trade names, and any other descriptive items are printed with artful graphic design in order to promote the sale. Recently, coded information on the label is becoming more important for inventory control and traceability function, and will be discussed later in the next section. The quantity of information on the label is increasing for whole distribution chain including retailers and consumers. The food label could be directly printed on the package surface, or attached by separate printed substrate, which is discussed in this section.

10.3.1 Label Types and Materials

A selection of labels depends on the desired appearance, printing type, resistance to moisture, or other conditions in the marketing chain, and labeling equipment. The substrate materials comprise paper, plastic film, and metallic foil. The labels may be produced with adhesive-coated or non-coated state: adhesive label is precoated with pressure-sensitive, heat-sensitive, or water-moistenable adhesive; nonadhesive labels include those applied with wet-glue, hot-melt, shrink sleeve, stretch wrap, or in-mold at time of attachment. Figure 10.10 shows the typical classification of labels.

Single side coated papers are widely used with wet glue application for large volume products such as beer, wine, and canned foods. The papers should have good whiteness, smoothness, and gloss for good quality printing. Uncoated papers are used to a limited extent for special effects of colored or embossed papers. Humidity resistant paper labels are required for beverages that are likely to be exposed to humid conditions. Paper labels coated with water-moistenable gum on the reverse side are used for corrugated boxes and glass bottles. A variety of papers are used as the substrates carrying pressure-sensitive or heat-sensitive adhesive.

Plastic films, single layer or laminate structures, are either decorated directly or used as printed labels. Plastic film labels with proper adhesive can work even in underwater conditions. Films such as polyester and oriented polypropylene can serve as the carrier web for pressure-sensitive labels. The plastic films can also be used as over-laminate on the paper label. However, the widely used form of plastic labels is in shrink/stretch wraps. Sleeves of polyethylene, polypropylene, and polyvinyl chloride are usually placed as partial or whole wraps on the lateral surface of bottles, jars, or cans. Reverse printing is usually used to prevent scuffing [5]. However, metallized film is printed on outer surface. Both opaque and transparent printings are applied depending on the product and the desired decoration effect. The shrink/stretch wrap labels can provide a large surface area of decoration and tamper-evidence.

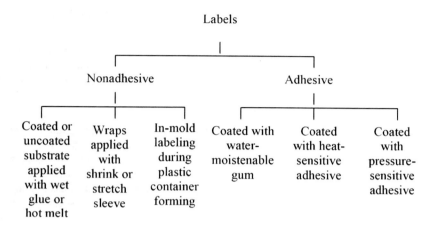

Figure 10.10 Types of label

Metallic foil is usually laminated onto paper with some difficulties in drying of adhesive solvent, labeling operation, and printing on the foil layer. However, it can give excellent appearance and bright reflective surface feel.

The label materials can be supplied in roll-stock or cut sheet. Roll-stock labels are more effective for high-speed long runs. Particularly plastic labels tend to be reel-fed with widening usage. Cut labels are useful for large sizes.

Pressure-sensitive labels consist of face material, pressure-sensitive adhesive, and release liner from which pressure-sensitive labels are detached: a silicon release coating is coated on the liner; the adhesives are based on acrylic formulations or rubber-resin blends. The label sticks to the surface or substrate when the adequate pressure is applied. Market for pressure-sensitive labels is steadily expanding these days [6]. Typical use includes wine bottles and yogurt containers. Pressure-sensitive labels have advantages of being retained stably on the surface of plastic bottles of carbonated beverages, which may expand and present a stress to cause the label to rip or become detached. When the bottle creeps due to the internal pressure, the label will slip on the glue but remain attached to the bottle.

Heat-sensitive labels can be divided into two types: instantaneous and delayed action [7]. Instantaneous action type is adhered onto the product by direct application of heat and pressure, while delayed action type becomes an active pressure-sensitive one by the application of heat. When applied to a container by pressure, it becomes a permanent label. Delayed action type is advantageous in that no direct heat is applied to the product package at the time of application. Instantaneous action type is used for end seals on biscuits and some bandings tolerant to heating.

In-mold labels are created during molding process of plastic containers. The labels are heat-sealed onto wall of the semi-rigid plastic containers being fabricated through blow molding, injection molding, or thermoforming. The in-mold label through a thermal-sealing process actually becomes a part of the plastic bottle or container during the molding process.

Holographic labels carry hologram featuring a three dimensional appearance. A hologram is created by embossing or volume methods. In the volume method, each hologram is exposed to a photosensitive medium and then developed according to the supplied data. An embossed hologram is created by passing a plastic web through the embossing device applying heat, pressure, and radiation treatment. The function of holographic label includes security and marketing tools.

Leaflet label is a booklet attached or held to the package by suitable means. It is used when there is a need to provide a large volume of information.

10.3.2 Labeling Process

The labeling operation is basically the process of adhesion between the printed substrate and package wall through the action of adhesive. Surface property is considered in selecting the adhesive system and is often modified by oxidation treatment for better bonding. Porous surfaces can be bonded together easily with hardening adhesives. Tacking adhesive is required for smooth surfaces and sometimes needs a preliminary primer coating. The adhesive may be applied, as wet or melted state, in line with label attachment to the package, or previously have been coated on the label substrate. The operation may be manual by using simple handgun or dispenser, or be fully automatic depending on the label type and volume. Labels should be stored under the atmospheric conditions close to those of the labeling operation.

The adhesives used for labeling include starch-based, casein-based, synthetic-resin emulsions, hot melt, and heat-seal coatings. Starch pastes are used for paper labels to a limited extent as wet glue labeling. Dextrin dispersions derived from starch by chemical treatment have properties of good transfer and fast drying in adhering labels; dextrins are cheap and easily cleanable with warm water; they are sometimes combined with polyvinyl acetate to improve their spreading property. Glass bottles are frequently labeled with wet starch-based adhesive; starch-based adhesive may also be applied as a remoistening adhesive for gummed paper label. Casein-based adhesive is particularly used for wet glue labeling of glass bottles; it provides the resistance to cold water immersion and removability in alkaline wash, which are especially good for returnable beer bottles. Aqueous synthetic-resin emulsions are mostly based on polyvinyl acetate, have the advantage of fast setting, and are used for labeling nonreturnable plastic bottles and coated glass; the emulsions usually include water-soluble colloids such as polyvinyl alcohol or 2-hydroxyethyl ether with optional additives of plasticizers, fillers, solvents, deformers, and preservatives; adhesion property can be improved by copolymerization with ethylene or acrylic esters. Hot melts are mostly of ethylene vinyl acetate copolymer (EVA) or low molecular weight polyethylene solids, which melt with heating to 120-200°C and set instantaneously on cooling; hot melt adhesives are 100% solid and are environmentally favored without any solvent use. They are suitable for plastic and glass bottles. However, they do not work for wet bottles. Use of solvent-based adhesives in the label is decreasing due to concerns on emission of volatile organic compounds: rubber-resin solutions are still used to a limited extent as a pressure-sensitive adhesive. In recent years, water based adhesives are increasingly used for environmental reasons.

Wet glue labeling is the least expensive and the most widely used system. The basic operation step consists of label feeding, adhesive coating, positioning of the label on the package surface, checking the label location, pressing the attached label onto the surface, and removal of labeled package. Precut labels supplied from the magazine are transferred to the container directly or indirectly; the operation may be in straight line or rotary movement with label pickup using vacuum gripper, mechanical finger, or glued surface. Wet glue labeling is widely used for canned foods, beer, and wine.

The adhesives on the pressure-sensitive label stick instantaneously to the container by sole application of adequate pressure. The adhesives precoated on the substrate include water-based acrylic emulsion, solvent-based blend of natural rubber and acryl resin, and hot-melt type. The adhesives are permanently tacky in dry state and bond to the contacting surface without help of water, solvent, or heat. Therefore, a backing sheet or film with release coating is placed as a cover contacting the adhesive layer in order to save the adhesive for the labeling process. In the operation, the label is peeled away from the backing sheet and fed to contact the containers by pressing, while the backing sheet is rewound onto a take-up reel. The pressure-sensitive labeling is clean without producing wet glue waste and easy to control with minimum cleanup and change-over time. These labels are widely used for cartons and cases.

Heat-seal labeling uses adhesives that become sticky when heated. Heat seal coatings are a wide variety of thermoplastic adhesives applied onto paper materials: solvent based vinyl acetate/vinyl chloride copolymers, water-based acrylics, and EVA are some examples. Instantaneous action type is activated by heating at the time of labeling and sets up instantaneously. Delayed action type is maintained as tacky for some period after heating treatment. Delayed action type has an advantage to separate the heating plate from the label application step. The heat-sensitive labels offer a high degree of adhesion for bottles subjected to high humidity conditions. They save clean-up time giving high productivity.

Sleeves of shrink labels are transferred to be positioned loosely over the container on the conveyor, and then shrink-fit in a heated tunnel. The sleeve labeling is usually done by rotary system, which die-cuts a sleeve from a rollstock tube of stretchable film, and mechanically applies it over the bottle using grippers. These sleeves offer a tamper-evident feature, and when applied onto glass bottles, covering a large part of the bottle body, the sleeves are also protective, reducing the damaging effects of possible impacts and containing the fragments due to an eventual breakage (see Section 7.5).

Injection molding, thermoforming, and blow molding can be used for in-mold labeling. The labels are printed on one side of a plastic material via gravure printing or other method with their backside containing a heat-sensitive adhesive. The label is embedded into the container wall by the action of the adhesive during the molding step of the manufacturing process. The labels of plastic film or paper fused to the plastic cannot be peeled off readily and are scratch resistant.

10.4 Coding

Traditionally thermally processed metal cans and glass containers were coded in closed format for carrying information such as production plant, product type, pack date, and processing batch. The coded information works as a safeguard useful in case of product recalls due to processing error or extraneous contamination. In later years, nutritional information came to be printed as consumer-readable format on the food package label in most countries. Starting from the end of 2006, for instance, packaging traceability law came into force in Europe as part of EU food safety policy for foods and food contact materials. Open dating of shelf life or pack date became mandatory for many products. These days, bar codes are added on the package to help to handle information in manufacturing, distribution, and retailing food products. Computers with telecommunication functions are involved in generating, printing, and decoding the codes with information transfer and management. These tools can constitute the major part of production management, inventory control, liability handling, and obedience to legal responsibility. In this section, coding is understood as packaging component consisting of a system of symbols or electronic signal set for information transfer through the whole channel of food processing, distribution, and final consumption.

10.4.1 Code Imprinting

Any code consisting of alphanumeric characters or symbols can be printed by conventional and digital printing methods described above. Some mention is, however, needed to cover a whole spectrum of code printing methods. Codes on metal can ends have often been marked in mechanically indented characters. The code is imprinted on the end before closing operation. Recently laser beam emission is used for marking on the package of glass, plastic, and paper [4]. The mechanism of marking is based on alteration of surface reflectivity, thermal color change, surface etching, or removal of layer coating. Usually CO_2 lasers are used to print format of dot-matrix or image mask. Most of code marking may be done before filling the container or after closing it. With increased demand for just-in-time coding, the marking process tends to be incorporated into in-line process of fill and seal.

Radio frequency identification (RFID) technology is recently designed to be combined with labeling for tracking and surveillance applications. Any RFID system basically consists of antenna, transceiver, and RFID tag in various forms. RFID system can read the information in the tag embedded inside the label at a distance without any contact and scanning on sight-line. A more detailed discussion will be given Chapter 15 (see Section 15.6.5).

10.4.2 Bar Code

Bar code mostly consists of a series of varying-width parallel bars and spaces to represent data of numbers, letters, and signs. The structure of a bar code is encoded according to the rule of symbology. The encoded data in the bar code is read optically and decoded into computer-compatible digital data by a scanner. As a scanning device illuminates the bar code symbol, the reflected width pattern of the bars and spaces is analyzed to extract the original encoded data. Visual images of bar codes should be printed in scanner-readable quality by proper printing methods.

There are many different bar code symbologies defining the rule for encoding the data (Figure 10.11). They differ both in the range of data type and the way of encoding the data. Even though there exist more than two hundred bar code symbologies, only a few of them are widely used. The most common linear symbologies started after 1970s are EAN/UPC, Code 39, Interleaved 2-of-5 (ITF), and Code 128. Most linear bar codes have a human-readable interpretation line directly below the symbol. Since 1990s, two-dimensional symbologies have been developed to encode larger amount of data: one type is multirow structure while the other one is matrix symbology in form of two-dimensional graphics.

In 1973, the grocery industry in USA established UPC as the standard bar code symbology for product marking. EAN code format, similar to UPC, has been adopted in Europe and other countries since 1976. EAN International (more than 100 countries, inside and outside Europe) and Uniform Code Council (the North American Society managing UPC codes) worked together (managing the EAN-UCC System) till 2005, when the organization changed its name to GS1. GS1 is nowadays a leading global organization aiming the implementation of global standards to improve the efficiency of supply and demand chains globally.

Figure 10.11 Examples of various bar code symbologies

All the former EAN/UPC codes are now collected under the GTIN umbrella of GS1, where GTIN, stands for 'Global Trade Item Number'. These are structures that may be 8, 12, 13, or 14 digits long. GTIN can be constructed using four different numbering structures: EAN, UPC, ITF, and Code128 (today GS1-128), depending upon the exact application. GTIN-8s will be encoded in an EAN-8 bar code. GTIN-12s may be shown in UPC-A, ITF-14, or GS1-128 bar codes. GTIN-13s may be encoded in EAN-13, ITF-14, or GS1-128 bar codes, and GTIN-14s may be encoded in ITF-14 or GS1-128 bar codes.

ITF-14 and GS1-128 are widely used for secondary packages, shipping containers, and logistic units, whereas items to be sold at retail level are usually marked with EAN-8, EAN-13, UPC-A, or UPC-E bar codes

UPC symbology, actually, has five versions, A, B, C, D, and E, while there are two versions of EAN (EAN-13 and EAN-8). UPC and EAN symbols have fixed length and encode only number data.

UPC version A commonly used for grocery items (now GTIN-12) encodes 10 digits plus two overhead digits while EAN-13 (now GTIN-13) does 12 digits and one overhead digit. The first overhead digit of a UPC A symbol is a number related to the type of product (0 for groceries, 3 for drugs, etc.). The next 5 digits are the UPC manufacturer's code, and the next five digits are the product code. The last digit is the check digit for validating the code. Any bar and space has widths of one, two, three, or four *modules*: the thinnest bar or space can be seem as one module wide. The bar code is guarded by starting and ending codes, both of which consist of longer bar-space-bar each in one module width. There are also long center guard bars in the center of the symbol (space-bar-space-bar-space in single modules) which divide the digits into two six-digit parts. Each digit is coded as a sequence of two bars and two spaces to occupy 7 module width. The left side digit combination starts with a space, while the right digits start with a bar.

The width of the narrowest bar or space (one module) is referred to as 'X dimension'. The X dimension adopted determines the width of all other bars and spaces, and ultimately the total length of the bar code. Nominal X dimension is 13 mils (1 mil=1/1000 inch). Nominal UPC-A symbol has width of 113 modules. The specifications for UPC A version state that it can be printed in 80% to 200% of nominal size: printable range of X dimension becomes 10.4 to 26 mils. The greater the X dimension, the more easily a bar code will be read with a scanner but with the higher cost of bigger labels. For proper scanning, most bar codes have a quiet zone of usually about 10 times the X dimension, i.e., clear space, at both ends. The height of the bars in the nominal size symbol is 22.85 mm (0.9 inch). The height of the symbol is desired to be at least half the length of the symbol. However, it is sometimes shortened to fit into the package label design. This reduced symbol height affects the ability to scan the symbol in any orientation.

UPC version E is a zero suppression version of UPC with a wide use next to UPC-A. By dropping out zeros, it has shorter length and is thus used on packaging that has a small surface for printing.

The EAN System is identical to UPC except for the number of digits. Standard EAN called EAN-13 has 13 numeric digits. According to the most updated specification, starting from 2002, the first 9 digits (from the left side) constitute a company prefix, of which the first 3 digits are 'flag' characters representing the country of the EAN international organization issuing the number or certain category of product or use. Before 2002, only the first 7 digits, with the flags of 2 or 3 characters, were used to identify the company prefix. The next digits refer to item and the last digit is for validation check. In all other respects, it is identical to UPC version A. For compatibility with UPC, flags 000 through 139 are assigned to the United States. Thus the UPC system may be seen as a subset of the more general EAN code. EAN-8 has a left-hand guard, two parts of four digits separated by a center guard, and a right-hand guard bars. An EAN-8 bar code has two flag digits, five data digits, and one check digit.

The check digit positioned at the end of the code validates that all characters have been decoded correctly. The checking procedure is unique for all GTIN (8, 12, 13, and 14) structures, provided that the digit numbering starts from the right end. It consists of summing up the value of all the digits at even positions, multiplying the sum by three, adding this value to the summation of all the digit values in odd positions (obviously excluding the control digit), and finally finding the value which can make a multiple of 10 by adding the previous total. If we take an example of UPC-A in Figure 10.11, total before finding multiple of 10 is 109, i.e., $(0 + 2 + 4 + 6 + 7 + 9) \times 3 + (1 + 3 + 5 + 8 + 8) = 109$. To make multiples of 10 from 109, the value of 1 should be added, which becomes the check digit.

With wider application of linear bar codes in product packaging and distribution, the needs and demands to encode more information in a smaller space increased, which was the driving force for the development of two-dimensional bar codes. 2D bar codes can fulfill the function of product identification and work as a database even while taking up significantly less space. Two types of 2D bar codes are currently used, which are stacked codes and matrix codes (Figure 10.11). The advantage of 2D codes is containment of more information in smaller space with higher security.

Stacked bar code is a stacked set of smaller bar codes. Typical stacked bar code, PDF417 developed in 1990 can encode a maximum of 2725 data characters. These features enabled decoding from scan paths that span multiple adjacent rows while incorporating error detection and correcting techniques. In PDF417 the basic data unit is cord word consisting of four bars and four spaces. Every code word has the same physical length 17 module wide. Any bar or space has module number from 1 to 6. A complete PDF417 symbol consists of at least 3 rows of up to 30 code words and may contain a maximum of 928 code words per symbol.

A two dimensional matrix code confers higher data densities than stacked codes in most cases, as well as orientation-independent scanning. A matrix code consists of a pattern of cells having shape of a square, hexagon, or circle. Data is encoded via the relative positions of these light and dark areas. Data Matrix, one of matrix codes can encode up to 3116 characters from the entire 256 byte ASCII character set.

The bar code is printed, desirably in black ink against white background, on the package using a variety of conventional and digital printing techniques. Code with static information on a large volume is usually printed in a different place by conventional process, while that with variable information on a small quantity is printed on site by digital printing. The printed bar code should be of good quality for the scanner to read the symbol for the first time and every time thereafter. Excessive ink spread should be avoided because of the possibility that the scanner is unable to read the code. For nonperfect planar surfaces, like corrugated board, the bigger and less critical structure of an ITF-14 code is advisable. When the digital printing methods are used, X dimension should be set to have an integer multiple of printer dots: X dimension is related to the printer's resolution expressed in dpi (dots per inch) by the following equation:

$$X = N/dpi \qquad (10.1)$$

where X is dimension or a module, N is integer number of dots per module, and dpi is the printer's resolution. In case of UPC-A code, X dimension can be varied from 80% to 200% of nominal 13 mils. Typical X dimensions for PDF417 bar code are 10 and 6.6 mil. Most laser printers today have resolutions of 300-1200 dpi, while the average ink jet prints have 360 or 720 dpi, and thermal transfer printers typically have 200-600 dpi.

10.5 Sealing of Plastic Surfaces

Flexible or rigid plastic packages are usually sealed at least once to function as an integrated unit. Preformed pouches are sealed on three sides (bottom and sides) prior to filling operation, after which the remaining side is sealed. Rigid or semi-rigid containers formed by injection, blow molding, or thermoforming are often hermetically sealed by bonding of the body either directly to the cover or to an inner liner inside the cap. The sealing of plastic surfaces is a process of creating adhesion between two interfaces in nature (see Section 2.4.5). The adhesive bonding of the surface may be formed from heating the polymers themselves, dissolving the interfaces by a solvent, or coating the surfaces by an adhesive. Usual method of sealing involves heated pressing of the interfaces followed by cooling, which is called as heat sealing. Adhesive sealing without applying heat is called cold sealing. Seals should be strong enough to withstand sealing and provide strength enough for the package to remain in integrated state through the distribution channel. Line speed and aesthetic appearance of the seal are other important factors for choosing the sealing method for a packaging system.

10.5.1 Heat Sealing

Heat sealing is a process that involves applying heat and pressure to melt two layers of heat sealants and press them together for a proper period of time to form a heat seal.

Figure 10.12 illustrates the steps in heat sealing and their explanations at the polymer chain level. In this figure, (a) shows that the process begins with two compatible heat sealant layers, (b) shows that heat and pressure are applied to melt the layers and press them together, (c) shows that time is allowed for the polymer chains to intermingle or diffuse across the interface between the two layers to form a polymer melt, and (d) shows that the polymer melt is cooled to form the heat seal. Scientific knowledge discussed in earlier chapters, such as properties of polymer and principle of heat transfer, is useful for understanding heat sealing.

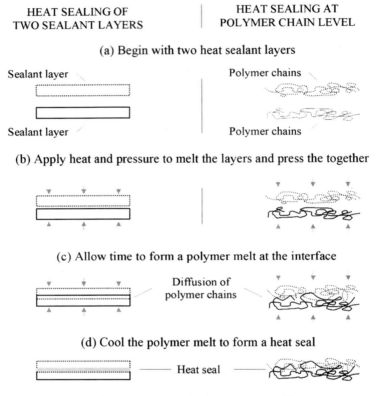

| HEAT SEALING OF TWO SEALANT LAYERS | HEAT SEALING AT POLYMER CHAIN LEVEL |

(a) Begin with two heat sealant layers

Sealant layer

Sealant layer

Polymer chains

Polymer chains

(b) Apply heat and pressure to melt the layers and press the together

(c) Allow time to form a polymer melt at the interface

Diffusion of polymer chains

(d) Cool the polymer melt to form a heat seal

Heat seal

Figure 10.12 Mechanism of heat sealing

The heat required to melt the sealant layers may by supplied by means of heat conduction, electromagnetic induction, ultrasonic friction, etc. To achieve proper seal strength, the temperature at the interface between the two sealant layers is an important factor. This temperature is usually a few degrees higher than the melting point of the sealant layers [8]. While heat seal should be strong enough to protect the product against possible abuses during distribution, it should not be too strong in situations where peelability of the seal is a desirable feature. Therefore, it is sometimes necessary to strike a balance between the protection and convenience functions in designing the seal.

Heat sealers are equipment used to form heat seals. As implied in Figure 10.12, all heat sealers must perform the basic functions of pressing the sealant layers firmly together, melting them, and allowing sufficient time (typically 0.5-3.0 s) to form a heat seal at the interface of the two layers. These functions are controlled by adjusting three important operating variables of the heat sealers: pressure, temperature, and time (or dwell time). In this section, several common types of heat sealers and their principles of operation are described.

Hot bar sealing (Figure 10.13) applies pressure through two metal bars, connected to a high pressure tank, to heat seal plastic bags, pouches, or containers. At least one of the bars is heated to provide sufficient temperature to melt the sealant layers. The folded plastic film to be sealed is placed between the flat bars under controlled conditions of temperature, pressure, and dwell time (Figure 10.13).

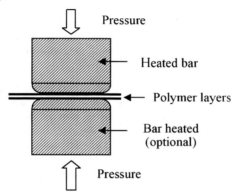

Figure 10.13 Bar sealing

Rounded edges on the bars are preferred to avoid puncturing of the contacting film surface [9]. One of the bars is often made of resilient surface mostly of silicon rubber, which applies uniform pressure onto the seal area when pressed. The bar surfaces are often designed as serrated jaws in order to remove wrinkles on the seal area and give higher seal strength. Probability of tiny seal leaks may increase with improper use of serrations. The sealed area is often flattened before pressing the two films between the bars. The heating for the melting of sealant is accomplished by conduction heat transfer from the bar to the sealant. The heating may be transferred from both bars or single bar maintained at constant temperature. The molten layers are fused and become solidified with cooling to ambient temperature. Optimal combination of temperature, pressure, and dwell time depends on film structure and sealer construction. Under practical operating conditions, pressure around 5 N cm^{-2} or slightly higher is usually sufficient and much higher level of pressure does not give beneficial effect on the seal properties [8]. Dwell time in bar sealing is usually controlled long enough to attain the desired interfacial temperature of melting under the used bar temperature condition. High production speed in bar sealing is due to a short dwell time, usually less than a second.

The heat sealant layers may be made of the same or different materials depending on the situation. A single layer polymer film, such as a film made of PE and PP, may be used as a heat sealant layer. A laminate may also used in heat sealing where the heat sealant is one of the layers. For example, the laminate structure PE/foil/PET is often used in hot bar sealing where the PE layer is the sealant.

The sealant layers should be located on the interfaces contacting each other, and have thermoplastic property of melting at a certain temperature for welding together. The sealant layer may be supported in lamination by another plastic layer or a nonthermoplastic layer such as paper and metal foil, which is resistant to high temperature. The sealant on the interface should be melted and fused together with another sealant layer at heat seal temperature, while the nonthermoplastic layer should remain intact without structure loss or excessive softening. Unsupported films of single or multiple thermoplastic layers are required to be heat-sealed with careful control to avoid sticking of the film surface onto sealer.

A variant of bar sealing uses the moving hot rollers between which the films are transferred. Another type of bar sealing is also used for applying the lid or cover on the cup or tray container. The sealing part of the container rim is pressed between the bars shaped exactly to fit the rim. Bar sealing is the most widely used sealing for flexible and semi-rigid plastic packages.

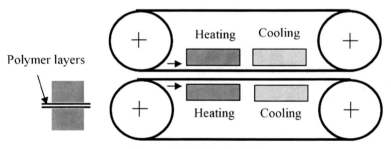

Figure 10.14 Band sealing

Band sealing is an improved version of continuous bar sealing, which transfers the matched films between two moving bands contacting the heated bars (Figure 10.14). Thereafter the bands also move the sealed films through the cooling bars to finish the sealing. The moving bands are usually made of stainless steel covered with Teflon® tape. The bands also play a role of pressing the seal area between the heated bars to attain the fused seal. Flattening the matched seal area is required before gripping by bands to attain the seal without wrinkles. Band sealing is mainly used for hermetically closing the pouches filled with food.

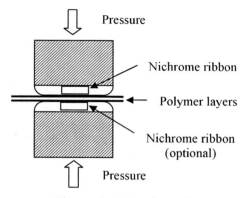

Figure 10.15 Impulse sealing

Impulse sealing applies heat to the matched films by the nichrome ribbons attached on one or both of the sealing jaws (Figure 10.15). The sealing jaws are usually covered with resilient material, on which the nichrome ribbon is covered by a Teflon® tape. The nichrome ribbons are heated only when electric current is passed through it. The time for electric current flow is controlled by a timer, which usually starts with closing the jaws to press the seal area. After the heating phase of electric current flow, the seal area is kept for some time under retained pressure to provide controlled cooling which helps to attain good solidified fusion on the seal area. The cooling is mostly controlled naturally by heat transfer from the seal through the resilient jaw surface. Sometimes the cooling can be accelerated by water flow for quick cooling. Shaped impulse sealer surfaces fitted to the rim of the containers are used for sealing covers on the cups and trays.

Impulse sealing allows stable seal even for unsupported films with simple construction, but is generally slower and has a narrower seal width than the bar and band sealing methods. The Teflon® coated release tape on the nichrome should be replaced often because of tearing and contamination. The nichrome ribbon also needs frequent replacement and maintenance because it experiences high mechanical stress due to very frequent cycles of heated expansion and cooled contraction.

Hot wire or knife sealing accomplishes the jobs of sealing and cutting at the same time by pushing the hot wire or knife through the matched film pair. The film parts becoming apart are fused on the ends by the heat. This type of sealing cannot produce hermetic consistent seal and is used for making coarse bags by simple dispensing at high speed. Polyethylene bags used for self-packing fresh produce are an example.

Hot gas sealing uses hot air, nitrogen, or gas flame to melt the thermoplastics on the surfaces. The heated gas or flame is directed at the interfaces to be sealed. The surfaces are then clamped together between cooled jaws. This method of heat sealing is useful for sealant-coated paperboard and thick plastic sheet having high resistance to conductive heat transfer. A polyethylene coated milk carton can be sealed faster by this method than by bar sealing [10].

Ultrasonic sealing uses heat generated from high-frequency mechanical motion. High-frequency electrical energy is converted into high-frequency mechanical motion, which in turn produces frictional heat under pressure at the interfaces for melting. A horn generates vibration at frequency of 15-70 kHz onto the interfaces for predetermined time period. The vibration increases the temperature at the interface for the plastic materials to melt and fuse together. After holding period for cooling, the system ejects the sealed container or materials. Ultrasonic sealing is characterized as the fact that heat is generated near the interface not to affect other part of material, which is valuable for thick materials having high resistance to conductive heat transfer. It is also useful for heat sensitive material like highly oriented films which cannot endure long hold time of heating. Ultrasonic sealing is used mostly for sealing the same series of plastic materials such as the pairs of polystyrene/high impact polystyrene (PS/HIPS) and acrylonitrile-butadiene-styrene copolymer (ABS)/styrene acrylate. Friction sealing physically moves one of two surfaces under pressure, creating heat through surface friction to melt and weld the surfaces together. Compared to ultrasonic sealing, friction sealing operates at much lower frequencies, higher amplitudes, and much greater pressure. Friction sealing can join a variety of thermoplastics including materials with high filler content.

High frequency (radio frequency) or dielectric sealing welds the polar thermoplastic surfaces placed between cold electrodes or dies of a high frequency electrical generator. The polar plastic materials experience the dielectric loss under oscillating high frequency electric field usually in the range of 50-80 MHz, which produces heat due to tendency of the molecules to orient according to the electric field. The polar plastic material is sealed between two dies which are usually mounted in a pneumatic or hydraulic press to provide pressure: one die is usually made of copper and has the shape of the required seal image; the other metal die has the same shape or a flat surface. The quality of a seal is determined by the electric current, time, and pressure. The usual heating period ranges from one to four seconds. This method of sealing is ideal for polar plastics such as polyvinyl chloride (PVC) and nylon. It can also seal with some liquid contamination on the seal area.

Induction heat sealing is a non-contact process that applies an electromagnetic field to heat seal a container with a screw cap which has a heat sealable foil innerseal (Figure 10.16). The innerseal may include a pulpboard, a wax layer, an aluminum foil, and a sealant layer. This method is typically used for making hermetic or tamper-evident seals on plastic bottles or jars. The operating principles are as follows: (1) the innerseal is pressed tightly against the lip of the bottle by the screw cap, (2) the bottle is then placed under the induction heat sealer for an appropriate period of time where the induction coil head creates an eddy current in the aluminum foil to generate resistance-type heating, (3) heat conduction from the aluminum foil causes the sealant layer to melt and the pressure from the cap causes the melt to stick onto the lip of the bottle, and (4) the bottle is removed from the induction heat sealer and the melt is allowed to cool and form a heat seal. The heat also causes the wax to stick onto the pulpboard, releasing the foil from the pulpboard.

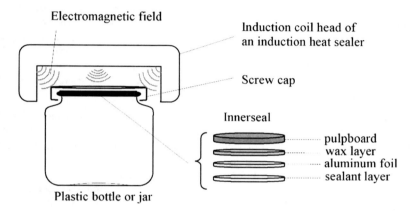

Figure 10.16 Induction heat sealing

10.5.2 Cold Seal

Seal between plastic interfaces attained without heat application is called cold seal. It is widely used in situations where heat sealing is impossible or damages the packaging material and/or the food product. Chocolates and candy products are examples of heat-sensitive food products. The cold seal can be achieved by dissolution of interfaces by solvent or use of adhesive.

If a solvent added on the interfaces can dissolve the plastics there, there would be intermingling or mixing of the components. When the solvent is removed from the interface by evaporation or migration, the intermingled part will solidify to form a weld. Pressure is usually applied to assist the mixed bonding of the interfaces. Mostly organic solvents are used for the solvent sealing.

Cold seal adhesives are applied to join the seal interfaces without heat. Natural rubber in latex (suspension in water) is commonly used as an adhesive [9]. Adhesives dissolved in organic solvent are also used to make a bond between the plastic blister and paperboard. Pressure is needed for activating the adhesives or assisting the bond. Hot melt adhesive timely applied can also make the seal without exposure to the heat. When organic solvents are used in cold seal process, their residual content and migration should be carefully controlled not to affect the food product quality. Oriented polypropylene and paper-backed foil are the flexible materials often sealed with hot melt or a cold-seal lacquer for packaging confectionery, bakery, and snack products.

10.5.3 Seal Strength Test

Seal quality can be examined visually to find any defects and wrinkles. More elaborate nondestructive techniques such as ultrasonic imaging, infrared laser, machine vision, X ray transmission, and magnetic resonance imaging have been developed for on-line detection of seal defects, but have not reached full commercialization yet. Any of the techniques are useful to a limited extent for detecting certain defects or flaws on the seal under standardized conditions, but do not work universally for variable conditions. Package integrity testing will be discussed in Chapter 11 (see Section 11.4.3).

The simplest and most widely used measurement method of seal strength is tensile test on the seal area which is called peel test: the specimens of the seal area with width of 10-30 mm (mostly 1 inch) are cut from the sealed part, stretched by the separate clamps and pulled away to record the force vs. strain curve (Fig. 10. 17a). The maximum peak force divided by the specimen width is termed commonly as seal strength. True seal strength is obtained when peeling of the bonded interfaces occurs with tensile stress. But the maximum force in tensile test does represent the yield stress of the film material itself, not the stress from peeling of the seal part, when the tearing or breakage takes place on the film part other than the seal area: therefore the occurrence of tearing failure in the peel test usually means that the seal provides enough strength achievable by the film.

Bursting test of a fully sealed pouch is also known to correlate well with the peel test. A sealed pouch is blown with air entering through a needle between the two separate flat plates: the compressed air in the package works to supply the force required to peel the seal interfaces (Fig. 10.17b). Yam et al. [11] reported that there is a good agreement between peel and burst tests when the peeling times for the two tests are equal: the seal strength is described by the product of bursting pressure and half-plate separation. The measured seal strength is a function of strain rate and generally higher with faster peeling. It needs to be mentioned that the high sealing strength of a seal does not mean a perfect seal without any risk of leaks but just the seal strong enough for mechanical handling. A seal with a microleak may show adequate seal strength [12]. However a seal strength test can be used to find optimum sealing condition for a package/sealer system. Package integrity can be confirmed by other testing methods, which will be discussed in Section 11.4.3.

High performance seals are also related to the so called 'hot tack' property of the sealing material. Hot tack is defined the capability of a heat seal joint to resist when stressed, while still hot from the sealing operation. A high hot tack is quite often a major requirement for demanding application such food packaging operations. Some EVA grades and ionomer polymers present quite good hot tack values and are therefore suggested as sealing layer for very fast packaging processes.

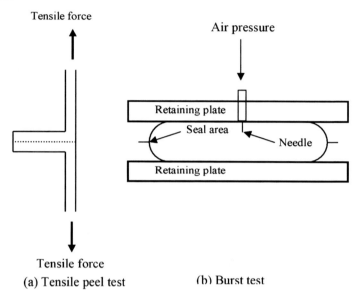

(a) Tensile peel test   (b) Burst test

Figure 10.17 Tensile peel and burst tests for measuring seal strength

One last interesting property of the sealing materials, very important in food packaging application, is the capability of sealing in the presence of dirt or contamination on the sealing interfaces. This is a quite common situation where powdered, oily, or liquid products are to be closed in a flexible package. Also in this case special grades of sealing material must be selected in order to have effective and fast seals.

10.6 Case Study: Finding Sealing Conditions for an LLDPE Film

Meka and Stehling [8] investigated the effect of heat-sealing process variables (seal bar temperature, dwell time, and pressure) on seal properties (seal strength, seal elongation, and seal energy) of a linear low density polyethylene (LLDPE) film 50 μm thick. Heat seals 9.5 mm wide were made by a laboratory heat sealer. The testing ranges of the independent variables were dwell time of 0.15-10 s, pressure of 15-1000 N cm^{-2}, and platen temperature of 105-170°C. The sealed films were cut into the width of 25.4 mm and mounted onto the tensile testing machine with initial jaw separation of 50.8 mm. The film samples were strained at a rate of 508 mm min^{-1} with measurement in maximum force. Deformation around seal area was also determined by measuring the thickness of the sealed samples across the two seal edges.

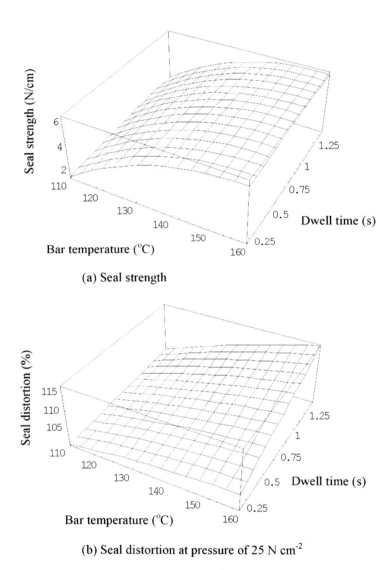

(a) Seal strength

(b) Seal distortion at pressure of 25 N cm^{-2}

Figure 10.18 Sealing properties as a function of bar temperature and dwell time. Seal distortion was at pressure at 25 N cm^{-2}.

From the regression analysis of the authors' reported data, seal strength was described as following equation:

Seal strength (N cm^{-1}) = -42.5532 + 0.61398T + 2.24063D - 0.001935T^2 - 0.37704D^2 - 0.00890TD

where T is bar temperature (°C), and D is dwell time (s).

Seal strength did not depend on the pressure in the range of 15-1000 N cm^{-2} used and was predominantly dependent on the bar temperature under practical operating range of dwell time (Figure 10.18a).

Seal distortion expressed as thickness ratio of the seal edge to initial film structure was dependent on bar temperature, dwell time, and pressure following the regression equation:

Seal distortion (%) = 73.083 + 0.4120T - 13.4247D - 0.1331P - 0.0014T^2 - 0.3186D^2 + 0.0009P^2 + 0.1502TD + 0.0006TP - 0.0176DP

where P is bar pressure (N cm^{-2}).

Seal distortion was higher with higher platen temperature, dwell time, and pressure (Figure 10.18b).

From consideration of seal strength and distortion as a function of process variables, feasible sealing condition were selected to have adequate seal strength and low distortion of the seal. A combination of high temperature and very short dwell time may have a potential risk of imprecise control for the interfacial temperature due to very rapid heat transfer in the film. The conditions of longer dwell time and plate temperature slightly higher than melting point of the polymer are desirable under the practical limitation of sealing time. However, too long a dwell time can cause distortion and should be avoided. Final melting temperature of the used film in differential scanning calorimeter analysis was measured as 126°C. Because increased pressure does not give beneficial effect on seal properties in usual range, pressure slightly higher than 5 N cm^{-2} is generally acceptable with optimum level depending on the sealer. Therefore a dwell time of 1 s at 135°C of bar temperature under the small pressure is an appropriate choice for this film.

Discussion Questions and Problems

1. Look around a supermarket to find the open dating of shelf life on the label of food packages. List the printing methods used for the open dating of the food products.

2. Find an orange juice package in a supermarket and list all the information on the label. Discuss the label information in relationship with regulations of your country.

3. Compare flexography and gravure for printing packages of dry snack foods. Discuss the advantages and disadvantages for both methods.

4. How can the environmental friendliness of the package printing be achieved?

5. Take a snack package in a supermarket and interpret its bar code symbol based on UPC or EAN symbology.

6. Think about the mechanism of cold-sealing plastic films and then list the possible process variables affecting the seal strength of the cold seal.

7. How can you decorate the semi-rigid plastic packages of soft drinks? Give your choice and discuss its advantages and disadvantages in comparison with other alternatives.

Bibliography

1. W Soroka. Fundamentals of Packaging Technology. Herndon, Virginia: Institute of Packaging Professionals, 1995, pp. 61-88.
2. H Speirs. Introduction to Printing and Finishing. 2nd ed: Pira International, 2003, pp. 1-175.
3. TA Turner. Canmaking: The Technology of Metal Protection and Decoration. London: Blackie Academic & Professional, 1998, pp. 1-131.
4. M Chamberlain. Marking, Coding and Labelling. Surrey, UK: Pira International, 1997, pp. 1-49.
5. M Hanlon, RJ Kelsey, HE Forcinio. Handbook of Package Engineering. Boca Raton, Florida: CRC Press, 1998, pp. 399-433.
6. IH Hall. Labels and Labelling. 2nd ed. Leatherhead, UK: Pira International, 1999, pp. 1-60.
7. MC Fairley. Labels and labeling machinery. In: M Bakker, ed. The Wiley Encyclopedia of Packaging Technology. New York: John Wiley & Sons, 1986, pp. 424-430.
8. P Meka, FC Stehling. Heat sealing of semicrystalline polymer films. I. Calculation and measurement of interfacial temperatures: effect of process variables on seal properties. J Appl Polym Sci 51:89-103, 1994.
9. RJ Hernandez, SEM Selke, JD Culter. Plastics Packaging. Munchi: Hanser Publishers, 2000, pp. 157-183.
10. WE Young. Sealing, heat. In: M Bakker, ed. The Wiley Encyclopedia of Packaging Technology. New York: John Wiley & Sons, 1986, pp. 574-578.
11. KL Yam, J Rossen, X-F Wu. Relationship between seal strength and burst pressure for pouches. Packag Technol Sci 6(5):239-244, 1993.
12. CL Harper, BA Blakistone, JB Litchfield, SA Morris. Developments in food packaging integrity testing. Trends Food Sci Tech 6:336-337, 1995.

Chapter 11

Food Packaging Operations and Technology

11.1 Food Packaging Line..313
11.2 Filling of Liquid and Wet Food Products...............................317
    11.2.1 Filling to Predetermined Level ································318
    11.2.2 Filling to Predetermined Volume ·······················321
11.3 Filling of Dry Solid Foods...322
    11.3.1 Filling by Count···323
    11.3.2 Filling by Volume··324
    11.3.3 Filling by Weight···326
11.4 Closure and Closing Operation..327
    11.4.1 Closures ···327
    11.4.2 Closing Operation···333
    11.4.3 Package Integrity Testing ··································333
11.5 Methods of Wrapping and Bagging...336
    11.5.1 Wrapping ···336
    11.5.2 Bagging··337
11.6 Form-Fill-Seal 339
    11.6.1 Vertical Form-Fill-Seal····································340
    11.6.2 Horizontal Form-Fill-Seal ·······························342
11.7 Various Forms of Contact and Contour Packaging 344
    11.7.1 Skin Packaging ··345
    11.7.2 Blister Packaging ···345
    11.7.3 Shrink Packaging ···346
    11.7.4 Stretch Wrapping··348
11.8 Case Studies 349
    11.8.1 Performance of a Piston-type Filler ···················349
    11.8.2 Pressure Differential System for Nondestructive Leak Detection ··351
Discussion Questions and Problems 352
Bibliography 353

11.1 Food Packaging Line

A food packaging line consists mainly of the packaging material and food product combined into a unit of package. The objective of the packaging operation is to fill the package safely with a desired volume or weight. To accomplish this objective effectively, adequate techniques of weighing, portioning, conveying of product and packaging material, sealing, etc. are linked together systematically. Hygienic guidelines should be followed for attaining the desired functions of food packaging. Appropriate tests of cleanability and decontamination should be commissioned for the filler itself and filler zone [1]. Especially the filler and its air supply system should be free of microbial contamination. CIP (clean in place) techniques should be used where possible, and the parts not cleanable by CIP should be able to be easily dismantled and cleaned fast.

The structure of operation machinery is very diverse depending on the product type, package type, desired accuracy and speed of filling, and space allowance. A typical packaging line in a food plant consists of depalletization, decasing, cleaning, filling, closing, coding, optional thermal processing, fill checking, labeling, casing, and palletization. All the operations should be constructed to have harmony among the individual machines. Filling equipment is generally the most expensive part of the line and thus is speed-controlling in the whole operation [2]. To attain maximum efficiency in the packaging machinery line, operations before and after filler are set to run faster than the filler (Figure 11.1) and thus to provide accumulation between machines so that minor stoppage before the filler can be accommodated without stopping the filler [3]. The faster postfill operations push up the speed of the line further from the filler. Figure 11.1 shows relative speed in the food packaging line.

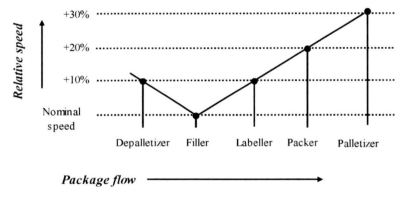

Figure 11.1 Relative speed of individual machines in typical packaging line. From Ref. [3] with permission.

Packaging lines for conveying packages through filling and sealing are classified into two types of straight-line and rotary configuration. In straight-line type, the containers transferred through straight line mode are filled by the filling valves under the stop cycle of intermittent-motion or are filled continuously by the filling heads moving together with them. The intermittent-motion filler is the one widely used and usually has low capacity: maximum operating speed is about 150 units per minute [4]. The number of filling stations or valves may be increased to enhance the speed. Dual-lane straight-line fillers allow more efficient utilization of filling stations: the filling valves, in a row, fill in one lane while the containers are being moved on the other lane. Rotary type filler feeds the containers continuously to the rotating turret, where the rotating valve fills the container on the rotational move (Figure 11.2). Containers on the conveyor are fed to rotating turret through starwheel, being spaced by timing screw. The filled containers on the turret are discharged by the other starwheel. Capacity of rotary filler is determined by the valve number and required filling time. As shown in Figure 11.2, the angle 'a' is for the filling step while angles 'b₁' and 'b₂' are, respectively, for raising and lowering containers before and after filling. The angle 'c' is the span for simple turn over. The capacity of the rotary filler in units per minute (c$_r$) can be described as function of angle 'a' (degree) in the filling [5]:

$$c_r = \frac{(a/360) \cdot n_v}{t_f} \qquad (11.1)$$

where n$_v$ is valve number, and t$_f$ is the time required for filling a container (min).

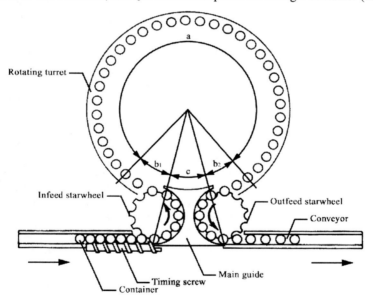

Figure 11.2 Configuration of rotary filler

The larger angle 'a' can be obtained with the larger diameter of the filler, which gives a higher speed of filling. The wider the diameter of the turret, the higher the number of the filling valves and thus the greater the speed. The number of valves ranges from 8 to 120. The speed of the rotary filler is commonly about 80 units per minute, but may reach 2000 units per minute [4].

Container should be designed to give stable feeding before filler even when there are a file-up of containers in the line [6]. Containers with large base area, low center of gravity, and parallel contact points are generally stable with no tendency to topple. Cross section of the containers should be such as to provide straight contact between them avoiding the possible jams and breakage on the conveyor. There is usually installed a device or arrangement stopping the product flow from product tank when there is no supply of containers [7]: filling valve mechanism or container detection system can work for this purpose.

Selection of a filling system should consider the characteristics of food and container types. Products may be liquid, dry solid, or wet paste. Liquid products are varied in viscosity and homogeneity with or without carbonation. Dry products may be powders free flowing or clogging. Filling is based on weight or volume depending on the product characteristics. Some products consist of solid and liquid parts whose ratio should be controlled for consistent quality. The product with heterogeneous mixture of ingredients is usually filled sequentially in separate stations because consistent fill of premix cannot be ensured by a single operation. Rigid glass containers and metal cans can be filled with application of high pressure or vacuum. However, there should be limitations on the level of pressure and vacuum for the semirigid plastic containers. Flexible pouches are hard to be filled to a specified level under fixed dimension.

Filling accuracy is also one of the most important considerations in selecting a filler. Underfills raise legal problems and overfills cause economic loss. Legal requirements in fill weight or volume should be satisfied without causing excessive overfill or 'give-away'. The accuracy of the fill is represented by standard deviation of fill weight or volume. Assuming normal distribution of fill amount dictates that 99.73% of the containers stay within the limits of three standard deviations from mean fill weight or volume as shown in Figure 11.3. Average fill weight is set at least three standard deviations above the declared fill on the label [4, 8]: thus less than 0.13% of the packages have the probability to have lower fill than declared on the label. The more accurate the filler system, the narrower the distribution of the fill amount and the less the 'give-away'. In order to attain product filling with the least overfill and underfill, statistical process control is adopted [9]. Mean and range control charts are the most often used tools to evaluate and correct the filling process.

The overall design of a food packaging line should consider hygienic conditions, harmony between individual steps, capacity, prevention of waste and spill, precision, flexibility, and coordination between packaging and other food processing operations [10]. To accomplish this objective, these days, computer-aided automated machinery and robots are increasingly used in food packaging lines.

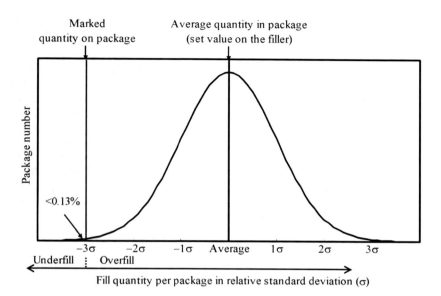

Figure 11.3 Setting of filling amount considering its distribution in the filler

11.2 Filling of Liquid and Wet Food Products

Variation in the product viscosity needs to be considered in selecting the type of liquid food fillers: low viscosity products of 1-10 centipoise (cp) such as wine, drinking water, and vinegar; medium viscosity products of 10-2000 cp such as sauce and vegetable oil; high viscosity products with more than 2000 cp such as ketchup and mayonnaise [5]. Semi-liquid with extremely viscous characteristics may have viscosity of up to 10000 cp and may be called semi-solid or paste. It needs to be noted that food viscosity can change with temperature change encountered before the filling: examples are some edible oils or tomato sauce. The carbonated beverages contain carbon dioxide dissolved in aqueous phase. Typical levels of carbonation in the beverages ranges from 1.5 of citrus soft drink to 5 volumes of soda water [11]. One volume of dissolved CO_2 amounts approximately to 2 g L^{-1} and produces 1 atm at room temperature. With carbonated beverages, equilibration in the pressure should be made between the internal bottle space and product by introduction of carbon dioxide before the start of filling [12]. Pressure resistance is a requirement of carbonated beverage packaging, which thus excludes the use of flexible pouches and paper cartons.

The filling of liquid and wet foods is accomplished based on a predetermined level in the container or predetermined volume. The former is commonly used particularly for glass bottles. The filling operation may be achieved under the conditions that container opening is sealed with filling devices, or the containers are left open to the atmosphere. Volumetric fillers are usually adopted for the latter type of unsealed system.

The liquid filling operation may be done by top filling or bottom-up filling. In top filling, the filling tube is located just on the neck of the container for the product to drop from top downward or run down the side wall with minimum turbulence and air entrapment. In bottom-up filling, the filling tube is inserted down to the bottom of the container and gradually ascend upward as the product covers the bottom part of the container [4, 13]: the filling tube may be submerged under the rising liquid or remain just above the surface of the liquid being filled. The bottom-up filling protects the product from frothing, aeration, formation of air pocket in semi-solid, and undue vaporization. Filling valves may be operated pneumatically, hydraulically, or electrically, and their operation may be actuated by an electronic or pneumatic control system.

11.2.1 Filling to Predetermined Level

In the predetermined level filling, the fill volume is determined by the height of liquid in the container. Consistency in the container dimension and geometry is a prerequisite to attain the constant volume in the container by the level filler. Most liquid fillers use gravity to cause the product to flow. The gravity flow can be made under vacuum or pressurized conditions. The gravity flow can also be accelerated or assisted by imposing overpressure in the product supply tank or suctioning vacuum in the filled container. The volumetric rate of filling, $\dot{V}$ (m³/s) can be understood well by Equations 11.2 (Hagen-Poiseuille's equation) and 11.3 for laminar and turbulent flows of Newtonian fluid, respectively :

$$\dot{V} = \frac{\pi \cdot \Delta p \cdot D^4}{128\mu \cdot L} \tag{11.2}$$

$$\dot{V} = \sqrt{\frac{\pi^2 \cdot \Delta p \cdot D^5}{32\rho \cdot f \cdot L}} \tag{11.3}$$

where Δp is pressure difference (Pa), D is pipe diameter (m), μ is viscosity (Pa s), L is pipe length (m), ρ is density (kg m⁻³), and f is Fanning friction factor depending on flow rate.

The essence of Equations 11.2 and 11.3 is that flow rate in fill pipe is proportional to pressure differential across the pipe length for low speed laminar flow and to square root of the pressure differential for turbulent flow. Imposing high driving pressure on the route or increasing the tube diameter would increase the fill rate.

Pure gravity filling allows the liquid food to flow by gravity from the supply tank through a spring-loaded sleeve valve, which contains a concentric inner tube within a tube: the outer one for product flow and the inner one for vent air (Figure 11.4). The valve is opened when the container is placed and pushed up into sealed position against spring pressure. The air in the container is exhausted through the vent tube above the product level in the supply tank. When the air vent tube is covered by rising liquid in the container, the flow stops to determine the fill level because no air can escape and thus no further liquid can flow into the container. Then the container is disengaged from the filler and moved toward the closure application step. Gravity filling offers very accurate fill level with relatively simple design and easy maintenance. It is used for most free flowing liquids such as juices and edible oils.

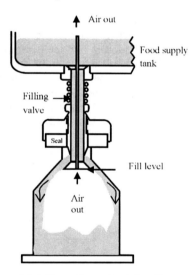

Figure 11.4 Operating principle of pure gravity filling

Gravity filling under vacuum achieves filling by gravity in a low level of vacuum (0.05-0.15 bar below atmospheric pressure). This low level pressure is maintained in the supply tank and extended to the container after an airtight seal is created between the fill valve and the container; other filling procedure and control of the level are undertaken as same as in pure gravity filling. Good control of fill level can be attained for still viscous liquids. It can also avoid filling a cracked container: vacuum cannot be attained in the container and thus container pressure is maintained higher than the supply tank to prohibit the flow of filling. For plastic bottles, collapse or deformation under vacuum may occur to make it hard or impossible to obtain consistent volume by filling to a predetermined level. Sometimes timed pulsating vacuum is used to have the thin plastic container walls flex in and out, helping the foam move up the vents [7].

Gravity filling under positive pressure called counter pressure filling is used for carbonated beverages. The supply tank is maintained under CO_2 pressure ranging from 1 to 8 bars above atmospheric pressure [13]. When the container is sealed to the valve, the vent tube opens to permit CO_2 pressure in the supply tank to be equilibrated in the container. Then the valve is actuated for the product to flow into the prepresseurized container gravimetrically. The product is ultimately filled to seal off the vent tube in the bottle filler or ball check valve in the can filler, which determines the fill level [12, 14]. Recently contact probes or conductivity probes have been developed and used for accurate level control [15]. Usual accuracy with a standard deviation of about 1 mm is obtained for the counter pressure level fillers. When the equilibrated stable conditions between the product tank and the container were ensured, the container is released and the closure is immediately applied. The product should remain as quiescent as possible to control the fill level. Operation at low temperature close to 0°C can keep foaming of beers and soft drinks at a minimum and filling speed at a maximum [16]. The normal stress on the plastic bottle by the filling valve may change its dimension and may not ensure the required fill volume by filling to predetermined level. Modern fillers for PET bottles use neck lifter with neck support ring to avoid this problem [12]. Care should be given to the point that the high CO_2 pressure in the bottle during filling operation may also cause temporary volume expansion of the plastic bottles resulting in a lower fill height than designed. Recently single or double pre-evacuations of bottles have been applied to the filling of carbonated beverages sensitive to oxygen in order to flush out residual oxygen [15].

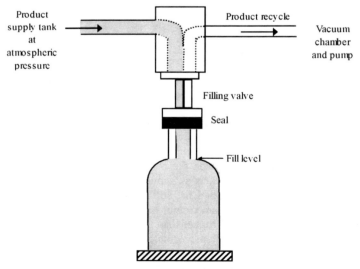

Figure 11.5 Operating principle of pure vacuum filling

Pure vacuum fillers use preliminary application of high vacuum in the container (0.85 or 0.95 bar below atmospheric pressure). The supply tank is normally maintained at atmospheric pressure. The vacuum in the container sucks down the product from the supply tank through the filling valve that opens after sealing against the container (Figure 11.5). The flow continues until the rising liquid blocks the vacuum port in the filling valve, which establishes the fill level. Overflow is recirculated to the vacuum chamber and recycled to supply tank. Vacuum filling is somewhat faster than gravity filling, and prevents a cracked container from being filled [13]. It is normally used when liquid is added to solid ingredients already positioned in the containers. Air trapped in the solid ingredients such as peach halves may be eliminated by the vacuum filler [7].

Pure pressure fillers apply pressure on the product by overpressure on the supply tank and direct pumping to make product flow from the tank to the container. The product is pumped into the container through the filling valve which is sealed against the container opening. The same mechanism as in pure vacuum filters is undergone to control the fill level and recirculate the overflow. The pressure filling is desirable for the product such as alcoholic beverages and hot filled juices, for which vacuum cannot be applied because of the requirement to maintain the constant composition or liquid state without boiled flashing [7]. It is also used frequently for lightly carbonated beverages.

Level-sensing fillers sense the fill level by ultra-low pressure air flow and fill to the predetermined level in the container. There is no need for sealing the container with the filling valve, and thus these fillers can be used for rigid and plastic containers. There is a sensing tube generating a weak air flow parallel to the fill nozzle. When the rising liquid blocks the air flow, backpressure triggers a control system to shut down the product flow into the container. The products should be free of foam which may block the sensing tube. Electronic or transonic sensing may also be used for level sensing [12, 15, 17]. An optical sensing technique may be used for transparent bottles.

11.2.2 Filling to Predetermined Volume

In the predetermined volume filling, the fill volume is measured before or during filling into the container. Accurate metering system should be provided with filler.

Piston fillers take the product from supply tank to a measured volume cylinder by rotary valve, and then discharge it into the container by push-in of the piston (Figure 11.6). The rotary valve is actuated by a mechanical cam to comply with the piston cycle of draw and discharge in the cylinder. Piston fillers are commonly used for viscous products that are paste, semi paste, or chunky with large particulates. Diaphragm fillers work with premeasured volume chamber the same as the cylinder in piston filler except that the cycle of charge and discharge is achieved by pressurized diaphragm. The diaphragm provides an absolute seal without frictional contact between the seal and cylinder wall. For accurate control of fill volume in volumetric fillers with metering cylinder, transonic level probe of float incorporating magnet may be used in the cylinder with pressurized stroke system [15].

Volume-cup fillers transfer first the liquid product to a measuring cup of precise volume and make the liquid to flow into a container through the filling tube. They have simple configurations and are reliable at an inexpensive price [5]. The measuring cup can be exchanged for different fill volumes. This type of filler is suitable for low-viscosity liquids [7].

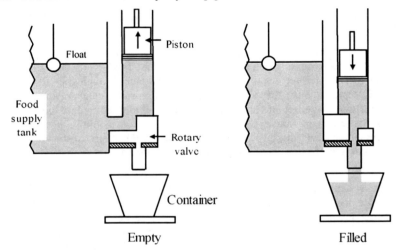

Figure 11.6 Piston filler

Timed flow fillers control the fill volume through the fill tube by time of the liquid flow. The liquid food under pressure is allowed to flow into the container for a controlled time. The pressure over the liquid in the supply tank should be maintained at a constant level for accurate fill, but some adoptions can handle minor variations in the pressure by a microprocessor control. This technique can be used for filling flexible bags [18].

Weighted filling used in rare cases for liquid or semi-solid foods weighs the fill weight by load cell in the filling station, which stops the filling at a predetermined weight. Usually gross weight is measured for the filling, but net weight may be checked with taring the container weight beforehand.

11.3 Filling of Dry Solid Foods

Filling of dry solid foods relies on measuring count, volume, or weight of the products. Dry food products may be divided into individually countable objects, free flowing powders, non-free flowing powders, fragile flakes, and sticky granules. Free flowing products fall freely under gravity and can be characterized by a cone shape with shallow repose angle when dumped on a flat surface. They have relatively consistent density and include rice, popcorn, sugar, beans, nuts, coffee, etc. Non-free flowing products do not flow readily under gravity and frequently clog the pipe. They are characterized by a cone shape with steep repose angle when dumped on a flat surface. They should be aided for feeding to a filler by an auxiliary device because of their tendency to agglomerate, clog, and entrap air. Wheat flour is an example of non-free flowing products. Sticky powders have the tendency to stick to metal or other surfaces: sugar and garlic powders become sticky when they absorb moisture. Fragile solids are easy to be injured by physical stress and include potato chips, popcorn, and breakfast cereals. Whatever the food material type is, variation in product properties such as density, flowability, and particle size should be minimized for good fill performance. To attain consistent and compact packaging in the container, high-frequency vibration and mechanical shock may sometimes be applied for the compression effect, particularly for the particulate materials with high ratio of length to width [10]. The nature of the product is required to be considered carefully in selecting the method of feeding and filling, which will be discussed below.

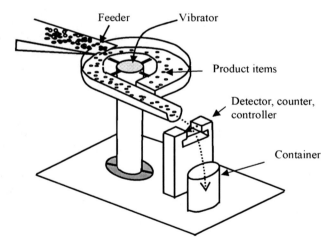

Figure 11.7 A counter filler with electronic counter mechanism

11.3.1 Filling by Count

Discrete items are often filled by counting (Figure 11.7). Electronic counters and perforated-disk counters are used for counting individual solid food items for filling into containers. Electronic counting of pieces is done by optical or non-optical systems: optical one counts the units of pieces passed by sensing the breakage of light or a specified shadow; the nonoptical one counts by electric or magnetic touch mechanisms [19]. A perforated-disk counter consists of a revolving metal or plastic plates having holes appropriate for tablets or pills [4]. The holes fed with products are allowed to drop the products into the discharge port. Counting fillers are used for filling fresh fruits and vegetables into net sacks, pieces of frozen meat and fish into trays, and processed food pieces into rigid containers [10].

The products should be aligned and fed into the detection zone in a controlled manner. Vibratory or V-belt conveyor systems are often employed. Recently computer-controlled robot systems with electronic camera have been introduced to handle, count, and place the delicate products in the containers [20].

11.3.2 Filling by Volume

Typical volumetric fillers are auger, cup, and vacuum fillers. All volume fillers depend on the bulk density of the product in attaining the required fill weight. Some recent machinery systems work in conjunction with automated weight checkers [20], which can correct the problem of density variation in the product. Auger fillers depend on the rotation of auger in the bottom of the product hopper (Figure 11.8). An auger is a long spiral wound-welded on a central spindle. With rotation of auger, the loose powder is compacted and moves out down the tube to fill the container. Rotation speed of the auger controls the rate of powder movement and the filling volume. The auger is driven by an electric motor on the top of the hopper. Auxiliary agitator may be added to facilitate the mixing and feeding in the hopper as shown in Figure 11.8. Auger fillers are suited for free-flowing and nonfree-flowing powders, granules, and flakes [8].

Volumetric cup fillers can handle free-flowing powders and granular materials (Figure 11.9). The product must be free flowing and have a relatively consistent density. A set of measuring cups are rotated beneath a fill hopper and the cup is filled. As rotation continues, the measuring cup is leveled off by a hopper brush. Rotation continues until the measuring cup is over the discharge hole, and the product is dispensed through a suitable chute via gravity. The sizing of the cups is based upon the volume required to fill a specified weight. Telescoping cups are used to make adjustments for different volumes as shown in Figure 11.9.

In vacuum filling, dry powders or granules are first sucked into a precise vacuum tube and then transferred to the container. This type of filler is normally used for rigid container. Pressure filling is hardly used for powder products due to its high risk of dust production and danger of explosion in packaging room.

Timed vibratory fillers use vibrators to achieve a controlled feed rate into the containers. Fill volume is controlled by the flow rate to the vibrator, frequency and amplitude of vibration, and feed time. These can handle powders, fragile products, and large discrete pieces [8]. Its operation is based on the assumption of constant product properties and filling speed, which cannot be satisfied perfectly and thus limits the filling accuracy somewhat.

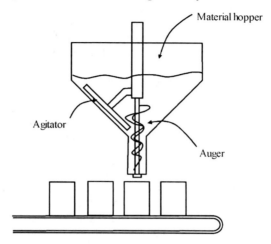

Figure 11.8 An auger filler. From Ref. [8] with kind permission of John Wiley and Sons.

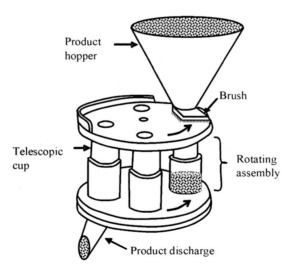

Figure 11.9 Volumetric cup filler

11.3.3 Filling by Weight

There are two principal types of weight fillers: gross weight and net weight fillers. Gross weight filling can be attained by operation of simultaneous fill and weigh, while net weight filling is usually operated with preweigh. Gross weighing is commonly used for the sticky products that tend to buildup on a weigh hopper or clog in the passage after weighing; for ensuring the filling accuracy of these products, the weight of the package should be checked after filling; filling accuracy based on gross weighing can be ensured by consistency and/or compensation of empty container weights. The net weighing system first weighs the product accurately and then transfers it to the container; net weight filling is the most accurate filling method for dry products, but it is slow.

Many different weighing techniques such as single- or double-action scale beams, head with compressed air, and electronic scales are employed for the weight filling. To have controlled weight, the supply of the product is divided into a main feed and a fine feed [6]. Initial bulk feeding fills about 80 or 90% of the required amount, and then fine feeding finishes the filling usually by dribbling up to an exact amount (Figure 11.10).

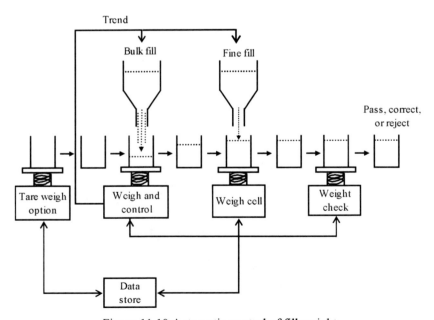

Figure 11.10 Automatic control of fill weight

In handling foods such as dry snacks and nuts, which vary widely in weight of individual pieces, computer-combination net weighing using multiple weigh buckets are employed so that each package contains a weight closest to the target one [21]. Each bucket is filled with a portion of the total net weight and a microprocessor assesses each bucket weight selecting the best combination of the buckets closest to the desired weight. Omar and de Silva [22] recently tried an optimized grouping of fish pieces into a package to make each combination as close as possible to the target weight of the package.

11.4 Closure and Closing Operation

Any filled package needs to be sealed or closed by an appropriate means. The sealing and/or closing operations complete a unitized separation of packaged product from the outside environment. Simple wire tie, tape, or clip may be applied for closing plastic or paper bags filled with foods such as breads and cookies. Tape closing is rarely used because of consumer's inconvenience in opening the bags. A wire tie is a metal core covered with plastic or paper, which gives dead-fold property in the twisted closing. Plastic clips usually of polystyrene (PS) with a printed message are used to close the bread bags. A zipper lock closure used for flexible plastic bags provides a convenience of easy opening and resealing; a zipper lock device located inside the heat-seal area offers resealing after opening the package. Most of the plastic pouches and trays are hermetically sealed by application of heat or adhesive: the sealing methods described in Section 10.5 can be applied to the pouch sealing.

Rigid containers of metal, glass, and plastics are mostly joined with closures to provide hermetic separation of the product from the outside. Metal cans filled with foods are usually closed with can end through double-seaming. Double-seaming ensures hermetic sealing of the can and makes it possible to heat-process the canned product. The properties of the double-seam have already been dealt in Section 8.6.5. Glass and plastic bottles are covered with a separated closure and joined by an appropriate mechanism. There are a variety of closures whose use is dependent on the product and container types. This section mostly deals with closing of glass and plastic containers, but the testing of closure and package integrity may also apply to the other types of packages.

11.4.1 Closures

Closure provides the hermetic sealing of the package and also allows the access for consumer's opening the package. So the closure, being an integral part of the package, performs functions of protection, convenience, and communication. There are so many types of closures used for different foods. Closures can be classified in various ways according to different criteria (Table 11.1). Threaded closures are applied and taken off by screwing on the container mouth. Friction closures are deformed with pressurized friction between cap material and container mouth. The closure material is usually made of plastic or metal. Tinplate and tin free steel are used for making crown and screw caps. Roll-on closures are made of aluminum. Plastic closures are replacing metal closures with improvement in the functionality of plastic ones. The most widely used plastic is polypropylene (PP) followed by polystyrene (PS), high density polyethylene (HDPE), and low density polyethylene (LDPE). Plastic closures can easily incorporate a dispensing or convenience function. Natural cork is still widely used for wine bottles. Choice of closure is also dependent on the internal pressure or condition. For example, high internal pressure of carbonated beverage requires stronger seal between the closure and the bottle mouth for attaining the package integrity, which restricts the window of applicable closures for this type of product. Specialty closures adopting a variety of specialized design are available for satisfying a particular function or requirement, which includes tamper-evidence, easy and controlled dispensing, gas venting, etc.

Table 11.1 Classification of closures based on different criteria

Criteria	Type and typical examples
Engagement method	Thread-engagement: screw cap, lug cap, roll-on, press-twist Friction-fit: crown, snap-fit, stopper, cork
Material	Steel: crown, lug cap, Al roll-on, screw cap Plastic: snap-fit, screw cap Wood: cork
Internal package atmosphere	Atmospheric: snap-fit Vacuum: lug cap, press-on/twist-on Pressurized: crown, roll-on
Container type	Glass: lug cap, crown Plastic bottle: roll-on, screw cap Metal: screw cap
Special function	Tamper-evident, dispensing convenience, child resistant, gas venting, etc.

Closure structure consists of a panel of top flat center area, a skirt of side vertical round, a shoulder connecting panel and skirt, thread (or lug), and tamper-evident band (Figure 11.11). Hermetic seal of closure onto the mouth of container is attained by the close contact between the closure and the mouth. The seal contact is often augmented by a resilient liner compressed between the closure and the container mouth, providing a tighter and secure fastening. Liner usually consists of backing material and facing material. The backing must be soft and elastic to provide cushioning and compression in the seal joint. The widely used backing materials include ethylene vinyl acetate (EVA) and cork. The facing should have an adequate barrier property and abrasion resistance.

Typical facing materials include Al-foil and laminated plastic film. Sometimes a lubricant is added on the surface of the facing. A plastisol of polyvinyl chloride (PVC) is applied as a molten state unto underside of the metal cap and then cured to work as an integral liner. In recent years, foamed plastic material laminated on single or both sides have been used for liners. The liner may simply be pushed into the cap and be free to rotate in the closing operation. Or the liner may be glued into the cap not to rotate freely. The fixed liner needs lubricant for closing and opening. In recent years the entire thermoplastic cap with integrated seal liner is molded in a single operation and is sometimes called 'linerless closure'.

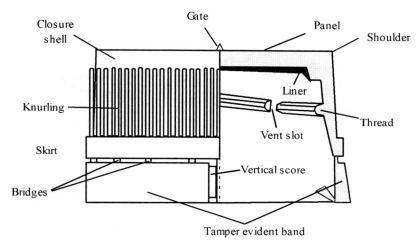

Figure 11.11 Typical design structure of continuous thread closure with its nomenclature. From Ref. [23] with permission.

Continuous thread (CT) closure is a screw cap made of metal or plastic (Figures 11.11 and 11.12). The metal materials cover tinplate, tin-free steel, and aluminum. The plastic materials include injection-molded thermoplastics such as PP and HDPE as well as thermosetting resins of phenol-formaldehyde and urea-formaldehyde shaped by compression molding. Thermoset closures have higher resistance to high temperature and solvent, are stiffer and more expensive than thermoplastic ones. PP is stiff enough, resistant to the moderately high temperature used in food processing, and also flexible to be formed as an integral linerless closure providing acceptable seal contact with the container mouth finish. Polyethylene closures are used where greater flexibility is required and/or use temperature is lower. Thermoplastic closures are replacing the metal and thermoset closures because of their economy and easiness of forming into a variety of shapes. However, plastic closures are used to a limited extent for packages heat-preserved after closing.

Figure 11.12 shows major dimensional codes used for CT closures. The CT closures are designated by diameter in mm followed by finish series number. Thus the designation of '22-400' means a closure with diameter of 22 mm and a finish type of a shallow continuous thread. The CT closure attains a seal with container by engagement of its thread with the corresponding threads on the container finish. The cap threads are designed to fit a specification of bottle threads (Figure 11.13). L-style thread with 30° angle is the one developed for glass bottles for which sharp corners are difficult to achieve. For plastic containers, M-style thread of 10° angle is generally used because it allows a tighter and better grip with closure thread. Rotation of 360° or more is usually required for a satisfactory seal from a CT closure.

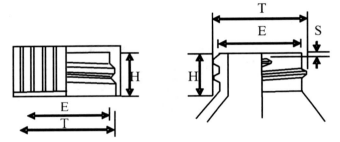

Figure 11.12 Closure dimension code. T=Outside diameter of the thread on bottle finish or dimension of thread root inside closure; E=Diameter of thread root on the bottle finish or inside dimension of thread in closure; H=Vertical dimension from top of finish to bead on bottle finish or dimension from inside top to skirt bottom on closure; S=Vertical dimension from top of finish to thread start on bottle finish or dimension from top inside to thread start on closure.

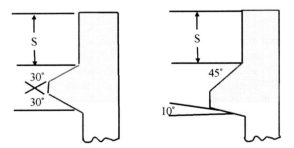

Figure 11.13 Styles of finish thread contour

With technology development CT closures have been widely used for most food products packed under vacuum, atmospheric, and pressurized conditions. Earlier versions of CT closures did not have the independent device of tamper evidence, but most of recent versions tend to be equipped with breakable band for this (Figure 11.11). Lower part of cap skirt is linked through perforations to the cap body and separated when the cap is twisted for opening. There are two types of tamper evident bands, a drop-down band remains on the bottle neck, and 'pigtail' band is removed from the neck staying on the cap [23]. The plastic closures for carbonated beverages have vent slots for ensuring the adequate gas release and preventing the closure from being propelled away (blow off of tail end) at the moment of opening the bottle.

Lug caps engage the noncontinuous thread on the container finish by 3, 4, or 6 lugs protruded inward on the bottom edges of the closures. The lug is formed raised internal impressions. The final engagement is achieved through the interlocking between the lugs and horizontal thread, and works to tighten the close contact of bottle finish with the liner on the cap. The stress in the contact is high to confer hermetic good seal, which makes the lug caps popular for heat-processed vacuum packs and restricts their construction material to the thick tinplate. The sealed lock is formed with fewer degree of rotation (less than a quarter turn) than CT closures, which gives the food packer the speedy capping and the consumers the convenience in removal of the cap at opening of the bottle. The lug caps are extensively used for vacuumed glass containers and hardly used for plastic containers because of their hard stress on the thread. They are available in sizes ranging from 27 to 100 mm.

Press-on/twisted-off metal closure attains a seal thread of plastisol by steamed vacuum capping of glass container. The pressure from the capping operation and the heat in steam sterilization form a thread on side gasket in the cap contacting the thread of the glass bottle finish, which assists the opening of the closure by the consumer. The press-on/twisted-off closure is mostly used for heat-preserved vacuum packages of baby foods, sauces, and juices. The closure size varies from 27 to 77 mm. The center part of top panel is embossed in the shape of ring, which becomes flexible a little to move down under vacuumed conditions of the package. With opening the package and breaking of the vacuum, the depressed ring pops up with sound. The held-down state of the ring indicates the soundness of vacuum condition. Thus, the embossed ring has a function of tamper evidence and is often called 'safety button'. The safety button device is also sometimes adopted for lug caps.

The crown cap attains a friction-fit sealing by short skirts with 21 flutes crimped onto the bottle mouth. The crown is fabricated by stamping tinplate or tin-free steel followed by addition of liners. The liner may be a preformed type of plastic or cork, or a plastisol directly flow-formed. Majority of crowns are 26 mm and are used for glass bottles of beer and carbonated beverages. A disadvantage of crowns is the inconvenience in opening requiring an opener. There are twist-off crowns with easy-opening function, but they provide less protection against tampering. The crimping operation of twist-off crowns is deigned to shape the metal material into the bottle thread.

A roll-on closure is formed into final shape by the pressurized rolling of the preformed aluminum shell onto the bottle threads. Simultaneously there exists a top pressure assuring the seal between the container finish and lining compound of the cap. Roll-on closures mostly have the breakable skirt part for tamper evidence device, which is similar to that of CT closures. Most common size of aluminum roll-on closures is 28 mm, with other variations of 14-38 mm. Al roll-on closures have extensively been used for packaging beverages carbonated and noncarbonated. These days the roll-on caps have been replaced to a great extent by plastic CT closures.

Plastic snap-fit closures are made mostly of low or medium density polyethylene flexible for application onto the container mouth. The cap is deformed for the engagement bead to fit onto the container mouth finish with passing the external bead on it. Opening of container deforms the cap again to snap back it from the mouth. The snap-fit closures can be applied to the container at high speed. Typical applications of snap-fit closure are syrup and noncarbonated mineral water that do not require vacuum or pressure inside the package.

There is also a plastic-metal combination closure, which includes a flat metal panel inserted into a threaded or beaded plastic ring closure. The metal panel contains the lining compound for hermetic vacuum sealing. The plastic ring offers the tamper evident function and reclosability.

Cork is compressible and elastic material which is highly impervious to air and water penetration, and has low thermal conductivity [24]. It is deformed with tight fit to the inside of the bottle mouth when pushed into it. The cork is used for stoppers of wine bottles. With diminishing production of cork, there have been attempts to replace natural cork with a plastic cork of polyethylene.

Sometimes the container for closing operation has an inner seal under the main cap or closure. The inner seal may be obtained by a glassine paper or foamed polystyrene with the help of adhesive. Or induction heat sealing is applied as an instantaneous inner seal at the capping operation (see Section 10.5.1). The inner seal works as a tool of tamper evidence.

As with some tamper evident tools already mentioned above, several other devices are added to the cap for providing tamper evident function. Shrink band or capsule over the cap, tape over the seal, and plastic tear tab attached onto the cap are the examples. Foil capsule, wrap, and wax seal are used for aesthetical preference and tamper evidence. Special convenience functions are often appended to the closure: ready and multiple accesses to the products are provided by a variety of designs such as spout, plug-orifice, and dispenser; some fitment devices are inserted to regulate the flow of liquid, powder, and flake foods; plastic over cap is applied for protecting an inner seal and/or reseal function. As another type of specialized closure, venting closure is applied for the food products producing gases during storage in order to maintain a stabilized pressure inside the package: the closure mostly has a few small notches, pinholes, valves, or microporous membrane in its structure for the gas to escape the package without disturbing the required package integrity. An oxygen absorbing liner may be inserted into the cap for the oxygen sensitive products such as beer.

11.4.2 Closing Operation

Closing operation of the filled container is commonly achieved by a chuck that grips the closure and turns or pushes it into the desired place. A roller or belt is also sometimes used to apply the cap to the container. The closing operation of container is closely integrated with the filler and constitutes a station with straight line or rotary configuration; the closure is presented at the right angle to the moving container, which captures the closure and is transferred to a sealing or chucking station. An appropriate mechanical action of the capping head tightens the closure to the container finish. The capping head tightens the closure until the desired torque is reached. A quality closure seal can be obtained by a good combination of closure, container, and capping machinery. A tamper-evident shrink band may be placed over the capped container and then exposed shortly to heat for secure shrinking.

Severity in twisting effect of the capping head is represented as torque, which is defined as the force multiplied by the perpendicular distance between the line of action of the force and the center of rotation at which it is exerted. Torque is measured at application or removal of a threaded or lugged closure. Application torque is a measure of tightening given by the rotating capping head. Removal torque is a measure of force required to loosen and remove the closure from the container mouth. It is influenced by several factors such as packaging material, package design, and process conditions. Removal torque usually decreases with time after application of the closure until it stabilizes at a certain level after several weeks: generally the closure loses about half of its torque in one day. Storage temperature also affects the change of the torque. Removal torque too low presents a risk of accidental loosing of the closure, while too high torque causes the difficulty in opening the container. The larger the cap diameter, the higher the required torque: removal torque of 1.2-1.7 N m is ideal for a 28 mm closure, and the torque for different diameter closure can roughly be given proportionally to the diameter ratio. The opening torque of the food products should stay in proper level suitable for their relevant consumers: different groups in age and behavior may be different in their opening torque or strength [25].

11.4.3 Package Integrity Testing

The food packages closed with a closure or sealed by application of heat or adhesive should have adequate integrity to preserve and deliver the contained food throughout the distribution system for the required shelf life. Ingress of microorganisms, dust, and gases should be prevented or minimized to keep the food safety and quality. Therefore, the integrity of the food packages should be assured by proper testing method before the products leave the manufacturing plant. In particular, there has been great emphasis on the package integrity for thermally processed foods of high moisture because their safety and stability after proper processing are threatened by post-process contamination through leaks.

Conventionally there have been specifications on the double-seam of the metal cans and closing of the glass bottles. The packaged products have been sampled randomly according to statistical quality control plan and examined in the seam or closure. For example, the double-seam of the can is cut and examined under magnified projection to see whether the specification is satisfied (see Section 8.6.5). If the requirements are not met for certain samples, the product lot from which the samples are taken is discarded or repacked. Level of vacuum is measured by piercing vacuum gauge for the food cans and glass jars thermally processed as an indication of the package integrity. These kinds of destructive testing have been extended to plastic packaging. A simple compression test may be used to have the general idea on the seal resistance against the stress. A filled and sealed food package is placed on flat surface with subsequent application of pressure and then occurrence of leaks is observed. A burst test of the empty package and tensile test of the seal area are valuable for assessing the soundness of heat-seal, and have been discussed in Section 10.5.3. A bubble test applies air pressure to a closed package located under water to find visually leaks emitting bubble streams. In dye penetration test, dye or ink solution is applied to inside surface of a cleaned package at the seal or suspected location of failure, and the possible penetration toward outside is visually observed after peeling up the test location: dye solutions of fluorescein, rodhamine B, and safarin O powder in isopropanol are most commonly used. An electrolytic test measures the electric current flowing across the packaging barrier which might have a pinhole or leak. The presence of electric current indicates the presence of leaks in the package. Light passed through the pinhole or defect area on the emptied opaque package is also used as an indication of loss in package integrity and commonly applied also for confirming the absence of pinhole on the Al-laminated film.

The critical defects causing the loss of hermetic plastic package integrity are found in the forms of channel leak, seal wrinkle, air void on the seal, incomplete weak seal, and contaminated seal. There have been extensive studies of biotest to determine the threshold size of the defect against bacterial penetration. Immersion biotest submerges the representative packages in a high concentration of active bacteria. Aerosol biotest uses a spray cabinet which sprays directly the bacterial dispersion onto the package. The bacterial penetration to the packaged product is detected by swollen packages or direct measurement of the live bacterial growth in the product after incubation for a couple of weeks. The most frequently used organisms are *Enterobacter aerogenes* for foods with pH >5.0 and *Lactobacillus cellobiosis* for foods with pH <5.0. Threshold size for bacterial contamination found from several researchers goes generally down to below 10 µm and is dependent on exposure condition, channel length, and type of the defect [26].

Even though destructive testing based on statistical control is useful, a lot of time and money should be spent for its management. Destructive testing prohibits testing of all the packages, which is another drawback. Therefore there has been a growing interest in nondestructive testing of integrity of every package on line. As an earlier version of the nondestructive testing, metal can or glass jar packed in vacuum is tapped by a metal rod to measure the frequency of vibration tone. The frequency different from normal range tells the losses of vacuum and thus seal integrity. Since the introduction of retortable plastic packages and semi-rigid plastic containers for thermal-processed foods, commercial package integrity testers targeting 100% on-line inspection of the packages have been developed and are available. There also have been a lot of technological trials and developments in this area that do not reach the commercialization stage. Table 11.2 lists recommended methods of package integrity testing depending on package types.

Table 11.2 Test methods recommended for detecting microleak of different types of food packages

Package type	Recommended test method
Paper board package	Compression, dye penetration, electrolytic test, incubation, sound, visual inspection
Flexible pouch	Burst, compression, dye penetration, incubation, tensile test, visual inspection
Plastic package with heat-sealed lid	Burst, dye penetration, incubation, sound, tensile test, vacuum, visual inspection
Plastic can with double-seamed metal end	Incubation, proximity sensor, scope projection, sound, visual inspection
Metal double seamed can	Dye penetration, incubation, proximity sensor, scope projection, sound, vacuum, visual inspection
Glass jar with screw closure	Dye penetration, incubation, proximity sensor, scope projection, sound, vacuum, visual inspection

Extracted from Ref.[27].

Most on-line package inspection systems are based on a stimulus-response principle as shown in Figure 11.14 [28]. Testers based on using pressure differential across the packaging barrier are the most popular on commercial scale. External pressure or vacuum is applied outside the package located inside the test chamber with measurement of resultant changes in structure and decay of pressure or vacuum in the package and/or the chamber: failure of the expected change indicates the presence of leakage. Gas injected at filling stage (such as helium and hydrogen) may be used as a tracer indicating leakage for evacuation test. Internal pressure can be created by mechanical means of squeezing or heating, and the subsequent structural and/or internal pressure changes are detected by appropriate sensor such as proximity sensor and load cell. An example of a pressure differential test is given in the case study below. There are other innovative techniques in the development stage, which applies machine vision, infrared imaging, X-ray, ultrasound, magnetic resonance imaging, etc. For modified atmosphere packages, O_2 and CO_2 color indicators responding to the gas concentration changes have been suggested as visible indicators for leaks [29].

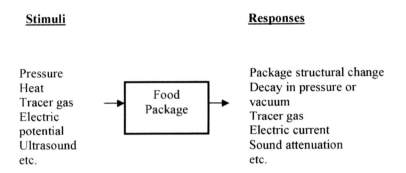

Figure 11.14 Principles of package integrity testing

11.5 Methods of Wrapping and Bagging

As with flexible packaging of foods, there are a variety of wrapping and bagging methods by paper, plastic film, metal foil, or combination materials. The method of wrapping and bagging is determined by required appearance, protection degree, economics, and speed.

11.5.1 Wrapping

Wrapping or overpackaging is used for protection of individual food product from mechanical, chemical, and biological stress, or for grouping together individually packaged items. The wrapped film may be a type of plain, stretch, or shrink form. The operation of wrapping may be done by several ways: product may be positioned first on or beneath the film, which is subsequently folded upon it; the product may be pushed horizontally into the wrapping film [30, 31]. The wrapped package may be stretch wrapped, heat-sealed, twist-folded mechanically, heated for shrink form, or cold-sealed with a layer of latex.

Figure 11.15 shows some basic patterns of wrapping foods. Parcel wrap of rectangular shape has an overlapped long seam with folded ends. Bias or oblique wrap has underfold seam and its foil application is often used for chocolate block in conjunction of paper sleeve wrap, which is a simple band with exposed ends. Twist wrap is a twisted cylinder and mostly used for hard candies. Roll wrap is cylinder form with long side seam and end-lock folds, which is used for mints, candies, cookies, and biscuits. Bunch wrap is used for holding irregular-shaped product such as chocolate, lettuce, and pizza with a series of folds and tucks in aluminum foil or stretch film. Bread wrap has long overlap on bottom with end-lock folds.

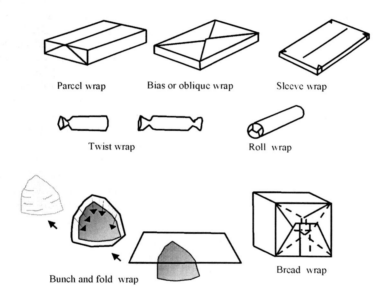

Figure 11.15 Main types of wrapping food

11.5.2 Bagging

Bags for packaging foods can be preformed or formed in line with filling and sealing. Preformed bag or pouch may be one of two styles: the open-mouth and valve bags. The former is open at one end, which is closed or folded after filling of food; closing of bags can be attained by twisting, folded taping, heat-sealing, clipping, tying, or sewing. Valve bags with closed ends are filled through a valve. The valves in heavy-duty bags for dry powders or granules are often designed to self-close by the internal pressure of filled product. But the closure of this type valve is not completely sift-proof and is sometimes augmented by hot melt sealing for integrity and sift-proofness. Small size valve bags are used for packaging drink products with function of easy dispensing, convenience, and reclosurability. Paper bags or sacks are manufactured in various premade forms by sewing, adhesive, or heat sealing of inner laminated layer. They can have a flat or gusseted structure with simply sewn, pasted, stepped, or satchel bottom end.

The simplest method of plastic bag making is bottom-sealed tube. The tubular blown film with or without gusset is sealed by hot-wire or impulse sealer on bottom, giving an open end. The bottom-sealed tubular bags usually of LDPE are often supplied on a roll and used for wrapping fresh fruits and vegetables in a supermarket. Other uses are industrial liners. Premade bottom-sealed plastic bags or pouches are often filled with foods and sealed to have two welds on bottom and top (Figure 11.16). Two weld pouches are often used for packing foods on small scale. Three weld pouches with lengthwise side seam are used for variety of foods as pillow packs, which are usually formed as tubular form in line; the side sealing by heat may be done in two ways: fin seal and lap seal (Figure 11.17). Fin seal is made by heat-welding between same sides of the film: only single surface with heat-sealability suffices for the sealing. Lap seal covers the outer surface of the film by the inner side for heat sealing: both surfaces should be heat-sealable. Three weld bags with bottom fold are formed by folding the film on its center lengthwise and subsequent filling and sealing on three sides. Four-weld pouches were based on joining of two films, which may come from identical or different webs. Stand-up pouch with gusset base is another variant of three or four weld one, and gives the function of standing by itself. The gusset base may be from a different sheet of stress-resilient film. Another recently developed stand-up pouch with attached recloseable opening valve is formed from three or four panels: front and back facings with two side gussets or a bottom gusset.

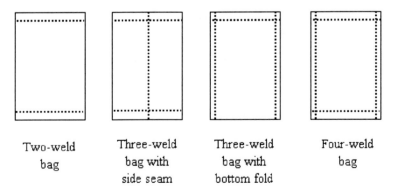

Two-weld	Three-weld	Three-weld	Four-weld
bag	bag with	bag with	bag
	side seam	bottom fold	

Figure 11.16 Typical types of sealed plastic bags. Dotted line represents sealing part.

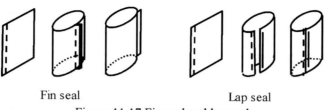

Fin seal   Lap seal

Figure 11.17 Fin seal and lap seal

Recently pouches are often equipped with zipper-type device enabling secure reclosure and multiple opening. There also appeared a pouch bag zipper closure extended inside from tamper-evident heat-seal. Bag-in-box packaging is another application area of specialized bag with closable spout [32]. The spout-sealed tough film of high oxygen barrier is combined by another film to form a bag with tapped closure. The bag filled with liquid food is located in a rigid box, and the food is dispensed in small amount conveniently through the spout whenever needed. The outer box is usually of corrugated board, but sometimes of metal or rigid plastic containers. The spout is usually used for filling the bag and also for dispensing the food. Many sophisticated designs of spout and closure are available depending on the end-user requirements. Major uses of bag-in-box are beverage packaging for restaurants and institutional food services with fill size ranging from 10 to 40 L. Small size down to 2 L for retail and large size of 1,000 L for bulk transportation are also available. With dispensing of the product, the bag collapses not to allow the admission of air, providing quality protection against oxidation.

11.6 Form-Fill-Seal

Form-fill-seal means an integrated operation of forming a package, filling the food product into the package, and sealing it usually in a single machinery or linked automated system. The steps of package forming, product filling, and package sealing are synchronized to have consistent fill and integrated sealing. Form-fill-seal provides high production rate at economical and reliable mechanism. Recent developments make the structure more compact and the speed higher with thanks to the intelligent use of electronics and various control components. Flexible bags and semi-rigid trays based on paper, plastic, and aluminum in combination are produced. Solid, powder, liquid, and viscous foods can be handled with form-fill-seal system. A variety of fillers are combined into the forming compartment to have a speedy handling of the products. Heat sealing is usually applied, but some high speed system adopts cold sealing technique particularly for confectionery products [20]. The form-fill-seal machines are divided into vertical and horizontal types depending on the flow of the film, sheet, or container. The vertical form-fill-seal machine transports the film and package from top to bottom direction in the course of forming, filling, sealing, and separation of individual package. The horizontal machine feeds the film and container on horizontal line or plane.

11.6.1 Vertical Form-Fill-Seal

The vertical form-fill-seal machine is characterized by forming a vertical tube from a film web reel. The film material unwinds from web reel and is transferred through guiding rollers. The bag forming may be of pillow type, three-weld pouch, or four-weld pouch (Figure 11.18). The bag forming is aided by forming shoulder or triangle device engineered optimally for bag type and material property.

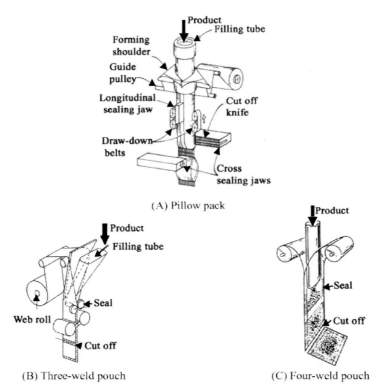

Figure 11.18 Typical types of vertical form-fill-seal operation

Pillow type bags are general type of vertical form-fill-seal. The film moves down the forming tube to have overlapping, on which side fin or lap seal is made. The film transport may be either intermittent or continuous. Intermittent movement is attained by clamping the material with the cross-seal jaw, which experience the cycle of up and down strokes. Figure 11.19 explains the principle of intermittent operation in vertical form-fill-seal pillow packaging. The intermittent movement of the film tube is provided by an arrangement with a reciprocating clamp suctioning it. Continuous film movement is attained by drive belts contacting the film (Figure 11.18A). The belt driven mechanism is a versatile design for high speed and simple operation [33]. Vacuum may be drawn through perforated drive belt(s) to lock the film firmly; however this design imposes limitation on machine speed and may produce noise. With film advancement cycle in lengthwise direction, the contents are filled into the formed pouch and the end portion is cross-sealed with top and bottom weldings. The filled package is then cut off crosswise between the two sealings by a cutting device and released away from the machine. The bottom sealed bag on the tube is ready for another cycle of filling. Pillow type vertical form-fill-seal machinery can be applied for modified atmosphere packaging of granular and free flowing products such as coffee, nuts, and snacks. In this case two concentric tubes are used for bag forming and filling of product and gas: inner tube for product filling and space between two tube walls for gas introduction [34].

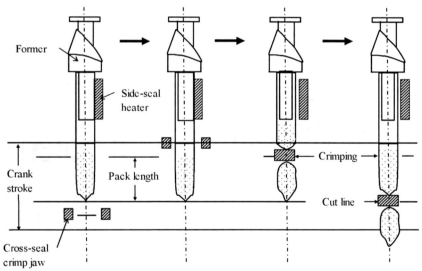

Figure 11.19 Intermittent operation in vertical form-fill-seal pillow packaging

Three-weld pouch in vertical form-fill-seal is formed by folding over a triangle from the film supplied from a web reel (Fig. 11.18B); the flat folded film is sealed crosswise and lengthwise with vertical movement, and then filled with product. Four-weld pouch in vertical form-fill-seal is made from combining two flat films (Fig. 11.18C); downward film advance is done either by vertical seam welders or by cross-seal jaws. The mechanism of cross sealing and cutting of vertical form-fill-seal in three-weld and four-weld pouches is the same as in pillow type.

Vertical form-fill-seal system has been improved to handle aseptic, cold, and hot-fill pumpable products with vacuumed conditions. Recent developments incorporate fast pack size change, printer, and labeler [35].

11.6.2 Horizontal Form-Fill-Seal

Horizontal form-fill-seal can handle various types of pouch and tray packages: three-weld and four-weld bags, pillow bag with side seam, thermoformed tray, etc. Bag forming in horizontal machine is similar to that in vertical one, except that the film movement is horizontal in the former (Figure 11.20). The movement of folded film and pouch may be intermittent or continuous. Filling is accomplished either by filling tubes fixed or moving together with pouch. Heat sealing may also be accomplished by band sealer or moving sealing bars. The design varies in the sequence of pouch cutting: pouch may be cut from the film fold before or after filling. The line of pouch movement during filling in pouch standing vertically may be linear or rotary. Pouch opening device or suction grip is often employed before filling step.

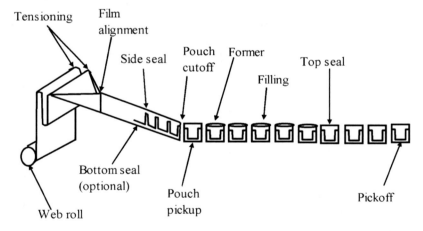

Figure 11.20 Typical flow diagram for horizontal form-fill-seal of bag package. From Ref. [36] with kind permission of John Wiley and Sons.

Pillow type pouch package can be formed and used for wrapping a wide variety of solid foods such as cookies, candy bars, and cheese by horizontal form-fill-seal machine. The film from single reel is passed through a folding box to form a horizontal tube, two edges of which are heat-sealed together under pressure usually on the bottom-side of the package (Figure 11.21). The product on the infeed or conveyor is passed into this tube. In some machines the film tube is formed around the product moving on the conveyor. Some types of machines adopt the feeding of the film below the product and forming a tube with longitudinal sealing on the upper side of the package; this type of machine called inverted type allows the product to be loaded onto the film, which is advantageous when handling sticky products such as cheese and sliced meats; the inverted pillow pack machine can also eliminate contamination on longitudinal seal area which may be caused by debris from crumbly products. Cross-sealing and cutting-off of the sealed package complete the cycle of the form-fill-seal. Some machines achieve the cross-sealing by orbital-motion sealer, which provides enough sealing time. Recent developments use low temperature seal coating for achieving faster sealing and avoiding film shrinkage [35]. Horizontal form-fill-seal machines for pillow type package are widely used for modified atmosphere packaging [34, 37]. A lance protruding into the tube of packaging film is used for the passage of the flushing gas. Shrink or vacuum packaging is also sometimes combined into the horizontal form-fill-seal machine.

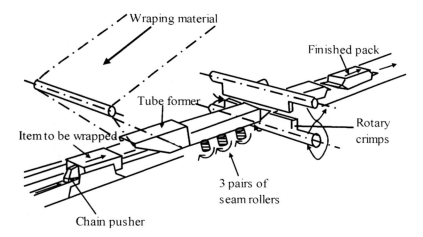

Figure 11.21 Horizontal form-fill-seal of pillow type package. From Ref. [20] with kind permission of Springer Science and Business Media.

A series of thermoforming, filling, and sealing also takes the concept of horizontal form-fill-seal. A thermoplastic web sheet is thermoformed into a rigid or semi-rigid tray, filled manually or by automatic method, lidded, and sealed with top web (Figure 11.22). Separation of filled container by cutting completes the cycle of form-fill-seal. A die train forming the tray also plays a role to move the bottom web, which has been formed into desired shape. Mechanical clip often aids the movement of the web film. The thermoform-fill-seal machines usually operate by intermittent motion, but some machines work continuously.

A variety of liquid, semi-solid, and wet solid foods are packaged by this method: the popular examples of foods are cured meats, cheeses, baked goods, and fresh pasta products. The thermoform-fill-seal machines are usually used in combination with vacuum or modified atmosphere packaging; the loaded trays are moved with lidded top layer to the chambers, in which evacuation and gas flushing are achieved with heat-sealing; evacuation of air and subsequent reintroduction of modified atmosphere can accomplish a target atmosphere with a low residual oxygen level. Recent innovations include the expanded polystyrene container with a high barrier film. Some developments are the combination with vacuum-skin or shrink packaging [34]. Some systems operate in semi-aseptic conditions for improved storage stability. Recently mechanical cold-forming of aluminum laminated film has been applied to form-fill-seal concept for retortable pouch packaging.

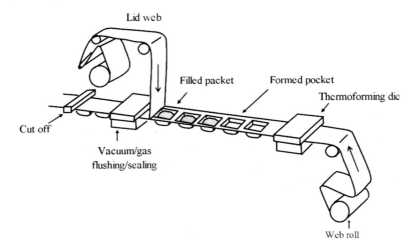

Figure 11.22 Thermoform-fill-seal system. From Ref. [38] with kind permission of John Wiley and Sons.

11.7 Various Forms of Contact and Contour Packaging

Transparent plastic film or form that is shaped or clung around the packaged product, gives an aesthetic look and may eliminate oxygen in the headspace when in close contact with the product. It can also help in handling of the packaged products. Proper application of contact and/or contour packaging thus may improve the appearance, protection against outside stresses, and preservation of food quality. Contour packaging has an emphasized function of sales promotion.

11.7.1 Skin Packaging

In case of nonfoodstuffs such as dishes and utensils, the product is placed on the heat-seal coated porous paperboard, over which the heated film is dropped down on the product (Figure 11.23). A vacuum is drawn through the porous paperboard to cause the film to drape over the product and the board. The film is heat-sealed onto the paperboard and the package is separated by cutting. An intimate contact between the product and film is attained to give a good contour wrap. Usual cycle time for skin packaging is about 40 seconds with line speed of 5-6 m/min [39]. Vacuumizing acts to form the intimated contact shape in the shaping, but low oxygen level is not maintained inside the package because of the porous structure of the paperboard. The paperboard should be compatible with printing and heat-seal coating, and is sometimes perforated at 3-4 holes of 0.15 mm diameter per cm^2 in order to increase the porosity. Films for skin packaging should have good properties in hot melt strength, tensile strength, adherence to paperboard, and clarity. The thin films of LDPE, PP, PVC, and ionomer in thickness around 100 µm are widely used for the skin packaging [40].

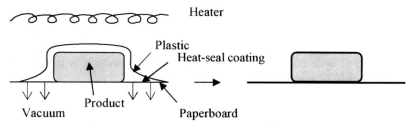

Figure 11.23 Principle concept of skin packaging

Skin packaging for foodstuffs uses a high barrier sheet as the bottom layer. It applies the heated top film on the printed bottom in the vacuum chamber with followed heat sealing, which can retain low oxygen level inside the package [30]. This technique called vacuum skin packaging is applied to sliced meat, cured meat, pate, and fish products [41]. Top films with proper permeability characteristics are used depending on the product. For retail fresh meats, gas permeable film is tightly wrapped on the product for the bright red color (see Section 17.3). For cured meats and fishes, multilayered high barrier film is used as a top layer for quality preservation.

11.7.2 Blister Packaging

Original meaning of 'blister' is concerned with a raised skin patch filled with watery matter, which may be caused by burning. The preformed blister is loaded with food product, and then was over-laid with heat-seal coated paperboard, which is heat-pressed to the former for sealing (Figure 11.24). Blister packaging has been another type of carded packaging like skin packaging, offering the good effect of display on the shelf. Recent development adopts form-fill-seal for blister packaging: blister is thermoformed, loaded with product, and sealed by top layer of coated aluminum foil or laminated film instead of paperboard. As blister, sheets or films of regenerated cellulose, PS, PP, polyethylene terephthalate (PET), and PVC with thickness of about 0.2 mm are used due to excellent clarity, thermoforming property, and heat sealing property [40].

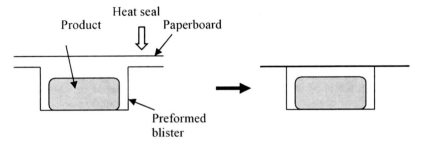

Figure 11.24 Principle concept of blister packaging

Blister packaging applications in food include chocolate, sliced ham, Frankfurter sausage, and sandwich. Tablets of boiled sweets and chewing gums are often packaged in blister sealed with bottom layer of sealant-coated aluminum foil with 20-30 μm thickness; this kind of package is called press-through-pack (PTP) and offers convenience in opening.

11.7.3 Shrink Packaging

Shrink packaging also provides tight wrap around the product. The shrink films are manufactured to shrink under heat by stretching a heated film (at a temperature close to its softening point) and cooling rapidly. These films have a tendency to shrink to the original dimension when heated. Shrink ratio is defined as the ratio of reduction in film length compared to original length when exposed to heat. Uniaxially oriented film has the heated shrinking in the direction of its orientation, while biaxially oriented one shrinks in both directions. Biaxially oriented film may be balanced in both directions, or preferentially balanced with different axial shrink ratios. In application, the shrink film is loosely wrapped on the product and then passed through a heated tunnel shrinking in tight contact between the product and the film (Figure 11.25). Heat exposure of the wrapped product for a few seconds does not ordinarily affect the product, but does let the film to shrink tightly around the product. Heat tunnel may be replaced by hot water bath or radiant heater. The film for wrapping may be in a form of center fold, sleeve, or bag. Usual form is in center fold, with which side and end of the wrapped film are sealed by L shaped bar sealer-trimmer. U shaped bar sealer-trimmer is used for four-side sealing on two web films (Figure 11.25). Sleeve wrap with open ends has often been used for tamper-evident neck-bend, printing decoration, or multipackaging of rigid containers. Simple sleeve wrap places the blown-extruded open tube over the container for tight fit. Large one may have one back sealing.

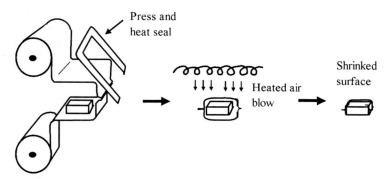

Figure 11.25 Principle concept of shrink packaging

Shrink packaging is used both for individual food item of irregular shape and for assembled unit of cans and bottles. In case of unitized package, shrink packaging is may be applied in sleeve wrap or all round wrap: sleeve wrap allows the two end sides of package open and all round one over-wraps whole collection with only small perforations for air escape. Sleeve wrap can save the material and the empty hole parts on the package sides may be used as handles for convenient transfer. The holes on side may also allow the accelerated moisture drying from the packaged cans coming from sterilization or pasteurization process. All round wrap acts to protect the contamination of water and dust from outside environment. Shrink wrap recently tends to combine with corrugated or paperboard tray in bundling packages of cans and bottles [42]. There has also been improvement in speed with development of automatic form-fill-seal equipment for shrink packaging [43].

Common uses for shrink packaging as primary packaging are for vacuum packaging of cured meats, poultry, fish, and frozen sub-primal cuts of meat, and for individual seal packaging of retailed fresh fruits and vegetables. Cured meats packaged under vacuum by barrier plastic film are immersed in hot water of 75-90°C to attain tight shrunk contact of the film. Citrus fruits are individually wrapped with HDPE film of 10 μm thickness and is treated by blowing hot air for tight contact seal, which reduces the water loss and blemish, delays physiological senescence, and blocks spreading of microbial spoilage infected by surface contact [44]. This individual seal-packaging acts to provide the water-saturated microatmosphere around the produce and also modify the gas exchange rate with respect to environment (see also Section 17.8).

Important attributes of shrink film are shrink ratio, shrink temperature, shrink tension, sealability, transparency, toughness, and slip. Shrink ratio is measured by the shrunk length when the film is exposed to hot water bath of controlled temperature for 1-2 seconds. Lower shrink temperature is generally better for high productivity and can damage less the heat-sensitive quality of foods such as fresh fruits and vegetables. High shrink tension is good for rigid packages while it may be detrimental to weak ones such as paper pack. LDPE, linear low density polyethylene (LLDPE), EVA, ionomer, PP, PS, PET, and PVC are the films widely used for shrink packaging. These days multi-layer shrink films are produced by coextrusion to have required barrier and mechanical properties [43]. Table 11.3 lists the properties of typical shrink films.

Table 11.3 Properties of typical shrink films

Property	Film		
	LDPE	PP	PVC
Thickness range (μm)	30-300	12-40	13-200
Tensile strength ($kg_f \cdot m^{-2}$)	160-300	1100-1400	400-1100
Maximum shrink (%)	70-80	25-45	60
Clarity (%)	40-65	75-82	75-80
Shrink temperature ($^{\circ}$C)	80-130	120-165	65-150
Heat seal temperature ($^{\circ}$C)	125-170	170-210	140-200

Adapted from Ref. [21, 45, 46].

11.7.4 Stretch Wrapping

A technique for tight contour packaging without heating is stretch wrapping. The stretch wrapping is applied to bundles, pallets of packaged goods, and single items to provide the protection, aesthetic appearance, and integrity of the items. The film is stretched around the product and then heat-sealed (Figure 11.26). The stress remained under stretched strain gives a tight tension on the contour wrap. A major advantage of stretch wrapping is that heat is not applied. Some machines adopts a step of prestretch before application of the load, making the usable film longer [42, 47]. Typically 60-100% stretch is obtained, but 250% elongation may be obtained with a certain machine [48]. Prestretch can save the film, but decrease the strength of the film, and therefore is used for loads requiring light tension; undesirable neck-in in stretching can also be reduced.

The desired properties of stretch film are proper elongation, minimal neck-in when stretched, high tensile strength in machine direction, high puncture strength, low stress relaxation, heat-sealability, and good cling [49]. Stretch wrapping is based on the ability of the film to retain the holding force when film is pulled to a certain length and stick to the load and film. All films relax in stress and then equilibrate after elongation takes place. The stress relaxation should not be too great to lose the tightness of the wrapping. The equilibrated stress retention is in the range of 50-65% for LDPE-, EVA-, and LLDPE-based film, whereas PVC can retain about 20% [48]. However, temperature increase makes the film under tension relax and thus breaks the original shape of package. Cling property allows the film to adhere to each other and holds the product together. Cling property usually increases with temperature and humidity. Widely used stretch films are of LDPE, LLDPE, EVA, PVC, PP, and nylon. EVA improves the stickiness of the film.

Stretch wrapping is most widely used for prepackaging of fresh produce and meat in retail marketing, providing the package with neat and transparent impression. Stretch wrapping of individual fruits and vegetables also works to improve their keeping quality by mechanisms similar to those in individual shrink packaging [44]. Wraps on pallet or shipping loads are another area for wide use of stretch wrapping.

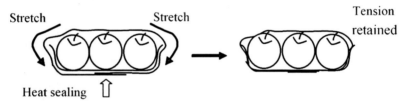

Figure 11.26 Principle concept of stretch wrapping

11.8 Case Studies

11.8.1 Performance of a Piston-type Filler

Filler speed and quality are affected by the rheological properties of foods. Some liquid foods may splash and contaminate the rim of the plastic container, which cause seal defects. Some types of semi-solid foods may heap high and cause difficulty in closing the container. Rao et al. [50] reported safe range of filling speed for given food viscosity and several problems encountered in the filling process, when a piston-type filler (FMC model PN010) was used. The filler consisted of filler tube in which liquid or semi-solid food was supplied by piston from a plenum chamber and discharged into a container by a plunger. Three types of containers were used which are cylindrical cup of 82 mL (height 4.0 cm, bottom diameter 6.0 cm), oval container of 218 mL (height 3.7 cm, axial diameters of 11.5 and 6.4 cm), and a rectangular one of 295 mL (height 3.4 cm, bottom of 12.2 and 9.2 cm). Foods with a variety of rheological properties were tested to find problems in the filling at several filling speeds, which include water, corn oil, apple juice, guar gum, tomato paste, pork and beans, apple sauce, crushed pineapple, and dog-food with different solid contents.

Figure 11.27 shows critical limits of filling rate vs. viscosity of Newtonian fluid food for different containers. Filling rate above the limits was not acceptable because of splashing or overflow. Higher viscosity food increased the threshold values for safe filling, which enabled faster filling. The critical limit for safe filling was also higher with lower volume container. Splashing was found to take place by two mechanisms: over flow of low-viscosity foods discharged at high velocity and break-up of fluids into droplets that land on the rim of the container.

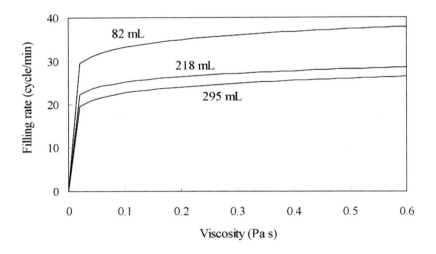

Figure 11.27 Critical upper limit for safe filling without splashing for different container volumes. Constructed from the data of Ref. [50].

In case of crushed pineapple with particles in the juice, phase separation was found to be a main cause for splashing. Splashing occurred both when the food hit the container bottom and when the plunger actuating final filling step hits the top of the heaped food in the container.

In case of foods with high contents of suspended insoluble solids, air pockets were introduced with the foods with high yield stress. Optimal design in cross section of a rotor blade and its rotation in the tank was suggested for eliminating the tunneling of the food. These products might also cause heaping in the container and difficulty in closing it due to the contact between the filling plunger surface and food. The contact may allow the food stuck on the plunger to fall on the rim of another container. This heaping problem occurred with foods of yield stress above 29 Pa. The authors recommended the use of small circular containers with deep height and mechanical vibration of the container to level the food surface.

11.8.2 Pressure Differential System for Nondestructive Leak Detection

Pascall [51] subjected semi-rigid polymeric tray packages of 355 mL (16 x 11.5 x 3.2 cm, tray body of polypropylene/ethylene vinyl alcohol/ polypropylene, flexible lid of 103 μm thickness) to a laboratory leak detection unit based on the principle of pressure differential (Figure 11.28A). The package was located onto the load cell in the sealed test chamber and subsequently evacuated to the pressure of 0.53 bar in 12 seconds. The package expanded in response to the vacuum was designed for the top lid to touch a plate at fixed height and exert a force on the load cell. During the subsequent stabilization phase of 30 seconds, there was an equilibration in the chamber temperature and creep expansion. In the final stage, the force exerted by the package on the load cell was monitored and used as a guide for integrity loss. Figure 11.28B shows the force-time response of the packages subjected to the test. Good package without leak would not cause the pressure or force to fall below a preset lower limit. The package with leakage would lose the pressure or force because of air loss through the leakage. The force decay was used as an indicator for the package defects. There was a positive correlation between the force decay and leak sizes for both pinholes and channel leaks. The correlation was better for pinhole leaks than for channel leaks. The author concluded that this unit was capable of detecting pinholes ≥10 μm and channel leaks ≥40 μm in diameter. This detection level seems to provide sufficient safety assurance in practical terms [28, 29].

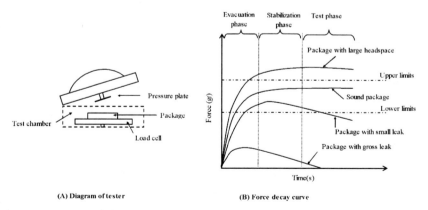

Figure 11.28 Response from a pressure differential tester. From Ref. [51] with kind permission of John Wiley & Sons.

Discussion Questions and Problems

1. Review the methods of bottom forming in plastic bags.

2. Compare the thermoform-fill-seal system with other form-fill-seal systems in food packaging.

3. List and compare potential benefits for skin packaging, shrink packaging, and stretch wrapping for packaging the cured meats.

4. A preliminary trail of filling machine for cereal packages showed the data of performance as given in Table 11.4. Discuss the soundness of the filling operation in satisfying the legal requirement and avoiding overfill. You may construct mean and control charts for the analysis.

Table 11.4. Net weight data of filling a cereal package (nominal weight is 400g)

Sample subgroup	Fill weight of each package					Mean	Standard deviation	Range
1	404	401	400	401	400	401	1.5	4
2	400	402	408	402	404	403	3.0	8
3	397	404	398	401	401	400	2.8	7
4	404	402	400	407	398	402	3.5	9
5	396	405	400	409	401	402	5.0	13
6	396	404	402	400	399	400	3.0	8
7	398	401	400	407	405	402	3.7	9
8	405	397	402	408	404	403	4.1	11
9	397	401	405	401	403	401	3.0	8
10	402	403	398	402	400	401	2.0	5
11	406	409	400	403	402	404	3.5	9
12	407	399	403	402	402	403	2.9	8
13	397	404	401	402	405	402	3.1	8
14	409	411	403	402	402	405	4.3	9
15	400	398	399	404	397	400	2.7	7
16	407	420	401	388	404	404	11.5	32
17	396	398	417	402	395	402	9.0	22
18	416	405	391	407	403	404	9.0	25
19	406	401	405	403	403	404	1.9	5
20	409	411	404	407	405	407	2.9	7

5. List the essential elements of fill technology for carbonated beverages and give some typical types of fillers used.

6. Compare crown and plastic CT closure for closing the bottle packages of carbonated soft drinks.

7. What kinds of physical properties are needed for stretch wrap films? How can those properties be identified? You may refer to Chapter 3 for the test methods.

8. What would be the differences between pinhole and channel leaks in package integrity testing? Discuss the difference based on physical principle and practical application of the pressure differential method.

Bibliography

1. Anonymous. Challenge tests for the evaluation of hygienic characteristics of packing machines for liquid and semi-liquid products. Trends Food Sci Tech 12:244-248, 2002.
2. WH Clifford, SW Gyeszley. Food packaging. In: AW Farrall, ed. Food Engineering Systems. Vol. 1. Westport Connecticut: AVI Publishing, 1976, pp. 431-473.
3. RAW Lea. Processing and packaging. In: PR Ashurst, ed. The Chemistry and Technology of Soft Drinks and Fruits Juices. Sheffield, UK: Sheffield Academic Press, 1998, pp. 85-102.
4. W Soroka. Fundamentals of Packaging Technology. Herndon, Virginia: Institute of Packaging Professionals, 1995, pp. 445-471.
5. M Honda. Fillers: liquid, semi-solid. In: T Yano, ed. Handbook of Food Packaging (in Japanese). Tokyo, Japan: Japan Association of Packaging Professionals, 1988, pp. 810-815.
6. FA Paine, HY Paine. A Handbook of Food Packaging. Glasgow, UK: Blackie & Sons, 1983, pp. 83-145.
7. FB Fairbanks. Filling machinery, liquid, still. In: AL Brody, KS Marsh, ed. The Wiley Encyclopedia of Packaging Technology. 2nd ed. New York: John Wiley & Sons, 1997, pp. 390-397.
8. RF Bardsley. Filling machinery, dry-products. In: M Bakker, ed. The Wiley Encyclopedia of Packaging Technology. New York: John Wiley & Sons, 1986, pp. 295-300.
9. NP Grigg, J Daly, M Stewart. Case study: the use of statistical process control in fish product packaging. Food Control 9:289-297, 1998.
10. GD Saravacos, AE Kostaropoulos. Handbook of Food Processing Equipment. New York: Kluwer Academic, 2002, pp. 575-623.
11. AL Griff. Carbonated beverage packaging. In: M Bakker, ed. The Wiley Encyclopedia of Packaging Technology. New York: John Wiley & Sons, 1986, pp. 121-124.
12. D Steen. Processing and handling of beverage packaging. In: GA Giles, ed. Handbook of Beverage Packaging. Sheffield, UK: Sheffield Academic Press, 1999, pp. 272-335.
13. WK Clarke. Filling, liquids products. The 1983 Packaging Encyclopedia. Newton, MA: Cahners Publishing, 1983, pp. 234-240.

14. WR Evans. Filling machinery, liquid, carbonated. In: M Bakker, ed. The Wiley Encyclopedia of Packaging Technology. New York: John Wiley & Sons, 1986, p. 300.
15. FG Vickers. Developments in bottling & can line technology - flexibility in filling technology. Brewer 82:549-554, 1996.
16. WR Evans. Filling machinery, liquid, carbonated. In: M Bakker, ed. The Wiley Encyclopedia of Packaging Technology. 1st Ed. New York: John Wiley & Sons, 1986, p. 300.
17. A Freidinger. Emballages et controle de remplissage dans l'industrie des boissons. Bios 14(3):35-38, 1983.
18. P Ronichon, JP Savina. Technology of bottling beverage. In: G Bureau, JL Multon, ed. Food Packaging Technology. Vol. 2. New York: VCH Publishers, 1996, pp. 105-126.
19. D Madison. Filling machinery, by count. In: M Bakker, ed. The Wiley Encyclopedia of Packaging Technology. New York: John Wiley & Sons, 1986, pp. 294-295.
20. JH Hooper. Confectionery Packaging Equipment. Gaithersburg, Maryland: Aspen Publishers, 1999, pp. 85-161.
21. JF Hanlon, RJ Kesley, HE Forcinio. Handbook of Package Engineering. 3rd ed. Boca Raton, FL: CRC Press, 1998, pp. 550-556.
22. FK Omar, CW de Silva. Optimal portion control of natural objects with application in automated cannery processing of fish. J Food Eng 46:31-41, 2000.
23. K Pitman. Closures in beverage packaging. In: GA Giles, ed. Handbook of Beverage Packaging. Sheffield, UK: Sheffield Academic Press, 1999, pp. 207-245.
24. JF Nairn, TM Norpell. Closures, bottle and jar. In: M Bakker, ed. The Wiley Encyclopedia of Packaging Technology. New York: John Wiley & Sons, 1986, pp. 172-185.
25. DA Carus, C Grant, R Wattie, MS Pridham. Development and validation of a technique to measure and compare the opening characteristics of tamper-evident bottle closures. Packag Technol Sci 19:105–118, 2006.
26. CL Harper, BA Blakistone, JB Litchfield, SA Morris. Developments in food packaging integrity testing. Trends Food Sci Tech 6:336-340, 1995.
27. GW Arndt, J Charbonneau. Nondestrcutive electroconductivity testing of packages for hermetic sealing. In: BA Blakistone, CL Harper, ed. Plastic Package Integrity Testing. Herndon, Virginia: Institute of Packaging Professionals, 1995, pp. 125-136.
28. KL Yam. Pressure differential techniques for package integrity inspection. In: BA Blakistone, CL Harper, ed. Plastic Package Integrity Testing. Herndon, Virginia: Institute of Packaging Professionals, 1995, pp. 137-145.
29. E Hurme. Detecting leaks in modified atmosphere packaging. In: R Ahvenainen, ed. Novel Food Packaging Techniques. Cambridge, UK: Woodhead Publishing, 2003, pp. 276-286.

30. C Behra, J Guerin. Technology of packaging under film. In: G Bureau, JL Multon, ed. Food Packaging Technology. Vol. 2. New York: VCH Publishers, 1996, pp. 127-142.
31. K Tamai. About Designing Flexible Packages (in Japanese). Tokyo, Japan: Process Technology Research Group, 1989, pp. 190-207.
32. J Arch. Bag-in-box, liquid product. In: AL Brody, KS Marsh, ed. The Wiley Encyclope-dia of Packaging Technology. 2nd ed. . New York: John Wiley & Sons, 1997, pp. 48-51.
33. GR Moyer. Form/fill/seal, vertical. In: M Bakker, ed. The Wiley Encyclopedia of Packaging Technology. New York: John Wiley & Sons, 1986, pp. 367-369.
34. MJ Hasting. Packaging machinery. In: RT Parry, ed. Principles and Applications of Modified Atmosphere Packaging of Foods. London: Blackie Academic & Professional, 1993, pp. 41-62.
35. C Rowan. Fast and convenient: form fill seal technology. Food Engineering & Ingredients 25 (4):30-32, 2000.
36. RF Bardsley. Form/fill/seal, horizontal. In: M Bakker, ed. The Wiley Encyclopedia of Packaging Technology. New York: John Wiley & Sons, 1986, pp. 364-367.
37. K Seko. Horizontal pillow form-fill-seal machine. In: T Yano, ed. Handbook of Food Packaging (in Japanese). Tokyo, Japan: Japan Association of Packaging Professionals, 1988, pp. 364-367.
38. LD Starr. Thermoform/fill/seal. In: M Bakker, ed. The Wiley Encyclopedia of Packaging Technology. New York: John Wiley & Sons, 1986, pp. 664-668.
39. Japan Packaging Institute. Handbook of Packaging Technology (in Japanese). Tokyo, Japan: Daily Industry News, 1983, pp. 1481-1535.
40. JM Gresher. Carded packaging. In: M Bakker, ed. The Wiley Encyclopedia of Packaging Technology. New York: John Wiley & Sons, 1986, pp. 124-129.
41. BPF Day. Chilled food packaging. In: C Dennis, M Springer, ed. Chilled Foods. New York: Ellis Horwood, 1992, pp. 145-159.
42. Anonymous. Wrapping, shrink and stretch. Packaging 32(5):216-219, 1987.
43. F Calmes. Films, shrink. Packaging 32(5):53-54,, 1987.
44. S Ben-Yehoshua. Individual seal-packaging of fruit and vegetables in plastic film. In: AL Brody, ed. Controlled/Modified Atmosphere/Vacuum Packaging of Foods. Trumbull, Connecticut: Food & Nutrition Press, 1989, pp. 101-117.
45. K Oki. Monolayer plastic film and sheet. In: T Yano, ed. Handbook of Food Packaging (in Japanese). Tokyo, Japan: Japan Association of Packaging Professionals, 1988, pp. 464-490.
46. CR Jolley, GD Wofford. Film, shrink. In: AL Brody, KS Marsh, ed. The Wiley Encyclopedia of Packaging Technology. 2nd ed. New York: John Wiley & Sons, 1997, pp. 431-434.

47. Anonymous. The latest trends in stretch and shrink wrap. Modern Materials Handling 42(2):65-67, 1987.
48. RL Caire. Films-stretch. In: M Bakker, ed. The Wiley Encyclopedia of Packaging Technology. New York: John Wiley & Sons, 1986, pp. 338-341.
49. BW Griggs. Film, stretch. Packaging Encyclopedia. Newton, MA: Cahners Publishing, 1988, pp. 49-50.
50. MA Rao, HJ Cooley, C Ortloff, K Chung, SC Wijts. Influence of rheological properties of fluid and semisolid foods on the performance of a filler. J Food Process Eng 16:289-304, 1993.
51. MA Pascall. Evaluation of a laboratory-scale pressure differential (force/decay) system for non-destructive leak detection of flexible and semi-rigid packaging Packag Technol Sci 15:197-208, 2002.

Chapter 12

Thermally Preserved Food Packaging: Retortable and Aseptic

12.1 Introduction 357
12.2 Thermal Destruction of Microorganisms and Food Quality 358
    12.2.1   Kinetics of Microbial Inactivation ··································358
    12.2.2   Kinetics in Thermal Degradation of Food Quality ··················362
12.3 Basics of Thermal Processing Design 362
    12.3.1   Selection of Processing Conditions ································362
    12.3.2   Determination of Thermal Process Condition ···················365
12.4 Hot Filling 367
12.5 In-container Pasteurization and Sterilization...368
    12.5.1   Heat Penetration into Packaged Foods ···························369
    12.5.2   Pasteurization ··371
    12.5.3   Sterilization for Shelf-Stable Low Acid Foods···············372
    12.5.4   Containers for heat preserved foods ····························376
12.6 Aseptic Packaging 379
    12.6.1   Continuous Heat Processing of Fluid-Based Foods···············381
    12.6.2   Sterilization of Packages and Food Contact Surfaces···········384
    12.6.3   Aseptic Filling and Packaging System ··························386
    12.6.4   Modified Version of Aseptic Packaging························390
12.7 Case Study: Design of a Thermal-Processed Tray-Set Containing High-
and Low-Acid Foods 392
Discussion Questions and Problems 393
Bibliography 394

12.1 Introduction

Heat is the most important means for food preservation. A typical thermal process consists of heating, holding, and cooling of a food product. Sometimes, the aim is to use heat to inactivate enzymes such as in blanching in which raw materials, mainly fruits or vegetables, are immersed in hot water or exposed to steam for a short time followed by rapid cooling. More often, the aim is to use various degrees of heat to preserve foods by inactivating foodborne microorganisms. Pasteurization is a mild heat process (below 100°C) for producing shelf-stable acid foods or refrigerated food products. Commercial sterilization is a severe heat process (higher than 110°C) for producing shelf-stable low acid foods. This chapter presents mainly the packaging related to the microbial inactivation.

Thermal processes may be classified based on whether the inactivation of microorganisms occurs before or after the food is filled inside the container. The former includes hot filling and aseptic packaging, in which a heat processed food is filled and packed in a clean environment. In hot filling, a high acid food is first heated to an adequate temperature to inactivate microorganisms, and the hot food is then filled into the container and remains for short time before cooling to disinfect the package's internal surface: target organisms are usually vegetative bacteria, yeasts, and molds. In aseptic packaging, the food and the package are sterilized (by heating and subsequent cooling) before the filling operation and the recontamination of the heat processed foods is prohibited completely by maintaining the environment and containers aseptic. In-container sterilization (also called as retorting) or pasteurization should usually be based on the proper level of microbial inactivation in geometrical center, which is normally assumed as a slowest heating point (cold point) or the location experiencing the typical temperature history of the container. Therefore the heat penetration into the geometrical center is measured or evaluated carefully to determine the process conditions of temperature and time.

The next two sections address the microbial kinetics of thermal destruction and the basic principles of the thermal processing, which are helpful for understanding the packaging of thermally preserved foods. The discussions are necessarily concise due to the space limit of this book. The remaining sections are dedicated to their packaging and process operational aspects. It should be recognized that thermal processing and packaging are closely related operations, and the selections of the processing condition and package design are mutually dependent.

12.2 Thermal Destruction of Microorganisms and Food Quality

Same reaction kinetic principle can be applied to microbial inactivation and quality changes caused by heat: first order kinetic equation holds for simplified explanation of thermal destruction of microorganisms, enzymes, nutrients, colors, and other quality attributes. The principle is useful to estimate the degree of microbial inactivation and quality loss under certain conditions of heat processing.

12.2.1 Kinetics of Microbial Inactivation

The rate of thermal destruction of vegetative microbes and spores is known to follow the first order reaction kinetics:

$$-\frac{dN}{dt} = k\,N \tag{12.1}$$

Upon integration, the equation becomes

$$\ln\left(\frac{N}{N_o}\right) = 2.303 \log\left(\frac{N}{N_o}\right) = -k\,t \tag{12.2}$$

where t is thermal processing time, k is rate constant, and N and N_o are the numbers of surviving microorganisms at $t = t$ and $t = 0$, respectively. The equation may be rearranged into a more convenient form:

$$\log\left(\frac{N}{N_o}\right) = -\frac{t}{D} \tag{12.3}$$

where D is known as decimal reduction time (or D value). Note that D is equal to 2.303/k and has a unit of time.

The D value has the physical meaning of being the thermal processing time (usually expressed in minutes) required to achieve a one log-cycle microbial reduction; that is, reducing the microbial number to 1/10 of its initial value. It is an important parameter that characterizes the heat resistance of a microorganism: the higher the D value, the more heat resistance the microorganism. The D value of a particular microorganism at a given temperature may be determined experimentally by first measuring N as a function of t and then plotting the data as log N versus t. According to Equation 12.3, the plot should follow a straight line and the D value may be estimated from its slope as shown in Figure 12.1.

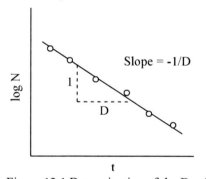

Figure 12.1 Determination of the D value

The D values of typical microbial strains are listed in Table 12.1. Note that the D value is temperature dependent, and thus each value is associated with a reference temperature.

The D value is known to decrease logarithmically with temperature according to

$$\log\left(\frac{D}{D_{T_f}}\right) = \frac{T_f - T}{z} \tag{12.4}$$

where z is known as thermal resistance constant (or z value); D and D_{T_r} are D values at temperature T and reference temperature T_r respectively. The unit of z is usually expressed in °C or °F.

Table 12.1 Thermal resistance data for microorganisms of major concern in thermal preservation of packaged foods

Microorganism	Reference temperature (°C)	D value (min)	z value (°C)
Proteolytic *Cl. botulinum*	121.1	0.1-0.2	9-12
Nonproteolytic *Cl. botulinum*	90	1.1	9
Cl. perfringenes	121.1	0.15	10
Cl. thermosaccharolyticum	121.1	3-4	7-15
B. stearothermophilus	121.1	4-5	7-12
B. coagulans	121.1	0.07	10
B. polymyxa	100	0.50	9
Cl. pasteurianum	100	0.5	9
B. cereus	90	1-36	9.7
Mycobacterium tuberculosis	82.2	0.0003	6
L. monocytogenes	70	0.3	7.5
Escherichia coli	70	0.001-0.04	3-7
Salmonella	70	0.001-0.01	
Yeasts and molds	82.2	0.0095	7

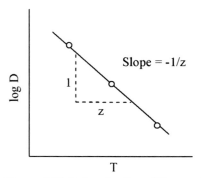

Figure 12.2 Determination of the z value

The z value has the physical meaning of being the temperature difference required to reduce the D value to 1/10 of its initial value. It is an important parameter that characterizes the sensitivity of the microorganism to temperature: the higher the z value, the less sensitive the microorganism to temperature changes. The z value of a particular microorganism may be determined experimentally by first measuring D as a function of T and then plotting the data as log D versus T. According to Equation 12.4, the plot should follow a straight line and the z value may be estimated from the slope as shown in Figure 12.2. Note that the procedure of determining D described earlier should be conducted for at least three temperatures in order to obtain a good estimate of the z value. Typical z values are shown in Table 12.1.

The effects of processing time and temperature on microbial reduction are illustrated in Table 12.2 for a hypothetical microorganism with $D_{110}=1.0$ min and $z =10$ C. At the reference temperature of this D value ($110°C$), the number of surviving microorganisms (N) will be reduced by one log (or 90%) per every minute of processing time. At $100°C$ being $10°C$ below the reference temperature, the z value dictates that the thermal processing time needs to increase by 10 times in order to achieve the same microbial reduction.

Table 12.2 Microbial reduction for $D_{110}=1.0$ min, $z =10°C$

Time (min) at various temperatures			N
100°C	110°C	120°C	
0	0	0	100000
10	1	0.1	10000
20	2	0.2	1000
30	3	0.3	100
40	4	0.4	10
50	5	0.5	1
60	6	0.6	0.1

The thermal death time F is defined as a multiple of D value:

$$F = n D \qquad (12.5)$$

where n is the required log-cycle of microbial reduction. The F value is the time required to reduce the microbial population by a certain number of log-cycles at a given temperature. For example, F=6D is equivalent to six log-cycle reduction or a 99.9999% reduction in microbial population. It is customary to use F_T^z to denote the thermal death time to reduce the population of a microorganism with z value at temperature T. For reference purpose, F_o is defined as $F_{121.1}^{10}$, where T and z are 121.1 and 10°C, respectively.

Based on the kinetics of microbial inactivation, the probability of survival in food can never reach absolute zero even after a long heating time. For spores of mesophilic proteolytic *Clostridium botulinum*, F=12D is commonly used, which corresponds to 3 minutes at 121.1˚C. *Cl. botulinum* is an anaerobic pathogenic bacterium capable of producing heat resistant spores. Some spores of nonpathogenic organisms are more heat resistant than *Cl. botulinum*. For these microbial strains such as *Bacillus stearothermophilus* and *Cl. thermosaccharolyticum*, lower degree of microbial reduction (n) of 4-6 is usually applied to obtain the F value. The desired degree of microbial destruction may differ with initial microbial load, tolerable probability level of nonpathogenic spoilage, and storage temperature conditions.

12.2.2 Kinetics in Thermal Degradation of Food Quality

Besides microbial destruction, thermal processing may also cause physical and chemical changes of the food such as nutrient loss and textural degradation. As with microbial destruction, the first order kinetics is applicable to many of these quality changes although their rates are significantly slower and less temperature dependent. Therefore Equation 12.6 similar to Equation 12.3 works to describe thermal destruction of food quality:

$$\log\left[\frac{c}{c_o}\right] = -\frac{t}{D_c} \qquad (12.6)$$

where D_c is decimal reduction time of the quality, and c_o and c_t are concentrations of the quality index at time 0 and t, respectively.

Change of D_c with temperature is also described by Equation 12.4 with using temperature dependence parameter z in range of 25-47°C. Similar to the F value, a cooking value (C value, usually at reference temperature of 100˚C) is used to describe the degree of quality change in heat processing:

$$C = n\,D_c \qquad (12.7)$$

It needs to be mentioned that C value as quality index is usually evaluated for the whole volume while F value is done mostly for the slowest heating point.

12.3 Basics of Thermal Processing Design

12.3.1 Selection of Processing Conditions

The thermal processing condition is aimed at assuring the lethality of the target microbial strain of a given food. This condition is usually satisfied by heating the food with integrated severity higher than the F value of the microbial strain. As shown in Table 12.3, the selection of the target microbial strain and the F value depends on the pH, water activity, storage temperature, and shelf life of the food.

Table 12.3 Classification of foods for thermal process design

Food type	Criteria of thermal process	Storage condition and shelf life
Shelf-stable foods		
Low acid foods with pH ≥ 4.5 and water activity ≥ 0.92	Minimum botulinum process of F_o=3, usually more severe process for sufficient stability and cooking	3-24 months at ambient storage depending on barrier properties of packaging
Moist acid foods with pH of 3.7-4.5	*Bacillus coagulans, B. polymyxa, B. macerans, Clostridium butyricum*, and *Cl. pasteurianum* as target organism, generally 90°C at cold point is attained	3-24 months at ambient storage depending on barrier properties of packaging
Moist high acid foods with pH < 3.7	Vegetative bacteria, yeasts and molds as target, 80°C at geometric center is usually sufficient	3-24 months at ambient storage depending on barrier properties of packaging
Chill-stored products		
Cook-chilled and *sous vide* packaged foods with pH > 5.0	6D inactivation of *Listeria monocytogenes* (70°C, 2 min)	Less than 10-14 days at below 8°C; storage at 0-3°C is desirable
Cook-chilled and *sous vide* packaged foods with pH > 5.0	6D inactivation of non-proteolytic *Cl. botulinum* (90°C, 10 min)	Less than 40 days at below 8°C; storage at 0-3°C is desirable
Milk products	Inactivation of *Coxelliae burnettii*, typically 63°C for 30 min or 72°C for 15 sec	7 days below 6°C

For shelf stable low-acid foods with pH ≥ 4.5 and water activity ≥ 0.92, *Cl. botulinum* is considered first as one to be destroyed and minimum botulinum process with F_o of 3 is usually satisfied [1, 2]. However, for the many kinds of packaged foods, more severe heat process is applied because of the considerations in variations of commercial manufacturing conditions, sufficient cooking for adequate sensory quality, and spoilage concerns due to more heat resistant nonpathogenic spores. In this case 5D inactivation of *B. stearothermophilus* may be selected as an alternative process condition. In alcoholic beverages such as beer and wine *Cl. botulinum* and any other heat resistant organisms are of no concern due to preservative action of contained ethanol component.

For the shelf stable moist foods with pH < 3.7 to be sold at ambient temperature, vegetative bacteria, yeasts, and molds are considered because spore formers cannot germinate and grow in these foods. These organisms are very sensitive to heat and therefore controlled by thermal process below 100°C. For these high acid products with pH < 3.7, reaching 80°C at geometric center is usually sufficient conditions for pasteurization.

For the shelf stable moist foods with pH between 3.7 and 4.5, spoilage spore formers such as *Bacillus coagulans*, *B. polymyxa*, *B. macerans*, *Clostridium butyricum*, and *Cl. pasteurianum* may be used as a reference for the thermal process design. From the consideration of these organisms, 10 and 5 minutes of $F_{93.3}^{8.3}$ have been suggested for the products with pH of 4.3-4.5 and 4.0-4.3, respectively [3]. F_o of 0.7 was suggested as another guide for the product with pH between 4.0 and 4.5 [4]. Attainment of 90°C by cold point is sometimes regarded as sufficient for the normal ambient storage of these acid foods.

For the cook-chilled and *sous vide* packaged foods with pH > 5.0 to be stored and distributed in chilled temperature below 10°C, the target organism for inactivation differs with shelf life. The product with short shelf life less than 10-14 days is processed based on the 6D inactivation of *Listeria monocytogenes*, which is attained typically by 2 minutes at 70°C [5]. The product with longer shelf life is usually based on destroying the nonproteolytic *Cl. botulinum* by 6 log cycle, which corresponds to 10 minutes at 90°C. The *sous vide* products are generally based on the process consisting of placing the precooked ingredients in high barrier bags, vacuumed sealing of the package, cooked pasteurization, and rapid chilling to less than 3°C. From the consideration of microbial safety and spoilage, the *sous vide* products are stored under refrigerated temperature (preferred range: 0-3°C) for limited period of time (at most 40 days).

Another category of foods to be pasteurized and stored under chilled condition for limited period is milk products. The typical pasteurization conditions are 63°C for 30 minutes or 72°C for 15 seconds, which is based on the inactivation of *Coxelliae burnettii*, the most heat resistant pathogen found in milk (9).

Application of other hurdles such as reduced water activity, added preservatives, and modified atmosphere may change or alleviate the heat process conditions. Wise choice of appropriate target microbial inactivation can ensure the food safety and save the physical, chemical, and sensory qualities.

12.3.2 Determination of Thermal Process Condition

Sometimes it may be assumed for processes like aseptic processing and hot filling of the liquid foods that the foods are heated instantly to a certain temperature, held there for a specified period, and then cooled instantly to low temperature. In this case the required thermal process time at different temperatures can be calculated by using the temperature dependence parameter of z value (Equation 12.4, Figure 12.2). If we take Table 12.2 as an example of a hypothetical destruction achieving the survivor reduction from 100000 to 0.1 (6 log reduction), process times of 6 minutes at 110°C, 0.6 minutes at 120°C, or 60 minutes at 100°C may be easily chosen and applied to give the same degree of sterility. However, in many usual situations such as in-container thermal processing, the temperature at slowest heating location during the process does not remain at constant level but changes with time. The different temperatures experienced during the course of heating and cooling stages need to be evaluated in terms of microbial lethality and integrated on the same basis. This is usually achieved by measuring the temperature history at slowest heating location (usually geometrical center in sterilization after packaging) and calculating the lethal effect of each temperature through the process by Equation 12.8:

$$L = 10^{(\frac{T-T_{ref}}{z})} \tag{12.8}$$

where L is lethality (min) based on the reference temperature, T_{ref}. The concept of lethality may be easily understood by adopting Table 12.2 as an example: if we take reference temperature as 110°C, 1 minute at 100°C will amount to 0.1 minute at T_{ref} in the microbial inactivation effectiveness, or 1 minute at 120°C will amount to 10 minutes at T_{ref} in the microbial inactivation effectiveness; by this way we can evaluate the effectiveness of different temperatures in terms of one at reference temperature.

Then the summation of all the effectiveness of different temperatures experienced by cold point during the whole process gives the integrated sterilizing value, which needs to be higher than the target thermal death time of F value. The integrated sterilizing value is also called as F value and may be defined as:

$$F = \sum L \cdot \Delta t = \int L \cdot dt \tag{12.9}$$

As with evaluation of microbial sterilizing value, quality changes described in C value (Equation 12.7) can be integrated similarly but with using different reference temperature (usually 100°C) and parameter z_c:

$$C = \int 10^{(T-T_{ref})/z_c} \cdot dt \tag{12.10}$$

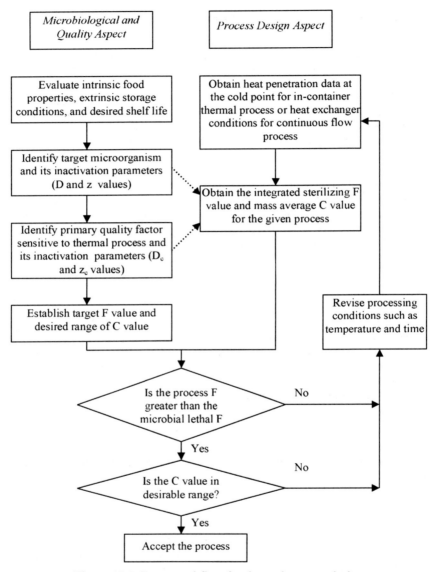

Figure 12.3 Conceptual flow for thermal process design

The design of thermal process is based on the objective of achieving the required level of microbial inactivation simultaneously with adequate cooking, i.e., desired range of C value: for certain heat-sensitive component, minimum C value is usually desirable. As mentioned above, evaluation of C value as quality deterioration index needs to be made for the whole volume or mass. Figure 12.3 presents a simplified logical flow for thermal process design.

If C value is known for a process condition, it can be related to the quality retention as follows:

$$C = \log(c_o/c_t)\, D_{c,ref} \qquad (12.11)$$

where $D_{c,ref}$ is D_c value of the relevant quality index at reference temperature.

Example 12.1 Determination of lethal degree in a thermal process

In a hypothetical example of Table 12.2, F_{110}^{10} value of 6.0 is assumed as desired for heat-processing a certain food. Is the hypothetical heat process consisting of 10 minutes at 100°C, 5 minutes at 105°C, 7 minutes at 116°C, and 6 minutes at 101°C based on the cold point would be enough for the required sterility? It is assumed that temperatures in the other time periods are so low as to be disregarded in the lethality calculation.

Lethal effectiveness of the different temperatures was obtained by using Equation 12.8 and then summed up according to Equation 12.9 in the table below:

Temperature, T (°C)	Time span, Δt (min)	Lethality based on 110°C, L (min)	L Δt
100	10	0.100	1.00
105	5	0.316	1.58
116	3	3.981	11.94
101	6	0.126	0.76
		Σ L $\Delta t =$	15.28

The integrated F value of 15.28 is greater than 6.0 of the target value, and thus the process at least satisfies the required sterility for the food. However, if it gives too much sterility, there might be concerns about the loss of sensory and nutrient qualities. In this case the less severe process may be tried and tested.

12.4 Hot Filling

Wide range of acid food products such as fruit juices, jams, sauces, and tomato ketchups can be preserved by filling at hot temperature above 85°C, closing/sealing of the package under a steamed vacuum, and holding there for short period time before cooling [6]. The residual heat from the hot filled product is thought to inactivate the vegetative bacteria, yeasts, and molds which are sensitive to heat. Vacuum in the container is also attained by the immediate closing and sealing after hot filling. The filled containers are sometimes held upside down or rotated to ensure the pasteurization on internal surfaces of the container including the closure or cap. The atmospheric environment should be maintained as clean as possible or in semi-aseptic condition in order to reduce the risk or probability of recontamination. In certain instances, the filled packages are immersed in a hot water chamber for short time to ensure the pasteurization effect on the container's inner surface.

When glass containers are used in hot filling process, precautions are required in order to avoid the thermal shock leading to glass fracture. The abrupt temperature rise in excess of 55-60°C may cause the breakage and should be avoided (see also Section 7.6). Preheating of the container may solve the breakage problem.

With plastic containers, heat resistant materials such as high density polyethylene (HDPE), polypropylene (PP), polyethylene terephthalate (PET), and polycarbonate (PC) are usually used for the hot filled products.

12.5 In-container Pasteurization and Sterilization

Traditional way of thermal processing is application of heat after filling food into container and subsequent closed sealing of the open end. The containers such as metal cans, glass jars, and preformed pouches may be delivered to food factory, filled with foods, and then subjected to heat processing. The plastic pouches and trays may be formed, filled, and sealed in food processing plant before thermal process. Through the filling operation, suitable means for air removal from the container should be provided. Hermetic closing is inspected and confirmed for the absence of leaks.

The heat in the thermal processing may be applied by steam, hot water, steam-air mixture, flame, or microwave energy. Whatever heat source is used, the process should be designed to ensure that any part in the container should not have risk of microbial spoilage and contamination. Temperature history during thermal process must be examined for this purpose. Any possible chance of recontamination after heat processing should be eliminated through the whole process. The concept of in-container thermal processing is explained in Figure 12.4 of retort pouch example. Application of high pressure on flexible packages of acid foods can also attain the consequence of microbial inactivation similar to heat pasteurization and is used to a limited extent in Japan. Proper selection of the pressurization conditions may reduce the intensity of thermal processing for better quality preservation, but the high-pressure technology has many problems to be overcome for wide commercial application and thus is not discussed here.

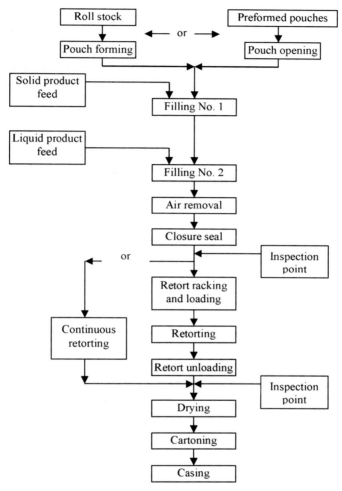

Figure 12.4 Flow diagram for production of foods preserved in retort pouch. From Ref. [7] with kind permission of Elsevier.

12.5.1 Heat Penetration into Packaged Foods

Whether the packaged foods are processed in a batch retort or continuously, the heat penetration into the slowest heating location determines the adequacy of the thermal process. As shown in Figure 12.5 the temperature profiles during heating and cooling stages are usually analyzed by plotting the temperature difference between food and medium of heating or cooling (ΔT) vs. time in semi-log scale. From this type of curve, two important parameters, j and f are derived. The j value is called a lag factor which is defined as

$$j = \frac{\Delta T_{po}}{\Delta T_o} \qquad (12.12)$$

and f value is defined as time for the temperature difference to decrease to 1/10. The j and f for heating are written as j_h and f_h, and those for cooling as j_c and f_c. The linear portion after t_l in Figure 12.5 may be described:

$$\Delta T = j\Delta T_o \cdot 10^{-t/f} \qquad (12.13)$$

For the curve-linear portion before t_l, Hayakawa [8] proposed some useful relationships to show the temperature change with time as function of f and j values, which are very useful for thermal process design (Table 12.4).

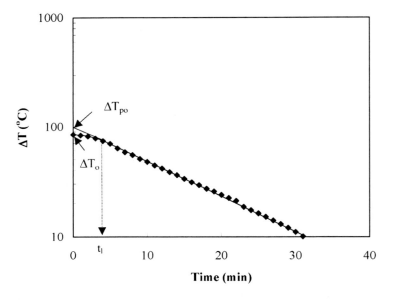

Figure 12.5 A typical temperature history curve for heating or cooling process of food product. ΔT is the temperature difference between food and medium of heating or cooling.

As long as the mechanisms of heating and cooling do not change, the f and j values may be considered to be relatively constant when exposed to a different temperature condition. Therefore these parameters may be useful for estimating temperature change of the foods exposed to various heating or cooling conditions. The f value is independent of location while j is dependent on the location inside the package.

Table 12.4 Formula to describe the curve-linear portion of temperature history exposed to constant environmental temperature condition

Range of j	t_1	ΔT
$0.045 \leqslant j < 0.4$	$t_1 = 0.3f$	$\Delta T = \Delta T_0 \times 10^{-(t/B)^{1/n}}$ where $n = (0.3 - \log j)/0.3$; $B = 0.3f\,(0.3 - \log j)^{-n}$
$0.4 \leqslant j < 1$	$t_1 = 0.9\,f\,(1-j)$	$\Delta T = \Delta T_0^{\cot(Bt - \pi/4)}$ where $B = \dfrac{1}{t_1}[\arctan\{\dfrac{\log \Delta T_0}{\log j \Delta T_0 - t_1/f}\} - \dfrac{\pi}{4}]$
$1 < j \leqslant 3$	$t_1 = 0.7\,f\,(j-1)$	$\Delta T = \Delta T_0^{\cos(Bt)}$ where $B = \dfrac{1}{t_1}\arccos[\dfrac{\log j \Delta T_0 - t_1/f}{\log \Delta T_0}]$

For some foods which heat quickly during initial stage and then slow down in the heating rate because of gelling of the components such as starch, there are two f values with broken heating curves. Many complicate mathematical analyses were reported, which are beyond the scope of this book, and textbooks on thermal processing and thermobacteriology should be referred to for detailed information.

12.5.2 Pasteurization

For acid foods with pH less than 4.5, pasteurization is usually carried out with temperature $\leq 100^\circ C$. Hot water bath, hot water spray, or steam may be applied as heating medium. Because low pH inhibits pathogenic proteolytic *Cl. botulinum* and limits the growth of heat-resistant spoilage microbial strains, the product may be shelf stable with well designed pasteurization process. For the product with pH between 3.7 and 4.5, attainment of center temperature at $90^\circ C$ is usually regarded as sufficient for the normal ambient storage. For the product with pH < 3.7, center temperature at $80^\circ C$ is generally sufficient.

In the pasteurization process with hot water bath, the products are located in the crates or racks, which are subsequently immersed in the hot water bath. The product-loaded crates, or racks are then moved to cooling stage with immersion into the cool water. The procedure may be operated batch wise in water tank(s) or continuously with movement of the containers on the conveyor belt through the heating and cooling tanks. Hot water bath pasteurizers are mostly used for canned products and foods packed in plastic pouch and trays.

Water spray pasteurization is usually carried out in continuous mode for the glass jars and cans of beer and acidic food products. The packages are usually moved in chain or on a conveyor through the consecutive zones of different temperatures. Glass containers should experience the gradual temperature rise and drop: for the heating stage the temperature difference between food and heating medium should be below 21°C and should not exceed 38°C under any conditions; for the cooling stage the temperature difference should be desirably below 10°C and should not exceed 21°C under any conditions [9].

In the cooling of the pasteurized products, the containers are cooled to about 40°C, which helps the evaporation of water remained on the package surface. However, in case of *sous vide* packaging, the products are cooled as quickly as possible to 3°C and stored below 3°C because of concerns on the microbiological spoilage.

Example 12.2 Estimating pasteurization time to reach certain temperature in sous vide packaging

Sous vide package of 2 kg spinach (dimension of 26.5 × 26.5 × 3.4 cm) heated or cooled in water has the heating and cooling parameters for geometric center as follows: j_h=1.14, f_h=30.6 min, j_c=0.93, f_c=66.3 min. The spinach is blanched in 1% NaCl solution and vacuum-packed in barrier plastic pouch. The pasteurization is undertaken under pressurized hot water of 97°C based on the time for center temperature to reach 75°C in heating phase. Cooling after the heating cycle is attained to reach 5°C. Initial temperature is 23°C and cooling water temperature is 0°C. Estimate the time spans required for heating and cooling.

The time to reach the required temperature is soundly to be on linear portion of heating and cooling curves. Therefore Equation 12.13 was used to estimate the time for the geometric center to reach 75°C.

$$(97 - 75) = 1.14 \times (97 - 23) \cdot 10^{-t/30.6} \text{ and thus t=17.9 min for heating phase}$$

For the cooling phase to reach 5°C,

$$(5 - 0) = 0.93 \times (75 - 0) \cdot 10^{-t/66.3} \text{ and thus t=75.9 min}$$

Total required time for the process would be 93.8 minutes, being sum of heating and cooling times.

12.5.3 Sterilization for Shelf-Stable Low Acid Foods

For the low acid foods, the thermal sterilization process requires high temperature of 110-130°C as heating medium temperature for inactivation of *Cl. botulinum* spores and other heat resistant organisms. The minimum process is taken as minimum botulinum process of F_o=3.0 minutes, and more severe conditions with F_o=6 minutes or higher are usually applied to control spoilage organisms [2]. The conditions of time-temperature relationship required for commercial sterility should be determined based on the heat penetration study and intrinsic characteristics such as initial microbial load and spoilage microorganisms of concern (Figure 12.3).

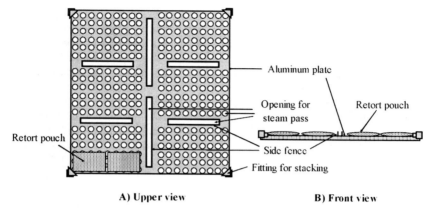

Figure 12.6 Sterilization rack for efficient heat transfer to retort pouches

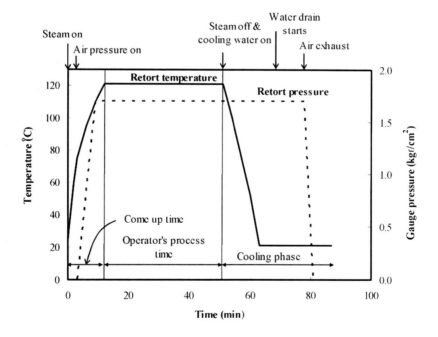

Figure 12.7 A typical pressurized sterilization process

Wide range of products packaged in glass jars, metal cans, and plastic pouches and trays are processed in a batch style or continuous retorting systems. Water, steam, or water/steam mixture is applied as a heating medium for the high temperature processing. In case of flexible pouches or semi-rigid plastic containers, the package is positioned in a rack in order to provide even exposure to heating medium of the retort (Figure 12.6). In the process batch or continuous, the heating medium temperature is controlled in a well-programmed manner to ensure the achievement of the required sterilizing value at the package center. After heating stages, the containers are cooled to about 40°C by using cooling water. For most cases, the pressure inside the retort is controlled by using the air overpressure. Especially for glass and plastic containers which are fragile and weak for the pressure differential between the package inside and retort, the balanced retort pressure is important for container integrity. The risk of package breakage due to bad pressure operation is higher for the initial cooling stage when the pressure inside the package becomes much higher than the retort pressure without overpressure. The overpressure applied by air ranges from 0.5 bar to 2.8 bar depending on the process temperature, initial vacuumized level, and headspace volume. Figure 12.7 shows a typical pressurized sterilization profile at 121.1°C: please note that absolute pressure is obtained by adding 1.03 kg_f/cm_2 to gauge pressure.

In case of glass packages, temperature of heating and cooling media should be changed stage-wise or slowly in order to avoid the breakage of the containers due to thermal shock. The pressure profile also needs to change in accordance with the temperature change pattern. Pressure in the food package being sterilized can be estimated by using Dalton's law of partial pressure: the total pressure of air-steam mixture is sum of partial pressures of air and water steam, which is given by following relationship:

$$P_T = P_a + P_s \tag{12.14}$$

where P_a and P_s are the respective partial pressures (bar) of air and steam in headspace. The air pressure at product temperature T (K), is given from the ideal gas law as Equation 12.15,

$$P_a = \frac{P_i V_i^h T}{T_i \{V^h - \alpha V_i^f (T - T_i)\}} \tag{12.15}$$

and the steam pressure at the temperature can be obtained as follows [10]:

$$P_s = 0.006112 \cdot \exp(53.479 - \frac{6808}{T} - 5.09 \cdot \ln T) \tag{12.16}$$

where P_i is the air pressure when the product is initially filled or sealed at temperature T_i (K), α is thermal expansion coefficient of food (K^{-1}) V_i^f is initial volume of food, V_i^h is the initial headspace volume, and V^h is the headspace volume at temperature T (K). Ideally the retort pressure adjusted to the Equation

12.14 would give no stress on the package from the pressure differential across the package wall.

In heating stage by steam or hot water, saturated water vapor pressure already exerts in the retort, and thus the air overpressure would be added to it in order to override the total required pressure. However, the total counter pressure needed in the cooling stage should be provided solely by the air pressure.

It was analyzed that higher internal pressure is built with lower headspace for rigid containers: so small headspace is too dangerous due to enormous pressure build-up inside the package. Thus headspace in glass jars should usually be greater than 7% to provide space for the required vacuum and some allowance for food expansion during thermal processing [6]. In case of metal cans with slight structural flexibility, filling level is usually controlled to have 6-9 mm thick headspace on the top of the container with less than 10% of total can volume [4]. For the semi-rigid plastic packages, higher headspace slightly increases the internal pressure because of relative flexibility of the container [11]. Intimate contact between food surface and top lid without headspace therefore, in certain designs, is suggested as desirable for minimum oxygen content and better heat transfer by good contact between wall and food contents in these containers [12]. For the retort pouches, higher degree of volume expansion is tolerable in the heat processing, but high air overpressure is needed in the cooling cycle due to the high pressure difference between inside and outside of the package.

Example 12.3 Estimating internal package pressure subjected to retorting

A noncarbonated beverage of 300 mL is filled in 320 mL glass bottle at the temperature of 55°C under the vacuum of 50 cm Hg. The product is going to be sterilized at 115°C. What would be the maximum internal package pressure? Volumetric thermal expansion coefficient of the liquid food may be assumed as $6.2 \times 10^{-4}\ K^{-1}$.

It can be assumed that the food temperature nearly reaches the retort temperature of 115°C. The headspace volume is 20 mL. The initial pressure of the container is given as:

$$P_i = (76\text{-}50)\ cm\ Hg \times (0.01333\ bar\ cm\ Hg^{-1}) = 0.347\ bar$$

The glass container is assumed to be in constant volume. The pressure of the container at this temperature is obtained by using Equations 12.14-12.16:

$$P_T = \frac{P_i V_i^h T}{T_i \{V^h - \alpha V_i^f (T - T_i)\}} + 0.006112 \cdot \exp\left(53.479 - \frac{6808}{T} - 5.09 \cdot \ln T\right)$$

$$= \frac{0.347\ bar \times 20\ mL \times 388\ K}{328 \times \{20 - 6.2 \times 10^{-4} \times 300 \times (388 - 328)\}} + 0.006112 \cdot \exp\left(53.479 - \frac{6808}{388} - 5.09 \cdot \ln 388\right)$$

= 0.929 + 1.638 = 2.567 bar

The pressure of the retort needs to be maintained around this level for some appropriate periods of heating and cooling phases.

12.5.4 Containers for Heat Preserved Foods

Traditionally metal cans and glass jars have been used extensively for packaging heat preserved foods. Their high temperature stability and excellent barrier to gases and moisture are the main characteristics allowing their use in thermal processed foods. Cans in tinplate and tin-free steel are used for both pasteurization and sterilization processes, while aluminum cans are mostly for pasteurization products of acidic, alcoholic, or carbonated beverages such as soft drinks and beer.

Table 12.5 Comparison of process time between can and retort pouch

Food	Container	Process time at 121.1°C (min)	Process lethality (F_o)
500g of pumpkin puree (conduction heating type)	Metal can	53.2	3.6
	Retort pouch	19.8	3.6
312 g of peas packed in brine (convection heating type)	Metal can	16.7	11.0
	Retort pouch	22.7	11.0

Extracted from the data of Ref. [13].

Retortable plastic packaging has appeared from early 1980s and have grown steadily due to consumers' acceptance of its convenience. The plastic packaging for heat preserved foods usually has the flat form of flexible pouch or semi-rigid tray. Pouch types may be formed into flat four-seal design or standing pouch with gusseted bottom. In-line formed and premade pouches are usually filled vertically with solid and/or liquid foods. A horizontal form/fill/seal method uses forming on a horizontal bed into several adjacent cavities, into which placeable products are filled with the seal areas shielded in the operation [14]. Bowl type plastic containers are used for some soups. The shallow slab form of plastic container achieves faster heat penetration into the geometric center of conduction-heating foods and thus makes the heat process time shorter compared to fat cylindrical containers such as metal cans and glass jars (Table 12.5): less severe heating may attain better nutrient retention and sensory quality. However, it is not so for foods of convection heating mechanism. Individual retortable plastic packages are mostly repacked in individual cartons to prevent damage or puncture stress during distribution. Recently Tetra Pak® developed a retortable square carton package of paperboard laminate material (Tetra Recart®), which can give logistic advantage by high efficiency of staking on pallet. Table 12.6 lists typical examples of retortable plastic package types.

Table 12.6 Typical examples of retortable plastic and paper packages

Package type	Material combination	Closing method
Pouch	PET/Al/PP PET/nylon/Al/PP PET/nylon/PP Nylon/PP PET/PVDC(or EVOH)/PP SiOx PET/nylon/PP	Heat seal; gusset part for stand up pouches may be constructed of more resilient material because of its exposure to high mechanical stress
Tray	PP/ PVDC(or EVOH)/PP	Heat sealing with same material of body or double seamed with Al can end
Carton	PP/ paperboard/PP/Al(or EVOH)/PP	Heat seal with easy open feature; efficient stacking due to brick shape

Because the heat processing procedure undergoes severe conditions of high temperature and some pressure differential, the containers for in-container heat preservation should have the strict requirements for package integrity and desired shelf life (see also Sections 10.5.3 and 11.4.3). Double seam between wall part and cover is used exclusively for hermetic sealing in metal cans and in plastic containers with metal ends; inspection of double seaming should be subjected to routine inspection including visual evaluation and destructive tear-down test. For the glass containers, cap closing is examined for security sealing. Vacuum level in the package as another indicator for seal integrity, is usually controlled and tested by a piercing vacuum gauge. In case of heat-sealed plastic containers, burst, peel, or leak detection test is conducted for seal integrity. Measures to prevent defective seals should be employed [7]: sealing conditions such as temperature, time, and pressure should be controlled for the sealant in use; contamination on the seal area should be avoided.

In terms of heat stability, metal can and glass are durable enough to pass the temperature of steam or hot water up to 130°C. However, glass containers should not be exposed to abrupt temperature change as mentioned before in consideration of preventing heat shock damage. For the plastic containers to be heat processed, the tolerable range of materials should be selected. Usually PP is used as structural layer for physical strength, and ethylene vinyl alcohol copolymer (EVOH) and polyvinylidene chloride copolymer (PVDC) are used as barrier layer (Table 12.6). PP also serves as food contact layer and outer surface, and acts as heat seal area. Aluminum foil is applied as a barrier layer against oxygen and light in case of flexible retort pouch. Crystallized polyethylene terephthalate (CPET), polyethylene naphthalate (PEN), amorphous polyamide (AMPA), and polycarbonate (PC) are being used or considered for use in retortable packaging because of their excellent heat stability. Figure 12.8 is useful for selecting the materials in heat-preserved packaging.

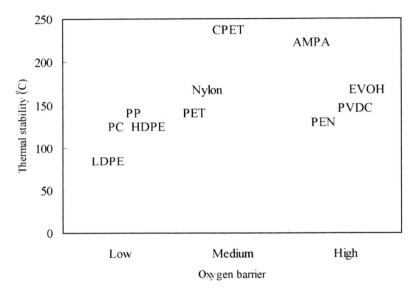

Figure 12.8 Heat stability and gas barrier of plastic materials with reference to use in heat-preserved packaging (based on the property in dry state)

As with package temperature tolerance, gas barrier properties are needed to protect the food product from oxidation. Heat resistant polymers with excellent barrier properties such as EVOH and PVDC are used as a barrier layer embedded in the multilayered structure (Table 12.6). Silica oxide (SiO_x) coated PET films and bottles have been developed for improvement in the gas barrier property. The oxygen barrier properties classified in consideration of thermally processed food packages are also given in Figure 12.8. For the cover lid on the package, aluminum foil may be used as a laminated film. Or flexible retort pouches are fabricated by multilayer film laminated with PP, aluminum foil, PET, and/or nylon layer. PET or nylon serves as surface layer with mechanical strength against outer stress. Precautions are needed in the use of EVOH and nylon as gas barrier layer, because oxygen permeability of these polymers increases under high humidity conditions (see Section 4.6.3). Table 12.7

compares packaging materials for sterilization process. As for the protection against the oxidative quality deteriorations, low levels of residual air or high vacuum is desired, which is also beneficial to lower the internal pressure of the package during thermal processing. Higher vacuum or low residual air is obtained by hot filling, heat exhausting, or mechanical exhaust by vacuum pump. Desired vacuum level in pasteurized glass jar is in the range of 30-45 cm Hg and that for sterilized product is 45 cm Hg [9]. For the retortable pouches of consumer size, residual air is suggested to be below 10 cm^3 or 2% of the food volume. Air removal methods employed for retortable pouch are steam flushing, drawing of vacuum, or squeezing the pouch between two vertical plates.

Table 12.7 Comparison among various food packaging materials for thermal processing

Property	Metal cans	Glass jars	Plastic pouches or trays
Usual geometry	Cylinder	Cylinder	Slab
Gas and moisture barrier	Excellent	Excellent	Variable
Transparency	Opaque	Transparent	Variable
Weight	Light	Heavy	Light
Production speed	Fast	Medium	Slow
Heat penetration	Slow	Slow	Relatively fast
Physical strength	Strong	Strong but weak to impact	Weak
Hermetic closing	Double seamed end	Cap	Heat-sealing or double seam
Outer individual package	No	No	Yes
Ease of transportation and handling	Not as good	Poor	Good
Convenience	Not as good	Good	Excellent
Recycle	Good	Good	Poor

12.6 Aseptic Packaging

Aseptic packaging usually implies sterilizing the container and food separately before the filling and sealing operations under aseptic environment (Figure 12.9). The term aseptic packaging has been defined as "all the means implemented during the phases of packaging, cleaning, and sterilizing which limit initial contamination of the container, its closure, the material used, and the atmosphere so that a sterile product will remain sterile during packaging, storage, and distribution" [15]. In order to achieve this, stringent means to destroy microorganisms and protect atmosphere from any contamination must be provided.

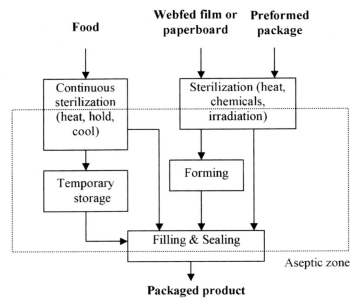

Figure 12.9 Typical aseptic packaging of liquid-based foods

Generally the technology has been applied to acidic liquid foods in the early development and extended to the low acid foods containing particles. Currently aseptic processing and packaging is used commercially in dairy products and for fruits juices, pea soups, sauces, tomato pastes, etc. Heat transfer to fluid before filling makes possible continuous heating and cooling, which can be carried out with short time at high temperature and thus provides better quality retention of the product. The benefit of HTST (high temperature and short time) processing is obtained from the difference in z values of microbial inactivation and degradation of nutrient and sensory qualities: lower z value of microbial destruction compared to that of quality degradation results in higher sensitivity of the microbes to high temperature (Figure 12.10). At higher temperature, microbial destruction is much faster than the destruction of nutrients and other components. Thus, aseptically processed foods are generally better in color and higher in retained nutrient (lower C value) than those manufactured by conventional canning.

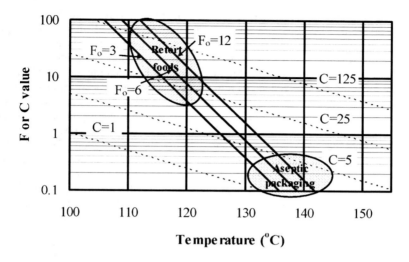

Figure 12.10 Comparison of retorting and aseptic packaging in terms of temperature-time relationship and product quality (C value based on 100°C with z_c of 33°C)

The theoretical background of aseptic food processing and packaging is the same with in-container sterilization in providing the heat treatment of required F value based on the target microorganism. However, the way of attaining commercial sterility differs between the two methods: in-container thermal processing is mostly controlled by monitoring the heat transfer to the geometric center of the food and the integrity of seal part; for the aseptic packaging, various potential sources of microbial contamination such as food, air, equipment, personnel, and package should be controlled to assure the aseptic conditions. Therefore aseptic packaging should include the validation procedure by physical, chemical, and microbiological means for all the steps in processing the foods.

12.6.1 Continuous Heat Processing of Fluid-Based Foods

The heat processing of liquid food is usually accomplished by heating product rapidly to 130-145°C, holding for a required time, and then cooling rapidly. Mechanical components of aseptic food processing, therefore, consist of metering pump, product heater, holding tube, product cooler, and back pressure valve. The condition of time and temperature is determined by the product characteristics, storage condition, and shelf life. Heat exchanger, indirect or direct, is used for heating the product [16]. Indirect heating relies on the heat transfer through a surface separating the product and the heating media. Tubular, plate, or scraped surface heat exchangers are applied for indirect heating. Tubular heat exchanger consists simply of coils and provides faster heat transfer to liquid food. Plate heat exchanger consisting of gasketed plates gives high energy efficiency with low pressure operation. Scraped surface heat exchanger with inner rotating screw is suitable for viscous products or fluids with solid particles. Direct heating involves the direct contact between the food and heating steam: injecting pressurized steam into the product (steam injection) or injecting the product into the steam (steam infusion). Direct heating methods give practically instantaneous heating compared to indirect heating or in-container heat processing (Figure 12.11). In direct heating the water condensed from steam should be removed or accounted for in the formulation. Recently the passage of electric current through the food has been adopted in aseptic processing. Such an electroresistive heating called ohmic heating, results in fast heating, in particular for the particles in the fluid food. Pulsed electric field has also been proposed for aseptic processing of fluid foods with less damage on food components, but does not reach the state of commercial application.

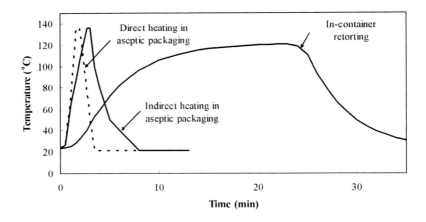

Figure 12.11 Comparison of different heating methods in thermal processing of foods

Following the heating section, the product enters holding section in which the desired temperature is maintained for the required period of time to accomplish the commercial sterility. The holding tube is deigned to have upward slope in order to eliminate the entrapment of air within the product. The product flows at a specified rate through a series of insulated pipes attached to the heating section. A back pressure sufficient to prevent product boiling or flashing should be applied and maintained during operation. Boiling may decrease the temperature and residence time of the product in the hold tube and endanger the attainment of required sterility. A metering pump located upstream from the holding section should be operated to maintain a required flow rate invariably. The controlled flow rate of the product is a critical factor to obtain the consistent sterilization value.

From the relationship between the flow rate and the volume of the holding section, the holding time or residence time is determined. The holding section volume divided by the flow rate gives the residence time when the velocity inside the tube is assumed constant, which applies to the case of plug flow. The assumption of plug flow can rarely apply to real flow situation. Turbulent flow is attained when the Reynolds number (Re) defined by Equation 12.17 exceeds 4000.

$$Re = \frac{D\rho \bar{v}}{\mu} \qquad (12.17)$$

where D is diameter of the tube (m), ρ is the density (kg m^{-3}), $\bar{v}$ is the average velocity (m s^{-1}) obtained from dividing the flow rate by cross sectional area of the tube, and μ is the viscosity (kg m^{-1} s^{-1}). In the turbulent flow there is small velocity variation inside the flow: the hold time is calculated conservatively based on the fastest moving component, which is usually assumed to be about 1.25 times the average velocity, i.e., shorter hold time by a factor of 1.25. When the Reynolds number is below 2100, laminar flow with parabolic flow profile is assumed, which has the velocity of the fastest moving component being twice the average velocity: therefore conservative hold time based on the maximum velocity is shorter by a factor 2 compared to that based on average velocity. Worst case of the hold time estimation is used as a conservative design of the aseptic processing.

For the liquid product containing particles, the process design should consider the microbial lethality at the slowest heating location in the particle. In normal heat exchangers heat is transferred first from the heating medium to the fluid and then to the particles. Heat transfer coefficient from the fluid to the particle may determine the temperature in the particle. Conservative approaches are used in the estimation of the heat transfer coefficient and residence time distribution of the particle. In ohmic heating particle heats faster than the fluid part because of the different mechanism of heat transfer [17].

The product removed from holding section enters another heat exchanger for cooling which regenerates heat for preheating the product entering the heater. Direct heating systems usually transfer the processed food to vacuum chamber evaporating a required amount of water, and then to a subsequent heat exchanger for final cooling. After the cooler, the processed food is routed to the product filler or aseptic surge tank, before which an automatic flow diversion device is located and operated reliably to divert the product flow away from the line when there occurs an improper event in holding section. An aseptic surge tank of 575-22700 liters may be located in an aseptic system to hold sterilized product after processing and before filling into package [18]. The product may be transferred to the surge tank and then delivered to the filler at a proper rate. The surge tank provides a buffer between processing and packaging systems, which is useful when the flow rate of the product sterilization system is not compatible with the filling rate of the packaging unit. A sterile air or sterile food-grade gas supply is necessary in order to maintain a protective positive pressure within the tank and to displace the contents.

The sterilization process should be validated in terms of selecting appropriate F value, achieving the required F value through mathematical model and experimental measurements, examining the critical factors and conducting the inoculated pack test. For acid foods, yeasts, lactobacilli, and molds are considered as test organisms. Aciduric bacilli may be warranted for certain uses; *Bacillus circulans* and *B. cereus* were found to be dominating spoilage organisms in aseptically packaged liquid dairy products which were pasteurized and stored in chilled conditions [19]. For the low acid foods, nontoxic microorganisms with higher resistance than that of *Cl. botulinum* are considered for the validation tests. Typical aseptic thermal processing conditions for low-acid foods with pH $\geq$ 4.5 are 138-149°C for 2-5 seconds. Those for acid foods with pH < 4.5 are 93-96°C for 15-30 seconds [18].

12.6.2 Sterilization of Packages and Food Contact Surfaces

Packaging materials get contaminated with microorganisms mainly by air and/or contact with machines or men. Usual range of total microbial contamination of the package surface is of 2-100 microorganisms per 100 cm^2 while more than 1000 organisms per 100 cm^2 were found in poor hygiene conditions [15, 20]. Small percentage of package may be contaminated as high as 10^5 organisms per package. So decimal reduction of 10^{-5} is usually applied as a target. Heat, chemicals, and irradiation have been used, alone, or in combination, for sterilizing the packaging materials and packaging machinery in aseptic packaging (Table 12.8). Saturated steam under pressure has been used for the sterilization of metal cans and lids and also for thermostable plastic cups. Plastic materials have limited thermal stability and thus have narrow window of temperature range applicable. Superheated steam is used for sterilizing the metal cans. Sufficiently high surface temperature should be attained to achieve the required sterility. The saturated steam frequently heats the materials to 121-129°C, while superheated steam to 176-232°C [21]. Superheated steam has an advantage of attaining high temperature at atmospheric pressure without pressurization but has a disadvantage that bacteria are more heat resistant under it. Hot air has similar pros and cons similar to those of superheated steam.

Table 12.8 Package sterilizing methods in aseptic packaging

Method	Container type	Remarks
Saturated steam/superheated steam	Metal cans or drums, composite cans, glass bottles, heat-resistant plastic cups	High temperature applied prohibits use of heat-sensitive plastics; saturated steam is more effective in microorganisms than superheated steam
Dry hot air	Metal or composite containers	High temperature at atmospheric pressure; weaker sterilization effect than in saturated steam in
Hydrogen peroxide	Plastic containers or webs, laminated foil or paper, preformed cartons	More effective at higher temperature; often combines with high temperature condition; combination with UV increases the effectiveness
Heat from extrusion process or blow molding	Plastic containers	No chemicals used; may be aided with other method for low acid food application
Irradiation	Plastic bags and containers, bag-in-box system	γ-ray is usually used for heat sensitive packaging materials; effective at ambient temperature but expensive; treatment is usually possible only at the place of irradiation source

The residual heat from the fabrication such as coextrusion, blow molding, and thermoforming may be used for the sterilization of the plastic containers. In a system of coextruded plastic cup, a superficial layer of thermoplastic multilayered film sheet is delaminated under aseptic conditions to expose the sterile surface, which is then thermoformed into cup container [18]. The lid film delaminated similarly from the coextruded film is used to cover and seal the container body which is filled with aseptically processed food.

Most aseptic packaging systems use hydrogen peroxide with concentrations of 30-35% in combination with heat for sterilizing packaging materials and other food contact surfaces. The packaging materials are dipped in the hydrogen peroxide solution or sprayed or rinsed with it. Subsequent hot air drying is applied to decrease the residual hydrogen peroxide to legal level, which is 0.5 ppm in United States [21]. It was shown that sterilizing the polypropylene and polyethylene surfaces by hydrogen peroxide did not influence the polymer composition and migration into foods [22]. Other chemicals such as food acids (citric, lactic, and acetic acids), ethanol, ethylene oxide, peracetic acid, and ozone are being used or examined for use in sterilizing the equipment and packaging materials.

Ionizing radiation produced from cobalt-60 source inactivates microorganisms on the pouches or bags. A dose below 60 kGy is usually approved by US FDA for the sterilization of the packaging materials with maximum does being dependent on the material. Ultraviolet (UV) ray is sometimes applied alone for partial decontamination of packaging surface or for helping the effectiveness of hydrogen peroxide treatment.

Table 12.9 Test organisms for aseptic packaging systems

Sterilization medium	Organisms
Saturated steam	*Bacillus stearothermophilus*
	Clostridium sporogenes
Super heated steam	*B. stearothermophilus*
	B. polymyxa
Dry heat	*B. stearothermophilus*
H_2O_2 + UV or heat	*B. subtilis*
Peracetic acid	*B. subtilis*
Heat of formation	*B. stearothermophilus*
Gamma radiation	*B. pumilus*

Adapted from Ref. [20, 23].

The sterilization treatment of packaging materials should be validated by microbial test. Known concentrations of test organisms are applied on the package surface, which is then treated by physical or chemical means. Containers are filled with a growth medium or product, incubated, and examined for the evidence of growth or spoilage. In the other method the survivors are directly determined on the treated surface by swabbing technique. Table 12.9 shows the test organisms for the aseptic system.

12.6.3 Aseptic Filling and Packaging System

Before start of the production of aseptically processed food products, equipment with food contact surfaces is sterilized by circulation of hot water or saturated steam through the food transferring system, which includes aseptic pump, aseptic surge tank, the tubes, and valves. Thirty minutes at 121°C or equivalent is a usual minimum requirement of the system for low acid foods, while 30 minutes at 104°C is suggested as a minimum for acid foods [21]. Air supply system including air filters and air supply pipe lines should be assured to be in sterile condition by presterilization. The aseptic zone covering entrance of presterilized container or materials and exit of the sealed packages, should be operated and maintained sterile during subsequent production stage. The aseptic zone should be constructed with adequate physical barriers between the sterile and nonsterile areas. The aseptic zone is usually maintained under positive pressure of sterile air. The sterile air is also used in managing the aseptic surge tank. An overpressure of sterile air or other inert gas is introduced and maintained to facilitate the product removal and prevent the contamination from the environment. If the sterile air is produced by high-efficiency purified air (HEPA) filters, they should be presterilized.

Various fillers are used depending on product characteristics and packaging system applied. Sine fillers are used mostly for pouch form-fill-seal machines, piston fillers are for cup form-fill-seal machines and bottle filling lines, and weight filling is for bulk metal drum packaging [18, 24]. The filling operation may be controlled to eliminate headspace air and maintain vacuum condition inside the package, which can be assisted by steam injection prior to closed sealing or seaming. On the other hand, the package headspace may be flushed with sterile air or nitrogen, which prevents excessive vacuum and distortion of the rigid container. The headspace of the cartons helps to open the package and pour the content food without spilling. Headspace is also advantageous for preventing contamination on the seal area. Nitrogen gas introduced in liquid form can retard potential oxidative deteriorative reactions in the food and strengthen the bottle dimension.

For validating the sterility of the aseptic zone, strips or discs inoculated with appropriate microorganisms and inoculum level are placed throughout the aseptic zone. After the sterilization cycle, the strips or discs are tested in the microbial survival.

Aseptic filling and packaging system may be classified according to the type of packaging materials and operation methods in package fabrication and sterilization. A variety of materials are adopted and used in aseptic packaging: metal cans and drums sterilized by steam or hot air, webfed paperboard and plastic film sterilized by hydrogen peroxide or alcohol, preformed paperboard and rigid plastic containers sterilized by hydrogen peroxide and heat, plastic containers fabricated by thermoform-fill-seal, glass and plastic bottles sterilized by hydrogen peroxide, and bag-in-box sterilized by gamma ray. Other variants exist with a variation in sterilization mode and container types.

Table 12.10 gives general overview on the package types and operation systems employed in aseptic packaging. Packaging materials for aseptic packaging do not have to withstand high temperature of thermal process unless thermal sterilization of packaging is not used. Therefore wide range of materials can be used in the aseptic packaging system. Currently laminated paper cartons and plastic containers are dominant form of aseptic packaging [25]. For the paperboard and plastic packages, a gas barrier layer such as aluminum foil, PVDC, EVOH is incorporated into the laminated or coextruded structure; the inner layer of polyethylene or polypropylene provides heat-sealing property; paperboard or thick plastic layer gives mechanical strength. The shelf life of aseptic packages is dependent on their permeability to oxygen and light. Sizes of aseptic packages range from about 100 mL for consumer pack to 300 L for institutional packs; as a special purpose of bulk storage, large container up to 4000 L capacity is used for aseptic processing and packaging [26]. Each aseptic packaging system has its own advantages and disadvantages. The selection of system is based on the product characteristics, shelf life, package type, cost, and convenience.

Table 12.10 Typical aseptic packaging systems

Package type	Packaging material	Package sterilization	Container closing in aseptic zone	Applications	Commercial system
Can	Metal	Superheated steam	Double seam	Wide range of low acid fluid foods	Dole
	Composite	Hot air, H_2O_2	Double seam	Fruit juice, coffee, sake	Dole, Cartocan
Bottle	Glass	Saturated steam, dry heat	Cap	Milk, fruit juice	Serac, Remy
	Plastics (HDPE, PP, PET), preformed or form-fill-seal	H_2O_2	Heat seal, jawed cap	Juice, dairy based beverage	Stork, Shibuya, Ampack-Ammann
Cup and tray	HIPS or PP buried with PVDC or EVOH, preformed or form-fill-seal; preformed Al cup	H_2O_2, extrusion heat	Heat seal	Pudding, cream, dairy products	Bosch, Conoffast, FreshFill, Gasti, Serac
Pouch and bag	Plastic laminates	H_2O_2 with UV radiation	Heat seal	Milk, sauces, tomato products	Prepack, Bosch
Carton	Paperboard-Al laminate coated with PE, prefabricated or form-fill-seal	H_2O_2 with UV radiation	Heat seal, ultrasonic or induction seal	Milk, juice, wide range of liquid foods	Tetra Pak, Combibloc, International Paper, Liqui-Pak
Bulk pack	Metal drum	Saturated steam	Metal cap	Fruit and vegetable concentrates, purees	Rheem, FranRica, Cherry-Burrel
	Bag-in-box of premade plastic bag	Saturated steam, ionizing radiation	Tightened plastic cap aided with heat seal	Tomato paste and sauce, fruit purees	Scholle

Compared to in-container heat processing, aseptic processing and packaging depends more on the proper quality assurance program for its success. The raw materials, designed thermal process, filling, and packaging operations should be controlled in every aspect of microbial contamination, absence of defect, and process variables. Critical factors to control potential hazards need to be identified and established in the operation. Principles of hazard analysis critical control point (HACCP) program should be rigorously adopted to help this. Table 12.11 presents potential failure modes in aseptic processing and packaging of foods.

Table 12.11 Types of potential failure modes and microbial profile in aseptic processing and packaging of food

Type	Failure mode	Microbial profile of concerns
1	Incoming raw ingredients, handling, storage, and batching	Incipient psychrotrophs, thermophilic spores
2	Equipment preparation and setup, clean in place and sanitation, and presterilization	Anaerobic spores and vegetative cells
3	Thermal process design & delivery – heating cycle	One type of spore – mesophilic or themophilic
4	Thermal process design & delivery – cooling cycle including surge tanks	Anaerobic and aerobic spores, vegetative cells
5	Incoming packaging material and its sterilization	Vegetative cells
6	Aseptic zone integrity and environmental load	Spores, vegetative cells
7	Container seal integrity	Mostly vegetative cells

Adapted from Ref. [27].

Once the sterilization processes for product and equipment have been established, microbial inoculated pack test should be conducted for proper system operation. At least 100 packages should be collected for each test batch, then incubated, and monitored for spoilage. The automatic controls and safety devices should also be challenged to test for proper function. After passing all these tests, it is suggested that at least four commercial production runs of uninoculated product be carried out and checked by incubation with 100% examination for spoilage and inspection on package seals. And even after start of commercial production, evaluation of the process should be continued on routine basis for successful operation of aseptic food packaging. Some nondestructive tests using ultrasound, redox potential, and pH indicators (placed on package lids), colorimeter, etc. have been developed and proposed for monitoring the microbial growth in incubated samples aseptically processed.

Example 12.3 Estimating lethality in aseptic processing

A chocolate-flavored milk is processed and packaged aseptically through tubular heat exchanger. The milk is heated to 141°C, flowed through hold section and then cooled to 5°C. The flow rate is maintained at 1.36 L min^{-1} in the hold tube with a diameter of 3.6 cm and length of 0.24 m. The density and viscosity of chocolate-milk are 1030 kg m^{-3} and 2.1 × 10^{-3} kg m^{-1} s^{-1}, respectively. Estimate the minimum F$_o$ value.

Average velocity is given by

$$\bar{v} = \frac{\text{Volumetric flow}}{\text{Crosssectional area}} = \frac{1.36 \times 10^{-3} \text{m}^3 \text{min}^{-1}}{(1/4)\pi(0.036\text{m})^2} = 1.34 \text{ m min}^{-1} = 0.0223 \text{ m s}^{-1}$$

and thus

$$Re = \frac{D\rho \bar{v}}{\mu} = \frac{0.036 \times 1030 \times 0.0223}{2.1\text{x}10^{-3}} = 394$$

Because Re<2100, the flow is laminar.

The retention time based on the flow rate, i.e., average velocity is given as

$$\text{Residence time} = \frac{\text{Volume of hold tube}}{\text{Volumetric flow rate}} = \frac{(1/4)\pi(0.036 \text{ m})^2 \times 0.24 \text{ m}}{1.36\text{x}10^{-3}\text{m}^3\text{min}^{-1}}$$

= 0.18 min = 10.8 s

However, the fastest moving fluid part would have the velocity twice the average one and thus have the half of the residence time based on the average velocity. Therefore the worst case hold time becomes 5.4 s, which gives

$$F_o = L \times \Delta t = 10^{(\frac{141-121.1}{10})}(5.4/60) \text{ min} = 8.8 \text{ min}$$

12.6.4 Modified Version of Aseptic Packaging

Sometimes the basic scheme of aseptic processing and packaging has been modified to accommodate the variety of food products, combine with other preservation mechanism, and attain an improved food quality. In the system the principles of aseptically handling the product are maintained while some arrangements to help the storage stability of the product are made. Sometimes the process is modified for easiness of the maintenance.

A variant of aseptic process combines the in-container heating and aseptic packaging: the prepared product is filled into the unsealed container, is heated to about 130°C in a chamber pressurized by steam, remains there for 3-6 minutes, and is sealed aseptically [6, 27]. Before coming out from the pressurized chamber or tunnel, the sealed container is cooled to about 100°C. A locking device or valve is needed for passage of the food through the pressurized chamber. Final cooling takes place under water spray. The whole process is similar to hot fill-hold-cool process but under pressured conditions at high temperature. This system is used for conventional containers such as cans and jars, in particular for large institutional-size cans. Solid foods can be handled in this system. Another advantage of this system is the improved quality coming from short heating time. Slightly different designs in heating method and overpressure medium are available depending on the product characteristics.

Another similar variant of aseptic packaging is for cooked rice, which was commercialized first in Japan [28]. The rice is washed to reduce the surface microbial load and then transferred to an ultraclean room maintained with HEPA filter. The soaked rice is heated for cooking and sterilization by steam at ultra high temperature and introduced into a barrier plastic trays which have been sterilized by UV ray. As rice is filled aseptically into the tray, steam is injected into the tray to evacuate the package and the package is sealed with headspace oxygen less than 0.3%. This system employs an integration of hot filling, hold, and aseptic packaging for the cooked rice. The product is distributed under ambient conditions with six month shelf life. It is reheated by microwave oven or immersion in the hot water when consumed.

Another example of using an ultraclean room is vacuum or modified atmosphere packaging of processed meat, cheese, pasta, and rice cakes, which has been called semi-aseptic packaging [25, 28]. The product is filled into the semi-rigid plastic container thermoformed aseptically, covered with coextruded barrier film, and then heat-sealed with vacuum or modified atmosphere. When the modified atmosphere is applied, 20-30% carbon dioxide with nitrogen in balance is used. The products are distributed at low temperature. In an absolute sense this system is not truly aseptic and may well be called a modified atmosphere packaging with low contamination. The contamination level of packaging atmosphere is controlled within class 10000, which represents a maximum number of particles greater than 0.5 μm per cubic foot of air. For helping the preservation quality, the product is smoked or added with salt and/or other preservatives. These combined hurdles prevent the product from the risk of *Cl. botulinum* but preserve the quality with minimum heat treatment.

Another successful semi-aseptic packaging is for the membrane-filtered beer [29]. The common spoilage microorganisms are yeasts and lactic acid bacteria. The membrane with micropores less than 0.45 μm is used for eliminating the spoilage organisms without heat treatment. The filtered beer is packaged to the glass or PET bottle under clean conditions. Even though very low levels of microorganisms exist in some instances, no spoilage is observed probably due to the preservative action of alcohol in beer. The beer produced from this technology gives fresh taste and flavor superior to the pasteurized one.

12.7 Case Study: Design of a Thermal-Processed Tray-Set Containing High- and Low-Acid Foods

Hayakawa et al. [30] analyzed the thermal processing for a tray-set which is served as a complete meal for military feeding. The set contains high- and low-acid food containers or compartments, which are different in cooking requirements. An example tray set consists of peach slices as high acid food, and white rice and chili con carne as low acid food (Table 12.12). Simultaneous thermal processing is desired for efficient process management with saving in labor and time. A variety of design variables such as retort temperature, insulation for peach slice package, and initial product temperature were examined for simultaneous heat processing. The heat process of 50 minutes at retort temperature of 120°C was found to give the desired lethality to all the foods, when the peach slice container was insulated by a 1.65 mm thick expanded plastic layer. The temperature history and cumulative lethality are given for all the three foods in Figure 12.12. For estimating the temperature change, Hayakawa's formulas with f and j values in Table 12.4 were employed. The respective F values of 0.057, 7.7, and 8.3 are attained for peach slices, white rice, and chile con carne, which are close to the target values in Table 12.12. The detailed calculation of this case study in MS Excel® can be found in a website given in the preface.

Table 12.12 A tray-set of high- and low acid foods in semi-rigid containers

	Peach slices	White rice	Chili con carne
Net weight (g)	227	227	326
Internal dimension of container (height × width × length in mm)	$30 \times 70 \times 118$ with insulation of 1.65 mm	$26 \times 92 \times 136$	$30 \times 70 \times 118$
Heat penetration parameters			
j_h	1.04	1.38	1.24
f_h (min)	74.0	28.3	26.5
j_c	1.06	1.43	1.21
f_c (min)	86.2	59.3	35.8
Target $F_{121.1}^{10}$ value (min)	0.05	7.5	7.5

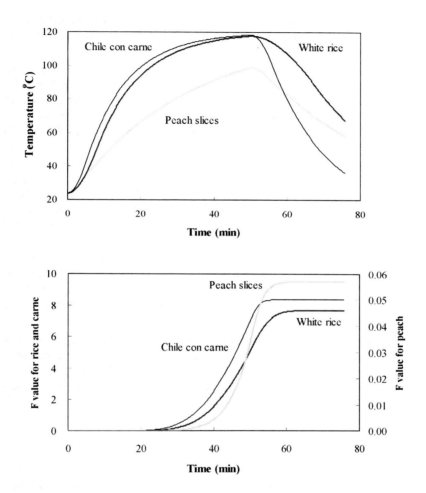

Figure 12.12 Profiles in temperature and cumulative lethality of low- and high-acid foods in a tray set being sterilized with process time of 50 minutes at 120°C. Refer to Table 12.12 for the package information.

Discussion Questions and Problems

1. How does headspace affect the processing and packaging operation? What would be the storage stability affected by oxygen included in the headspace and dissolved in the product?

2. What would be the advantages and disadvantages of packaging orange juice aseptically in paper-based laminate gable top package compared to pasteurization in metal can?

3. An acid food metal can with pH of 4.1 shows the heat-penetration parameters for geometric center as follows: j_h=1.82, f_h=26.3 min. The cold-filled product with initial temperature of 25°C is going to be pasteurized under hot water bath of 98°C. The process is designed that cooling starts when the center temperature arrives at 90°C. What would be the process time for heating cycle?

4. A hot filled retort pouch of 200g curry sauce is going to be sterilized in the steam/air retort at 121°C with a pressure of 1.8 bar absolute. The product has the temperature of 75°C at the filling with residual air of 8 mL. Volumetric thermal expansion coefficient of the food may be assumed as 5.8 x 10^{-4} K^{-1}. The density of the food is 1240 kg m^{-3}. What would be the maximum volume of the package during retorting?

5. A spinach puree is aseptically processed at 142°C for 5.3 s. What is the F_o value for this process. Estimate the C_{100} value based on the z_c value of 26.2 °C for destruction of chlorophyll a. If D_c at 121.1 °C for the chlorophyll component destruction is 13.6 min, what would be the retention of the chlorophyll component for this process?

6. An orange juice product is going to be packaged aseptically through a tubular heat exchanger. The juice is heated to 100°C, flowed through hold section, and then cooled to 20°C. The hold tube with a diameter of 4.7 cm and length of 10 cm is used to give F_{100}^6 of 0.3 min. The density and viscosity of orange juice are 1043 kg m^{-3} and 2.6 x 10^{-3} kg m^{-1} s^{-1}, respectively. What would be the required flow rate?

Bibliography

1. Microbiology Sub-Committee of the Heat Preserved Foods Panel. Guidelines for the Establishment of Scheduled Processes for Low-Acid Canned Foods. Chipping Campden, UK: The Campden Food Preservation Research Association, 1977, pp. 4-11.
2. SD Holdsworth. Thermal Processing of Packaged Foods. London: Blackie Academic & Professional, 1997, pp. 70-229.
3. KL Brown. Principles of heat preservation. In: JAG Rees, J Bettison, ed. Processing and Packaging of Heat Preserved Foods. Glasgow, UK: Blackie, 1991, pp. 15-49.
4. AC Hersom, ED Hulland. Canned Foods. Thermal Processing and Microbiology. Edinburgh, UK: Longman, 1980, pp. 87-102, 177-207.
5. GD Betts. Critical factors affecting the safety of minimally processed chilled foods. In: S Ghazala, ed. Sous Vide and Cook-Chill Processing for the Food Industry. Gaithersburg: Aspen Publishers, 1998, pp. 131-164.

6. J Larousse, BE Brown. Food Canning Technology. New York: Wiley-VCH, 1997, pp. 297-602.
7. K Yamaguchi. Retortable packaging In: T Kadoya, ed. Food Packaging. San Diego, CA: Academic Press, 1990, pp. 185-211.
8. KI Hayakawa. Estimating temperatures of foods during various heating or cooling treatments. ASHRAE J 14:65-69, 1972.
9. J Bettison. Packaging of heat preserved foods in glass containers. In: JAG Rees, J Bettison, ed. Processing and Packaging of Heat Preserved Foods. Glasgow, UK: Blackie, 1991, pp. 138-163.
10. CF Bohren, BA Albrecht. Atmospheric Thermodynamics. New York: Oxford University Press, 1998, pp. 55-56.
11. PN Patel, DI Chandarana, A Gavin III. Internal pressure profile in semi-rigid food packaged during thermal processing in steam/air. J Food Sci 56:831-834, 1991.
12. PJG Proffit. Packaging of heat preserved foods in plastic containers. In: JAG Rees, J Bettison, ed. Processing and Packaging of Heat Preserved Foods. Glasgow, UK: Blackie, 1991, pp. 164-186.
13. CJ Snyder, JM Henderson. A preliminary study of heating characteristics and quality attributes of product packaged in the retort pouch compared with the conventional can. J Food Process Eng 11:221-236, 1989.
14. MJ Kirwan, JW Strawbridge. Plastics in food packaging. In: R Coles, D McDowell, MJ Kirwan, ed. Food Packaging Technology. Oxford, UK: Blackwell Publishing, 2003, pp. 174-240.
15. JP Ranno, D Lemaire. Aseptic packaging. In: G Bureau, JL Multon, ed. Food Packaging Technology. Vol. 2. New York: VCH, 1996, pp. 35-50.
16. SP Emond. Continuous heat processing. In: P Richardson, ed. Thermal Technologies in Food Processing. Boca Raton, FL: CRC Press, 2001, pp. 29-48.
17. HJ Kim, YM Choi, TCS Yang, IA Taub, P Tempest, P Skudder, G Tucker, DL Parrott. Validation of ohmic heating for quality enhancement of food products Food Technol 50(5):253-261, 1996.
18. A Lopez. A Complete Course in Canning. Book II. 12th ed. Baltimore: The Canning Trade, 1987, pp. 160-211.
19. TW Dommett. Spoilage of aseptically pasteurized packaged pasteurized liquid dairy products by thermoduric psychrotrophs. Food Australia 44:459-461, 1992.
20. MA Cousin. Microbiology of aseptic processing and packaging. In: JV Chambers, PE Nelson, ed. Principles of Aseptic Processing and Packaging. Washington, DC: The Food Processors Institute, 1993, pp. 47-86.
21. KE Stevenson, KA Ito. Aseptic processing and packaging of heat preserved foods. In: JAG Rees, J Bettison, ed. Processing and Packaging of Heat Preserved Foods. Glasgow, UK: Blackie, 1991, pp. 72-91.
22. L Castle, AJ Mercer, J Gilbert. Chemical migration from polypropylene and polyethylene aseptic food packaging as affected by hydrogen peroxide sterilization. J Food Protect 58:170-174, 1995.

23. DT Bernard, A Gavin III, V Scott, BD Shafer, KE Stevenson, JA Unverferth, DI Chandarana. Validation of aseptic processing and packaging. Food Technol 44(12):119-122, 1990.
24. MN Ramesh. Food preservation by heat treatment. In: MS Rahman, ed. Handbook of Food Preservation. New York: Marcel Dekker, 1999, pp. 95-172.
25. M Yokoyama. Aseptic packaged foods. In: T Kadoya, ed. Food Packaging. San Diego, CA: Academic Press, 1990, pp. 213-228.
26. JD Floros. Aseptic packaging technology. In: JV Chambers, PE Nelson, ed. Principles of Aseptic Processing and Packaging. Washington, DC: The Food Processors Institute, 1993, pp. 115-148.
27. JRD David, RH Graves, VR Carlson. Aseptic Processing and Packaging of Food. Boca Raton, FL: CRC Press, 1996, pp. 3-194.
28. AL Brody. Integrating aseptic and modified atmosphere packaging to fulfill a vision of tomorrow. Food Technol 56(4):56-66, 1996.
29. GC Reid, A Hwang, RH Meisel. The sterile filtration and packaging of beer into polyethylene terephthalate containers. J Am Soc Brew Chem 48:85-91, 1990.
30. KI Hayakawa, J Wang, SG Gilbert. Simultaneous heat processing of high- and low-acid foods in semirigid containers-a theoretical analysis. J Food Sci 56:1714-1717, 1991.

Chapter 13

Vacuum/Modified Atmosphere Packaging

13.1 Basic Principles 397
    13.1.1 Gases for MAP ···399
    13.1.2 Packaging Operations for MAP Foods ·································404
    13.1.3 Safety of MAP foods ···406
13.2 Nonrespiring Products 410
13.3 Respiring Products 412
13.4 Case Study: Design of Modified Atmosphere Package for Blueberries 420
Discussion Questions and Problems 422
Bibliography 422

13.1 Basic Principles

Modification of packaging atmosphere from normal atmosphere (78% nitrogen, 21% oxygen) is widely used as effective modern packaging technologies which can significantly extend shelf life and improve quality level of most food products. The modification of internal package atmosphere may take place at the level of total pressure and/or partial pressures of component gas components (Figure 13.1). The composition or condition of the package atmosphere optimal for a food product is different with product type, and thus is created or built in consideration of interaction between atmospheric composition and packaged food.

When the total pressure inside the package becomes lower than that of outside environment as a consequence of a more or less evacuation of the internal atmosphere, it is commonly referred to as vacuum packaging or hypobaric packaging. The degree of residual pressure inside the package can vary greatly, according to the time and intensity of the evacuation step during packaging operation. Vacuum packaging is usually the condition with a negligible amount of nitrogen, oxygen, and carbon dioxide, and can be with vacuumed headspace (in rigid or semi-rigid packages) or tight contact between packaging film and food surface (in flexible packaging). It is quite common to call 'hypobaric packaging' the condition where the residual pressure is equal or higher than 2.6 kPa but at subatmospheric pressure. Typically packaging of moist foods in fact cannot reach the residual pressure down below the water vapor pressure at the corresponding temperature. The effectiveness of vacuum packaging for shelf life extension is mainly due to the low or very low amount of residual oxygen, as a consequence of the air extraction: oxidative reaction and aerobic respiration are reduced. Moreover, vacuumed package may be seen to be an effective means in eliminating possible biological and chemical contaminants from the space surrounding the food. The reduction of the package volume should also be noted as an important advantage of vacuum for flexile packaging.

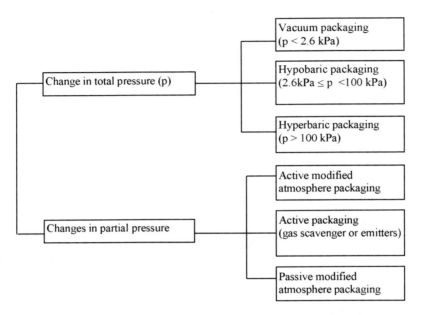

Fig. 13.1 Possible ways to modify the atmosphere inside the food package

On the contrary to hypobaric packaging, the generic term, 'hyperbaric packaging' is used for the case when the total pressure inside a rigid package (e.g., a bottle or a can) increases even higher than outside pressure as a consequence of the over pressure of carbon dioxide (added or produced by fermentation) or generated by any gas mixture added for protective aims.

As previously mentioned, the modification of package atmosphere may concern the partial pressures of its constituents; i.e., we can have a gas mixture inside the package whose composition is quite different from that of normal atmosphere under the condition that total pressure is equilibrated with outside atmosphere. This case is called as modified atmosphere packaging (MAP). MAP condition can be achieved actively (the initial substitution of normal air with a selected atmosphere) or passively (a change of internal atmosphere from the natural consequence of the balance between gas permeation across the package walls and gas production/consumption inside the package). A classification of the possible ways for changing the package internal atmosphere, for completeness, should also include active packaging, i.e., the possible gas scavenging (oxygen, carbon dioxide, ethylene, etc.) as well as the possible gas releasing (carbon dioxide, ethylene, etc.). In certain conditions, it may cause slight change in total pressure. Active packaging is discussed in detail later in Chapter 15.

Properly designed MAP can preserve the quality of a variety of packaged foods including fresh produce, meats, fishes, and bakery products. Shelf life extension is achieved by several different mechanisms that are mainly biological in nature (microbial or enzyme inhibition) but also chemical (the delay or reduction of chemical oxidation, the preservation of typical pigments) and even physical (staling reduction for bread and bakery products).

13.1.1 Gases for MAP

The gas mixture applied in MAP usually consists of nitrogen, oxygen, and carbon dioxide, but other gases can be usefully included in the modified atmosphere. In Europe, for instance, argon, nitrous oxide, and helium are permitted by the food law for use in packaging and debate is going on the use of small percentage of carbon monoxide and hydrogen. Hydrogen as well as helium is admitted or proposed as tracking gas, i.e., their presence outside the package in a nondestructive test could effectively reveal a leak in the material or a defect in the sealing. Nitrous oxide is mainly used as a propellant gas for liquid or semisolid foods (mounted cream, processed cheeses) but also some protective actions have been observed for this gas [1]. All the gases employed are, in different extent and modalities, useful for improving the preservation. The composition of the package atmosphere optimal for a product is different with product type. Properties of each gas component should be understood with reference to its effect on the quality degradation of packaged food. Its interaction with food component and packaging material also needs to be considered.

Table 13.1 shows the properties of the main gas components in terms of use for modified atmosphere packaging. In the perspective of food preservation, oxygen is not preferred gas and thus usually excluded from package headspace in food MAP. Oxygen attacks lipid components, colors, and flavors in foods enzymatically or nonenzymatically. Most food spoilage bacteria, yeasts, and molds are aerobic and require oxygen for growth leading to deterioration. Removal of oxygen from the package can slow down all these chemical reactions and microbial growth. Major beneficial effect of vacuum packaging is in the complete elimination of oxygen. However the degree of oxygen removal in vacuum packaging and MAP depends on the product type. Many processed meat products are evacuated completely for long term storage. Fresh fruits and vegetables need some level of oxygen supply to maintain aerobic respiration that is essential for live cell activity and freshness. In case of retail meats, on the contrary, a high level of oxygen is needed for the status of oxymyoglobin responsible for bright color, which consumers appreciate as high and fresh quality. Recently high oxygen concentration has been proposed to allegedly slow down the microbial deterioration and inhibit enzymatic browning for fresh produce.

Table 13.1 Properties of the main gas components for MAP of foods

Gas	Properties	Use in MAP
Oxygen	Moderately soluble in water and fat; moderately permeable through plastic film; supports aerobic spoilage microorganisms; oxidizes fat and other sensitive components; leads to oxidized forms of myoglobin (oxymyoglobin of bright red color and metmyoglobin of brown color); suppresses microbial growth at high concentration; maintains and accelerates aerobic respiration of fresh produce; inhibits strict anaerobic bacteria	Usually excluded from package atmosphere to improve the quality preservation; used to maintain minimum aerobic respiration for fresh produce; avoids anaerobic condition; blooms the meat color; slows down the microbial deterioration and inhibits enzymatic browning at high concentration under certain conditions
Carbon dioxide	Highly soluble in water and fat; most permeable through plastic film; suppresses aerobic spoilage microorganisms; reduces aerobic respiration of fresh produce; may cause discoloration and acid taste for certain foods at high concentration	Delays microbial deterioration mostly caused by Gram-negative bacteria and molds; produces tight contact contour package due to dissolution of CO_2 under 100%; reduces the physiological metabolism of fresh produce; reduces the bread staling under certain condition
Nitrogen	Inert, low solubility in water and fat; least permeable through plastic film	Displaces oxygen so as to prevent oxidation and aerobic microbe growth; inert filler gas to prevent package collapse and to dilute oxygen or carbon dioxide properly
Argon	Inert, more soluble in water and fat than nitrogen; higher density than nitrogen; it competes with oxygen due similar size and properties	Displaces oxygen so as to prevent oxidation and aerobic microbial growth

Carbon dioxide is unique in that it has high solubility in aqueous and/or fatty foods and antimicrobial activity against aerobic Gram-negative bacteria, molds and some yeasts. Suppression of microbial growth is observed in increased lag time and decreased growth rate as shown in Figure 13.2. Mechanism of microbial inhibition by CO_2 was not elucidated yet, but the proposed theories are related to (a) alteration of cell membrane function on nutrient uptake, (b) direct inhibition of enzymes, (c) intracellular pH change due to penetration of dissolved CO_2 through cell membrane, and (d) direct changes in protein properties [2, 3]. Whatever mechanism works, the level of dissolved CO_2 in foods is known to affect directly the degree of microbial inhibition (Figure 13.2).

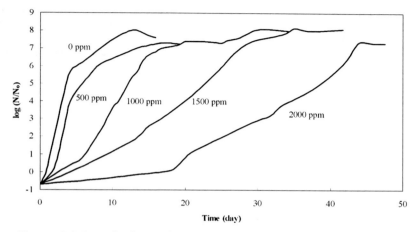

Figure 13.2 Growth of *Pseudomonas fluorescens* at 7°C for different CO_2 concentrations in a microbial broth medium, which are attained by different package atmospheres. Ratio of microbial count (N) to initial one (N_o), sketched according to the data of Ref. [4].

The solubility of CO_2 in aqueous foods increases with moisture increase and temperature decrease [5]. Therefore the effectiveness of CO_2 gas in microbial inhibition increases with higher moisture food at lower temperature. Dissolved amount of CO_2 in aqueous solution can be described by Henry's Law (Equation 13.1):

$$C_{CO2} = k_{CO2} \, P_{CO2} \qquad (13.1)$$

where C_{CO2} is concentration of CO_2 dissolved in aqueous phase (mg mL^{-1}), P_{CO2} is partial pressure of CO_2 in headspace (bar), and k_{CO2} is Henry's Law constant (mg mL^{-1} bar^{-1}). The k_{CO2} for water is dependent on temperature, CO_2 partial pressure and dissolved oxygen [5]:

$$k_{CO2} = 3.36764 + 0.07(1 - \frac{C_{O2}}{9}) - (0.014 - 0.00044C_{O2})P_{CO2} - 0.12723T$$
$$+ 2.8256 \times 10^{-3} \, T^2 - 3.3597 \times 10^{-5} \, T^3 + 1.5933 \times 10^{-7} \, T^4 \qquad (13.2)$$

where T is temperature (°C), and C_{O2} is concentration of O_2 dissolved in aqueous phase (mg L^{-1}).

Figure 13.3 shows the dependence of microbial growth parameters on temperature. Lag time decreases and specific growth rate (inverse of time for microbial count to increase 2.718 times initial value) increases with temperature for both air and CO_2 atmosphere; relative ratios of both parameters between two atmospheric conditions increase with temperature. It is noted that both relative ratios are aligned in parallel with concentration of dissolved CO_2. This implies again that dissolved CO_2 determines the relative efficacy of CO_2 atmosphere and maintaining low temperature is thus essential to attain the beneficial effect of CO_2 atmosphere in retarding microbial spoilage. In case of *Listeria monocytogenes*, the effect of growth inhibition by CO_2 at low temperature was observed mainly in extending the lag time (Figure 13.3).

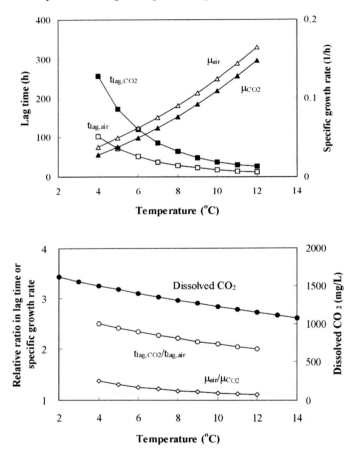

Figure 13.3 Solubility of CO_2 in water and growth *of Listeria monocytogenes* as function of temperature under CO_2 partial pressure of 0.5 atm. Solubility of CO_2 in water was from Equation 13.1 and growth model parameters *of L. monocytogenes* was calculated from Ref. [6] by assuming water activity of 1.0 and CO_2 solubility based on Equation 13.1. $t_{lag,air}$ and $t_{lag,CO2}$ are lag times for atmospheres of air and 0.5 atm CO_2, respectively. μ_{air} and μ_{CO2} are specific growth rates for the atmospheres of air and 0.5 atm CO_2, respectively.

On the other hand a modified atmosphere with very high CO_2 concentration may sometimes induce some deleterious effects in high moisture foods such as fresh fishes and meats [7, 8]. High dissolution of flushed CO_2 gas in the foods reduces the headspace volume in case of semi-rigid plastic package resulting in package collapse. High CO_2 dissolution may also cause the increased drip or exudates of flesh foods. Low pH resulting from dissolved CO_2 may cause discoloration on the surface and taste change of fresh muscle foods. All these problems may impose restriction on the maximum CO_2 concentration used for MAP. Package collapse due to CO_2 dissolution can be avoided by adding tablets of solid CO_2, injection of gas under slight overpressure, or pretreatment of foods with CO_2-saturated water or bicarbonate solutions [7].

Recently the combined treatment of high O_2 (80-90%) and 10-20% CO_2 has been suggested to provide adequate suppression of microbial growth, allowing a safe, prolonged shelf life of minimally-processed vegetables from the study on pure cultures of *Pseudomonas fluorescens, Enterobacter agglomerans, Aureobacterium* strain 27, *Candida guilliermondii, C. sake, Salmonella typhimurium, S. enteritidis, Escherichia coli, Listeria monocytogenes, Leuconostoc mesenteroides var. mesenteroides, Lactobacillus plantarum,* and *Lactococcus lactis* [9].

Nitrogen is an inert filler gas that has little antimicrobial activity on its own. Displacement of oxygen by nitrogen acts to block out the oxidative chemical and physiological reactions, preserving the food quality. Its low solubility in foods makes it possible to be used as filler gas preventing package collapse when highly soluble CO_2 is used at high concentration in MAP under semi-rigid containers. In any case, nitrogen in the proper amount is used for diluting oxygen or carbon dioxide (the most effective gases) to the right level.

Argon as well is an inert filler gas, quite largely present in the atmosphere where it represents the third air constituent (about 1% in volume). It has been suggested for modified atmosphere for its chemical and physical properties. Being more soluble and denser than nitrogen, it is able to remove oxygen from the reactive sites more efficiently than nitrogen, and thus it can inhibit oxygen-sensitive deterioration quite easily. Several studies in vitro demonstrated the positive effects of argon in protecting oxygen sensitive systems such as substrate mixture with malic dehydrogenase or phenoloxidase [10].

Carbon monoxide (CO) is not at all a common gas in MAP applications, but its possible use deserves some mentioning. Carbon monoxide is a colorless, odorless, and tasteless gas that is, as it is well known, poisonous and potentially lethal. Inhaled CO binds to hemoglobin in blood to form carboxyhemoglobin, reducing the oxygen-carrying capacity of hemoglobin. Carbon monoxide has approximately 250 times greater affinity to hemoglobin than oxygen. In the same way CO can also bind to myoglobin in the muscles, forming carboxymyoglobin. This product is much more stable than oxymyoglobin and has a bright red-pink color, and therefore the use of small amounts of carbon monoxide in MAP for meat and some fish is really attracting. Moreover, CO is able to inhibit several oxidase enzymatic systems, particularly those including copper, like tyrosinase which is widely found and responsible for many browning reactions in fruits and vegetables. Finally carbon monoxide also has antimicrobial activity, particularly against some specific spoilage bacteria like *Brocothrix thermosphacta*, a common contaminant in meat products [11]. Actually, the positive effects of carbon monoxide are largely exploited, not always knowingly, in the smoking techniques that are largely applied on fish and meat products with some CO production. CO is an unapproved food additive in the USA and Europe. In Japan, Canada, and other countries its use for fresh meats and fishes is prohibited. With quickly evolving science, however, different regulatory and commercial decisions might come. Controversial decisions in the USA and in other countries already admit the use of CO for treatments and liquid smokes. In Europe, Norway (not a EU country) strongly desires an approval for use of CO for meats, even if in 2001 the Scientific Committee on Food of the European Commission opposed to the use of carbon monoxide as component of food packaging gases in modified atmosphere packaging [12].

13.1.2 Packaging Operations for MAP Foods

The gases for MAP of foods can be supplied from the sources of compressed gas in cylinder, liquid gas, or air separation [8, 13]. The gas cylinder may be of single pure gas or premixed gas. The gases in different bombs may be combined to attain the desired composition just before use for packaging. Premixed gas is advantageous in supplying the right gas composition. The gas vaporized from liquid gas supply can be introduced to the packaging machine. Or liquid nitrogen is dispensed in droplets directly into cans, jars, bottles, or foil packs to flush out residual oxygen after evaporation. The examples include beers, beverages, peanuts, dried potatoes, nuts, and coffees. The rigid packages with liquid droplet system tend to be pressurized by evaporated nitrogen. Nitrogen and oxygen can be produced from air separating systems applying selective adsorption on molecular sieve materials or effusion through small hollow fiber: molecular sieves with controlled pore size can adsorb and separate reversibly particular gas molecules; gas molecules of smaller molecular weight in a mixed compressed gas pass through microchannels of hollow fiber at higher speed, which can be used for concentration of the gas. Purity of gas from air separating systems is inferior to that of cylinder gas or liquid source. The air separating systems are readily adaptable for maintaining the required level of gas composition and widely used for controlled atmosphere storage and transport of fresh fruits and vegetables [14].

Vacuum packaging operation includes removal of air in tray or pouch loaded with food. Air exclusion is attained and maintained by using vacuum pump and heat sealing. The retention of vacuum condition should be maintained by hermetic sealing and gas barrier packaging.

Table 13.2 Typical formats of MAP machinery

Format		Machinery type		Process	Typical rate (packs/min)
Semi-rigid tray pack	Horizontal form-fill-seal	Evacuation and reintroduction	Automatic	Trays thermo-formed in-line	90
Semi-rigid tray pack	Fill-seal	Evacuation and reintroduction	Semi-automatic	Preformed trays or bags	15
Pillow-pack	Horizontal or vertical form-fill-seal	Continuous gas flushing	Automatic	Formed from reel	70-100
Outer box or drum with inner MAP bag	Bulk	Evacuation and reintroduction	Manual or semi-automatic	Bag reinjected with gas within outer pack	Varies

Adapted from Ref. [15, 16].

Replacement of air in MAP operation may start with evacuation: vacuumized containers or bags are then reintroduced with gas mixture of desired composition [8, 13, 15]. This two-step operation is usually achieved in the chamber in which tray or pouch with food is located. On rare case the pouches or large wrapped boxes in open surroundings may be directly connected to snorkel achieving evacuation and reintroduction of mixed gas. Evacuation/gas flushing takes time in operation and gives restriction on the machine speed, but this type of operation is excellent in attaining the desired gas flushing. In the other type of operation in MAP the gas mixture adjusted in composition is continuously flowed through the lance into the package to replace the air: the pillow packages continuously formed in form-fill-seal machines are usually flushed with the premixed gas. Because there should be some mixing of entering and leaving gases in the package, there is a limit to the efficiency of this technique: typical residual O_2 level in the packages with this technique is 2-5%. Table 13.2 gives typical systems of MAP machinery. The flushing of package atmosphere with gas mixture different to air is called active modification.

In contrast to active modification, the package atmosphere can be modified by the respiring activity of food product. Fresh fruits and vegetables consume oxygen and produce carbon dioxide in the respiration, and thus can modify the O_2 and CO_2 concentrations in the hermetically sealed containers. These phenomena are also common for cheeses containing high load of live microorganisms (like Camembert, Taleggio, blue cheese, Gorgonzola). This type of atmosphere modification is called passive modification. Proper balance between gas permeation and respiration is important for attaining target atmosphere of package, which will be discussed later.

13.1.3 Safety of MAP foods

Even though vacuum packaging and MAP help to keep the freshness of the foods and extend their shelf life, it cannot be effective without hygienic and temperature control in production, distribution, and storage. Initial high quality product without contamination by any pathogenic microorganisms is prerequisite for successful MAP. Packaging in vacuum, nitrogen, or gas mixture without oxygen can slow down or inhibit aerobic bacteria, yeasts, and molds. Table 13.3 lists microorganisms in aspect of oxygen requirement. The gas mixture with CO_2 in excess of 5% can reduce the growth rate of most food spoilage bacteria. Higher level of CO_2 concentration has inhibitory action against many pathogenic bacteria. The effect of CO_2 gas on the growth differs with microbial species (Table 13.4). Generally aerobic microbes such as *Pseudomonas*, *Acinetobacter/Moraxella*, and molds are more sensitive to CO_2 gas than anaerobes such as *Lactobacillus* and *Clostridium*. Packaging fish and meat in the atmosphere with high CO_2 concentration transforms their microflora to be dominated by *Lactobacillus* spp [3, 17]. Gram-negative bacteria are more sensitive to CO_2 gas compared to Gram-positive ones. However it needs to be mentioned that the reported effects of CO_2 gas on microbial activity do not agree completely, and sometimes are contradictory in the literatures. It should again be emphasized that good hygienic practice with controlled temperature should be applied with MAP.

Even though vacuum and modified atmosphere packaging can be applied to the products at ambient temperature, its application is limited usually to shelf stable foods with low water activity or heat-preserved in commercial sterility, which do not present health hazard. Because vacuum and modified atmosphere packagings are effective at low temperature in retarding microbial growth of perishable foods, they are applied usually with the chilled storage for these foods: temperature abuse at storage and distribution may introduce the hygienic problems for them. There has been a great concern about potential risk for anaerobic *Clostridium botulinum*, particularly type E, to grow and produce toxin at refrigerated temperature under vacuum or modified atmosphere packaging without oxygen. There has been some reports warning that toxin by nonproteolytic *Cl. botulinum* may be produced before organoleptic spoilage symptoms appear [18, 19]. There have also been hygienic concerns for the pathogenic organisms such as *Clostridium perfringenes*, *Yersinia enterocolitica*, *Escherichia coli* O157:H7, *Listeria monocytogenes*, *Staphylococcus aureus*, *Aeromonas hydrophila*, *Salmonella* spp., and *Bacillus cereus*, which can grow under refrigerated and anaerobic conditions. *L. monocytogenes*, a typical psychrotrophic pathogen was reported to be able to grow in refrigerated MAP food products, but was inhibited in its growth under high CO_2 concentration especially in combination with additional hurdles such as lactic acid, salt, lowered water activity, and lowered pH [3, 6, 17]. The safety of MAP foods has been tested by simple inoculation test of a pathogen, relative rate of organoleptic spoilage versus toxigenicity, prediction of statistical probability of toxin formation, or relative ratio of spoilage progress versus pathogenicity [20]. MAP effect on the pathogenic bacteria is generally summarized not to increase the risks from *Cl. perfringenes, Y. enterocolitica, E. coli, L. monocytogenes, S. aureus, A. hydrophila, Salmonella* spp., and *B. cereus* [3, 7, 17].

Table 13.3 Classification of some food spoilage and pathogenic microorganisms based on oxygen requirement

Aerobes – require oxygen for growth
Acinetobacter/Moraxella, Bacillus spp.,
Micrococcus spp., *Pseudomonas* spp.,
Molds

Facultative organisms – grow in presence or absence of oxygen
Aeromonas hydrophila, Bacillus cereus,
Shewanella putrifaciens, Brocothrix thermosphacta,
Corynebacterium spp., *Escherichia coli,*
Listeria monocytogenes, Salmonella spp.,
Shigella spp., *Staphylococcus* spp.,
Vibrio parahaemolyticus, Yersinia enterolitica,
Yeasts

Microaerophiles – grow best in the presence of a small amount of free oxygen
Campylobacter jejuni, Lactobacillus spp.

Anaerobes – inhibited/killed by oxygen
Clostridium botulinum, Clostridium perfringens

Table 13.4 Relative effect of CO_2 on the growth of foodborne microorganisms

Organism	CO_2 effect			
	Growth unaffected	Growth inhibited	Growth weakly inhibited	Growth stimulated
Gram-negative				
Acinetobacter		+		
Aeromonas hydrophila		+	+	
Alteromonas spp.		+	+	
Campylobacter jejuni				
survival	+			+
growth		+	+	
Enterobacter spp.	+	+		
Escherchia coli		+	+	
Moraxella spp.		+		
Proteus spp.		+		
Pseudomonas spp.		+	+	
Salmonella spp.		+	+	+
Serratia spp.	+	+		
Vibrio spp.		+		
Yersina enterocolitica		+		
Gram-positive				
Bacillus spp.		+	+	
Brochotrix thermosphacta		+	+	
Clostridium botulinum	+		+	+
Clostridium PA3679		+		
C. sporogenes-survival		+		+
C. perfringens		+		+
Enterococci		+	+	
Corynebacterium spp.		+		
Lactobacillus spp.	+		+	+
Leuconostoc spp.		+		
Listeria monocytogenes	+	+		
Staphylococcus aureus		+	+	
Streptococcus spp. (besides Enterocci)	+			
Molds		+	+	
Yeasts	+		+	

Reprinted from Ref. [3] with kind permission of Journal of Food Protection.

Therefore the major concern in refrigerated vacuumed/MAP products is nonproteolytic *Cl. botulinum*. The characteristics of *Cl. botulinum* need to be consulted for avoiding or minimizing the risk (Table 13.5). Hurdles other than low temperature and modified atmosphere may be combined for assuring the safety from *Cl. botulinum*: pH < 5.0, water activity < 0.97 or salt content > 5% when stored below 10°C [17, 21]. Or the shelf life should be controlled to less than 10 days to take account of the time for germination of spores under chilled condition below 10°C. If longer shelf life is desired without the constraints on pH, water activity, and salt content, pasteurization based on inactivation of psychrotrophic, nonproteolytic *Cl. botulinum* can be applied, which usually corresponds to thermal death time of 10 minutes at 90°C. When temperature abuse may give rise to the potential risk of mesophillic, proteolytic *Cl. botulinum*, other hurdles in product composition (pH < 4.6, water activity < 0.94 or salt content > 10%) may be used. Or commercial sterility based on thermal death time of 3.0 minutes at 121.1°C (F_o of 3 minutes) can make it possible to have shelf stability without the risk from *Cl. botulinum* at ambient temperature (see also Section 12.3.1). Above all the storage of MAP products below 3.3°C will assure their safety with enhanced microbial stability and keeping quality.

Table 13.5 Growth characteristics of pathogenic *Cl. botulinum*

Properties	Group	
	Proteolytic	Nonproteolytic
Toxin type	A, B, F	B, E, F
Inhibitory pH	4.6	5.0
Inhibitory NaCl concentration	10%	5%
Minimal water activity	0.94	0.97
Minimum temperature for growth and toxin production	10°C	3.3°C
D_{100} of spores	25 min	0.1 min

If vegetative pathogenic bacteria are to be destroyed, pasteurization process based on inactivation of the most heat resistant *L. monocytogenes* is applied as 2 minutes at 70°C. Irradiation treatment is also sometimes combined with MAP to inactivate vegetative pathogenic bacteria, ensuring the food safety with extended shelf life. However, it is again emphasized that raw material quality and processing procedures determine the quality and safety of MAP and vacuum packed products. Proper temperature control in storage is required for benefits offered by MAP, particularly for microbial growth retardation by CO_2-enriched atmosphere.

Example 13.1 Estimating volume change of an MAP by CO_2 dissolution
Consider 300 g of fresh fish fillet packaged with 100% CO_2 under the conditions of gas-product ratio of 3.0 in high barrier plastic film bag. What would be the volume of the package located at 3°C? Estimate the dissolved CO_2 concentration

in the fresh fish. You may assume that 300g fish is similar to 300 mL water in dissolving CO_2.

We assume that the package initially consists of 300 mL aqueous food and 900 mL headspace. The flexible packaging with 100% CO_2 stipulates that CO_2 partial pressure stays at 1 atm. The mass balance of CO_2 dictates that initially flushed CO_2 is equilibrated in the food and gas phase as follows:

Total amount of flushed CO_2 = CO_2 dissolved in food + CO_2 in headspace

Using ideal gas equation to have mg unit for this relation:

$$\frac{101325 \text{ Pa} \times 900 \text{ mL} \times 0.044 \text{ kg/mol}}{8.314 \text{ J}/(\text{K} \cdot \text{mol}) \times 276\text{K}} = k_{co_2} \times 1.013 \text{ bar} \times 300 \text{ mL} + \frac{101325 \text{ Pa} \times V \times 0.044 \text{ kg/mol}}{8.314 \text{ J}/(\text{K} \cdot \text{mol}) \times 276\text{K}}$$

The k_{CO2} at 1 atm (1.013 bar) is calculated to be 3.066 mg mL^{-1} bar^{-1} with disregard to dissolved oxygen in Equation 13.2. Substitution of k_{CO2} to the above equation, headspace volume (V) is obtained as 420 mL. The dissolved CO_2 concentration in the fresh fish is estimated as 3.106 mg mL^{-1} or 3106 mg L^{-1} at 1.013 bar of CO_2 partial pressure from Equation 13.1. Initial free volume of 900 mL was estimated to be reduced to 420 mL, which shows the degree of potential package collapse in case of semi-rigid package. It is noted that this solution is based on some simplified assumptions but gives meaningful insight into the package situation.

13.2 Nonrespiring Products

The nonrespiring products such as meats, fishes, bakery products, and dry snacks are relatively static in terms of production and consumption of gases such as O_2, CO_2, and N_2, which are commonly applied in MAP. However it does not mean that these foods have no interaction with package headspace. Some amount of CO_2 is dissolved in tissues of meat and fish when gas with high CO_2 concentration is flushed into the package. Some degree of CO_2 production and O_2 consumption by contaminated microorganisms and flesh tissue may occur with fresh meat and fish products. However the contribution of these ingenious reactions to the changes in package atmosphere is not significant under normal practices of packaging and storage conditions. Therefore we might assume that package atmosphere is maintained constant when the products are packed with barrier packaging materials. Some types of fermented dairy products such as cheese and yogurt may be taken as respiring when they produce significant amount of CO_2 by lactic acid bacteria or molds. But they can usually be treated as nonrespiring because of insignificant level of their CO_2 production when the products are stored at low temperature close to 0°C within limited time of short shelf life.

Table 13.6 Typical examples of gas mixtures for MAP of nonrespiring foods

Product	Temperature (°C)	Gas composition (%)			Major benefits from MAP
		O_2	CO_2	N_2	
Fresh meat and pork	0-2	30-85	15-40	0-10	Inhibits microbial spoilage, maintain bright red color
Poultry	0-2	0	20-50	50-80	Inhibits microbial spoilage and oxidative rancidity
Processed meat	0-3	0	0-40 or vacuum packaged	60-100	Inhibits microbial spoilage and oxidative rancidity
Fish - white	0-2	30-50	40-50	30	Inhibits microbial spoilage; O_2 inclusion is for preventing the growth of *Cl. botulinum* and minimize color change
Fish - oily	0-2	0	40-60	40-60	Inhibits microbial spoilage and oxidative rancidity
Eggs	3-25	0	0-20	80-100	Keeps the CO_2 inside the shell, inhibits mold growth on surface
Cheese	0-3		10-40 or vacuum packaged	60-90	Inhibits microbial spoilage and oxidative rancidity
Cream	0-3			100	Inhibits aerobic microbial spoilage and oxidative rancidity without taste change from CO_2 dissolution
Fresh pasta	4	0	50-80	20-50	Inhibits microbial spoilage and oxidative changes
Bread	Ambient	0	40-70	30-60	Inhibits mold growth and may slow down the staling rate
Dried snacks	Ambient	0	0	100	Prevents the oxidative changes and damage of fragile structure
Combination products	0-3	0	40-60	40-60	Inhibits microbial spoilage and oxidative changes

Adapted from Ref. [3, 8, 13].

Hence MAP of the nonrespiring products consists of hermetically enclosing the foods under the desired gas or vacuum with the bags or containers of excellent gas barrier property. In the vacuum packaging headspace air is simply evacuated with followed sealing of package. In the MAP the gas mixture of controlled composition is flushed into the package headspace mechanically and then the containers are sealed. The gas composition applied differs with product type: a best combination of O_2, CO_2, and N_2 gases is selected for preserving the quality of the product, which should consider the modes of quality deterioration and their interaction with package atmosphere. Table 13.6 shows some examples of gas composition for typical nonrespiring food products.

Relative static nature in MAP system of nonrespiring foods requires high gas barrier materials for maintaining the optimal atmosphere initially flushed. O_2 permeation rate of 50 cm^3 m^{-2} day^{-1} atm^{-1} is suggested as maximum limit [13]. As a general rule moisture barrier property is also required for high moisture or low moisture foods. Metal cans and glass jars satisfy this requirement as absolute barrier against O_2, CO_2, and moisture, and are primarily used for shelf stable foods. Dry milk, nuts, and confectionery products are the examples packaged by cans or glass jars in MAP. Metal cans also provide with light barrier. However, majority of perishable MAP foods without respiration activity are contained in plastic pouches or trays of high barrier. Multilayered films or sheets based on nylon, polyethylene terephthalate (PET), polyvinylidene chloride (PVDC), and ethylene vinyl alcohol (EVOH) are the most frequently used materials: the structure may be laminated, coextruded, or coated one. Base wall in tray may sometimes consist of thick materials with medium barrier such as polyvinyl chloride (PVC) and high density polyethylene (HDPE): typical thickness is 400 μm in this type. Refer to Chapter 4 for information on gas and moisture permeabilities. In case of plastic MAP, seal integrity and transparency are other important properties besides permeability. Anti-fog treatment on the films improves the transparency.

Another important factor for MAP incorporated with CO_2 is ratio of gas to product. Because the effectiveness of CO_2 gas depends on the dissolved CO_2 amount, too low headspace gas volume per product leads to very low level of dissolved CO_2 resulting in little antimicrobial effect. Therefore proper gas volume should be flushed initially to have the desired beneficial effect of CO_2 gas. At least gas product ratio of 2:1 has been suggested for MAP of fish and meat products [7, 17]. Visual image of package in marketing is also considered for determining gas volume ratio.

13.3 Respiring Products

Modified atmosphere (MA) of reduced O_2 and elevated CO_2 concentrations applied to fresh produce packaging reduces respiration rate, ethylene production, and softening of the texture, and therefore keeps its freshness extending shelf life. Optimum gas composition to preserve the quality at best conditions differs with commodity (Table 13.7). Many fresh fruits and vegetables can benefit by the freshness keeping when located in proper MA. The package atmosphere should not go beyond the O_2 and CO_2 tolerance limits specific for commodity: O_2 concentration lower than the tolerance limit induces anaerobic respiration, formation of alcohols and aldehydes, and development of off-flavor and odor; CO_2 concentration higher than the tolerance limit causes unfavorable physiological disorders such as internal tissue breakdown. Table 13.7 shows some examples of the optimum MAP conditions for fresh produce. However it needs to be mentioned that a lot of variables such as variety, temperature, growing conditions, and harvest method may affect the location or window of the optimal atmosphere.

Table 13.7 Recommended optimal modified atmosphere conditions for fresh produce

Commodity	Temperature range (°C)	Relative humidity (%)	Modified atmosphere (%)	
			O_2	CO_2
Fruits				
Apple	0-5	90	1-3	1-3
Apricot	0-5	90	2-3	2-3
Avocado	5-13	90-95	2-5	3-10
Banana	13-15	90-95	2-5	2-5
Blueberry	0-5	90-95	5-10	15-20
Cherry, sweet	0-5	90-95	3-10	10-15
Orange	3-9	90-95	5-10	0-5
Peach	0-5	90-95	1-2	3-5
Pear	0-5	90-95	2-3	0-1
Persimmon	0-5	90-95	3-5	5-8
Strawberry	0-5	90-95	4-10	15-20
Vegetables				
Asparagus	0-5	95-100	air	5-10
Broccoli	0-5	95-100	1-2	5-10
Cabbage	0-5	95-100	3-5	3-6
Cauliflower	0-5	95-98	2-5	2-5
Corn, sweet	0-5	95-98	2-4	10-180
Cucumber	8-12	90-95	3-5	0
Lettuce	0-5	95-100	1-5	0
Mushroom	0-5	95-98	1-3	10-15
Green pepper	8-12	90-95	3-5	2-8
Spinach	0-5	95-98	7-10	5-10
Tomato, partly ripe	8-12	90-95	3-5	0-3

MAP of respiring produce cannot be attained just with flushing by gas of desired gas composition, because the respiration may change the package atmosphere toward outside of optimal window. Respiration process consumes oxygen and produces carbon dioxide as follows:

$$C_6H_{12}O_6 + 6O_2 \rightarrow 6CO_2 + 6H_2O + heat \qquad (13.3)$$

The molar ratio of CO_2 produced to O_2 consumed is called respiratory quotient (RQ) and is ideally 1 for glucose, but ranges from 0.7 to 1.3 depending on the substrate metabolized. The water produced is transpired to vapor by respiration heat increasing the humidity inside the package.

Therefore respiration should be balanced to permeation of oxygen and carbon dioxide in order to achieve a target package atmosphere modified from normal atmosphere: the optimal MA for a produce package is created and maintained by an intricate interplay between the respiration of the produce and the gas permeation of the packaging material. Figure 13.4 describes the dynamics of fresh produce MAP. In usual situation of MAP, CO_2 diffuses out of the package and O_2 enters into the inside of the package due to the concentration gradient across the permeable film. Transpired water vapor is also permeated through the film out of the package even though its gradient is not high in usual distribution and storage conditions. For some produces, particularly fruits, ethylene (C_2H_4) is produced by endogenous enzymes and accelerates their respiration and ripening process. Fast rate of its removal, by permeating out or adsorption on the adsorbent, would be desirable for preserving the freshness.

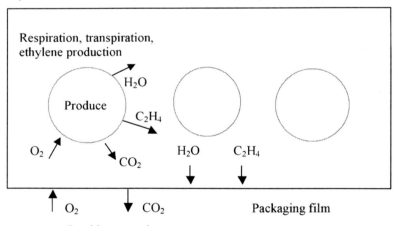

Figure 13.4 Dynamics of fresh produce MAP

As described in Figure 13.4, respiration rate is an important factor for MAP of fresh produce. The respiration rate is also an indication of the rate of catabolic changes and quality deterioration in fresh produce: the higher the rate, the shorter the storage life. The respiration rate is affected not only by internal factors such as commodity, maturity, and harvest time but also by external factors such as temperature and gas concentration of atmosphere [22, 23]. Mathematical model describing respiration as function of environmental variables is very useful for designing the fresh produce MAP. An enzyme kinetics type respiration model (Equation 13.4) has been developed to describe the respiration rate of fresh produce as a function of O_2 and CO_2 concentrations [24], and is widely used in designing MAP of many fresh commodities.

$$r = \frac{V_m[O_2]}{K_m + (1+[CO_2]/K_i)[O_2]} \quad (13.4)$$

where r is respiration rate in O_2 consumption or CO_2 production (mL kg^{-1} h^{-1}), $[O_2]$ and $[CO_2]$ are O_2 and CO_2 concentrations (%), respectively, and V_m, K_m and K_i are parameters that represent maximum respiration rate, O_2 concentration at half V_m and inhibition constant for CO_2, respectively. Table 13.8 shows the respiration model parameters for some commodities.

Table 13.8 Respiration model parameter values and respiration activation energies for some fresh produces

Commodity	Temp. (°C)	Respiration expression	Respiration model parameters			Activation energy (kJ mol^{-1})
			V_m (mg kg^{-1} h^{-1})	K_m (% O_2)	K_i (% CO_2)	
Blueberry 'Coville'	15	O_2 consumption	68.0	0.4	2.9	147.3
		CO_2 evolution	51.0	0.2	4.9	163.3
Broccoli	7	O_2 consumption	210.3	0.6	2.3	62.7
		CO_2 evolution	235.2	1.7	1.9	66.1
Cauliflower	13	O_2 consumption	133.7	1.7	3.0	21.2-48.2
		CO_2 evolution	134.4	1.4	3.1	21.2-48.2
Dry coleslaw	5	O_2 consumption	22.7	1.1	23.2	74.8
Green pepper	10	O_2 consumption	54.3	6.0	1.3	48.7-57.3
		CO_2 evolution	31.8	2.4	4.3	48.7-57.3
Sweet cherry 'Burlat'	5	O_2 consumption	10	11	>200	83.3

Adapted from Ref. [25-27].

The degree of film permeability to oxygen and carbon dioxide is another critical factor controlling the atmosphere inside the package. The permeation of these two gases is proportional to partial pressure gradient, surface area, inverse of film thickness, and permeability coefficient as described in Chapter 4 (see Sections 4.3 and 4.4). Respiration and permeation are combined to mass balance equations of O_2 and CO_2 to predict O_2 and CO_2 concentrations inside the package [28]:

$$\frac{d[O_2]}{dt} = 100\{\frac{S\overline{P}_{o2}(0.21-[O_2]/100)p}{VL} - \frac{Wr_{o2}}{V}\} \tag{13.5}$$

$$\frac{d[CO_2]}{dt} = 100\{\frac{S\overline{P}_{co2}(0.00-[CO_2]/100)p}{VL} + \frac{Wr_{co2}}{V}\} \tag{13.6}$$

where t is time (h), V is free volume (mL), L is thickness (μm or mil), W is produce weight (kg), S is package surface area (m^2), p is atmospheric pressure (atm), $\overline{P}_{o2}$ and $\overline{P}_{co2}$ are permeability coefficients to O_2 and CO_2 (mL μm m^{-2} h^{-1} atm^{-1} or mL mil m^{-2} h^{-1} atm^{-1}), respectively, and r_{o2} and r_{co2} are oxygen consumption and carbon dioxide evolution rates (mL kg^{-1} h^{-1}), respectively.

The differential Equations 13.5 and 13.6 can be solved numerically to give O_2 and CO_2 concentration changes with time for a given conditions of packaging. The r_{o2} and r_{co2} as function of O_2 and CO_2 concentrations inside the package can be taken into account by Equation 13.4. More exact but complicated equations have been proposed to give the changes in package volume or pressure [29], but usual cases of MAP design can be handled by Equations 13.5 and 13.6.

Whether the package atmosphere is modified actively or passively, the equilibrated one should be within or close to optimal gas composition, and should not go beyond tolerance limit of O_2 and/or CO_2 concentrations. The equilibrated modified atmospheric composition in the package ($[O_2]$ and $[CO_2]$) can be obtained by equilibration between respiration and permeation as follows ($d[O_2]/dt = 0$ in Equation 13.5; $d[CO_2]/dt = 0$ in Equation 13.6):

$$\frac{S\overline{P}_{o2}(0.21-[O_2]/100)p}{L} = W \cdot r_{o2} \tag{13.7}$$

$$\frac{S\overline{P}_{co2}([CO_2]/100-0.00)p}{L} = W \cdot r_{co2} \tag{13.8}$$

The r_{o2} and r_{co2} can be given as function of O_2 and CO_2 concentrations by Equation 13.4.

A simpler relation can be obtained when r_{o2} and r_{co2} are equal (RQ = 1): Equations 13.7 and 13.8 can be equalized to result in:

$$[CO_2] = \frac{1}{\left(\overline{P}_{CO_2} / \overline{P}_{O_2}\right)} (21 - [O_2])$$
(13.9)

The relationship of Equation 13.9 tells that the O_2 and CO_2 concentrations equilibrated inside the package is interdependent as described in Figure 13.5: the $[O_2]$ and $[CO_2]$ are located on the line whose slope is determined by the relative ratio in permeabilities of CO_2 and O_2, $\left(\overline{P}_{CO_2} / \overline{P}_{O_2}\right)$[30].

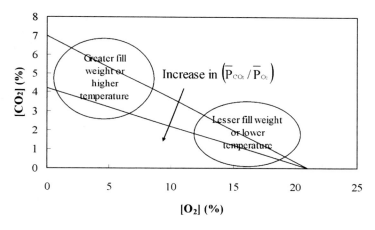

Figure 13.5 Equilibrated O_2 and CO_2 concentrations versus $\left(\overline{P}_{CO_2} / \overline{P}_{O_2}\right)$

When solution of the differential Equations 13.5 and 13.6 is examined, small free volume decreases the time to equilibration with no effect on the equilibrated gas concentration as shown in Figure 13.6. Therefore the equilibrated $[O_2]$ and $[CO_2]$ are independent of free volume in Equations 13.7 and 13.8, respectively. Active flushing can also attain quickly the equilibration in package atmosphere and improve the effectiveness of MAP in keeping freshness.

Designing an optimal package is to find an ideal combination of film type and package dimension to attain the recommended internal gas composition. Proper selection of packaging film should consider its permeabilities to O_2 and CO_2 gases. As with absolute values in gas permeabilities, the relative ratio of CO_2 permeability to O_2 permeability determines the possible range of attainable O_2 and CO_2 concentrations as shown in Figure 13.5. For a given film, increased produce weight increases CO_2 concentration and decreases O_2 concentration in equilibrated package atmosphere along a straight line. Table 13.9 lists typical gas permeability data for some plastic films. It is noted that most polymeric films have the $\left(\overline{P}_{CO_2}/\overline{P}_{O_2}\right)$ value of 3-6, which limits package atmosphere window attainable by plastic film package of fresh produce. In order to alleviate this restriction and attain the desired gas composition inside the package, different films and/or window of highly permeable membrane may be combined in a package [31] or some perforations may be built on the film to have high gas transfer with $\left(\overline{P}_{CO_2}/\overline{P}_{O_2}\right)$ close to 1 [25, 32].

Table 13.9 Typical gas permeability coefficients at 10°C and permeability activation energies for polymeric film

Polymeric film	Permeabilities (mL mil m^{-2} h^{-1} atm^{-1})		$\dfrac{\overline{P}_{CO_2}}{\overline{P}_{O_2}}$	Activation energy (kJ mol^{-1})	
	O$_2$	CO$_2$		O$_2$	CO$_2$
Polybutadiene	1118	9892	8.8	29.7	21.8
Low-density polyethylene	110	366	3.3	30.2	31.1
Ceramic-filled LDPE	199	882	4.4	36.8	28.4
Linear low-density Polyethylene	257	1002	3.9	-	-
High-density polyethylene	2.1	9.8	4.6	35.1	30.1
Cast polypropylene	53	151	2.9	-	-
Oriented polypropylene	34	105	3.1	-	-
Polyethylene terephthalate	1.8	6.1	3.3	26.8	25.9
Nylon laminated multilayer film	1.7	6.0	3.5	52.6	50.0
Ethylene vinyl acetate	166	985	5.9	48.4	37.0
Ceramic-filled polystyrene	116	630	5.4	34.5	26.2
Silicone rubber	11170	71300	6.4	8.4	0.0
Perforation (air)	2.44×10^9	1.89×10^9	0.8	3.6	3.6
Microporous film	3.81×10^7	3.81×10^7	1.0	13.0	3.7

From Ref. [25] with kind permission of Springer Science and Business Media.

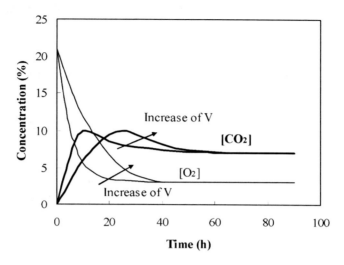

Figure 13.6 Effect of free volume (V) on the progress in internal atmosphere of a fresh produce package

A produce package able to attain an optimal package atmosphere at given temperature conditions will have a different one under temperature-abused conditions. Examination of Tables 13.8 and 13.9 shows that activation energy for respiration is generally higher than that of the film's gas permeability: relatively higher temperature effect on respiration than gas permeation. This leads to higher CO_2 and lower O_2 concentrations with temperature increase (Figure 13.5), which may go outside the tolerance limits of O_2 and CO_2 concentrations. This underlines the importance of temperature control for MAP fresh produce.

CO_2 scavengers may be used for preventing potential injurious CO_2 buildup in the produce package. Ethylene absorbers are sometimes used to delay the onset of senescence in bulk packaging. Active packaging techniques of gas absorption are discussed in Chapter 15.

Humidity control by moisture absorbent has also been tried by some researchers, because most of the plastic film packages result in undesirable saturated humidity with moisture condensation which may be harmful for some types of fresh commodities [33].

Example 13.2 Equilibrated atmosphere of a fresh produce package
Fishman et al. [34] showed that respiration of mango fruit (mL kg^{-1} h^{-1}) at 20°C could be described as function of oxygen concentration in air by $r = 0.92[O_2]$. If 0.61 kg of the fruits are packaged with 1 mil thick low density polyethylene

(LDPE) film of 0.03 m^2 surface area and free volume of 780 mL, what would be the equilibrated package atmosphere? Assume that O_2 consumption rate equals CO_2 production, i.e., RQ is 1.

Gas permeabilities of low density polyethylene at 20°C are obtained from Table 13.7 by using Arrhenius equation (see Section 4.6.3). Thickness and volume changes due to temperature increase are disregarded.

$$\overline{P}_{o2} = 110\,mL \cdot mil\,/(m^2 \cdot h \cdot atm) \times \exp(\frac{30200\,J\,/\,mol}{8.314\,J\,/(K \cdot mol)} \times (\frac{1}{283K} - \frac{1}{293K}))$$

$$= 170\,mL \cdot mil\,/(m^2 \cdot h \cdot atm)$$

$$\overline{P}_{co2} = 366\,mL \cdot mil\,/(m^2 \cdot h \cdot atm) \times \exp(\frac{31100\,J\,/\,mol}{8.314\,J\,/(K \cdot mol)} \times (\frac{1}{283K} - \frac{1}{293K}))$$

$$= 575\,mL \cdot mil\,/(m^2 \cdot h \cdot atm)$$

Applying the relationships of Equations 13.7 and 13.8,

$$\frac{S\overline{P}_{o2}(0.21 - [O_2]/100)p}{L} = W \times 0.92[O_2]$$

$$\frac{0.03 \times 170 \times (0.21 - [O_2]/100) \times 1}{1} = 0.61 \times 0.92[O_2]$$

$$[O_2] = 1.88\%$$

$$\frac{S\overline{P}_{CO2}([CO_2]/100 - 0.00)p}{L} = W \times 0.92[O_2]$$

$$\frac{0.03 \times 575 \times ([CO_2]/100 - 0.00) \times 1}{1} = 0.61 \times 0.92 \times 1.88$$

$$[CO_2] = 6.12\%$$

13.4 Case Study: Design of Modified Atmosphere Package for Blueberries

According to Song et al. [35] the respiration for 'Duke' blueberry at 15°C could be described by the model parameters in Equation 13.4: V_m, K_m, and K_i for O_2 consumption are 22.71 mL kg^{-1} h^{-1}, 7.63%, and 14.42%, respectively, and those for CO_2 production 17.64 mL kg^{-1} h^{-1}, 5.08%, and 11.99%, respectively. A low density polyethylene film with O_2 and CO_2 permeabilities of 7900 and 27200 mL μm m^{-2} h^{-1} atm^{-1} was applied for packaging 250 g blueberries with 413 mL free volume. The used film had a surface area of 0.069 m^2 with thickness of 33.9

µm. The respiration model was incorporated into Equations 13.5 and 13.6 as follows:

$$\frac{d[O_2]}{dt} = 100[\frac{0.069 \times 7900 \times (0.21 - [O_2]/100) \times 1}{413 \times 33.9} - \frac{0.25}{413} \cdot \{\frac{22.71[O_2]}{7.63 + (1 + [CO_2]/14.42)[O_2]}\}]$$

$$\frac{d[CO_2]}{dt} = 100[\frac{0.069 \times 27200 \times (0.00 - [CO_2]/100) \times 1}{413 \times 33.9} + \frac{0.25}{413} \cdot \{\frac{17.64 \times [O_2]}{5.08 + (1 + [CO_2]/11.99)[O_2]}\}]$$

Solving these two simultaneous differential equations by Runge Kutta method gives the estimation on O_2 and CO_2 changes in the package as shown in Figure 13.7. The solution of these differential equations can be undertaken by a computer program in Mathcad®, which can be found in a website given in the preface. O_2 and CO_2 concentrations in the package attained the respective equilibrated values of 6.7 and 3.9% in about 60 hours.

The equilibrated gas concentrations can also be obtained by using Equations 13.7 and 13.8, which were substituted with appropriate parameters or values:

$$\frac{0.069 \times 7900 \times (0.21 - [O_2]/100) \times 1}{33.9} = 0.25 \times \frac{22.71 \times [O_2]}{7.63 + (1 + [CO_2]/14.42)[O_2]}$$

$$\frac{0.69 \times 27200 \times ([CO_2]/100 - 0.00) \times 1}{33.9} = 0.25 \times \frac{17.64 \times [O_2]}{5.08 + (1 + [CO_2]/11.99)[O_2]}$$

The solution of these simultaneous equations can be undertaken also by another Mathcad® computer program, which can be found in the book website. O_2 and CO_2 concentrations of 6.5 and 3.8% were obtained from the solution, which is also exactly equal to the values after 80 hours in Figure 13.7.

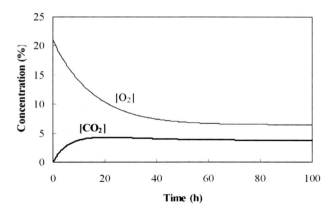

Figure 13.7 Estimated package atmosphere for the blueberry package

Discussion Questions and Problems

1. Compare the vacuum packaging and MAP for sausage in terms of quality preservation and packaging operation.

2. Consider 300 g of fresh fish fillet packaged with 50% CO_2 and 50% N_2 under the conditions of gas product ratio of 3.0 in high barrier rigid tray. What would be the CO_2 concentration in headspace of the package located at 3°C? Estimate also the dissolved CO_2 concentration in the fresh fish. You may assume that 300g fish is similar to 300 mL water in dissolving CO_2.

3. Hayakawa et al. [28] presented functional relationship of tomato respiration (mL kg^{-1} h^{-1}) at 21°C in the interested range of modified atmosphere (4% ≤ [O_2] < 12%, 0% ≤ [CO_2] <10%) as follows: r_{O2} = 2.01[O_2]; r_{CO2} = 18.5. What would be the equilibrium package atmosphere of O_2 and CO_2 concentrations for the package of 0.47 kg tomato with free volume of 843 mL and surface area of 0.034 m^2? The packaging film is their RMF-61 film with permeation property of $\overline{P}_{O2}/L = 1356\,mL \cdot m^{-2} \cdot h^{-1} \cdot atm^{-1}$ and $\overline{P}_{CO2}/L = 6983\,mL \cdot m^{-2} \cdot h^{-1} \cdot atm^{-1}$.

4. Using the respiration information of green pepper at 10°C in Table 13.8, predict the gas concentration changes of the permeable plastic package containing 110 g of peppers. The packaging film is 1 mil thick LDPE film with the dimension of 15 × 20 cm. From the packaging conditions, the package has an initial free volume of 230 mL.

5. Evaluate pros and cons for high CO_2 concentration MAP of fresh fish.

6. From the literature review, evaluate potential hygienic hazards of *Escherichia coli* O157:H7 in MAP of low O_2 and high CO_2 concentrations.

Bibliography

1. P Rocculi, S Romani, MD Rosa. Effect of MAP with argon and nitrous oxide on quality maintenance of minimally processed kiwifruit. Postharvest Biol Tec 35:319-328, 2005.
2. ME Stiles. Scientific principles of controlled/modified atmosphere packaging. In: B Ooraikul, ME Stiles, ed. Modified Atmosphere Packaging of Food. New York: Ellis Horwood, 1991, pp. 18-25.
3. JM Farber. Microbiological aspects of modified-atmosphere packaging technology - a review. J Food Prot 54:58-70, 1991.
4. F Devlieghere, J Debevere, J Van Impe. Concentration of carbon dioxide in the water-phase as a parameter to model the effect of a modified atmosphere on microorganisms. Int J Food Microbiol 43:105–113, 1998.
5. M Rammert, MHP Paderson. Die Loslichkeit von Kohlendioxid in Getraken. Brauwelt 131:488-499, 1991.

6. F Devlieghere, AH Geeraerd, KJ Versyck, B Vandewaetere, J Van Impe, J Debevere. Growth of *Listeria monocytogenes* in modified atmosphere packed cooked meat products: a predictive model. Food Microbiol 18:53-66, 2001.
7. HK Davis. Fish. In: RT Parry, ed. Principles and Applications of Modified Atmosphere Packaging of Foods. London: Blackie Academic & Professional, 1993, pp. 189-228.
8. RT Parry. Introduction. In: RT Parry, ed. Principles and Applications of Modified Atmosphere Packaging of Foods. London: Blackie Academic & Professional, 1993, pp. 1-18.
9. A Amanatidou, EJ Smid, LGM Gorris. Effect of elevated oxygen and carbon dioxide on the surface growth of vegetable-associated micro-organisms. J Appl Microbiol 86:429-438, 1999.
10. D Zhang, PC Quantick, JM Grigor, R Wiktorowicz, J Irven. A comparative study of effects of nitrogen and argon on tyrosinase and malic dehydrogenase activities. Food Chem 72:45-49, 2001.
11. DS Clark, CP Lentz, LA Roth. Use of carbon monoxide for extending shelf life of prepackaged fresh beef. Can Inst Food Sci Tec J 9:114-117, 1976.
12. WS Otwell, M Balaban, H Kristinsson. Use of carbon monoxide for color retention in fish. In Conference Proceedings First Joint Trans-Atlantic Fisheries Technology Conference-TAFT 2003 33rd WEFTA and 48th AFTC meetings. 2003. Reykjavik, Iceland.
13. BPF Day. Guidelines for the Good Manufacturing and Handling of Modified Atmosphere Packed Food Products. Chipping Campden, UK: The Campden Food and Drink Research Association, 1992, pp. 26-70.
14. AK Thompson. Controlled Atmosphere Storage of Fruits and Vegetables. Wallingford, UK: CAB International, 1998, pp. 14-55.
15. DJ Cakebread. European market development and opportunities for MAP. In: AL Brody, ed. Modified Atmosphere Food Packaging. Herndon, Virginia: Institute of Packaging Professionals, 1994, pp. 47-67.
16. MJ Hasting. Packaging machinery. In: RT Parry, ed. Principles and Applications of Modified Atmosphere Packaging of Foods. London: Blackie Academic & Professional, 1993, pp. 41-62.
17. M Sivertsvik, WK Jeksrud, JT Rosnes. A review of modified atmosphere packaging of fish and fishery products – significance of microbial growth, activities and safety. Int J Food Sci Tech 37:107-127, 2002.
18. AL Brody. Microbiological safety of modified/controlled atmosphere/vacuum packaged foods. In: AL Brody, ed. Controlled/Modified Atmosphere/Vacuum Packaging of Foods. Tumbul, Connecticut: Food & Nutrition Press, 1989, pp. 159-174.
19. CB Hintlian, JH Hotchkiss. The safety of modified atmosphere packaging. A review. Food Technol 40(12):70-76, 1986.
20. JH Hotchkiss. Experimental approaches to determine the safety of food packaged in modified atmospheres. Food Technol 42(9):55-64, 1988.

21. SJ Walker. Chilled food microbiology. In: C Dennis, M Stringer, ed. Chilled Foods, A Comprehensive Guide. New York: Ellis Horwood, 1992, pp. 165-195.

22. D Zagory, AA Kader. Modified atmosphere packaging of fresh produce Food Technol 42(9):70-77, 1988.

23. WD Powrie, BJ Skura. Modified atmosphere packaging of fruits and vegetables. In: B Ooraikul, ME Stiles, ed. Modified Atmosphere Packaging of Food. New York: Ellis Horwood, 1991, pp. 169-245.

24. DS Lee, PE Haggar, J Lee, KL Yam. Model for fresh produce respiration in modified atmospheres based on principles of enzyme kinetics. J Food Sci 56:1580-1585, 1991.

25. KL Yam, DS Lee. Design of modified atmosphere packaging for fresh produce. In: ML Rooney, ed. Active Food Packaging. London: Blackie Academic & Professional, 1995, pp. 55-73.

26. CP McLaughlin, D O'Beirne. Respiration rate of a dry coleslaw mix as affected by storage temperature and respiratory gas concentrations. J Food Sci 64:116-119, 1999.

27. P Jaime, ML Salvador, R Oria. Respiration rate of sweet cherries: 'Burlat', 'Sunburst' and 'Sweetheart' cultivars. J Food Sci 66:43-47, 2001.

28. KI Hayakawa, YS Henig, SG Gilbert. Formulae for predicting gas exchange of fresh produce in polymeric film package. J Food Sci 40:186-191, 1975.

29. PC Talasila, AC Cameron. Prediction equations for gases in flexible modified atmosphere packages of respiring produce are different than those for rigid packages. J Food Sci 62:926-930, 1997.

30. JD Mannapperuma, RP Singh. Modeling of gas exchange in polymeric pack-ages of fresh fruits and vegetables. In: RP Singh, FAR Olivera, ed. Minimal Processing of Foods and Process Optimization. Boca Raton, FL: CRC Press, 1994, pp. 437-458.

31. P Veeraju, M Karel. Controlling atmosphere in a fresh fruit package. Modern Packaging 39:168-254, 1966.

32. DS Lee, JS Kang, P Renault. Dynamics of internal atmosphere and humidity in perforated packages of peeled garlic cloves. Int J Food Sci Tech 35:455-464, 2000.

33. Y Song, DS Lee, KL Yam. Predicting relative humidity in modified atmosphere packaging system containing blueberry and moisture absorbent. J Food Process Preserv 25:49-70, 2001.

34. S Fishman, V Rodov, J Peretz, S Ben-Yehoshua. Model for gas exchange dynamics in modified-atmosphere packages of fruits and vegetables. J Food Sci 60:1078-1987, 1995.

35. Y Song, N Vorsa, KL Yam. Modeling respiration–transpiration in a modified atmosphere packaging system containing blueberry. J Food Eng 53:103-109, 2002.

Chapter 14

Microwavable Packaging

14.1 Microwaves and Microwave Oven...425
14.2 Microwave/Food/Packaging Interactions...426
    14.2.1   Microwave/Heat Conversion Mechanism·····························427
    14.2.2   Dielectric Properties ···428
    14.2.3   Penetration Depth ···429
    14.2.4   Mathematical Equations and Models·································430
14.3 Challenges in Microwave Heating of Foods...431
    14.3.1   Nonuniform Heating ···432
    14.3.2   Lack of Browning and Crisping ····································433
    14.3.3   Variation in Microwave Ovens······································434
    14.3.4   Meeting the Challenges ···434
14.4 Microwavable Packaging Materials 435
    14.4.1   Microwave Transparent Materials·································435
    14.4.2   Microwave Reflective Materials····································436
    14.4.3   Microwave Absorbent Materials ···································437
14.5 Interactive Microwave Food Packages...439
    14.5.1   Surface Heating Packages···439
    14.5.2   Field Modification Packages ··440
    14.5.3   Steam Cooking Packages···441
14.6 Case Study: Effect of Metal Shielding on Microwave Heating Uniformity 442
Discussion Questions and Problems 443
Bibliography 443

14.1 Microwaves and Microwave Oven

After development and introduction of microwave oven in the 1950s, it has become a popular household appliance currently having the penetration level higher than 90% of the US households. With recent lifestyle changes favoring convenience and quick preparation of food, the microwave oven provides a convenient means for the consumer to cook or reheat food quickly and easily. Consumers can quickly microwave a ready-to-heat meal in a container and enjoy the meal from the same container. A variety of microwavable food products have been developed by food industry and are now ubiquitous in the supermarket. The packaging industry has also developed new technologies and packages that are compatible with microwave heating. Because the heating mechanisms of food in the microwave oven is quite different from conventional oven, development of successful microwavable food products requires the developer to have a good understanding of the microwave heating of food, as well as careful considerations of the package design and consumer expectation.

Microwaves are short electromagnetic waves located in the portion of electromagnetic wave spectrum between radio waves and visible light. The energy is delivered in the form of propagating sine waves with an electric field and a magnetic field orthogonal to each other. Microwaves are relatively harmless to humans because they are a form of nonionizing radiation, unlike the much more powerful ionizing radiation (such as X-ray or gamma ray) that can damage the cells of living tissue. Microwaves are used in daily applications such as cooking, radar detection, telecommunications, and so on.

Most microwave ovens for food applications operate at two frequencies. The household microwave oven operates at 2450 MHz (2.45×10^9 cycles per second), and the industrial microwave oven operates at 915 MHz (9.15×10^8 cycles per seconds). The wavelengths associated with those frequencies are 0.122 and 0.382 m, respectively, when the microwaves are assumed to travel at the speed of light (3×10^8 m s^{-1}). Microwaves travel at approximately the speed of light in air, but they travel at a lower speed inside a food material. The relationship between frequency and wavelength is expressed by the equation:

$$v = f \lambda \qquad (14.1)$$

where v is velocity (m s^{-1}), f is frequency (Hz), and λ is wavelength (m) of electromagnetic wave.

In the microwave oven, microwaves are generated by an electronic vacuum tube known as magnetron located outside the oven cavity. The microwaves then travel through a hollow metal tube called 'waveguide' to the oven cavity. The food in a microwave oven is heated by absorbing the microwaves, while the air in the oven and the oven wall stay cold by transmitting and reflecting them, respectively. The colder air exchanges heat with the food surface evaporating some water into the air. The evaporated water in the air is flowed away by a mechanical means. To improve the heating uniformity, the microwave oven is often equipped with a stirrer or a turntable. The stirrer is a fanlike set of spinning metal blades used to scatter the microwaves and disperses them evenly within the oven. The turntable rotates the food during the microwave process. The history, features, standardization, and safety matters relating to the microwave oven are discussed by Decareau [1].

14.2 Microwave/Food/Packaging Interactions

There are three possible modes of interaction when microwaves impinge upon a material: absorption of microwaves by the material, reflection of microwaves by the material, and transmission of microwaves through the material. The material may be a food or a packaging material. The food must absorb a portion of the microwave energy in order for heating to occur. Most foods do not reflect microwaves, and thus all the remaining unabsorbed microwave energy is transmitted. Some packaging materials, such as susceptors, absorb microwave energy and become hot. Metals, such as aluminum foils, reflect microwaves. Paper, plastics, and glass are transparent to microwaves. To optimize the microwave heating of food, it is necessary to consider the reflection, absorption, and transmission of microwaves by the food and the package.

14.2.1 Microwave/Heat Conversion Mechanism

Microwave energy is not heat energy. In order for microwaves to heat food, they must first be converted to heat. There are two mechanisms by which this energy conversion can occur: dipole rotation and ionic polarization. The two mechanisms are quite similar, except the first involves mobile dipoles while the second involves mobile ions. Both dipoles and ions interact only with the electric field, not the magnetic field.

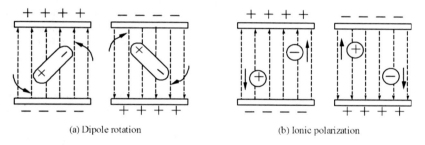

(a) Dipole rotation   (b) Ionic polarization

Figure 14.1 Microwave/heat conversion mechanisms. The dashed lines denote the alternating electric field at the frequency of microwaves. (a) A dipole rotates back and forth. At high frequencies (such as 2450 MHz), there is not sufficient time for the dipole to rotate 180°, and thus the actual rotation angle is much smaller. (b) A positive ion and a negative ion oscillate in an alternating electric field.

Figure 14.1a illustrates the dipole rotation mechanism of a polar molecule. In the presence of an electrical field, the polar molecule behaves like a microscopic magnet, which attempts to align with the field by rotating around its axis. As the polarity of the electric field changes, the direction of rotation also changes. The molecule thus absorbs microwave energy by rotating back and forth billions of times at the frequency of microwaves. Since the molecule is often bound to other molecules, the rotating action also causes it to rub against those other molecules. The rubbing action disrupts the bonds between the molecules, which in turn causes friction and heat dissipation.

Water molecule is the most abundant polar molecule in food. The water molecules in liquid water are quite mobile, and they readily absorb microwave energy and dissipate it as heat through dipolar rotation. In contrary, the water molecules in ice are much less mobile due to the confined crystal structure, and they do not absorb microwave well. The distribution of moisture and the state of water (liquid water or ice) are often two critical factors that determine the behavior of microwave heating of foods.

Figure 14.1b illustrates the ionic polarization of a positive ion and a negative ion in solution. In the presence of an electric field, the ions move in the direction of the field. As the polarity of the electric field changes, the ions move in the opposition direction. The ions absorb microwave energy by oscillating themselves at the frequency of microwaves. The oscillating action in turn causes heat dissipation through friction. The common ions in food are those from salts such as sodium chloride. Since ions are less abundant than water molecules in most foods, ionic polarization often plays a less important role than dipole rotation.

14.2.2 Dielectric Properties

While dipole rotation and ionic polarization provide a qualitative understanding of the microwave/heat conversion mechanisms, the dielectric properties provide a quantitative characterization of the interactions between microwave electromagnetic energy and food. The dielectric properties, along with thermal and other physical properties, determine the heating behavior of the food in the microwave oven.

An important dielectric property is dielectric loss factor ($\varepsilon"$), which indicates the ability of the food to dissipate electrical energy. The term 'loss' refers to the loss of energy in the form of heat by the food. It is useful to remember that a material with a high $\varepsilon"$ value (also known as a lossy material) heats well, while a material with a low $\varepsilon"$ value heats poorly in the microwave oven. The dielectric loss factor is related to two other dielectric properties by the equation:

$$\tan \delta = \varepsilon" / \varepsilon' \qquad (14.2)$$

where $\tan \delta$ is loss tangent, and ε' is dielectric constant.

The dielectric properties (ε' and $\varepsilon"$) are functions of frequency, temperature, moisture content, and salt content. Values of dielectric properties for foods and other materials can be found in the literature [2-4]. Examples of $\varepsilon"$ and ε' values at 2450 MHz are shown in Table 14.1. Although the literature values can be used as guidelines, actual measurements are often required because of the variability of composition of the materials.

The dielectric properties provide a quick indication of how well a material heats in the microwave oven. For example, the $\varepsilon"$ value of water (12.5) at 25°C is several orders of magnitude higher than that of ice (0.0029) at -12°C. This means water heats far better than ice in the microwave oven. Ice is almost transparent to microwaves because its molecules are tightly bound and do not rotate easily through the mechanism of dipolar rotation. The dramatic increase in $\varepsilon"$ value is also observed, when ice changes to water, during the thawing of frozen foods including beef, potato, and spinach (Table 14.1). Plastics and paper have low ε" values because they are almost transparent to microwaves.

Table 14.1 Dielectric constant (ε'), dielectric loss factor (ε"), and penetration depth (D$_p$) of various foods and materials at 2450 MHz

Materials	ε'	ε"	D$_p$ (cm)
Ice (-12°C)	3.2	0.0029	2400
Water (1.5°C)	80.5	25.0	1.4
Water (25°C)	78	12.5	2.8
Water (75°C)	60.5	39.9	0.8
0.1 M NaCl (25°C)	75.5	18.1	1.9
Fat and oil (average)	2.5	0.15	41.1
Raw beef (-15°C)	5.0	0.75	11.7
Raw beef (25°C)	40	12	2.1
Roast beef (23°C)	28	5.6	3.7
Boiled potatoes (-15°C)	4.5	0.9	9.2
Boiled potatoes (23°C)	38	11.4	2.1
Boiled spinach (-15°C)	13	6.5	2.2
Boiled spinach (23°C)	34	27.2	0.9
Polyethylene	2.3	0.003	1970
Paper	2.7	0.15	42.7
Metal	∞	0	0
Free space	1	0	∞

14.2.3 Penetration Depth

The speed of microwave heating is due to the deep penetration of microwaves into the food, and the dielectric properties can be used to determine the extent of penetration. When microwaves strike a food surface, they arrive with some initial power level. As microwaves penetrate the food, their power is attenuated since some of their energy is absorbed by the food. The term penetration depth (D$_p$) is defined as the depth at which the microwave power level is reduced to 36.8% (or 1/e) of its initial value, which can be estimated using the equation:

$$D_p = \frac{\lambda_o}{\sqrt{2}\pi\left[\varepsilon'\left(\sqrt{(1+\tan^2\delta)}-1\right)\right]^{1/2}} \qquad (14.3)$$

where λ_o is wavelength in free space. At 2450 MHz, $\lambda_o = 12.24$ cm and thus

$$D_p = \frac{2.76}{\left[\varepsilon'(\sqrt{1 + \tan^2 \delta} - 1)\right]^{1/2}} \qquad (14.4)$$

where D_p is in cm. The penetration depth is a 'visual term' that describes how well a food absorbs microwaves: the shorter is the penetration depth, the more the food absorbs microwaves.

The meaning of penetration depth is further illustrated in Figure 14.2. At the first D_p, 36.8% of the initial power remains, while 63.2% of the power is absorbed. At the second D_p, $(0.368)^2 = 13.5\%$ remains and 86.5% is absorbed. At the third D_p, $(0.368)^3 = 5.0\%$ remains and 95% is absorbed. The penetration depth depends on the composition of material, the frequency of microwaves, and temperature.

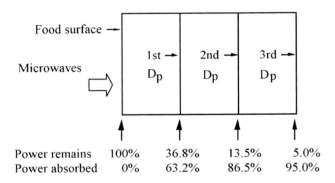

Figure 14.2 Power levels at various penetration depths

Typical values of D_p for various materials at 2450 MHz are also shown in Table 14.1. As mentioned earlier, liquid water absorbs microwaves far better than ice. The D_p of water at 25°C is 2.8 cm, but D_p of ice at −12°C is 2400 cm! Frozen foods have longer penetration depths than unfrozen foods. For example, the D_p values for frozen beef and unfrozen beef are 11.7 cm and 2.1 cm, respectively.

14.2.4 Mathematical Equations and Models

Besides Equation 14.3, other simple equations can also provide researchers and food product developers with a better understanding of the microwave heating process. For example, the microwave power absorption can be estimated using the equation:

$$P = k \, \varepsilon'' \, f \, E^2 \qquad (14.5)$$

where P is power absorption (W cm^{-3}), k is constant (5.56 x 10^{-13} farad cm^{-1}), ε'' is dielectric loss factor of food (dimensionless), f is frequency of microwaves (Hz), and E is electric field strength of microwave (V cm^{-1}). The power absorption is directly proportional to the dielectric loss factor, indicating again that a lossy material (which has high ε'' value) is a good absorber of microwave energy.

The rate of temperature increase of the food can be estimated using the equation:

$$\frac{dT}{dt} = \frac{k\,\varepsilon''\,f\,E^2}{\rho\,C_p} \qquad (14.6)$$

where T is average temperature of food ($^\circ$C), t is time (s), ρ is density of food (g cm^{-3}), and C_p is specific heat of food (J g^{-1} $^\circ$C^{-1}). Equation 14.6 is an energy balance equation, which assumes the microwave energy absorbed is balanced by the heat gain of the food. Note that this is a relatively simple equation, and it assumes that temperature is evenly distributed within the food.

Mathematical models based on heat-and-mass transfer principles are also available to provide more sophisticated information during the microwave heating of food [5, 6]. A typical model consists of a set of partial differential equations with the proper initial and boundary conditions. The models can be used to predict the temperature and moisture distribution histories of foods during microwave heating. To use the models, values for dielectric properties, thermal properties, density, electrical field strength, and product dimensions are required. Models for frozen food are more complicated than those for unfrozen food, because the microwave heating behavior changes greatly from the frozen state to unfrozen state.

The models can simulate what-if scenarios and thus can help to minimize the number of experiments and shorten the product development time. However, the models are limited mostly to the predictions of temperature and moisture content, and they do not deal with other important factors such as taste and texture. Most models are also limited to foods that are homogenous and have regular shapes.

14.3 Challenges in Microwave Heating of Foods

While microwavable food products with the benefits of speed cooking and convenience have made significant inroads in the past two decades, there still remain many challenges for the developer and manufacturer. Besides convenience, other factors such as taste and texture are also important to the consumer. Quite often, the consumer perceives food heated by the microwave oven do not taste as good as that heated by the conventional oven. This is because the heating mechanisms of food in the microwave oven and conventional oven are indeed quite different. To develop successful microwavable food products, the developer must have a good understanding of the microwave heating of food, as well as careful considerations of the package design and consumer expectation.

The challenges in the development of microwavable food products arise from the need to deal with the many variables relating to the food, package, and microwave oven. For the food, there are variables of food composition, shape, size, specific heat, density, dielectric properties, and thermal conductivity. For the package, there are variables of shape, size, and properties of packaging material. For the microwave oven, there are variables relating to the design of the oven. A related and more important challenge is to solve the problems of the consumer. From the consumer's point of view, the most noticeable problems are those associated with nonuniform heating, lack of browning and crisping, and variation in microwave ovens.

14.3.1 Nonuniform Heating

Nonuniform heating is a major problem associated with microwave heating. Uneven temperature distribution in the food causes nonuniform cooking and microbial inactivation, often accompanied with uneven moisture distribution affecting product texture. The problem is especially noticeable for frozen food. It is not uncommon for a frozen food heated in a microwave oven to boil around the edges while the center remains frozen. The problem is caused by the differences in microwave energy absorption of liquid water and ice.

In frozen foods, the water molecules on the surface are relatively free to move compared to the water molecules inside the food. When a frozen food is microwaved, heating begins at the surface where the water molecules are more ready to absorb microwave energy. This causes the adjacent ice crystals to melt and the surface temperature to rise, while the inside temperature is still little unaffected. As more liquid water is available, the heating of the surface becomes more rapidly. This can lead to 'runaway heating', in which heating is excessive at the surface while the inside is still frozen. To minimize runaway heating during thawing, microwave energy should be delivered at a slow rate, which allows more time for heat to conduct from the surface to the inside.

Irregular shape of the food can also cause nonuniform heating. The thin parts tend to overcook, while the thick parts tend to undercook. This situation also occurs in conventional cooking but is less pronounced because the cooking is slower. Shapes such as cylinder and sphere often concentrate microwave energy at the center of the product, while the slab shape products attain the highest temperature along the edges and corners. Relatively uniform heating of cylindrical and spherical foods is attained when the food diameter is in the range of 2-2.5 times the penetration depth [7]. The corner heating effect in slab is more severe with the shorter penetration depth food and is reduced by oval or circular shapes. An annulus or doughnut shape is optimal for reducing the corner effect. However, the degree of hot spotting at center or corner varies with dielectric food properties and dimensional conditions. Another cause of nonuniform heating is that different foods have different dielectric and thermal properties. When a microwave meal consists of two or more items, it is possible that the items heat at different rates. For example, when microwave is heating a frozen meal consisting of meat and vegetable, the vegetable often becomes overheated and dried out before the meat reaches the serving temperature.

Inclusion of holding time in microwave heating is a simple method used to reduce the nonuniform temperature distribution. The duty cycle (fraction of cycle time during which the power is on) can be varied by the consumers to improve the temperature uniformity of microwave-heated foods and is also used for controlling the power output of the microwave oven with low cost. Selective metal shielding on some part of packaged foods can also contribute to minimizing the uneven heating.

14.3.2 Lack of Browning and Crisping

Another problem is that, unlike the conventional oven, the microwave oven is not able to produce foods that are brown and crisp. This is because the heating mechanisms of the conventional oven and the microwave oven are quite different.

In the conventional oven, the food is heated by hot air in the oven, and if the heating element is not shielded, the food is also heated by radiated heat. Heating is concentrated on the food surface by means of heat convection and radiation. The inside of the food is also heated, at a slowly rate, by means of heat conduction. The heating causes the moisture on the food surface to evaporate rapidly, and later, browning and crisping to occur. Although the moisture inside the food tends to migrate to the surface, the rate is not sufficiently fast to prevent browning and crisping. As a result, the food surface becomes brown and crispy while its inside remains moist and soft.

In the microwave oven, there is no hot air, and heating is mostly due to the interaction between microwaves and water. Microwave heating is not concentrated on the food surface, but it is distributed within the food depending on the penetration depth. The heating on the food surface is no longer sufficiently intense to cause browning and crisping. Unless the food is microwaved for a long time to remove all or most of the water in the food (which is not desirable because the food quality may no longer be acceptable), browning and crisping either do not occur at all or are inadequate.

Browning formulations have been developed for various meat and dough products [1]. Commercial steak sauces, barbecue sauces, soy sauces, and the like are brushed on meat before microwave heating. Reusable browning dishes are also available for browning food surfaces in the microwave oven. Most of the commercial browning dishes are made of glass-ceramic substrate with tin oxide coating on the underside. The packaging industry has also developed a disposable browning and crisping material, known as susceptor, discussed later in this chapter.

14.3.3 Variation in Microwave Ovens

Yet another problem is the large variation of performance in different microwave ovens. Microwave ovens are available in different powers, oven cavity sizes, with or without a turntable, with or without a stirrer (to distribute microwaves more evenly in the oven). Consequently, different microwave ovens may produce greatly different results, even if the same cooking instructions are used. To accommodate the differences, the food manufacturer can only place vague microwave heating instructions on their packages. For example, a package may contain vague instructions such as "heat between 4 to 8 minutes, depending on the microwave oven".

14.3.4 Meeting the Challenges

There is no easy solution to deal with the complex process of microwave heating. In developing a microwavable food product, the scientist or technologist has to rely on the somewhat useful but incomplete scientific knowledge described in the previous sections, as well as trial-and-error or empirical methods.

From the discipline point of view, there are three approaches to deal with the challenges. The first is the food chemist's approach, in which food ingredients are modified and browning formulations are added to make the food more microwavable. The compositional content of water, fat, and protein can be tailored for better performance in microwave oven: water and fat attract microwave for faster heating [8]. Sugar-based formulation may be coated on the food surface to generate golden brown surface by reaction with protein in the heating. Typical browning formulations proposed include salt, reducing sugars, amino acids, shortening, and yeast extract [1]. The second is the packaging engineer's approach, in which the package is modified to enhance the performance of microwave heating. Generally foods in round or oval shape containers perform better than those in containers with sharp corners. The third is the microwave engineer's approach, in which new and useful features are added to the microwave oven. Other modes of heat transfer such as infrared radiation and hot-air jet-impingement heating increase the spatial uniformity of heating simultaneously with increased rate of heating [9]. Ideally, these approaches should be integrated into a system to deliver the highest quality of microwavable foods to the consumer. As another way to finetune the microwave oven for cooking a food item, Samsung Electronics recently launched a cooking control system (Smart Oven®) that reads and carries out cooking instruction written in two dimensional bar code on food package (see Section 15.8): simply swiping the package's bar code on the oven scanner transfers the cooking instruction information to the oven, which will be operated in a programmed manner; information sharing between food company and oven maker in

microwave sensitive food item(s) can be shielded so that the entire meal can be heated more even.

Aluminum is also used as an electromagnetic field modifier to redirect microwave energy in a manner to optimize the heating performance [11]. Aluminum can intensify the microwave energy locally or redirect it to places in the package that otherwise would receive relatively little direct microwave exposure. This approach has been used to redirect microwave energy from the edges to the center for frozen food products such as lasagna.

When aluminum foils are used in the microwave oven, precautions are necessary to prevent arcing, which can occur between foil packages and the oven walls, between two packages, across tears, wrinkles, and so on. Arcing can be prevented by following simple rules of separating metal objects, using a ceramic plate or turntable and using shallow format foil dishes: the foil must not touch other metal, such as a metal turntable, the oven sides, or another container; the container should preferentially be used singly, and placed on a ceramic plate if the turntable is metal; the board lids lined with foil should be removed and replaced by a suitable plastic cover or film; the shallow trays should be at least two thirds full. In the package design, any foil components should be receded from the edge of the package to avoid arcing with the oven walls. Aluminum tray laminated by PP or PET can also be used in microwave oven without arcing. Recently metal cans coated with electric insulator have also been attempted for use in microwave oven. The plastic-coated shallow metal can removed of metal end and covered with plastic lid can be microwaved without arcing. In addition to following those rules, it is also necessary to thoroughly test the package/product to ensure that the package is safe to use.

14.4.3 Microwave Absorbent Materials

Microwave absorbent materials used for food packaging are commonly known as susceptors. The major purpose of susceptors is to generate surface heating to mimic the browning and crisping ability of the conventional oven. For the concentrated microwave energy absorption on the surface, a thin layer material of high conductivity is bonded to a support structure of plastic film. Although many types of susceptors such as ferrites, metal oxides, salt hydrates, and titanium nitride have been invented, the only commercially available type is the metallized film susceptor (Figure 14.4). This type of susceptor consists of a metallized PET film (typically about 12 μm thick) laminated to a thin paperboard. The paperboard serves the task of dimensional stability at high temperature of around 200°C. The metal layer is a very thin (less than 100 angstroms, usually 30-60 angstroms), discontinuous layer of aluminum, which is responsible for generating localized resistance heating when exposed to microwaves. The thicker aluminum layer decreases electric resistivity resulting in higher portion of power reflection and lower portion of power transmission, while the thinnest layer transmits all the energy (Figure 14.5). Maximum absorption occurs at a film resistivity corresponding to half value of free space, which is used for the susceptor heating [7]. The heating can cause the susceptor to reach surface temperatures over 200°C within seconds.

Figure 14.4 Metallized film susceptors (Courtesy of Graphic Packaging Inc.)

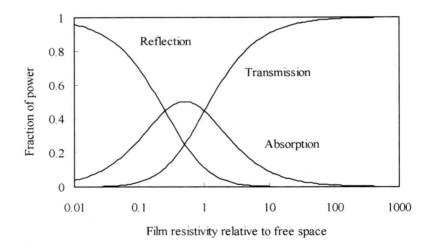

Film resistivity relative to free space

Figure 14.5 Fractional power of reflection, absorption, and transmission as a function of electric resistivity of a susceptor film

In order to avoid the runaway heating, susceptors may be provided with the function which self-limits the level of heating. Weakening of bond between the metallized film and the base support substrate is utilized for the purpose. With weakening of the bond the metallized film shrinks and is distorted to change the conductivity characteristics of the metal layer reducing its ability to generate heat energy. The bond characteristics need to be tuned for the required pattern of weakening with temperature increase by selecting optimal variables of the film, metal surface, adhesion process, etc. A regular pattern with small nonmetallized areas or grid of metallized lines may create concentrating paths for current flow, which disables the susceptor at high temperature and limits the heating. Printable coatings of conductive materials in an ink-like polymer matrix have also been tried in developmental stage for easy fabrication and self-limiting performance of susceptors.

14.5 Interactive Microwave Food Packages

As mentioned above, requirements of most microwavable packaging materials are transparency to microwave and resistance to high temperature. The materials should preferably have a low dielectric loss factor less than 0.05. Dual-ovenable packaging materials should be able to withstand the temperature of 200-230°C and include PET-coated paperboard products, CPET, TPX®, copolymer of cyclohexanedimethanol and terephthalic acid (PCTA, or Thermx® of Eastman Chemical), polyetherimide, and polycarbonate. Microwave-only materials usually cover the temperature below water-boiling temperature of 100°C and include PP, polyphenylene oxide/polystyrene blend, and HDPE. Food products with microwavable characteristics and packaging are often marked as 'microwavable' on their label. High-fat food containing little water may lead to extremely high temperature posing a risk of consumer burning. As with microwave-transparent materials, metal foil of the reflective material and metallized susceptor of absorptive nature may be used to modify the microwave field and absorption pattern. All the package designs should look for an elaborate combination and harmony of proper roles between transparent, reflective, and absorptive materials, which should be based on the consideration of desired food quality characteristics. Recently more elegant designs of active packaging concepts have been introduced in microwave packaging: elaborate use of susceptor, field modification technique, and steamed pressure cooking with self-venting device [12, 13]. As a way to improve the flavor of the microwaved food, aroma is incorporated in the polymeric structure and released at heating in the microwave oven [8]. Active microwave packaging may sometimes become unfavorable for the dual ovenability. It needs to be mentioned that any active microwave package design should be tested in both normal usage and abusive conditions for heating performance and safety considerations.

14.5.1 Surface Heating Packages

The rationale for surface heating is to generate thermal energy in close proximity of food surface, bringing about early surface drying, browning, and/or crisping [11]. Two approaches are used for surface heating function: microwave-absorbent susceptors are the means predominantly applied, while fringe effects using the principle of localized field concentration at edges of the electrical conductive materials are proposed in some patent. As discussed in Section 14.4.3, metallized plastic film induces a localized and concentrated heating for puffing the popcorn, browning the surface, and creating the baked flavor (Figure 14.4).

Susceptors have been used for products such as frozen pizza, frozen French fries, frozen waffles, frozen hot pies, and popcorn with instant microwave popcorn being the single largest application. Susceptors are available in the forms of flat pads, sleeves, and pouches. The flat pads are suitable for products (such as pizza) that require heating only on one surface. The sleeves and pouches are suitable for heating on multiple surfaces. Matching the active heating area to the shape of food product is desired for efficient and balanced heating producing better browning and crisping performance. Complicated design providing close contact between food surface and susceptor can help the transfer of generated heat to the food. Susceptors are also available in various patterns, in which portions of the metallized layer are deactivated [11]. The patterns are designed to provide more control of heating. A company uses a printed checkerboard pattern to generate various levels of heating based on the size of the check. There are some designs which facilitate the surface heating on wider area by raising and suspending the susceptors and the food over the oven floor.

Fringes or edges of the conductive material can generate high energy around them, which can be used for producing surface browning. The concept has been developed as slot line structure of paperboard press tray with a limited commercial usage [11]. Combination of fringe effect heater with metallized susceptor structure is seen to have a good potential of the effective surface heating.

There is a public concern of migration of mobile compounds from the susceptor to the food, because the susceptor can reach high temperature. The high temperatures may result in the formation of significant numbers of volatile chemicals from the susceptor components and loss of barrier properties of food-contact materials leading to rapid transfer of nonvolatile adjuvants to foods. The FDA has issued voluntary guidelines regarding the safe use of the susceptor for the packaging to follow. Appropriate migration test at the actual use condition of high temperature is required in order to demonstrate the safe use.

14.5.2 Field Modification Packages

Modification of microwave field by microwave reflective materials such as aluminum foil thicker than 6 μm can be used to redirect the microwave energy in a predictable way to optimize the heating performance: even heating to avoid overheating at edges and undercooking at food center as well as providing different heating rates to different components of multi-component foods. The foil is usually laminated to supporting carrier such as PET film and/or paperboard.

Shielding around corners or edges of the food packages alter energy distribution and can protect those parts from being overheated. Microwave energy supply to a specific part of the food can be restricted by using the shielded package. Close contact of the shield with food surface is helpful in most cases. Patterned shields with some etched holes or demetallized areas are often adopted to attain uniform temperature distribution and eliminate hot spotting (Figure 14.3). Package of multi-component meals may be designed to have different size apertures for each component on the metal foil sleeve and thus different exposures to microwave. This approach can provide the desired heating profiles for all the components whose dielectric properties and target temperatures of cooking differ with component.

Focusing the microwave energy onto the target area is used for concentrated heating. Field intensification elements for focusing energy include lenses using controlled-size apertures, one-way mirrors trapping and reflecting the power, and conductive loops with electric lengths equal to integer multiples of energy wavelength [11]. Recently, a series of antennae structures made of aluminum foil (6-7 µm thick) are mounted on plastic film or tray to promote even heating, surface browning, crisping, and selective heating [12]. The antennae work to collect the energy, transmit it to the desired area, and dissipate it usefully. A variety of component designs are used: large antennae affect bulk heating properties, small ones induce surface heating, and transmission devices redistribute the power. The antenna design can be used in combination with susceptor for better performance. An innovative example of field modification technology is MicroRite® of Graphic Packaging International.

14.5.3 Steam Cooking Packages

There have been attempts to use the power of steam in microwave heating and cooking. Microwaving produces steam from moisture contained in wet food or moisturized paper pad. The steam then penetrates into the food and package headspace, builds the pressure, and expands the package. The package is equipped or attached with self-venting valve or lid responding to pressure increase. Small vents, slits, or holes may be located in the seam or film surface. This type package provides the steam-cooker effect and saves the consumers the job of package opening. The concept of steamed pressure cooking has been applied to microwave products of fresh vegetables, fishes, and chicken. Typical example of steam-cooking microwavable package is Thermipack® of Cryovac/Sealed Air.

14.6 Case Study: Effect of Metal Shielding on Microwave Heating Uniformity

Ho and Yam [14] studied how metal band arrangement patterns could affect the heating uniformity and power absorption of the cylindrical model food containing 3% agar. The food sample was contained in 600 mL glass beaker and had the dimension of 84 mm in diameter and 74 mm in height. Aluminum foil bands of varying width were attached on the beaker's lateral outer surface in various combinations of horizontal, vertical, and slanted patterns with different spacing. Unshielded (open) sample was used as a reference for comparison. Relative temperature uniformity (RU) and power absorption (RP) in percentage defined as Equations 14.7 and 14.8, respectively, were determined as a function of design variables of shield number, width, spacing, orientation, and exposed area.

$$RU = \frac{SD}{SD(open)} \times 100 \qquad (14.7)$$

$$RP = \frac{P}{P(open)} \times 100 \qquad (14.8)$$

where SD is standard deviation of linear heating rates measured at various locations of the model food (Figure 14.6a), and P is power absorption described as sensible heat increase rate of 400 mL distilled water placed in the beaker. Lower RU value means more uniform heating rate distribution and higher RP means faster heating. Desired heating comes along with the good heating uniformity and power absorption, which would result in lower RU and higher RP values.

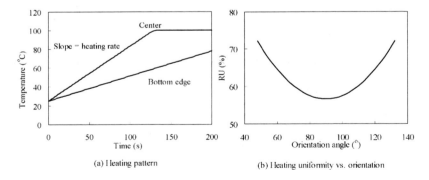

(a) Heating pattern

(b) Heating uniformity vs. orientation

Figure 14.6 (a) Heating pattern in microwave heating of model food and (b) temperature uniformity as function of orientation angle

It was found possible to design shielding patterns that greatly reduce RU without greatly increasing the heating time. RU value was significantly affected by orientation and width of shield, but correlated little with shield area. The lowest RU value was obtained with 1.5 cm shield width and 90° orientation angle which is vertical, while RP value was not changed with the shield width and orientation. The investigators formulated a simplified relation to express RU value as function of orientation as shown in Figure 14.6b.

Discussion Questions and Problems

1. Discuss the difference between microwave heating and infrared radiation heating.

2. Discuss hygienic safety aspects of microwave susceptor. Find the ways to avoid or reduce the potential risk related to it.

3. Discuss the potential problems of microwave thawing of frozen foods. Give your ideas to ensure their safe defrosting.

4. What would be the dimensional option to attain uniform reheating in microwavable food package design? You can take a real food package as an example for the analysis.

5. Provide some food products which can have benefits from microwave packaging using steamed pressure cooking. What would be the rationale for adopting the technology in the aspect of convenience and food quality?

6. Compare roast beef and boiled spinach in microwave heating. Refer to Table 14.1 and other resources for food properties.

Bibliography

1. RV Decareau. Microwave Foods: New Product Development. Trumbull, Connecticut: Food & Nutrition Press, 1992, pp. 1-127.
2. NE Bengtsson, PO Risman. Dielectric properties of foods at 3GHz as determined by a cavity perturbation technique II. Measurements of food materials. J Microwave Power 6:107-123, 1971.
3. SO Nelson. Electrical properties of agricultural products (A critical review). T ASAE 16:384-400, 1973.
4. AK Datta, E Sun, A Solis. Food dielectric property data and their composition-based prediction. In: MA Rao, SSH Rizvi, ed. Engineering Properties of Foods. New York: Marcel Dekker, 1995, pp. 457-494.
5. KG Ayappa. Modeling transport processes during microwave heating: a review. Rev Chem Eng 13(2):1-68, 1997.
6. LA Campanone, NE Zaritzky. Mathematical analysis of microwave heating process. J Food Eng 69:359-368, 2005.
7. CR Buffler. Microwave Cooking and Processing, 1993, pp. 69-97.

8. K Bertrand. Microwavable foods satisfy need for speed and palatability. Food Technol 59(1):30-34, 2005.
9. AK Datta, SSR Geedipalli, MF Almeida. Microwave combination heating. Food Technol 59(1):36-40, 2005.
10. GJ Huss. Microwave packaging and dual-ovenable materials. In: AL Brody, KS Marsh, ed. The Wiley Encyclopedia of Packaging Technology. New York: John Wiley & Sons, 1997, pp. 642-646.
11. TH Bohrer, RK Brown. Packaging techniques for microwaveable foods. In: AK Datta, RC Anantheswaran, ed. Handbook of Microwave Technology for Food Applications. New York: Marcel Dekker, 2001, pp. 397-469.
12. N Waite. Active Packaging. Leatherhead, UK: Pira International, 2003, pp. 85-90.
13. J Fowle. Developments in Microwaveable Packaging. Leatherhead, UK: Pira International, 2005, pp. 45-64.
14. YC Ho, KL Yam. Effect of metal shielding on microwave heating uniformity of a cylindrical food model. J Food Process Preserv 16:337-359, 1992.

Chapter 15

Active and Intelligent Packaging

15.1 Introduction 445
15.2 Active Packaging – Absorbing System 448
    15.2.1 Oxygen Absorbers 449
    15.2.2 Moisture Absorbers 451
    15.2.3 Carbon Dioxide Absorbers 452
    15.2.4 Ethylene Absorbers 453
    15.2.5 Other Absorbers or Removers 453
15.3 Active Packaging – Releasing System 454
    15.3.1 Carbon Dioxide Emitters 456
    15.3.2 Antimicrobial Packaging Systems 456
    15.3.3 Antioxidant Releasers 459
    15.3.4 Other Releasers 460
15.4 Active Packaging – Other System 460
    15.4.1 Self-Heating Systems 460
    15.4.2 Self-Cooling Systems 461
    15.4.3 Selective Permeation Devices and Others 461
15.5 Intelligent Packaging Framework 462
15.6 Smart Packaging Devices 464
    15.6.1 Time Temperature Integrator or Indicator 465
    15.6.2 Leak, Gas, or Other Indicators 466
    15.6.3 Freshness Indicators 467
    15.6.4 Bar Code 468
    15.6.5 Electronic Identification Tags 468
15.7 Legislative and Human Behavior Issues 470
15.8 Case Study: Intelligent Microwave Oven with Barcode Reader 471
Discussion Questions and Problems 472
Bibliography 473

15.1 Introduction

Traditional basic functions of packaging include protection, communication, convenience, and containment (Figure 15.1, see Section 1.4). These functions are not totally exclusive but are connected with each other; moreover extra enhanced functions have been sought by the food packaging sector for meeting the consumer demands for minimally processed foods with fewer preservatives, increased regulatory requirements, market globalization, and concern for food safety. Active packaging and intelligent packaging are the main areas in which most of recent innovative ideas have been applied to satisfy these needs, broadening and redefining the function of food packaging.

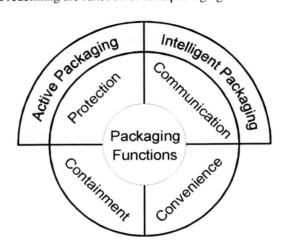

Figure 15.1 Model of packaging function. From Ref. [1].

The concept of *active packaging* started with a shift in the protection function of packaging from passive to active. Previously, primary packaging materials were considered as 'passive', meaning that they functioned only as an inert barrier to protect the product against oxygen and moisture. Recently, a host of new packaging materials have been developed to provide 'active' protection for the product. Active packaging has been defined as a system in which the product, the package, and the environment interact in a positive way to extend shelf life or to achieve some characteristics that cannot be obtained otherwise [2]. A multinational European study, Actipak has also defined it as a packaging system that actively changes the condition of the packed food to extend shelf life or to improve food safety or sensory properties, while maintaining the quality of the packaged food [3]. In this context, the function of active packaging technologies is related mostly to enhancing the protection of package and thus active packaging was placed above the protection function in Figure 15.1 to reflect this concept. This placement, however, does not prevent active packaging from performing other functions: a single active packaging device can provide several functions at the same time as will be seen later in this chapter.

All the active packaging technologies involve some physical, chemical, or biological action for altering the interactions between the package, the product, and the package headspace to achieve certain desired outcome [4, 5]. They can be divided into three categories of absorber, releasing system, and other system

[3]. Absorbing system [5, 6] is a group of technologies that use packaging films or sachets to remove undesired gases and substances (such as oxygen, carbon dioxide, moisture, ethylene, and taints) from the package so that a favorable internal package environment and food condition are achieved. Releasing system is a group of technologies that actively add or emit desired or active compounds (such as carbon dioxide, ethanol, antimicrobials, antioxidants, enzymes, flavors, and nutraceuticals) to protect and enhance food quality. Most attention of controlled release concept has been focused on antimicrobial packaging [7, 8] and antioxidant packaging [9, 10]. Other active packaging system may include the tasks of self-heating, self-cooling, microwave susceptor, and selective permeable film.

The term *intelligent packaging* has also been used with ambiguous and vague meanings, and sometimes is used interchangeably with *smart packaging*. European Actipack project defined intelligent packaging as one that monitors the conditions of packaged foods to give information about the quality of the packaged food during transport and storage [3]. Recently Yam et al. [1] define intelligent packaging as a packaging system that is capable of carrying out intelligent functions (such as detecting, sensing, recording, tracing, communicating, and applying scientific logic) to facilitate decision making in order to extend shelf life, enhance safety, improve quality, provide information, and warn about possible problems. In this definition, the ability to communicate has been stressed as a uniqueness of intelligent packaging: since the package and the food move constantly together throughout the supply chain cycle, the package is the food's best companion and is in the best position to communicate the conditions of the food. According to the definition, a package is 'intelligent' if it has the ability to track the product, sense the environment inside or outside the package, and communicate with human. For example, an intelligent package is one that can monitor the quality/safety condition of a food product and provide early warning to the consumer or food manufacturer.

Accordingly, intelligent packaging was placed above the communication function in the proposed model of Figure 15.1, in which intelligent packaging is a provider of enhanced communication and active packaging is a provider of enhanced protection. Thus, in the total packaging system, intelligent packaging is the component responsible for sensing the environment and processing information, and active packaging is the component responsible for taking some action (e.g., release of an antimicrobial) to protect the food product. Note that the terms intelligent packaging and active packaging are not mutually exclusive; some packaging systems may be classified either as intelligent packaging or active packaging or both, but this situation does not detract the usefulness of these terms. In appropriate situations, intelligent packaging, active packaging, and the traditional packaging functions work together to provide a total packaging solution.

Table 15.1 Active packaging systems of absorbers with material types, their working principles, and applications

Active task type	Material	Working principle	Desired effect and applications	Commercial status and variations
Oxygen absorber	Reduced iron; ascorbic acid; sulfites; oxidases; polydienes; aromatic nylon	Oxidation reaction of Fe, ascorbic acid, sulfites or polymer component; enzymatic catalysis of oxidation of glucose or ethanol	Prevention or reduction of aerobic microbial growth on bakery and prepared food products; prevention or reduction of oxidative quality deterioration in fatty foods and dry food products	There are insert types and modified polymers for absorbing headspace oxygen or enhancing O_2 barrier; commercial absorbers based on iron are in different varieties of moisture-sensitive and self-reactive products
Moisture absorbers	Silica gel; polyacrylate; CaO; propylene glycol; sugar; inorganic salt	Moisture adsorption; hydration reaction	Retention of low moisture content leading to crisp texture and minimum quality loss of dried products; removal of excess water on surface of flesh foods and fresh-cut produces	Propylene glycol of viscous state is packed in polyvinyl alcohol pouch, polyacrylate of super-absorbent polymer is in sheet form and others are mostly packed in spunbonded Tyvek sachet
Carbon dioxide absorbers	$Ca(OH)_2$; Na_2CO_3; zeolite; silica gel	Chemical reaction with CO_2; physical adsorption	Prevention of package blooming or bursting caused by CO_2 from lactic fermented kimchi or roasted coffee beans	Physical absorbents of zeolite and silica gel are reversible adsorbents, while the others of chemical absorbents are irreversible in reaction nature
Ethylene absorbers	$KMnO_4$ on aluminum oxide or silica gel; SiO_2; zeolite; active carbon	Oxidation of ethylene to acetic acid and alcohol; physical adsorption	Retarding ripening and senescence of fresh produce	$KMnO_4$ changes color from purple to brown with oxidation of ethylene
Other absorber	Immobilized enzymes; molecular sieve; ferrous salt with organic acid	Catalyzing the transformation reactions to remove naringin, lactose, or cholesterol; adsorption or oxidation of volatile flavors	Reduction of bitterness in grapefruit juice; removing of lactose or cholesterol in milk products; reduction and delayed production of off-flavor in fish and oily foods	Off-flavor absorption should be used not to hide or mask the microbial spoilage and hygienic problem

15.2 Active Packaging – Absorbing System

Some components present in the headspace or in food such as oxygen, moisture, and ethylene often cause undesirable reaction or give unpleasant quality changes.

Those components may exist in the packaged containment from the time of food production, may be produced from physiologically active food, or permeated into the food package. The actively responding devices to remove these components can improve the storage stability of both food and package, and also sometimes enhance the chemical, physical, and sensory quality of the contained food (Table 15.1).

15.2.1 Oxygen Absorbers

As stated in Section 13.1.1, oxygen is usually not a gas preferred in the perspective of food preservation. Lipid components, colors, and flavors in foods are deteriorated by their reactions with oxygen. Oxygen supports the growth of aerobic bacteria, yeasts, and molds which may cause food spoilage. Removal of oxygen from the package can slow down all these oxidation reactions and microbial growth, attaining beneficial effect of food quality preservation.

Currently available oxygen absorbing systems can be divided into two types: insert type and reactive polymer structure type [11]. The insert type includes sachet, self-adhesive labels, and adhesive devices which are placed in the package or attached onto the appropriate internal surface of the package. The reactive polymer structure type covers monolayer and multilayer materials and closure liners for bottles. Oxygen absorbers are an area which was most intensively investigated and reached the commercialization of wide market. Extensive list and information on commercial oxygen scavengers can be found in the literatures [4, 11-13].

The most successful oxygen scavengers of sachet form on commercial scale are based on iron. Powdered iron is contained alone or with other catalysts in oxygen permeable film pouch. The basic oxidation reaction used by iron for absorbing oxygen can be described as:

$$4\ Fe + 3\ O_2 + 6\ H_2O \rightarrow 4\ Fe(OH)_3 \qquad (15.1)$$

Because the rusting or oxidation reaction of iron requires water, the moisture is supplied as a vapor from the moist food or initially added in the absorbent. So the sachet containing water by itself is the self-reacting scavenger when exposed to air, while that without water starts to absorb oxygen when exposed to humid conditions [13]. Rough estimation of iron's capacity to absorb the oxygen is around 300 mL O_2 per gram of iron [3, 5]. Iron-based oxygen scavengers usually cannot pass through the metal detector and may have problem in microwave heating. They have some limitation of oxygen absorption under the atmosphere dominant with high CO_2 concentration [11]. Commercial brands of iron-based oxygen scavengers include Ageless® of Mitsubishi Gas Chemical Co., ATCO® of Standa Industrie, and FreshMax® of Multisorb Technologies.

The disadvantages of iron-based oxygen scavengers are overcome by organic type absorbers. Typical ones are of ascorbic acid, catechol, or polyunsaturated fatty acids. The system may contain small amounts of catalysts such as transition metal to control the oxidation of the substrates. A commercial example of ascorbic acid-based oxygen scavenger is OxySorb® originally developed by Philsbury Food Company [12]. Some oxidase enzymes catalyzing oxidation of glucose and ethanol have also been proposed as oxygen removal system. An example of commercial enzymatic O_2 scavenger is Bioka® of Bioka Ltd. [12]. Some organic oxygen scavengers may produce carbon dioxide from their oxidation reaction. Like iron based-scavenger, they also require moisture for the oxidation reaction except catechol, polyunsaturated fatty acids, and alcohol oxidase [3, 5]. Sachet form may pose the risk of its ingestion or objection by consumers, and thus adhesive label or card forms have been developed to avoid these problems. They are secured mostly onto internal surface of the package. Some insert type oxygen scavengers are added with capacity to produce the same volume of carbon dioxide as the removed oxygen: the produced carbon dioxide may compensate for the decrease of volume or pressure due to oxygen removal, alleviating the possible package collapse. CO_2 emitting mechanism will be discussed later in this chapter.

Modification of polymer composition and/or structure is used to give oxygen-absorbing capacity [11]. This approach can be used to offer barrier enhancement and/or headspace O_2 scavenging: oxygen barrier enhancement is attained by placing an oxygen-scavenging layer in the film or sheet structure used for making pouch, bottle, or jar; headspace O_2 scavenging is achieved by the polymer liner placed in the crown or closure of the bottle and jar. One simple example using the same idea of iron oxidation is polyolefin blended with iron compounds. The modified polymer blend is usually designed to be located as a barrier layer of multilayer structure. Because of the nature of iron oxidation mechanism, iron-based polymer scavenger system requires moisture for its activation. Commercial examples of iron-based polymer scavenger include Oxyguard® of Toyo Seikan and Shelfplus® of Ciba Specialty Chemicals. In a similar principle, plastic blended with oxidizable sulfites and ascorbate has been used as liner material to remove oxygen through the bottle closure [11]. An example of this type is Darex OST® of W.R. Grace [12]. Polymers bound with unsaturated functional groups have been developed for oxygen scavenging. Oxidation of polydienes and polyketones can scavenge oxygen from package headspace. The autoxidation reaction is triggered by ultraviolet light with aid of transition metal catalysts. OS® of Cryovac Sealed Air and Amosorb® of BP Amoco may be examples listed in this category.

Intensive researches and developments in oxygen scavenging polymer have been conducted in polyethylene terephthalate (PET) bottle particularly for beer, which is very sensitive to oxygen and has a large market volume. An oxidizable aromatic nylon with transition metal catalyst like cobalt was blended with PET or sandwiched between PET layers. An example of O_2 absorbing PET material is   Oxbar®   of   Constar   International   incorporating   MXD-6   nylon (polymetaxylylene adipamide or polymetaxylylene diamine-hexanoic acid) between PET layers [14].

Successful applications of oxygen scavengers cover whole categories of foods, with low to high water activity, chilled and ambient stored, retorted and minimally processed. Beneficial effects of reduced nutritional loss, better preserved color and flavor, desirable sensory quality, prevented insect infestation, and retarded microbial spoilage and rancidity have been observed in many reports and literatures. However, the package design with oxygen scavenger system should take into account the oxygen absorbing speed and capacity of the system in relation with the quality protection requirements of specific food item in order to have optimal results. The possible concerns of consumer's mishandling of the sachet, package collapse, and hygienic problems of anaerobic bacteria caused by depletion of oxygen should be dealt with.

15.2.2 Moisture Absorbers

Main use of moisture absorbent is to maintain low water activity of packaged dry foods whose physical qualities are lost at moisture adsorption on their surface. Desiccants of silica gel, calcium chloride, and calcium oxide are most widely used as sachet sealed with the moisture permeable spunbonded plastic film. Silica gel removes moisture by physical adsorption mechanism which can be reversible by temperature change. Most commercial moisture absorbers are based on silica gel. Calcium oxide reacts to remove water irreversibly as:

$$CaO + H_2O \rightarrow Ca(OH)_2 \tag{15.2}$$

The Reaction 15.2 proceeds slowly with heat production [12].

Polyacrylates and graft copolymer of starch called *superabsorbent polymer* can absorb liquid water and moisture vapor, and thus are used for absorbing liquid water from high water activity foods such as meat, poultry, and fish. The polymer particles are usually embedded in the porous drip sheet. Superabsorbent polymers having crosslinked networks of flexible chains accept water diffused into the structure with swelling. Another approach to remove the excess water on wet food surface is the humectant placed between two moisture permeable plastic film layers. One example of such system is Pitchit$^{®}$ of Showa Denko, viscous polypropylene glycol sandwiched between polyvinyl alcohol films. For buffering water activity inside the fresh produce package, sachets of hygroscopic sugar and inorganic salt have also been tried. The examined materials cover sucrose, xylitol, sorbitol, potassium chloride, calcium chloride and sodium chloride [15]. For proper control of relative humidity inside the package, the sorption isotherm of the absorbent materials and the sachet permeability to moisture need to be characterized.

15.2.3 Carbon Dioxide Absorbers

Carbon dioxide is produced from degradation reaction, microbial growth, and respiration. High amount of CO_2 produced in the roasting of coffee is released through the packaged storage, expanding the package. Nonpasteurized vegetables fermented by lactic acid bacteria produce large volume of CO_2 expanding the package or building up its internal pressure. Fresh produce consumes oxygen and produces CO_2 by aerobic respiration; excessive accumulation of CO_2 in the package may cause the physiological injury onto the fresh fruits and vegetables. Therefore, controlled removal of CO_2 from package headspace is desired for these kinds of packaging situations.

The most versatile commercial CO_2 absorber is calcium hydroxide, which reacts with carbon dioxide to produce calcium carbonate:

$$Ca(OH)_2 + CO_2 \rightarrow CaCO_3 + H_2O \tag{15.3}$$

Sodium carbonate can also absorb CO_2 under high humidity condition by following reaction [16] :

$$Na_2CO_3 + CO_2 + H_2O \rightarrow 2\ NaHCO_3 \tag{15.4}$$

Physical adsorbents such as active carbon and zeolite also have the capability to absorb carbon dioxide, however, they do not work when directly exposed to high humidity conditions because of their preferred adsorption of moisture vapor rather than CO_2 [17]. Sequential adsorption and desorption of CO_2 can be achieved by time-variable controlled exposure to moisture.

A patent assigned to Mitsubishi Gas Chemical Co. introduced a live fish transport system wherein is placed a bag which absorbs CO_2 by alkaline earth metal hydroxide and produces O_2 by solid peroxide and peroxide decomposition catalyst [18]. EMCO Packaging System produces a proprietary CO_2 absorber having O_2 generation function, which is useful for high O_2 modified atmosphere packaging (MAP) system of fresh-cut produce maintaining O_2 level above 40%. For the details of high oxygen MAP, consult Sections 13.1 and 17.8.

One-way check-valve, which is activated by pressure buildup, is widely used for roasted coffee packaging to release carbon dioxide outside the package. In Taiwan, a similar type of valve is used for packages of rice washed with supercritical CO_2. Even though the valve does not absorb carbon dioxide, it works to remove it helping to maintain the stable package structure. Similarly a pinhole channel was appended on the package of kimchi, a Korean lactic fermented vegetable to attain a beneficial internal atmosphere without pressure buildup [19].

15.2.4 Ethylene Absorbers

Ethylene (C_2H_4) produced by fresh produce and some microorganisms is a plant hormone which accelerates respiration rate and hastens ripening and senescence of the fresh produce. Even though the amount of ethylene production depends on the commodity, maturity, status of preparation, and temperature, any fresh produce package contains certain level of ethylene in package headspace. The presence of gaseous ethylene in the produce package is mostly deleterious, speeding the softening and reducing the shelf life. Therefore the removal of ethylene is desirable for keeping freshness of fresh produce, particularly for the commodity sensitive to ethylene.

Because ethylene is very reactive due to presence of double bond in its chemical structure, there are a variety of methods to degrade it. The most popular method is oxidation of ethylene by potassium permanganate ($KMnO_4$) adsorbed on an inert carrier with large surface area such as silica gel, alumina, and activated carbon [14, 20]. Typical content of $KMnO_4$ is about 4-6%. The overall equation of the oxidation is described as:

$$3\ C_2H_4 + 12\ KMnO_4\ \rightarrow\ 12\ MnO_2 + 12\ KOH + 6\ CO_2 \qquad (15.5)$$

The $KMnO_4$ based ethylene scavengers change color from purple to brown with their reduction reaction of Equation 15.5. They should not be in contact with food surface because of their toxicity.

Ethylene can also be adsorbed onto microporous media such as activated carbon, silica gel, and zeolite. Metal catalysts such as palladium may be impregnated with the adsorbents for further breakdown of ethylene. Polyethylene film filled with ceramic powder is claimed to have ethylene removing capability, but has not been proven rigorously in its power. However, it seems to be evident that inclusion of porous ceramic powder increases the permeability to ethylene contributing to reducing ethylene concentration in the fresh produce packaging.

As a promising development of ethylene removing packaging material, hydrophobic ethylene-permeable plastic films incorporated with electron-deficient dienes or trienes such as tetrazine have been suggested [20]. The candidate polymers for embedding these chemicals include silicone polycarbonate, polystyrene, polyethylene, and polypropylene.

15.2.5 Other Absorbers or Removers

As a way to improve the flavor of citrus fruit juices, cellulose acetate film immobilized with naringinase was developed, which can hydrolyze the bitter compound of naringin to nonbitter naringenin and prunin [21]. Other innovative packaging concepts incorporating enzyme include those for removing lactose and cholesterol from milk products. The packaging layer incorporating lactase to break down lactose to glucose and galactose can eliminate lactose intolerance problem for some people deficient in lactase. Cholesterol reductase in the inner packaging layer can reduce cholesterol to coprosterol, a harmless substance not to be absorbed by human intestine, which can enhance the healthiness of milk products. However, all these concepts of enzyme-immobilized packaging have not reached commercialization yet.

Malodorous volatile compounds such as aldehydes, amines, and hydrogen sulfides can cause consumer's rejection of the food product even at very low detectable level. The absorption of these off-odor compounds has been attempted by several principles: physical adsorption, chemical oxidation, etc. Molecular sieves such as silica gel, zeolite, and activated carbon can adsorb volatile components of low molecular weight. Ferrous salt and organic acid incorporated into polymer are claimed to oxidize amine and other oxidizable odor compounds [6]. Acid compounds in heat-seal polymer layer has been suggested as a means to remove the amine compounds. A patent claims that polyethylene imine embedded in polyolefin can scavenge undesirable aldehydes [4].

There are controversies about using flavor or odor absorption packaging to improve sensory quality of foods. Off-odor compound is a useful signal to show the food spoilage, and thus the removal of the odor signal may mislead consumers to eat the spoiled food. Therefore the use of odor absorption packaging should be carefully used to improve both hygienic and sensory quality of packaged food. Franzetti et al. [22] reported that an MAP system with absorbers for volatile amines and liquid led to the shelf life extension of chill-stored fish with delayed microbial growth and lower accumulation of headspace trimethylamine, however, they cautioned that some measures are needed to exclude the possibility that foods, microbiologically spoiled but sensorially acceptable, reach the consumers.

15.3 Active Packaging – Releasing System

Some gases or solute components are produced or delivered deliberately inside the package to cause desirable effects on package structure or food qualities. For the package with oxygen scavenger, carbon dioxide is often produced in proportion to the oxygen absorbed in order to prevent the package collapse (Table 15.2). Antimicrobials and antioxidants embedded in the packaging material can be released in controlled rate under the direct or indirect contact with food. The desired effect would be the improved storage stability and shelf life extension. Several concepts and applications of releasing active packaging system will be described below. However, it should be noted that the packaging system releasing certain compounds to the contained food is subject to food packaging regulations particularly in terms of safety and migration aspects. Even though any food-contact packaging materials should meet the requirements of hygienic standards set by corresponding country's legislation, there are more concerns on migration problem of the releasing packaging materials. For the regulations related to the hygienic safety and migration of the packaging materials, refer to Section 5.7.

Table 15.2 Active packaging systems of releasers with material types, their working principles, and applications

Active task type	Material	Working principle	Desired effect and applications	Commercial status and variations
Carbon dioxide emitters	$FeCO_3$; $NaHCO_3$; Na_2CO_3; ascorbic acid	Hydrolysis reaction of $FeCO_3$; reaction of $NaHCO_3$ or Na_2CO_3 with organic acid; oxidation of ascorbic acid	Inhibition or growth reduction of Gram-negative bacteria and molds on prepared food products; prevention of package collapse	$FeCO_3$ is very unstable not to be used without any halide metal catalysts
Antimicrobial packaging	Antimicrobials embedded or encapsulated in packaging materials; package surface intrinsically inhibitory to microorganisms	Release of antimicrobial agent onto the food surface or contacted suppression of microbial growth without migration; release of ethanol vapor to package headspace	Inhibition of microbial growth	Direct contact between packaging and food surface is mostly required for both migratory and nonmigratory antimicrobial packaging except one of volatile nature; vapors such as ethanol, allyl isothiocyanate, and ClO_2 in the package headspace may present an offensive smell at initial opening of package
Antioxidant releasers	Antioxidants (BHA, BHT, tocopherol) embedded in packaging materials	Release of antioxidants or consumption of oxygen	Retardation of oxidative quality changes and protection of the polymer	Antioxidants incorporated for polymer protection may work for food protection by their oxygen absorption and migration to food
Other releasers	Flavors incorporated in packaging matrix	Release of desirable flavor components	Improvement of food flavors	Release of desirable flavor should be used not to disguise the microbial off-flavors

15.3.1 Carbon Dioxide Emitters

There are some cases of package collapse due to volume reduction or pressure decrease of headspace atmospheric gases: oxygen scavenging in a peanut can may cause pressure decrease, and CO_2 dissolution from CO_2 enriched headspace into the flesh foods in the MAP products can cause partial vacuum or collapse of semi-rigid plastic trays. In these cases, production of CO_2 gas amounting to the disappeared gases can help to maintain package integrity and appearance. Creating high CO_2 concentration in the package can also contribute to retarding microbial growth and spoilage on the food.

Emission of CO_2 responding to O_2 gas absorbed can be described by following reaction:

$$4 \text{ FeCO}_3 + 6 \text{ H}_2\text{O} + \text{O}_2 \rightarrow 4 \text{ Fe(OH)}_3 + 4 \text{ CO}_2 \qquad (15.6)$$

Because ferrous carbonate is very unstable under ambient conditions, metal halide and other catalysts are incorporated with it to control the reaction rate of CO_2 production; by combining appropriate amount of iron powder to consume oxygen (see Equation. 15.1), the CO_2 emitted can be tailed to be equal to O_2 absorbed in some sachet products. Other oxidation of organic compounds like ascorbic acid used for O_2 scavenging can also emit some amount of CO_2.

Another carbon dioxide generation system consists of sodium carbonate and citric acid in the exudate absorber [23]. High and stable CO_2 concentration could be maintained by using this system for fish fillet MAP to reduce package deformation and extend shelf life.

15.3.2 Antimicrobial Packaging Systems

Because food spoilage usually starts on food surfaces due to the presence and growth of pathogenic or spoilage organisms, control of microbial growth on the food surfaces is thought to be effective way of improving the storage stability of the packaged foods and extending their shelf life. Controlled release of antimicrobial agents from food packaging material to the food surface has been considered to be able to inhibit or reduce microbial growth and spoilage by using minimum amount of antimicrobial agent on the consumed food. Therefore, dominant way of conferring antimicrobial activity onto food surface by packaging is through controlled migration of active compounds from packaging to food surface. Controlled release of antimicrobial substance from packaging material into the food may be an effective way of inhibiting the microbial growth, and can strengthen the efficiency of its direct addition to food [24, 25]. The migration may be achieved by direct contact between food and packaging material or through gas phase diffusion from packaging layer to food surface (Figure 15.2). Even though the former is the packaging situations usually met, the latter mode has been successful in commercialized antimicrobial packaging application because of its simple and wide applicability. The volatile antimicrobials can easily diffuse or penetrate the bulk food structure and surface which do not contact package surface.

Ethanol emitters are an old antimicrobial packaging system utilizing gas phase migration of the antimicrobial substance (Figure 15.2B). Ethanol is encapsulated or adsorbed onto SiO_2 or other medium, which is contained in a sachet of plastic film permeable to ethanol or perforated with a small pinhole. The sachet placed in food packaging is to release ethanol vapor at slow rate inhibiting mold growth on bakery and rice cake products. Typical product of ethanol emitter is Ethicap® of Freund Industrial Co. One drawback of ethanol emitter is strong ethanol aroma at time of opening the food package because ethanol vapor fills package headspace and is adsorbed on the food.

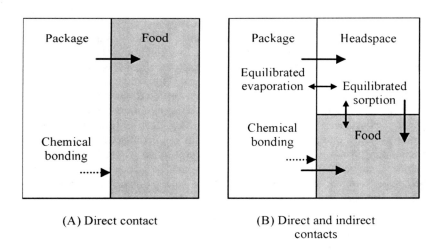

(A) Direct contact

(B) Direct and indirect contacts

Figure 15.2 Patterns of migration and action in antimicrobial food packaging system. Single side solid arrow denotes antimicrobial action by migration while single side dotted arrow denotes antimicrobial action on the food surface without migration. Double side solid arrow means equilibrium relation. Adapted from Ref. [26].

Gas phase migration is also used by another antimicrobial packaging of cyclodextrin encapsulating wasabi extract whose active volatile antimicrobial compound is allyl isothiocyanate (AIT). The AIT in the encapsulated powder is claimed to be nonvolatile in dry intact state, but become volatile when the AIT-cyclodextrin complex is exposed to high humidity condition after packaging of food. The evaporated AIT then moves to the food surface and inhibits the growth of aerobic bacteria and molds. The AIT-cyclodextrin complex powder has been incorporated in the packaging materials of drip sheets, polyethylene films, and tablets commercialized in Japan for use in rice lunch boxes, meat, and fresh produce. Another system releasing a volatile antimicrobial is chlorine dioxide (ClO_2) generation system from sodium chlorite and acid precursors embedded in polymer film such as low density polyethylene (LDPE) and paraffin. Atmospheric humidity and temperature affect the hydrolysis rate of an acid anhydride controlling the release of the acid, which then diffuses internally in the film to release the ClO_2 from the chlorite. An example of ClO_2 generation packaging is Microgarde® of Southwest Research Institute. CSIRO of Australia has also developed a system which gradually releases SO_2 from pads of sodium metabisulphite incorporated microporous material to control mold growth in some fruits [7].

Most usual types of antimicrobial packaging system are based on direct contact between packaging material and food surfaces (Figure 15.2A). The antimicrobial agents are incorporated into or coated on the packaging layer. Some antimicrobial functional groups may be attached to polymer backbone, or bioactive agents such as enzymes and other organic compounds may be immobilized in the packaging material matrix, maintaining their antimicrobial activity as themselves [7]. Some polymers such as chitosan and UV-irradiated nylon are intrinsically antimicrobial. Except these nonmigratory chemical substances bonded to packaging surface, most antimicrobial packaging systems available or investigated derive their microbial-inhibitory activity from the migrated substance. However, there is often no clear-cut distinction between migratory and nonmigratory antimicrobial packaging materials. Antimicrobial compounds proposed and/or tested for incorporation in packaging materials include silver ion, organic acids such as sorbate, propionate, and benzoate or their respective acid anhydrides, bacteriocins such as nisin and pediocin, plant extracts, chitosan, and enzymes such as lysozyme.

The most primitive form of fabricating antimicrobial plastic packaging is by extruding the polymer masterbatch mixed with an active substance. However, extrusion process applies high temperature for melting the polymer pellets and shaping into film and container. Therefore LDPE of low melting temperature is the most popular for extruded antimicrobial plastic materials. The commercially available products in Japan and Korea are LDPE films impregnated with silver zeolite which is thermally stable. Sorbates and their acid anhydrides have also been added in LDPE film fabrication. Usual concentration of incorporation is around 1%. Even though natural materials such as grapefruit seed extract and nisin have been used for extrusion or heat-setting of polyethylene and retained some antimicrobial activity after processing, their sensitivity to heat restricts the fabrication using thermal melting of plastics. Thus solution coating or solvent casting has been tried to incorporate these active substances into packaging matrix with better activity retention. When a preservative is incorporated or coated as a surface layer on the structure film by the coating or solvent casting rather than a whole film structure, the total amount of active substance and the corresponding material cost can also be reduced.

Antimicrobial packaging for extending shelf life and improving food safety should consider the target microorganism and food compositional properties [7]. Narrow antimicrobial spectrum, in a certain case, may be desirable for packaged foods of which microbial problem is caused by a specific type of strain. However, generally the packaging materials with a wide antimicrobial spectrum are desirable for universal use in a variety of foods. Several antimicrobial agents may be incorporated in a polymer matrix for this purpose. Because of high interests in the potential of antimicrobial food packaging, intensive researches are being conducted at present in this area and interested readers are suggested to refer to extensive reviews [7, 26, 27].

15.3.3 Antioxidant Releasers

Even though an oxygen scavenger in package prevents or retards oxidative deterioration of packaged food, antioxidants can be incorporated into or coated onto food packaging materials as to be released into the headspace or food for more effective control of the oxidative changes. The antioxidant incorporated in plastic packaging material may have dual roles of protecting the polymer and the packaged food from oxidation [6, 12]. Synthetic or natural antioxidants such as butylated hydroxytoluene (BHT) and α-tocopherol, incorporated in polyethylene (PE) and polypropylene (PP) packagings, were shown to protect cereals, milk, vegetable oils, and emulsions from oxidation [5, 6, 10]. BHT was shown to be transferred from packaging to solid food by evaporative diffusion. Waxed paper has once been used as reservoir for antioxidant release for cereal packaging. Recently α-tocopherol receives a great attention for incorporation into polyolefins because of its compatibility with polyolefins and thermal stability under polymer processing conditions. High density polyethylene bottles containing α-tocopherol gave a lower off-flavor and a higher acceptability for storage of drinking water by preventing the oxidation of polymer substrates when compared to those with BHT or Irganox [28, 29]. Wessling et al. [30] compared PE and PP as the media for incorporating α-tocopherol. The LDPE impregnated with α-tocopherol transferred the α-tocopherol to the packaged foods showing the potential of use as an active antioxidative packaging material to prevent oxidation of the foods. PP not showing the migration has been suggested to scavenge oxygen of the headspace of the package or at the surface of packaged foods [30].

Antioxidant may be incorporated into packaging material together with antimicrobial agent to preserve perishable foods that are sensitive to both microbial spoilage and oxidative deterioration. Antimicrobial and antioxidant packaging materials incorporating nisin and α-tocopherol, provided both antimicrobial and antioxidative effectiveness for a model emulsion system and milk cream at 10°C [31]. Another application combining ascorbic acid and silver zeolite has also been reported for antioxidative and antimicrobial activities [6].

15.3.4 Other Releasers

Controlled release of flavor components from plastic and paper packaging materials has been suggested as a method to improve the organoleptic perception of the package food during short period of initial opening [4]. Flavor release might also be used as a means to mask off-odors coming from the food or the packaging. Flavor components in liquid oil can be loaded in concentration of 25-40% onto the plastic pellets, which are then used for molding or extrusion. Other technologies of incorporating flavors include spiced interior coating, scented lacquering, scented printing ink, deposition of porous adsorbent, lamination with scented film, and microencapsulation. However, the flavor-release technologies should not be misused to mask the development of microbial off-odors and sell products that are below standard or even dangerous for the consumer [5].

There have also been suggested some concepts releasing other components such as enzymes, enzyme inhibitors, amino acids, and saccharides, or fumigants to result in some desired changes of the packaged foods. The released materials may induce specific reaction such as nonenzymatic browning, or kill the insects in packaging.

15.4 Active Packaging – Other System

Some active packaging concepts take the reactive modes other than absorption and releasing of compositional components in food or package headspace. Packaging components may actively control the heating or cooling speed. They may also be designed to change the gas permeability by responding to outside environment. Microwave susceptor is also a type of package element controlling heating rate and pattern, and has already been dealt with in microwave packaging of Chapter 14.

15.4.1 Self-Heating Systems

A typical reaction used for generating heat in food packaging is the hydration reaction of lime:

$$CaO + H_2O \rightarrow Ca(OH)_2 + 65.2 \text{ kJ} \tag{15.7}$$

One mole of solid calcium oxide reacts with one mole of liquid water to produce one mole of solid calcium hydroxide with heat evolution of 65.2 kJ at 25°C. Excess water (about 2 to 2.5 times the theoretical amount based on stoichiometric calculation) is usually used because hydrated lime absorbs some of water after hydration reaction. The lime and water are usually sealed in separated compartments located in contact with packaged food. There are made some devices such as depressing button, slitting band, pulling string, and puncturing lance to break the wall or membrane separating the compartments of lime and water at the time of usage. The heating rate and temperature can be controlled by the amount of lime and physical design of the package. The heating rate and temperature may be controlled by mixing the lime with some amounts of dolomite (CaO-MgO), magnesium oxide (MgO), magnesium sulfate ($MgSO_4$), or magnesium chloride ($MgCl_2$) which has lower heat of hydration. Another heating mechanism is oxidation of iron and magnesium by salt water [12]. Uniform heating rate can be obtained by homogenous blending of lime and water under condition of wide and close contact between the reaction mixture and food. Package material surfacing outside environment often consists of insulating layer.

Commercial applications of self-heating packaging system include packages of sake, coffee, and cooked rice. Most of the applications for liquid foods are based on direct conductive heat transfer from reaction mixture to food. Some applications for solid foods adopt initial generation of steam from the heat and subsequent steam flow into food container. Available packages are usually small size plastic tubes and metal cans of sizes mostly less than 300 mL. Usual heating time to the desired temperature of 60-75°C is in the range of 10-15 minutes.

15.4.2 Self-Cooling Systems

High demand from consumers' side for instantly cooled beverage cans evoked a global race to develop self-cooling packaging system. There appeared many innovative concepts of self-cooling packaging, but none have reached mature commercialization so far at the time of writing this book. However, some sachets are available which remove heat from surrounding air when exposed to high humidity conditions.

Proposed self-cooling mechanisms include endothermic hydration of mixture of dehydrated ammonium nitrate and powdered ammonium chloride, and instant release of liquefied gas such as carbon dioxide. Mixing of water and sodium thiosulfate can also cause an endothermic reaction used for Freddo Freddo® self-cooling cans in Italy [12]. Expansive evaporation of compressed liquefied gas can draw large amount of heat, but has a problem of high pressure maintenance in the container or package.

15.4.3 Selective Permeation Devices and Others

Because most plastic films have limitations in their gas permeability to provide optimal internal atmosphere in modified atmosphere packaging of high respiration fresh produce such as fresh-cut fruits and vegetables (see Sections 13.3 and 17.8), some means to enhance or modify the gas permeation properties of the packaging material have been devised. Very highly gas-permeable label may be attached as an adhesive patch onto a die-cut hole on the package to match the packaging's gas permeation to produce respiration. One simple approach is to have microperforations, pinholes, or channels on the packaging, which can increase the gas permeability and change the CO_2 to O_2 gas permeability ratio toward the value close to 1. Microperforations of 30-100 μm in diameter have been made by laser beam or other means. Commercial example of microperforated plastic film is P-Plus® bags marketed by Amcor and Sumimoto Bakelite. A US patent [32] used microporous plastic membrane as a part of fresh produce container to attain the desired package atmospheres for some particular produces. Yun et al. [33] also introduced the use of microporous earthenware sheet as a part of rigid package for fresh produce. A package with pinhole or channel is also reported to help build up the internal atmosphere desirable for quality of lactic fermented vegetable of Korean kimchi which contains live lactic acid bacteria [19]: the attachment of pinhole can also eliminate the pressure build-up problem and is commercially applied as a long channel on the seal surface area in small size packages.

Another innovative approach is to adjust the gas permeability responding to temperature change in distribution. Intellipac® film of Landec Corporation with fatty acid based side-chains undergoes the reversible physical change from crystalline to amorphous state with crossing over a switch temperature, which makes the film highly permeable to gases. This property can be utilized to prevent the formation of anoxic condition in fresh produce packages experiencing temperature abuse (see Sections 13.3 and 17.8).

15.5 Intelligent Packaging Framework

Intelligent packaging can play an important role in facilitating the flow of both materials and information in the food supply chain cycle. In Figure 15.3, the outer circles represent the supply chain cycle from raw material through manufacturing, packaging, distribution, product use, and disposal. The package, in one form or another (such as pouch, container, drum, pallet), is traditionally used to facilitate the flow of materials (represented by the arrows in the figure) from one location to another, by performing the basic functions of containment and protection of the product. Furthermore, the package can also facilitate the flow of information (represented by the communication links between the inner circle and outer circles), although this communication function has been largely overlooked. The package can indeed be a highly effective communicator—it can carry actual information in the direction of material flow (e.g., via truck, train, or ship), and it can transmit information visually (e.g., via an indicator) or electronically (e.g., via a barcode or the Internet) throughout every phase of the supply chain cycle.

A conceptual framework describing the flow of information in an intelligent packaging system is illustrated in Figure 15.4. The system consists of four components: *smart package devices, data layers, data processing, and information highway* (wire or wireless communication networks) in the food supply chain. The smart package devices are largely responsible for giving birth to the concept of intelligent packaging since they impart the package with a new ability to acquire, store, and transfer data. The data layers, data processing, and information highway are collectively referred here as the *decision support system.*

As shown in Figure 15.4, the smart package devices and the decision support system are designed to work together to monitor changes in the internal and external environments of the food package and to communicate the conditions of the food product, so that timely decisions can be made and appropriate actions taken. From the quality and safety viewpoint, the external environment can be further divided into the physical, ambient, and human environments (see Section 1.5), which are factors important for determining shelf life. However, the business environment is also an important factor; in fact, the development of smart package devices (especially data carriers) and the information highway are largely motivated by the desire to increase profit and operation efficiency. Presently, business data (such as product identification, quantity, and price) and *business models* (rules for processing information to maximize profits) are incorporated into the system to facilitate product checkout, inventory control, and product traceability. It is interesting to note that just a decade ago, intelligent packaging was not an attractive concept since package devices and computer networks were expensive and quite limited. Today, more powerful and affordable information technology has created a favorable environment for intelligent packaging to flourish.

Supply Chain Cycle

Figure 15.3 Material flow (→) and information flow (⚡) in the food supply chain. From Ref. [1].

A challenging question to the food packaging scientist or technologist is whether more efficient delivery of safe and quality food products can also be achieved by superimposing an additional layer of *food data* and *food models* (capitalized in Figure 15.4) on the information highway of the food supply chain, the goal of intelligent packaging. The food data refers to data that are indicative of food quality and safety (such as time-temperature history, microbial count, pH, water activity), and the food models refer to scientific principles or heuristic rules for processing the food data to enable sound decision making. The answer to this question is likely positive, although significant research and development is needed before the potential of intelligent packaging for enhancing food quality and safety could be fully realized.

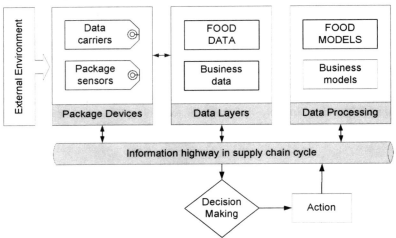

Figure 15.4 Conceptual framework of intelligent packaging system. From Ref. [1].

15.6 Smart Packaging Devices

Smart package devices are defined here as small, inexpensive labels or tags that are attached onto primary packaging (e.g., pouches, trays, and bottles), or more often onto secondary packaging (e.g., shipping containers), to facilitate communication throughout the supply chain so that appropriate actions may be taken to achieve desired benefits in food quality and safety enhancement. There are two basic types of smart package devices: *data carriers* (such as barcode labels and radio frequency identification (RFID) tags) that are used to store and transmit data, and *package indicators* (such as time-temperature indicators, gas indicators, biosensors) that are used to monitor the external environment and, whenever appropriate, issue warnings. As shown in Figure 15.4, these devices provide a communication channel between the external environment and other components in the system. These devices differ from each other not only in 'hardware' (physical makeup), but also in the amount and type of data that can be carried and how the data are captured and distributed. In a typical intelligent packaging system, multiple smart package devices are employed at several strategic locations throughout the supply chain. Table 15.3 lists categories of smart packaging devices.

Table 15.3 Categories of smart packaging devices with their functions

Category	Functions
TTI (Time temperature integrator or indicator)	Food quality monitoring
Leak or gas indicator	Food quality monitoring, temper-evidence
Freshness indicator	Food quality monitoring
Bar code	Tracking, logistic control
Electronic identification tags	Antitheft, brand protection, tracking, logistic control, information flow, tamper-evidence

15.6.1 Time Temperature Integrator or Indicator

Temperature is usually the most important environmental factor influencing the kinetics of physical and chemical deteriorations, as well as microbial growth in food products. Time-temperature indicators (TTIs) are typically small self-adhesive labels attached onto shipping containers or individual consumer packages. TTI changes its surface color responding to temperature history experienced during distribution and storage, which is particularly useful for warning on the exposure of chilled or frozen food products to temperature abuses. They are also used as 'freshness indicators' for estimating the remaining shelf life for perishable products. The color change of TTI comes from irreversible evidence of a physical or physiochemical change, embedded in the label or tag. It informs the supply chain and consumer that correct storage conditions have or have not been met. It can give reassurance and real information rather than fixed sell-by date of printed form to consumers.

There are three basic types of commercially available TTIs: critical temperature indicators, partial history indicators, and full history integrators [34, 35]. Critical temperature indicators indicate only whether the food package has been exposed to outside the allowed temperature range or remained in the desired temperature range: some indicators show irreversible color change above a threshold temperature and recently reversible thermochromic ink-print is used to show the visible sign of whether the beverage bottles or cans stay in desired cold temperature range. Partial history indicators respond and integrate the temperature effect only above (or below) a certain critical temperature. Full history indicators respond to temperature continuously and integrate the response to show the cumulative quality change effect dependent on temperature history. The responses of these labels are usually some visually distinct changes that are temperature dependent, such as an increase in color intensity and diffusion of a dye along a straight path. Their operating principles and performance have been reviewed extensively in the literatures [35-37].

On the commercial level of TTI, there are several products available. Fresh-Check® of Lifelines Technologies (Morris Plains, New Jersey, USA) is based on the polymer compounds that change color by polymerization of diacetylenic monomers whose rate is dependent on temperature. MonitorMark® of 3M (St. Paul, MN, USA) consists of blotter pack and track separated by a polyester film. Chemicals of specific melting points and blue dye are included into the paper blotter pad. The indicator is activated by the melting of the chemicals and the removal of the polyester film. The diffusion of melted chemicals with blue color onto the track is used as a measure of quality change. Vitsab® TTI (Belmont, NC, USA) is based on enzymatic hydrolysis of a lipid substrate, which causes pH decrease. The enzyme and substrate separated in different compartment are mixed by mechanical activation and reacted to produce fatty acid changing the color from deep green to bright yellow with dependence on temperature. Recently a couple of new TTI products have been introduced commercially: OnVu® TTI of Fresh Point based on color change of photochromic organic crystal activated by ultraviolet ray; TT Sensor® of Avery Dennison activated by attachment of another clear film.

15.6.2 Leak, Gas, or Other Indicators

The gas composition in the package headspace often changes as a result of the activity of the food product, the nature of the package, or the environmental conditions. For example, respiration of fresh produce, gas generation by spoilage microorganisms, or gas transmission through the packaging material or package leaks, may cause the gas composition inside the package to change. Gas indicators in the form of a package label or printed on packaging films can monitor changes in the gas composition, thereby providing a means of monitoring the quality and safety of food products.

Oxygen indicators are the most common gas indicator for food packaging applications, since oxygen in air can cause oxidative rancidity, color change, and microbial spoilage. Gas indicators for oxygen are used to detect a package leak, improper sealing, and quality loss of modified atmosphere packages. A number of oxygen indicators are designed to show color changes due to leaking or tampered packages. Ahvenainen et al. [38] and Smiddy et al. [39] used oxygen indicators to detect improper sealing and quality deterioration of modified atmosphere packages containing pizza and cooked beef, respectively.

Some indicators can tell whether the package has been subjected to unacceptable shock during its handling and distribution. Some laminates mostly consisted of liquid crystal or crystalline polymer change reflectivity in response to an electrical charge or mechanical movement [37]. They can show the record of tampering, bending, or mechanical shock.

15.6.3 Freshness Indicators

Freshness indicators indicate directly the quality of packaged food, mostly related to the microbial growth and metabolism [40]. An ideal freshness indicator is one showing the spoilage or freshness loss of the food contained in the package. The freshness indicator is based on the mechanism to detect the metabolites indicating the microbial quality of the food. In order to justify the indicator mechanism, sound evidence should be provided that there is close correlation between these metabolites and relevant food quality. The potential target molecules of freshness indication cover glucose, organic acids such as lactic and acetic acids, ethanol, volatile nitrogen compounds such as ammonia, dimethylamine, and trimethylamine, biogenic amines such as histamine, carbon dioxide, ATP-degradation products, and sulfuric compounds such as H_2S. The increase or decrease of the target compound is used as an indication for freshness. Most of the freshness indictors are based on color change caused by microbial metabolites produced from spoilage. Majority of the proposed concepts are based on color change of pH-dyes, which change color in gaseous contact with volatile compounds produced from food spoilage. A typical example is pH-dye based indicator responding to carbon dioxide which is produced from microbial spoilage of food. A similar example of pH-dye based indicator has been developed as polymeric label to detect and show the ripeness of a fermented vegetable, kimchi [41]. Gas indicators for water vapor, carbon dioxide, ethanol, and hydrogen sulfide, may also be used for indicating freshness. The detailed information on freshness indicator can be found in the literature [40].

Some innovative concepts of detecting pathogenic bacteria by immunochemical reactions have been proposed: the resultant color change in polymeric label or ink is induced by the presence of the bacteria or toxin. A commercial example is Toxin Guard® of Toxin Alert, which is built on polyethylene-based material and detects the pathogenic bacteria by the immobilized antibodies. Biosensors and electronic nose also have potential for use as freshness indictor, although commercial biosensors for intelligent packaging are not available presently. For example, SIRA Technologies (Pasadena, CA) is developing a biosensor/barcode called Food Sentinel System® to detect pathogens in food packages [42]. In this system, a specific-pathogen antibody is attached to a membrane forming part of the barcode; the presence of contaminating bacteria will cause the formation of a localized dark bar, rendering the barcode unreadable upon scanning.

15.6.4 Bar Code

Barcodes are the least expensive and most popular form of data carriers. The UPC (Universal Product Code) barcode was introduced in the 1970s and has since become ubiquitous in the grocery store for facilitating inventory control, stock reordering, and checkout [43]. Bar code performs the tracking function in itself. To enable barcodes to communicate with scanners and printers, many standards have been developed over the years into commonly accepted languages known as 'symbologies', although only a few of them are widely used today. Some symbologies are described in other part of this book (Section 10.4.2). To address the growing demand for encoding more data in a smaller space, new families of barcode symbologies are recently being introduced. As scanners are becoming more powerful and affordable, two-dimensional barcodes are also gaining popularity. They allow the encoding of additional information not possible with linear barcodes, such as nutritional information, cooking instructions, website address of food manufacturer, and even graphics. With the advent of wireless handheld barcode scanners, barcode scanning is finding new and innovative applications. For example, Bioett (Lund, Sweden) has developed a TTI/barcode system in which data may be read by a handheld scanner, displayed on a computer monitor, and downloaded into a database for analysis.

15.6.5 Electronic Identification Tags

Development of low-cost electronics has opened a brand new area of intelligent packaging including electronic article surveillance (EAS), electromagnetic identification (EMID), and radio frequency identification (RFID) systems. An electronic article surveillance (EAS) device identifies its presence in the reading zone of a reader to protect from theft. The EAS systems operate from a simple principle that a transmitter sends a signal at defined frequencies to a receiver in a surveillance area. EAS devise can reveal only an item's presence and has memory of 1 bit. Even though vast numbers of 1-bit transponders are used in EAS to protect goods in shops and businesses, there are only limited applications of EAS in food packaging. Electromagnetic identification (EMID) systems are magnetically neutral tags which can be easily coded with information by using magnetic encryption. The tag gives each piece of plastic packaging a unique, unalterable, and secure identity number that can be read from a distance of 20 mm by a hand-held scanner or by conveyor belt mounted, automatic readers. The data are either downloaded or fed directly into a process control or data logging system. An RFID system uses data communication using radio frequency wave between a sensing device and an RFID tag. A typical RFID consists of a host system and RFID equipment (interrogator and transponders). An interrogator often called reader emits radio waves to capture data from a transponder called as RFID tag, and the data is then passed onto a host computer (which may be connected to a local network or to the Internet) for analysis and decision making. The RFID tag is attached on the packages and contains internally a minuscule microchip connected to a tiny antenna. RFID has been available for many years for tracking expensive items and livestock, and has begun its broad application in packaging recently. Because of its great potential, there has been an overwhelming interest from major retailers, companies, government agencies, and researchers in using RFID tags in product packaging and distribution. In some forecasts, low cost RFID tags are expected to eventually move into the bar code region [37, 44]. Therefore this section will also describe the RFID in more detail.

In a typical RFID system, RFID tags may be classified into two types: *passive tags* that have no battery and are powered by the energy supplied by the reader, and *active tags* that have their own battery for powering the microchip's circuitry and broadcasting signals to the reader. The more expensive active tags have a reading range of 30 meters or more, while the less expensive passive tags have a reading range of up to 4.5 m. The actual reading range depends on many factors including the frequency of operation, the power of the reader, and the possible interference from metal objects. An RFID tag with antenna and basic features costs about \$0.10 when produced in tens of millions, while the cheapest chips with read-only capacity have the price of \$0.05 [37]. More expensive active tag may cost as much as \$75 each, but the cost is expected to fall considerably and rapidly with technology development and the production volume increase [1, 44].

Compared to the barcode, the RFID tag has several unique characteristics. Line-of-sight is usually not required: that is, the RFID tag does not need to be oriented towards the reader for data transfer to occur because radio waves travel through a wide array of nonmetallic materials. A significantly larger data storage capacity is available (up to 1 MB for high-end RFID tags), which may be used to store information such as temperature and relative humidity data, nutritional information, and cooking instructions. Read-write operations are supported by some RFID tags, which are useful in providing real-time information updates as the tagged items move through the supply chain. Multiple RFID tags may be read simultaneously at a rapid rate. An RFID tag may also be integrated with a time-temperature indicator or a biosensor to carry time-temperature history and microbiological data. KSW Microtec (Dresden, Germany) has developed a battery powered TTI/RFID tag using a technology in which thin-film batteries are printed onto a flexible substrate. Infratab (Oxnard, California) is also developing a battery powered TTI/RFID tag. Unlike the traditional TTI that is based on diffusion or a biochemical reaction, the TTI/RFID tag uses a microchip to sense and integrate temperature over time to determine the shelf life of a product. Advances in smart ink and printing technology are expected to allow the electronic tags to be printed as integrated circuit and display on the label: the polymer material dissolved in appropriate solvents may be printed like electronic ink to form the necessary structure [37].

RFID technology is still at its early stages of implementation, and at present the focus is on simple tasks such as product identification and tracking, and not on complicated matters that involve the application of scientific food principles. However, its potential in intelligent food packaging is enormous. The integration of food science knowledge will be required to develop the necessary decision support system for enhancing food safety and quality in rapidly changing food supply chain. Already several major retailers have started to ask their leading suppliers to use RFID tags on shipping crates and pallets. Marks & Spencer, a supermarket chain, has launched the rollout of RFID tagging of 3.5 million returnable food produce delivery trays [44]. When the RFID technology will become more established and be integrated with other technologies including food science, there will appear self-powered smart labels with polymer transistor circuits which help the logistic control, ensure food safety, and are compatible with environmental friendly packages.

15.7 Legislative and Human Behavior Issues

Adoption of active and intelligent packaging is so different with regional areas and countries. While sachets of active packaging such as oxygen absorbers and ethanol emitters have widely been accepted in Japan and other oriental countries, they have penetrated western market only to a limited extent mostly in modified form of label or card attached to internal package surface. On the contrary, TTI has been widely used in Europe for chilled foods, whereas it has only very limited introduction in oriental countries as Japan and Korea. Legislations and human behavior coming from cultural background affect the acceptance and use of active and intelligent packaging. In Japan and Korea, people do not have any objection against inedible sachet packed with food inside the package. There are little social concerns about the risk of ingesting the active packaging sachet. Use of active packaging can be covered in general food packaging regulation of those countries. In the USA and Europe, the food regulations related to packaging are more strict making active packaging difficult to be adopted [12, 45, 46]. In the USA, any active or intelligent packaging should pass the process of food additive petition and/or food contact substance notification in same way for passive packaging material. In Europe, substance potentially releasable from active packaging should be on the positive list and satisfy the specific and overall migration limits. For regulations with food packaging migration in the USA and Europe, please refer to Section 5.7. In those regions, consumers are more cautious about introduction of active packaging sachet, but are active to accept TTI label in consideration of food safety improvement [47].

In the regulations of Australia and New Zealand, there are provisions for using sachets of moisture absorber, ethanol emitter, and flavor releaser. In Europe, there have been attempts to incorporate active and intelligent packaging in the regulations after a multinational research project, Actipak [45]. Considerations given for active and intelligent packaging include extension of positive list, allowance of higher overall migration, appropriate migration test condition, compliance with food additive usage, inhibition of flavoring to hide food spoilage, preventive measure against swallowing the sachet, and change of labeling requirements on product weight, shelf life, effective function, and warning message for safety. In order to introduce successfully the new concepts of active and intelligent packaging in the market, substantial information campaign supporting their benefit and function may be required in parallel with legislative change [45].

15.8 Case Study: Intelligent Microwave Oven with Barcode Reader

The intelligent packaging concept has been applied to cooking appliances such as the intelligent microwave oven system shown in Figure 15.5 [48]. The uniqueness of this system is in its use of information sharing to enhance food quality and convenience. The PDF 417 barcode in the package carries data about the food product, and the data processing system generates the proper heating instructions for the microwave oven. Information exchange also occurs through the user interface (such as a touch screen or voice recognition system) and the Internet.

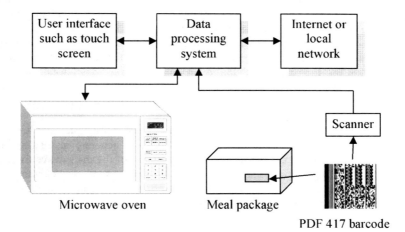

Figure 15.5 Intelligent microwave oven system

Since microwave ovens come in different sizes and power outputs, the heating instructions printed on microwaveable food packages are obviously vague in order to accommodate the many different ovens in the market. Using these vague instructions often does not allow achievement of good food quality. This problem may be overcome by scanning the bar code to enable the decision support system to match the microwave oven and the food package. To achieve a higher level of quality, temperature and moisture sensors may be placed inside the microwave oven to provide feedback for the data processing system. Part of this concept has been commercialized by Smart Oven® of Samsung Electronics whose scanner swipes the PDF 417 barcode on the food package to extract the cooking instruction: there has been collaboration between food companies and the microwave oven manufacturer for printing the bar code of optimal cooking mode on the food packages.

Scanning of the bar code eliminates the need for entering heating instructions. This is particularly useful for microwave/convective ovens that use complicated instructions. These combination ovens are capable of providing higher food quality than the conventional microwave ovens, but their heating instructions involve multiple steps since both microwave energy and convective heat are employed. Scanning of the bar code is also helpful for people who are visually impaired or have difficulties in understanding the language. The Internet connection provides convenient access to information relating to the packaged food such as the manufacturer's website, recipes, food allergen information, and product recall.

The development of the intelligent microwave oven system requires the application of scientific knowledge to design the data layers and the data processing system. The data layers should contain information relating to the food, the package, and the microwave oven. The architecture of the data layers could be rather complicated since data may enter the system in different ways. For example, the food manufacturer may encode the food and packaging information in the bar code, the oven manufacturer may store the oven information in a database connected to the information processing system, the consumer may enter his or her preferences through the touch screen, and information may be exchanged via the Internet. The data processing system should include algorithms that are based on heat transfer principles and heuristic rules relating to food quality and safety, in order to generate instructions for controlling the magnetron and turntable (if available) of the microwave oven. In addition, the algorithms may have other abilities such as obeying the requirement or preferences of the consumer, warning the consumer against food allergens, and tracking the dietary intake of the consumer.

Discussion Questions and Problems

1. Show how you can obtain about 300 mL O_2 per gram of iron as an estimation of iron's capacity to absorb the oxygen. Use the chemical reaction of Equation 15.1.

2. How can you estimate the capacity of calcium hydroxide to absorb the carbon dioxide? Refer to chemical reaction of Equation 15.3.

3. What type of active packaging would you like to use for fermented yogurt?

4. What kind of active packaging technology can be used in order to prevent the modified atmosphere packages of fresh-cut vegetables from becoming anoxic? What kind of gas indicator may be useful in having consumers recognize the anoxic packages?

5. What are the possible disadvantages and risks of intelligent packaging adopting RFID technology?

6. Find the most appropriate TTI product for packages of fresh fish fillets. You may first investigate the quality deterioration mode and rate of the food product in the literature, and then examine the rate and temperature dependence of TTI products currently available or developed by some researchers. Commercial TTI information can be found in Internet web sites of the producers, and new developments can be found in scientific literatures. Studying shelf life estimation principle in Chapter 16 would be helpful.

Bibliography

1. KL Yam, PT Takhistov, J Miltz. Intelligent packaging: concepts and applications. J Food Sci 70:R1-10, 2005.
2. J Miltz, N Passy, CH Mannheim. Trends and applications of active packaging systems. In: P Ackerman, M Jagerstad, M Ohlsson, ed. Food and Packaging Materials - Chemical Interaction. The Royal Society of Chemistry, 1995, pp. 201-210.
3. R Ahvenainen. Active and intelligent packaging, an introduction. In: R Ahvenainen, ed. Novel Food Packaging Techniques. Cambridge, UK: Woodhead Publishing Ltd., 2003, pp. 5-21.
4. AL Brody, ER Strupinsky, LR Kline. Active Packaging for Food Applications. Lancaster: Technomic Publishing, 2001, pp. 11-130.
5. L Vermeiren, F Devlieghere, M van Beest, N de Kruijf, J Debevere. Developments in the active packaging of foods. Trends Food Sci Tech 10:77-86, 1999.
6. ML Rooney. Active packaging in polymer films. In: ML Rooney, ed. Active Food Packaging. London: Blackie Academic and Professional, 1995, pp. 74-110.
7. P Appendini, JH Hotchkiss. Review of antimicrobial food packaging. Innov Food Sci Emerg 3:113-126, 2002.
8. P Suppakul, J Miltz, K Sonneveld, SW Bigger. Active packaging technologies with an emphasis on antimicrobial packaging and its applications. J Food Sci 68:408-420, 2003.
9. J Miltz, P Hoojjat, JK Han, JR Giacin, BR Harte, IJ Gray. Loss of antioxidants from high-density polyethylene: its effect on oatmeal cereal oxidation. In: JH Hotchkiss, ed. Food and Packaging Interactions. Washington D.C.: American Chemical Society, 1988, pp. 83-93.

10. C Wessling, T Nielsen, JR Giacin. Antioxidant ability of BHT- and α-tocopherol impregnated LDPE film in packaging of oatmeal. J Sci Food Agr 81:194-201, 2000.
11. ML Rooney. Oxygen-scavenging packaging. In: JH Han, ed. Innovations in Food Packaging. Amsterdam: Elsevier Academic Press, 2005, pp. 123-137.
12. N Waite. Active Packaging. Surrey, United Kingdom: Pira International, 2003, pp. 1-110.
13. JP Smith, J Hoshino, Y Abe. Interactive packaging involving sachet technology. In: ML Rooney, ed. Active Food Packaging. London: Blackie Academic and Professional, 1995, pp. 143-173.
14. L Vermeiren, L Heirlings, F Devlieghere, J Debevere. Oxygen, ethylene and other scavengers. In: R Ahvenainen, ed. Novel Food Packaging Techniques. Cambridge, UK: Woodhead Publishing, 2003, pp. 22-49.
15. A Shirazi, AC Cameron. Controlling relative humidity in modified atmosphere packages of tomato fruit. Hortscience 27:336-339, 1992.
16. DH Shin, HS Cheigh, DS Lee. The use of Na₂CO₃-based CO₂ absorbent systems to alleviate pressure buildup and volume expansion of kimchi packages. J Food Eng 53:229-235, 2002.
17. DS Lee, DH Shin, DU Lee, JC Kim, HS Cheigh. The use of physical carbon dioxide absorbents to control pressure buildup and volume expansion of kimchi packages. J Food Eng 48:183-188, 2001.
18. K Yoshida, Y Hiro, J Kokubo, C Nishizawa, S Watanabe. Oxygen generating materials, carbon dioxide absorbing materials, and transport system and transport method of live fishery products. US Patent 6612259, 2003.
19. DS Lee, HD Paik. Use of a pinhole to develop an active packaging system for kimchi, a Korean fermented vegetable. Packag Technol Sci 10:33-43, 1997.
20. D Zagory. Ethylene-removing packaging. In: ML Rooney, ed. Active Food Packaging. London: Blackie Academic & Professional, 1995, pp. 38-54.
21. NFF Soares, JH Hotchkiss. Bitterness reduction in grapefruit juice through active packaging. Packag Technol Sci 11(1):9-18, 1998.
22. L Franzetti, S Martinoli, L Piergiovanni, A Galli. Influence of active packaging on the shelf-life of minimally processed fish products in a modified atmosphere. Packag Technol Sci 14(6):267-274, 2001.
23. B Bjerkeng, M Sivertsvik, JT Rosnes, H Bergslien. Reducing package deformation and increasing filling degree in packages of cod fillets in CO₂-enriched atmospheres by adding sodium carbonate and citric acid to an exudate absorber. In: P Ackerman, M Jagerstad, M Ohlsson, ed. Foods and Packaging Materials-Chemical Interactions. Cambridge, UK: Royal Society of Chemistry, 1995, pp. 222-227.
24. D Chung, ML Chikindas, KL Yam. Inhibition of *Saccharomyces cerevisiae* by slow release of propyl paraben from a polymer coating. J Food Protect 64:1420-1424, 2001.

25. Y Chi-Zhang, KL Yam, ML Chikindas. Effective control of *Listeria monocytogenes* by combination of nisin formulated and slowly released into a broth system. Int J Food Microbiol 90:15-22, 2004.
26. JH Han. Antimicrobial food packaging. In: R Ahvenainen, ed. Novel Food Packaging Techniques. Cambridge, UK: Woodhead Publishing Ltd., 2003, pp. 50-70.
27. DS Lee. Packaging containing natural antimicrobial or antioxidative agents. In: JH Han, ed. Innovations in Food Packaging. Amsterdam: Elsevier Academic Press, 2005, pp. 108-122.
28. YC Ho, KL Yam, SS Young, PF Zambetti. Comparison of vitamin E, Irganox 1010 and BHT as antioxidants on release of off-flavor from HDPE bottles. J Plast Film Sheet 10:194-212, 1994.
29. PF Zambetti, SL Baker, DC Kelley. Alpha tocopherol as an antioxidant for extrusion coating polymers. Tappi J 78:167-171, 1995.
30. C Wessling, T Nielsen, A Leufven, M Jagerstad. Retention of a-tocopherol in low-density polyethylene (LDPE) and polypropylene (PP) in contact with foodstuffs and food-simulating liquids. J Sci Food Agr 79:1635-1641, 1999.
31. CH Lee, DS An, SC Lee, HJ Park, DS Lee. A coating for use as an antimicrobial and antioxidative packaging material incorporating nisin and a-tocopherol. J Food Eng 62:323-329, 2004.
32. HS Anderson. Controlled atmosphere package. US Patent 4842875, 1989.
33. JH Yun, DS An, KE Lee, BS Jun, DS Lee. Modified atmosphere packaging of fresh produce using microporous earthenware material. Packag Technol Sci 19:269-278, 2006.
34. RP Singh. Scientific principles of shelf-life evaluation. In: D Man, A Jones, ed. Shelf-life Evaluation of Food. Gaithersburg, Maryland: Aspen publishers, 2000, pp. 3-22.
35. PS Taoukis, TP Labuza. Time-temperature indicators (TTIs). In: R Ahvenainen, ed. Novel Food Packaging Techniques. Cambridge UK: Woodhead Publishing Limited, 2003, pp. 103-126.
36. JD Selman. Time-temperature indicators. In: ML Rooney, ed. Active Food Packaging. New York: Blackie Academic & Professional, 1995, pp. 215-237.
37. D Collins. Intelligent Packaging. Leatherhead: Pira International, 2003, pp. 1-72.
38. R Ahvenainen, M Eilamo, E Hurme. Detection of improper sealing and quality deterioration of modified-atmosphere-packed pizza by a colour indicator. Food Control 8:177-184, 1997.
39. M Smiddy, M Fitzgerald, JP Kerry, DB Papkovsky, CK O' Sullivan, GG Guilbault. Use of oxygen sensors to non-destructively measure the oxygen content in modified atmosphere and vacuum packed beef: impact of oxygen content on lipid oxidation. Meat Sci 61:285-290, 2002.
40. M Smolander. The use of freshness indicators in packaging. In: R Ahvenainen, ed. Novel Food Packaging Techniques. Cambridge, UK: Woodhead Publishing Ltd., 2003, pp. 126-143.

41. S-I Hong, W-S Park. Use of color indicators as an active packaging system for evaluating kimchi fermentation. J Food Eng 46:67-72, 2000.
42. CE Ayala, DL Park. New bar code will help monitor food safety. Louisiana Agri 43(2):8, 2000.
43. V Manthou, M Vlachopoulou. Bar-code technology for inventory and marketing management systems: A model for its development and implementation. Int J Prod Econ 71:157-164, 2001.
44. Anonymous. The RFID in Retail Manual 2005. Leatherhead: Plra International, 2004, pp. 1-14.
45. ND Kruijf, R Rijk. Legislative issues relating to active and intelligent packaging. In: R Ahvenainen, ed. Novel Food Packaging Techniques. Cambridge, UK: Woodhead Publishing, 2003, pp. 459-496.
46. YS Song, MA Hepp. US Food and Drug Administration approach to regulating intelligent and active packaging components. In: JH Han, ed. Innovations in Food Packaging. Amsterdam: Elsevier Academic Press, 2005, pp. 475-481.
47. L Lahteenmaki, A Arvola. Testing consumer responses to new packaging concepts. In: R Ahvenainen, ed. Novel Food Packaging Techniques. Cambridge, UK: Woodhead Publishing, 2003, pp. 550-562.
48. KL Yam. Intelligent packaging and the future smart kitchen. Packag Technol Sci 13:83-85, 2000.

Part Three

Packaging Food Science

The next two chapters are devoted to food science principles and knowledge most relevant to food packaging and their applications to practical problems. Chapter 16 discusses deterioration kinetics of foods and the role of packaging system to control or retard food deteriorations, thereby extending the shelf life of food. It elucidates conceptually and quantitatively the relationships between the shelf life of packaged food and important variables relating to the food, the package, and the environment. Chapter 17 discusses the stability of various food categories and their packaging requirements. When appropriate, knowledge from previous chapters is applied to suggest packaging strategies to extend the shelf lives of food products.

Chapter 16

Shelf Life of Packaged Food Products

16.1 Basic Concepts 480
    16.1.1 Definitions of Shelf Life ··480
    16.1.2 Factors Affecting Shelf Life of Packaged Foods ··············482
16.2 Food Factors Affecting Shelf Life...483
    16.2.1 Food Deterioration Modes···483
    16.2.2 Package Dependent versus Product Dependent Deteriorations ·····485
    16.2.3 Quality Indexes and Critical Limits································486
    16.2.4 Sensory Quality ···487
    16.2.5 Microbial Count ··488
    16.2.6 Losses in Nutrients and Pigments·····································490
    16.2.7 Production of Undesirable Components ·····························492
    16.2.8 Physical Changes and Processes·······································492
16.3 Kinetics of Food Deterioration..493
    16.3.1 Chemical Kinetics: Reaction Order and Rate Constant ···········493
    16.3.2 Microbial Growth Model···497
    16.3.3 Other Kinetic Model Associated with Physical Changes ···········498
    16.3.4 Temperature Dependence: Arrhenius and Shelf Life Plots ········499
    16.3.5 Moisture Dependence ··502
    16.3.6 Oxygen Dependence··506
16.4 Environmental Factors Affecting Shelf Life 507
    16.4.1 Ambient Environment ··507
    16.4.2 Physical Environment···510
    16.4.3 Human Environment ··512
16.5 Package Factors Affecting Shelf Life...512
    16.5.1 Packaging Parameters Important to Shelf Life ····················512
    16.5.2 Permeation versus Reaction Controlled Shelf Life·················513
    16.5.3 Package Interactions ··514
16.6 Shelf Life Studies 515
    16.6.1 Testing Under Normal Conditions ····································515
    16.6.2 Testing Under Accelerated Conditions·······························516
    16.6.3 Procedures for Shelf Life Studies ·····································517
16.7 Shelf Life Models...521
    16.7.1 Shelf Life of Chemical and Microbial Deteriorations ············522
    16.7.2 Shelf Life Models of Constant H_2O and O_2 Driving Forces ········524
    16.7.3 Shelf Life Model of Variable H_2O Driving Force ·················527
    16.7.4 Shelf Life Model of Variable O_2 Driving Force ··················531
    16.7.5 Package/Food Compatibility Dependent Shelf Life Model··········533

16.8   Case Study: Shelf Life of Potato Chip with Two Interacting Quality
Deterioration Mechanisms 534
Discussion Questions and Problems 536
Bibliography 540

16.1 Basic Concepts

While shelf life is a popular term in the food industry, its concept is seldom well understood. This chapter introduces the basic concepts of shelf life useful for evaluating existing or developing new food products. Matching the packaging to the desired shelf life is necessary for quality assurance, and food processors in many countries have a legal responsibility to ensure that the shelf lives on their package labels are accurate [1-3].

16.1.1 Definitions of Shelf Life

Shelf life may be defined in different ways depending on the specific purposes. Here are two definitions:

- Shelf life is the period of time from the production and packaging of a product to the point at which the product first becomes unacceptable under defined environmental conditions.

- Shelf life is the period of time during which a food retains acceptable characteristics of flavor, color, aroma, texture, nutritional value, and safety under defined environmental conditions. It also encompasses the time during which a product may reasonably be expected to meet its label claims for specific attributes, e.g., vitamin supplementation.

The first is a general propose definition which gives a brief description of shelf life. The second is a definition which provides richer technical information relating shelf life to specific food attributes. In these definitions, the key word is 'acceptable' or 'unacceptable'—selecting good criteria to determine the acceptability of a new food product is often subjective and challenging.

In the product development process, it is also useful to define two related shelf lives below:

- Required shelf life is the minimum shelf life that a food product must have in order to become viable in the marketplace. It depends primarily on marketing and logistics considerations, which in turn determine where and how the product is shipped and thus the distribution environments the product will likely encounter. This shelf life is typically the sum of the time for the product to travel from the manufacturing plant to the grocery store, plus the time for the product to display in the store, and the time for the product to stay at the consumer's home before consumption.

■ Product shelf life is the period of time during which the quality of the food product remains acceptable. This shelf life depends primarily on the deterioration kinetics and stability of the food product and the environment encountered by the food product during storage and distribution. While a product should be designed such that its product shelf life meets or exceeds the required shelf life, care should also be exercised not to over-package and unnecessarily increase the packaging cost.

What if the product shelf life is shorter than the required shelf life (for example, the shelf life expires before the product has been sold)? In situations where food deteriorations are driven by oxygen or moisture from the ambient environment, better packaging may be used to increase the product shelf life. This may be accomplished using better barrier materials to protect the food product from oxygen and moisture, or using packaging technologies such as modified atmosphere packaging or oxygen absorber to improve the internal package environment thereby extending shelf life. In situations where food deteriorations are not affected by oxygen, moisture, light, or airborne microbes in air, packaging is not an effective solution and other approaches should be sought. Reformulating the ingredients in the food product, or using different processing techniques or conditions, may sometimes improve product stability and thus extend product shelf life. Redesigning the distribution route and thus the ambient environment experienced by the food product is also a possible solution.

Shelf life dating for food package labels has become increasingly important due to the demands from the consumer and regulatory agencies. For perishable foods, sell-by-date or use-by-date is mandatory in many countries [3]. For nonperishable foods, many food processors voluntarily display best-use-before-date, best-quality-life, best-before-date, or date-of-minimum-durability on their package labels. For food product with sell-by-date, it should remain acceptable for a certain period of time after the sell-by-date to allow the consumer to use the product.

This chapter is focused on shelf life driven primarily by biochemical deterioration leading to undesirable microbial growth, off-flavors, textural change, and nutritional loss. Biochemical deterioration depends on biochemical stability of food product as well as environmental factors such as temperature, relative humidity, oxygen, and light. It is important to mention that a food product may also become unacceptable due to physical damages such as accidentally crushing the food product or breaking the package seal. The extent of possible physical damages depend on fragility of the product and its package, as well as physical environment related to shock, vibration, and handling; however, shelf life associated with physical damages is mentioned only briefly in this chapter.

16.1.2 Factors Affecting Shelf Life of Packaged Foods

A key concept shown by the straight arrows in Figure 16.1 is that shelf life of a packaged food product is affected by three major factors—the food, package, and environment. The dotted double-arrow curves indicate that these factors are sometimes related: for example, certain foods require certain packages and certain packages are suitable only for certain foods, food stability is affected by environmental conditions (relating to distribution logistics) and thus the environmental conditions may be manipulated to improve product shelf life, the type of package should match the type of distribution environment and vice versa.

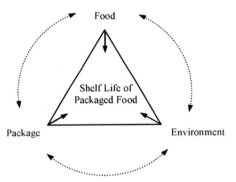

Figure 16.1 Factors affecting shelf life of packaged food

These three factors have already been introduced in Section 1.6.3. Their importance to shelf life is summarized below and will be further discussed later in this chapter.

- Food factor. The important considerations are how stable is the food product, what major deterioration modes cause the end of shelf life, and how fast those deteriorations occur under the distribution environment of interest. It is also important to identify quality indexes to measure the extent of food deterioration, as well as defining the levels at which those quality indexes are no longer unacceptable.

- Environmental factor. It is important to understand how environmental factors affect product shelf life. In Section 1.5, three package environments have been introduced— ambient environment, physical environment, and human environment. While all of them are important, this chapter is focused primarily on the effects of ambient environment on product shelf life, especially the effects of temperature and relative humidity on the kinetics of food deterioration.

- Package factor. The important consideration is to determine to what extent the package can affect product shelf life. For oxygen and moisture sensitive foods, the package is necessary to protect the food product from oxygen and water vapor from the ambient environment, and package parameters such as permeability, surface area, and thickness are important considerations. Packaging is also important to protect the food product from physical damages.

16.2 Food Factors Affecting Shelf Life

Among the three factors illustrated in Figure 16.1, the food product is most important. Below are some important questions to ask about the food product:

- What are the major deterioration modes which cause the end of product shelf life?

- What food products can benefit from packaging for improving shelf life?

- What quality indexes can be used to measure or quantify the rates of those deterioration modes?

- Below what levels the quality indexes are no longer acceptable to the consumer?

The information provided in the next few sections will help to answer these questions.

16.2.1 Food Deterioration Modes

The expiration of shelf life is frequently caused by some deterioration in the food product leading to unacceptable food quality and safety issues. Food deterioration modes can be categorized into physical, chemical, and microbiological [4]. Table 16.1 shows the typical deterioration modes and shelf lives of some food products, as well as environmental factors critical to food deterioration rates.

A food product, especially if it contains multiple ingredients, may have many deterioration modes, but it is not necessary to deal with all possible deterioration modes. Instead, efforts should be made to identify one or two major deterioration modes which occur most rapidly, because the product shelf life is limited by those fast deterioration modes, not by the slow ones. In some situations, two different deterioration modes may be related; for example, microbial growth may sometimes causes oxidative spoilage in food. The deterioration modes of a food product may be identified using a sensory panel or based on prior knowledge of the food product.

Table 16.1 Major modes of deterioration, critical environmental factors, and shelf lives of some food products

Food product	Deterioration modes (assuming intact package)	Critical environmental factors	Shelf life (typical)
Perishables			
Milk and dairy products	Bacterial growth, oxidized flavor, hydrolytic rancidity	Oxygen, temperature	7-30 days at 0 to 7°C
Fresh bakery products	Staling, microbial growth, moisture loss causing hardening, oxidative rancidity	Oxygen, temperature, humidity	2 days (bread) 7 days (cake)
Fresh red meat	Bacterial growth, loss of red color	Oxygen, temperature, light	3-4 days at 0 to 7°C
Fresh poultry	Bacterial growth, off-odor	Oxygen, temperature, light	2-7 days at 0 to 7°C
Fresh fish	Bacterial growth, off-odor	Temperature	3-14 days when stored on ice (marine fish)
Fresh fruits and vegetables	Respiration, compositional changes, nutrient loss, wilting, bruising, microbial growth	Temperature, relative humidity, light, oxygen, physical handling	
Semi-perishables			
Fried snack food	Rancidity, loss of crispness, breakage	Oxygen, light, temperature, relative humidity, physical handling	3-12 weeks
Cheese	Rancidity, browning, lactose crystallization, undesirable mold growth	Temperature, relative humidity	4-24 months
Ice cream	Graininess caused by ice or lactose crystallization, texture	Fluctuating temperature (below freezing)	1-4 months
Shelf stable nonperishables			
Dehydrated foods	Browning, rancidity, loss of texture, loss of nutrients	Relative humidity, temperature, light, oxygen	1-24 months
Nonfat dry milk	Flavor deterioration, loss of solubilization, caking, nutrient loss	Relative humidity, temperature	8-12 months
Breakfast cereals	Rancidity, loss of crispness, nutrient loss	Relative humidity, temperature, rough handling	6-18 months
Pasta	Texture changes, staling, vitamin and protein quality loss, breakage	Relative humidity, temperature, light, oxygen, rough handling	9-48 months
Frozen concentrated juices	Loss of cloudiness, yeast growth, loss of vitamin, loss of color or flavor	Oxygen, temperature and thawing	18-30 months

Table 16.1 Continued

Food product	Deterioration modes (assuming intact package)	Critical environmental factors	Shelf life (typical)
Frozen fruits and vegetables	Loss of nutrients, loss of texture, flavor, odor, color, and formation of package ice	Oxygen, temperature, temperature fluctuations	6-24 months
Frozen meats, poultry and fish	Rancidity, protein denaturation, color change, desiccation (freezer burn), toughening	Oxygen, temperature, temperature fluctuations	2-14 months
Frozen convenience foods	Rancidity in meat portions, weeping and curdling of flavor, loss of color, package ice	Oxygen, temperature, temperature fluctuations	6-12 months
Canned fruits and vegetables	Loss of flavor, texture, color, nutrients	Temperature	12-36 months
Coffee	Rancidity, loss of flavor and odor	Oxygen, temperature, light, relative humidity	9-36 months
Tea	Loss of flavor, absorption of foreign odors	Oxygen, temperature, light, humidity	18 months

From Ref. [2].

16.2.2 Package Dependent versus Product Dependent Deteriorations

What deterioration modes or types of foods can be protected against by packaging? To answer this question, it is useful to introduce the concepts of package-dependent deterioration and product-dependent deterioration. Some food products may involve both package-dependent deteriorations and product-dependent deteriorations. Packaging can only protect against deteriorations which are package-dependent.

Package-dependent deteriorations are food deteriorations driven by environmental factors that are controllable by the package or packaging technologies. Those environment factors include oxygen, moisture, microbes, and light from the ambient environment, against which the food product may be protected by using the package as a barrier to separate the product and the external environment. The use of packaging technologies, such as modified atmosphere packaging and oxygen scavenger, may also improve the internal environment of the package for shelf life extension.

Product-dependent deteriorations are food deteriorations driven by intrinsic stability of the food product, not by environmental factors or packaging. Intrinsic stability is dictated by food formulation and processing conditions. Consider, for example, a meat sandwich in which the meat inside the sandwich has a higher water activity than the bread outside. Due to the difference in water activity, moisture migrates from the meat to the bread leading the product to become unacceptable. Wrapping the food product with a high barrier packaging material cannot retard the internal mechanism of moisture migration. A possible solution is to reformulate the product to minimize the difference in water activity; another possible solution is to insert an edible barrier layer between the meat and the bread to retard moisture migration.

16.2.3 Quality Indexes and Critical Limits

Once the major deterioration modes are known, the next task is to identify quality indexes to quantify the extent of deterioration. Any quality factor should possess three characteristics: measurable, reproducible, and relevant. First, the quality index must be measurable by human or instruments. Preferably it should also be measured easily without labor intensive procedures or expensive instruments. Second, measurements of the quality index should have acceptable degrees of reproducibility suitable for the purpose of experiment. Third, the quality index should be relevant, meaning that it is an accurate indication of the deterioration mode and consumer acceptability. Identifying appropriate quality indexes is sometimes difficult for new food products.

Quality indexes may be either subjective (measurable by human) or objective (measurable by instruments). Subjective quality indexes such as sensory scores have the advantage of being most relevant to the consumer acceptability, but they also suffer from the disadvantage that measurements have relatively low degree of reproducibility and are labor intensive to obtain. Objective quality indexes such as amounts of volatile compounds measured by GC/MS have the advantage that measurements have good reproducibility, but they also have the disadvantage that the acquired data may not correlate well to consumer acceptability. In most cases, a combination of subjective and objective quality indexes is best for describing the quality of food product.

To define shelf life, it is also necessary to establish a critical limit for each quality index, below which the food product is no longer acceptable. This concept is illustrated in Figure 16.2, where the curves represent different environmental conditions (such as temperature) or packaging conditions (such as packages with different barrier properties), and t_{s1}, t_{s2}, and t_{s3} are corresponding shelf lives. With more protective packaging or preserving environment, longer shelf life can be obtained (from t_{s3} to t_{s1}). Defining the critical limit is not straightforward since there are many factors such as input from sensory panels, legal requirements, and market viability to consider. The critical limit of a food product may also vary from location to location, especially from country to country, as people from different locations have different preferences and circumstances.

In the next several sections, some common deterioration modes and quality indexes for shelf life determination are discussed.

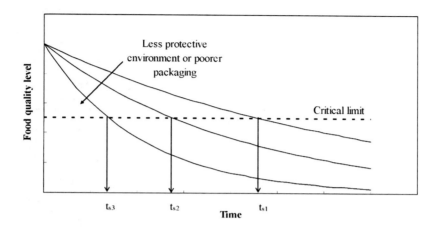

Figure 16.2 Concept of shelf life defined as time to reach the critical limit of quality

16.2.4 Sensory Quality

Sensory attributes are important quality indexes that people use to decide whether a food product is acceptable or not acceptable. Most chemical and physical changes in foods are associated with degenerated sensory quality [5]. Appearance changes are commonly observed on the storage of red meat (browning), fruit juices (darkening), dehydrated products (browning), jams (syneresis), dairy gels (syneresis), and emulsions (separation). Loss of aroma is a problem for products such as bread, spices, and coffee. Off-odor and off-flavor may be caused by microbial spoilage, oxidation, or migration of undesirable compounds from package to food. Undesirable textural changes may occur in fresh fruits and vegetables (softening and wilting), bread (staling), and snack foods (crispness loss).

Response of human senses can be systematically measured by a well-implemented sensory evaluation system, which includes properly selected panel and operation of suitable test method [5]. Sensory analysis in shelf life study is usually based on answering the questions about (1) how long the product can be stored without unacceptable changes and (2) how sensory attributes change with storage time [6]. Figure 16.3 shows some common sensory test procedures, which can be divided into analytical test in the laboratory and consumer test in real world. Discrimination tests are used to determine whether there is a difference between two samples, usually a control and a test sample. Quantitative tests are used for recording the perceived intensity of attributes or degree of likings during the storage. Line and hedonic scales are most frequently used to obtain numeric sensory data.

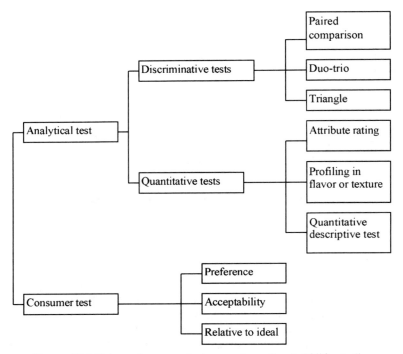

Figure 16.3 Types of sensory test procedures for shelf life studies

16.2.5 Microbial Count

Microbial safety is of prime importance for food products, especially fresh or minimally processed foods. Therefore some guidelines exist for the microbial standard for certain food products. In some countries, mesophilic aerobic bacterial count is set to be below 10^5/g immediately after processing and below 10^6/g at the end of shelf life for cook-chill products [7]. Sometimes there are guidelines dictating the absence of pathogenic microorganisms such as *Salmonella, Campylobacter, Listeria monocytogenes* and *Escherichia coli*. These guidelines differ with food category, country, and agency issuing them. Table 16.2 shows some examples of microbial norms for cook-chill products and fresh-cut vegetables. It should be mentioned that the microbiological norms specified can only be applied within the context of the original guideline and may be varied with time depending on the country and agency.

Table 16.2 Some examples of microbiological norms for cook-chill products and fresh-cut vegetables

Country	Microbiological norms
Cook-chill products	
Netherlands	At the end of shelf life
	For pathogens:
	Salmonella and Campylobacter: absent in 25 g
	Listeria monocytogenes: absent in 0.01 g
	Staphyloccocus aureus: $\leq 10^5$ g^{-1}
	Bacillus cereus: $\leq 10^5$ g^{-1}
	Clostridium perfringens: $\leq 10^5$ g^{-1}
	Mesophilic aerobic count: maximum=10^6 g^{-1}
UK	Immediately before reheating:
	Mesophilic aerobic count: $< 10^5$ g^{-1}
	E. coli: < 10 g^{-1}
	St. aureus (coagulase +ve): 10^2 g^{-1}
	C. perfringens: $< 10^2$ g^{-1}
	L. monocytogenes: absent in 25 g
	Salmonella spp.: absent in 25 g
Fresh-cut vegetables	
France	Salmonella: absent in 25 g
	Coliforms: $<10^3$ g^{-1}; number of positive results or results higher than 10 g^{-1} should be equal to or less than 2 from 5 samples drawn.
	Mesophilic aerobic count: maximum=5×10^7 g^{-1}
Germany	*E. coli*: $< 10^2$ g^{-1}
	Salmonella: absent in 25 g
	L. monocytogenes: $\leq 10^2$ g^{-1}
	Mesophilic aerobic count: maximum=5×10^7 g^{-1}

Adapted from Ref. [7, 8].

Apart from legal aspects, counts in aerobic bacteria, lactic acid bacteria, specific spoilage bacteria, or yeasts are often used to specify the limit of shelf life for the food product which spoils by microbial growth on its surface. Colony forming unit (cfu) per gram sample is measured by spread plate method or pour plate method for use in shelf life determination by microbial criteria. As an example, Figure 16.4 shows an example of the increase in total aerobic bacterial count with time for a seasoned food product. In many shelf life studies, 10^6 or 10^7 cfu/g of total aerobic bacterial count is used for deciding the end of the shelf life. But the limit criteria are different with food, storage conditions, etc. The range of 10^7 to 10^8 cfu/g is sometimes used for fresh-cut vegetables as given in Table 16.2. Apart from standard colony count method, optical density or conductance is measured for quickly quantifying the microbial growth on the food. As with other quality degradations, intrinsic food characteristics (pH, water activity, additives, etc.) and extrinsic environmental conditions (temperature, packaging, and humidity) affect the microbial food quality deterioration rate and also the microbial flora of significant concerns in aspects of spoilage and safety. The spoilage or pathogenic microbial species able to grow in range of pH, water activity, and temperature of the specific food product can be identified firstly and their counts may be used as quality index or criterion [4].

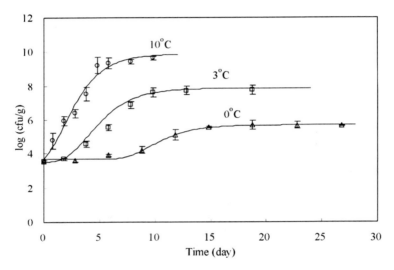

Figure 16.4 Total aerobic microbial count of a seasoned tofu product stored at chilled conditions. Solid lines are fittings by Gompertz equation (Equation 16.15).

16.2.6 Losses in Nutrients and Pigments

Many nutrients of vitamins and amino acids degrade with storage time [9]. Pigments such as chlorophylls, anthocyanins, and betalaines also lose their color and decrease in their concentration during storage of foods. These changes aggravate the nutritional and sensory qualities, and thus limit the shelf life. Compared to the usual processing steps employing thermal and physical operations, storage of packaged foods often has a small but significant effect on the nutrient and pigment retentions. Table 16.3 shows typical losses of thiamin in canned foods during storage. These losses are sometimes used as an index for determining the limit quality and shelf life for the food product. Several factors such as water activity, pH, packaging condition, and storage temperature affect the relative rates of the degradation in the nutrients and pigments. Table 16.4

shows the generalized summary for vitamin stability against intrinsic and extrinsic effects.

Table 16.3 Typical losses of thiamin from canned foods during storage

Food	Retention after 12 months of storage (%)	
	38°C	1.5°C
Apricots	35	72
Green beans	8	76
Lima beans	48	92
Tomato juice	60	100
Peas	68	100
Orange juice	78	100

From Ref. [10] with permission.

Table 16.4 Generalized vitamin stability to intrinsic and extrinsic conditions

Nutrient	Neutral	Acid	Alkaline	Air or oxygen	Light	Heat
Vitamin A	S	U	S	U	U	U
Ascorbic acid	U	S	U	U	U	U
Biotin	S	S	S	S	S	U
Carotenes	S	U	S	U	U	U
Choline	S	S	S	U	S	S
Vitamin B12	S	S	S	U	U	S
Vitamin D	S	S	U	U	U	U
Folate	U	U	U	U	U	U
Vitamin K	S	U	U	S	U	S
Niacin	S	S	S	S	S	S
Pantothenic acid	S	U	U	S	S	U
Vitamin B6	S	S	S	S	U	U
Riboflavin	S	S	U	S	U	U
Thiamin	U	S	U	U	S	U
Tocopherols	S	S	S	U	U	U

S: stable; U: unstable. From Ref. [10] with permission.

16.2.7 Production of Undesirable Components

Proteins may be oxidized to result in thiol-disulfide interchange reactions, cross-linking, and the formation of degradation products [9]. Some of these changes worsen the protein nutritive value. Proteins can also interact with aldehydes and sugars to form cross-links, and with lipids to form complexes resulting in deterioration of food texture and protein nutritive value. Volatile basic nitrogen (VBN) accumulated from protein degradation is used as a good index of quality deterioration in storage of protein-based foods, particularly for fish and fish products. 20-25 mg N/100g as VBN has been suggested as the quality limit of acceptability for fish products by Ababouch et al. [11]. Trimethylamine produced from trimethylamine oxide by bacteria on the marine fish is also used as an index of spoilage. Hypoxanthine is accumulated from the breakdown of adenosine triphosphate in dead fish and used as another chemical index for shelf life study.

Lipids, especially when unsaturated, are susceptible to oxidation by exposure to oxygen [9]. Oxidation of lipids produces hydroperoxides, dimerized fatty acids, and trans-fatty acids, resulting in loss of nutritive value and toxicological worry. Very small amount of oxygen can sometimes cause discernable sensory quality losses due to oxidation reaction products such as oxidized fatty acids and volatile flavor compounds. Peroxide value is well correlated to the oxygen absorption, thus to the early steps of chemical oxidation which has direct relation to oxygen permeability of the packaging. Thiobarbituric acid reactive substances expressed in the malondialdehyde content from the oxidized lipids is frequently used as a quality index for studying the shelf life of lipid-based foods. Free fatty acids are also produced by hydrolytic cleavage of triglycerides in the presence of moisture and result in rancid off-flavor. Hexanal is another index used for rancidity of cereal and nut products [1].

Carbohydrates with active carbonyl groups interacts with amino acids in proteins to form nonenzymatic browning pigments, which alter color, flavor, texture, and nutritional value [9]. Nonenzymatic browning measured in water extractable brown pigment concentration is widely used as a quality index for shelf life studies of many dried and semi-moist foods. An intermediate product for nonenzymatic browning pigment, hydroxymethylfurfural is also used as an index for potential discoloration.

16.2.8 Physical Changes and Processes

Sorption or desorption of moisture may occur and lead to unacceptable quality when a food is exposed to the humidity conditions different with equilibrium humidity of the food. Moisture sorption may cause caking for the powdered food and sticky soft texture for dry snack food. Moisture loss on the other hand induces hardening for breads and bakery products. Because the moisture transport should occur from the package or into the package in order for moisture to be sorbed or desorbed on the food, water vapor permeability of the packaging materials plays a key role determining the rates. In this context, shelf life of a food is determined by the permeability characteristics of the packaging materials, which emphasizes the importance of packaging design.

In case of bakery products and cooked rice with gelatinized starch, staling proceeds with increase in hardness of their texture, which is measured by a texturometer. The mechanism of staling is the change of starch from an amorphous state to a crystalline state. Staling rate decreases with temperature up to 55°C, while it increases with decrease in temperature up to 4°C and then decreases with further temperature decrease. Staling can be prevented if the products are stored above 55°C or frozen-stored below –18°C. Staling or retrogradation may be reversed to nearly fresh state by reheating to high temperature. The rate of staling is known to have little relation with packaging conditions.

Sugars crystallize in jams, icings, caramels, and cake decorations when moisture migrates between the component bases. In chocolate products, appearance of dull grayish spots called bloom occurs from sugar crystallization and fat separation when proper processing and storage are not applied. High level of drip is also unacceptable in stored animal tissue. Exudate formed from the meat deeply affects its tenderness and is related to the change in water holding capacity of the proteins that may change during the storage due to the pH modification (e.g., from the microbial growth) or enzymatic activity.

Surface color change measured by Hunter color difference meter is sometimes used as an index of quality change deciding the shelf life. For example, 'a' value in Hunter color scale indicating redness is often measured for quality evaluation of the meat products. 'L' value is also used for describing the discoloration of a variety of food products during storage. Sometimes some other combinations of the 'L', 'a', and 'b' values are used to describe the color attributes of the foods.

16.3 Kinetics of Food Deterioration

Quantification of quality changes and permeation of gas and vapor through the packaging materials can be described by the kinetic equations, which are useful for systematic determination and prediction of shelf life of packaged foods. Sometimes empirical or mechanistic kinetic equations are employed to express microbial quality of foods. Diffusional transport mechanism is used for the permeation of oxygen gas and water vapor through the package film or wall layer as discussed in Chapter 4.

16.3.1 Chemical Kinetics: Reaction Order and Rate Constant

Chemical kinetics is traditionally used to describe the rate of disappearance or generation of compounds in chemical reactions. As shown below, the chemical kinetic rate equation may be used to describe the loss of desirable compounds such as nutrients or pigments, or the generation of undesirable compounds such as hexanal from oxidation in food. The concentrations of these desirable or undesirable compounds may be used as quality indexes. Furthermore, the chemical kinetics rate equation is sometimes extended to describe other food deteriorations such as loss of sensory quality, although the validity of this extension must be confirmed by experimental data.

The rate of loss of compound B may be described by the chemical kinetics rate equation:

$$\frac{d[B]}{dt} = -k[B]^n \tag{16.1}$$

where [B] is the concentration of B and may be used as a quality factor, t is time, k is reaction rate constant, and n is reaction order. The negative sign denotes that the concentration decreases with time.

For the zero order kinetics (n=0), the loss of B is expressed by:

$$\frac{d[B]}{dt} = -k \tag{16.2}$$

Integrating the above equation yields a linear relationship between [B] and t:

$$[B] = [B]_o - k\,t \tag{16.3}$$

where $[B]_o$ and [B] is concentrations of quality index initially and at time t, respectively. If $[B]_c$ is the critical limit, then the shelf life t_s is given by:

$$t_s = \frac{[B]_o - [B]_c}{k} \tag{16.4}$$

If the experimental data follow the zero order kinetics, a plot of [B] versus t should yield a straight line and the rate constant k can be estimated from the slope. The shelf life t_s may also be found from the plot as shown in Figure 16.5. Zero order reaction may be applied to quality deterioration of frozen foods and sensory attribute intensity.

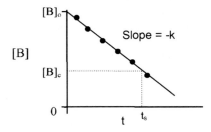

Figure 16.5 Zero-order reaction for quality loss

For the first order kinetics (n=1), the loss of B is expressed by:

$$\frac{d[B]}{dt} = -k\,[B] \tag{16.5}$$

Integrating the above equation shows that [B] is exponentially decaying with increasing t:

$$[B] = [B]_o\,e^{-kt} \tag{16.6}$$

which suggests that a plot of [B] versus t is nonlinear as shown in Figure 16.6. The above equation may also be expressed in the linearized form below:

$$\ln[B] = \ln[B]_o - kt \tag{16.7}$$

which suggests that a plot of ln [B] versus t is linear. If $[B]_c$ is the critical limit, then the shelf life t_s is given by:

$$t_s = \frac{\ln\{[B]_o/[B]_c\}}{k} \tag{16.8}$$

If the experimental data follow the first order kinetics, a plot of ln [B] versus t should yield a straight line and the rate constant k can be estimated from the slope. First order kinetics may be applied to food deterioration reactions such as destruction of nutrients and pigments.

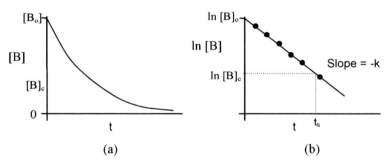

Figure 16.6 First-order reaction for quality loss

In shelf life studies, [B] is measured as a function of t, and the data are then used to determine the reaction order and rate constant. Although zero order and first order kinetics are usually tested, second order and fractional orders may also be attempted. For example, second order has been used to describe vitamin C destruction in canned infant food [26], and half order has been used to describe red pigment degradation in beet powder [27].

Typically, zero order and first order deviate significantly from each other only after [B] has decreased substantially as shown in Figure 16.7. Hence deciding suitable reaction order requires wise selection of experimental measurement schedule, which should cover a wide range of quality destruction [12, 13]. Conversely, it can be said that any order of zero or first may be used without significant error for practical purpose where quality loss to the limit is small.

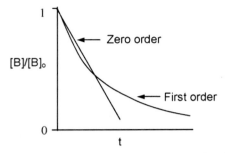

Figure 16.7 Zero order versus first-order kinetics

When food deteriorations result in the accumulation of undesirable compounds, the reaction rate equation is written as function of reaction product concentration, [P]:

$$\frac{d[P]}{dt} = k[B]^n \tag{16.9}$$

for the reaction scheme of B →P and $[B] = [B]_o - [P]$. For zero order kinetics, integrating the above equation yields:

$$[P] - [P]_o = kt \tag{16.10}$$

where [P] and $[P]_o$ are concentrations of reaction product at time t and 0, respectively. If $[P]_c$ is the critical limit, the shelf life is given by:

$$t_s = \frac{[P]_c - [P]_o}{k} \tag{16.11}$$

The zero order kinetics may be used to describe, for example, the accumulation of pigments from nonenzymatic browning in dry and semi-moist foods.

For first order kinetics, integration of the rate equation yields:

$$\ln(1 - \frac{[P]}{[B]_o}) = -kt \tag{16.12}$$

If $[P]_c$ is the critical limit, the shelf life is given by:

$$t_s = \frac{\ln(\frac{[B]_0}{[B]_0 - [P]_c})}{k} \qquad (16.13)$$

An important observation from the above equations for t_s is the relationship:

$$t_s \propto \frac{1}{k} \qquad (16.14)$$

which applies to both the zero order and first order kinetics. In fact, it can be shown easily that this relationship is applicable to nth or any order of kinetics.

16.3.2 Microbial Growth Model

Empirical kinetic models are available to describe the rate of microbial growth in foods [14, 15]. Two commonly used models are the Gompertz equation

$$\log N(t) = C_1 + C_2 \exp(-\exp(-C_3 (t-C_4))) \qquad (16.15)$$

and the logistic equation

$$\log N(t) = C_1 + \frac{C_2}{1+\exp(-C_3(t - C_4))} \qquad (16.16)$$

where $N(t)$ is microbial number (cfu/g) at time t (day), and C_1-C_4 are parameters.

For the Gompertz equation, lag time λ (day) and maximum specific growth rate μ (day^{-1}, the inverse of the time required for e (2.718) fold increase in microbial count) can be obtained from:

$$\lambda = C_4 - \frac{1}{C_3}\{1-\exp(1-\exp(C_3 C_4))\} \qquad (16.17)$$

$$\mu = \frac{2.303 C_2 C_3}{e} \qquad (16.18)$$

For the logistic equation, λ and μ can be obtained from:

$$\lambda = C_4 - \frac{2}{C_3}\{1 - \frac{2}{1+\exp(C_3 C_4)}\} \qquad (16.19)$$

$$\mu = 0.576 C_2 C_3 \qquad (16.20)$$

A microbial growth model based on mechanistic principles has been proposed recently by Baranyi and Roberts [16]:

$$\frac{dq(t)}{dt} = \mu q(t) \qquad (16.21)$$

$$\frac{dN(t)}{dt} = \mu(\frac{q(t)}{1+q(t)})(1-\frac{N(t)}{N_{max}})N(t) \qquad (16.22)$$

where q(t) is normalized concentration of unknown substance critically needed for cell growth and represents physiological state of the cell population at time t, and N_{max} are the maximum cell density (cfu/g).

The model of Baranyi and Roberts is integrated to obtain:

$$\log N(t) = \log N_o + \frac{\mu}{2.303}\cdot C - \frac{1}{2.303}\cdot \ln(1+\frac{e^{\mu C}-1}{10^{(\log N_{max}-\log N_o)}}) \qquad (16.23)$$

where N_o is the initial cell density (cfu/g), and C is defined by:

$$C = t + \frac{1}{\mu}\cdot \ln[\frac{e^{-\mu t}+q_o}{1+q_o}] \qquad (16.24)$$

with q_o being a parameter for the initial physiological state of the microbe cells.

The lag time (λ) is related to q_o and μ:

$$\lambda = \frac{\ln(1+\frac{1}{q_o})}{\mu} \qquad (16.25)$$

Because the model of Baranyi and Roberts is based on differential Equations 16.21 and 16.22, it has an advantage to be conveniently used for describing a microbial growth on the food exposed to a dynamically changing environment [17, 18].

16.3.3 Other Kinetic Model Associated with Physical Changes

In the composite chocolate products such chocolate-enrobed biscuits, migration of moisture and/or lipid between two phases causes the cocoa butter to recrystallize as bloom. The fat migration was described by an approximated solution of Fick's Second Law of Diffusion (see also Equation 5.7 in Chapter 5):

$$\frac{m_t}{m_\infty} = 2\sqrt{\frac{Dt}{\pi L^2}} \qquad (16.26)$$

or in a simpler approximation of

$$\frac{m_t}{m_\infty} = \frac{A\sqrt{D\cdot t}}{V} \qquad (16.27)$$

where m_t is the migrated moisture or fat within the food layer at time t, m_∞ is the migrated moisture or fat within the food layer at equilibrium, A is the contact surface between phases, D is diffusion coefficient, L is the thickness of the food layer, and V is volume of the source product containing originally the moisture or fat [19].

The similar form of equations can be applied for the taint migration from the packaging material and the scalping of flavor from the food, for these cases V in Equation 16.27 becomes the volumes of packaging material or food and L packaging material thickness. It needs to be mentioned that the Equations 16.26 and 16.27 are simplified forms of solution of Fick's Second Law, which are applicable to the relatively small initial time span ($m_t/m_\infty < 0.6$). For the detailed solution of the diffusion in migration and adsorption, please refer to Chapter 5 (Section 5.6).

For the staling of gelatinized starch-based products such as bread, tortilla, and cooked rice, Avrami equation is used to express its progress with time:

$$\theta = \exp(-k_c t^{n_a})$$ (16.28)

where θ is ratio of uncrystallized starch and may be defined as a function of hardness by $[(H_\infty - H_t)/(H_\infty - H_o)]$; H_o is hardness at time zero; H_t is hardness at time t; H_∞ is hardness after completion of staling or retrogradation; k_c is a rate constant; and n_a is Avrami power index.

16.3.4 Temperature Dependence: Arrhenius and Shelf Life Plots

For many food deteriorations, the temperature dependence of rate constant k may be described by the Arrhenius equation:

$$k = k_o \exp(-\frac{E_a}{RT})$$ (16.29)

where k_o is frequency factor, E_a is activation energy (J mol^{-1}), R is the gas constant (8.314 J K^{-1} mol^{-1}), and T is absolute temperature (K). E_a is a measure of the sensitivity of rate constant k to temperature: higher E_a results in more rapid increase in k with temperature.

The Arrhenius equation may be transformed into the linear form:

$$\ln k = \ln k_o - \frac{E_a}{RT}$$ (16.30)

Accordingly, a plot of ln k versus 1/T should yield a straight line with a slope of $-E_a/R$ as shown in Figure 16.8a.

In shelf life studies, k values at three or more temperatures are usually measured and the parameter E_a is then estimated from the slope in Figure 16.8a. Once E_a is known, any k values including those experimental values may be calculated using the Arrhenius equation. Alternatively, k values may be obtained by extrapolating the experimental data as shown in Figure 16.8a.

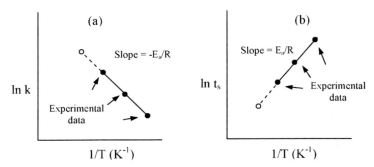

Figure 16.8 Arrhenius plots for rate constant and shelf life

Recall from Section 16.3.1 that shelf life t_s and rate constant k are related by the equation:

$$t_s \propto \frac{1}{k} \tag{16.14}$$

Substituting the Arrhenius equation into the above equation yields:

$$t_s \propto \frac{1}{k_0 \exp(-\frac{E_a}{RT})} \tag{16.31}$$

which may be rewritten as:

$$\ln t_s = \frac{E_a}{R} \frac{1}{T} + \text{constant} \tag{16.32}$$

An important implication from this equation is that shelf life data should be plotted as $\ln t_s$ versus $1/T$ as shown in Figure 16.8b, not as t_s versus T or other forms. Typically, this Arrhenius plot is generated using three or more t_s values measured at different temperatures. This plot may then be used to predict t_s values at lower temperatures as illustrated in Figure 16.8b—in fact, this is the kinetics basis for accelerated shelf life testing (ASLT).

It is somewhat surprising that the plots in Figure 16.8 may be approximated by the simplified plots in Figure 16.9. Note that the $1/T$ in Figure 16.8 (where absolute temperature K must be used) is now replaced by T in Figure 16.9 (which may be in °C or K). It can be shown mathematically that this is a good approximation so long as the temperature range being considered is small (say

ΔT is within 40°C), which is the case for most shelf life studies. The plot of ln t_s versus T is sometimes known as shelf life plot.

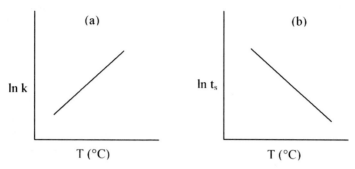

Figure 16.9 Simplified plots for rate constant and shelf life

Although activation energy E_a is well established in the scientific community, this parameter is difficult to understand by people with less scientific background. To overcome this difficulty, the more easily understood term Q_{10} is also used:

$$Q_{10} = \frac{\text{Rate constant at temperature (T + 10) °C}}{\text{Rate constant at temperature T °C}} = \frac{k_{T+10}}{k_T} \qquad (16.33)$$

where T must be in °C. Q_{10} is the familiar concept from general chemistry textbooks stating that for every 10°C increase in temperature, the reaction rates of many chemical reactions will double or triple (equivalent to Q_{10} of 2 or 3).

Since $k \propto (1/t_s)$, Q_{10} may also be expressed as:

$$Q_{10} = \frac{t_{s,T}}{t_{s,T+10}} \qquad (16.34)$$

When this relationship is applied to Figure 16.9b, slope of shelf life plot becomes $(-\ln Q_{10}/10)$. The table below illustrates the effect of Q_{10} on shelf life, assuming the shelf life at 45°C is 2 weeks.

T (°C)	Shelf life for different Q_{10}		
	$Q_{10} = 2$	$Q_{10} = 3$	$Q_{10} = 5$
45	2 weeks	2 weeks	2 weeks
35	4 weeks	6 weeks	10 weeks
25	8 weeks	18 weeks	50 weeks
15	16 weeks	54 weeks	250 weeks

Food products have a rather wide range of Q_{10} values; for example, typical Q_{10} values for canned foods are 1-4, those for dehydrated foods are 2-10, and those for frozen foods are 3-40. Q_{10} values are useful for the design of experimental plans for accelerated shelf life testing.

For the microbial food spoilage, maximum specific growth rate and/or reciprocal lag time has been expressed as Belehradek equation or square root model:

$$\sqrt{\mu} = b_{\mu}\,(T - T_{min,\mu})$$ (16.35)

$$\sqrt{1/\lambda} = b_{\lambda}\,(T - T_{min,\lambda})$$ (16.36)

where $T_{min,\mu}$ and $T_{min,\lambda}$ are notional minimum temperature for microbial growth, and b_{μ} and b_{λ} are constants [15]. Even though Arrhenius equation is also used for temperature dependence of maximum growth rate, square root model is more widely used for modeling the microbial growth.

16.3.5 Moisture Dependence

Many chemical, biological, and physical changes are dependent on moisture content or water activity in the foods, whose changes may be dictated by the packaging protection from the environment in the distribution channel. The concept of water activity is known to describe better the effect of the moisture on quality changes than the moisture content dry basis or wet basis. General dependence of many important chemical reactions and microbiological spoilage on water activity is given by Figure 16.10.

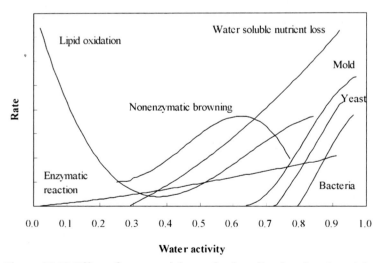

Figure 16.10 Effect of water activity on food quality deterioration. Adapted from Ref. [20, 21].

Table 16.5 Monolayer moisture value (M_m) for some dry foods at 25°C

Food	M_m (g water/g solid)	Food	M_m (g water/g solid)
Avocado	0.032	Whole milk	0.035
Banana	0.040	Skim milk	0.056
Barley	0.086	Mushrooms	0.049
Beans	0.070	Peas	0.044
Cabbage	0.091	Peppermint tea	0.068
Carrot	0.045	Pine apple	0.195
Celery	0.062	Potato	0.052
Cheese (Emmental)	0.033	Potato flakes	0.048
Cinnamon	0.061	Radish	0.054
Citrus juice	0.050	Rice	0.079
Collagen	0.096	Rough rice	0.070
Corn flours	0.035	Starch gel	0.087
Egg paste	0.057	Potato starch	0.060
Egg albumin	0.056	Sucrose	0.073
Fish protein concentrate	0.053	Wheat	0.064
Gelatin	0.085	Whey powders	0.024
Ginger	0.070	Active dry yeast	0.056

Adapted from Ref. [22].

Most bacteria can grow at $a_w > 0.9$, while some halophile yeasts and xerophilic molds can grow at a_w as low as 0.7. At $a_w > 0.6$ enzyme-catalyzed reactions such as hydrolysis and oxidation may occur. Destruction of water soluble vitamins and pigments also increases with water activity. These increases of the enzymatic and chemical reaction rates with water activity are explained by increased mobility of the substrates and reactants under high moisture content. Nonenzymatic browning shows a maximum around the water activity of 0.6-0.7; diffusion of the reactants such as reducing sugars and amino acids is accelerated at higher water activity to increase the browning rate, but further increase of moisture beyond a certain level reduces the rate by the dilution effect of the reactants and inhibition by water itself which is the product of the reaction. On the other hand, oxidation of fat, fat-soluble vitamins, and pigments is accelerated with decrease of water activity below monolayer value, which corresponds to monomolecular saturation of adsorption sites by water molecules (Table 16.5). The autoxidation reaction slows down with water activity increase due to the protection by hydrogen bonding between the reaction intermediates and water molecule. However at higher water activity, the oxidation is again enhanced by the mobility of the metal ions catalyzing the reaction.

Table 16.6 Maximum permissible water activity for some dry foods at 20°C

Food	a_w	Food	a_w
Baking soda	0.45	Sugars:	
Crackers	0.43	Pure fructose	0.63
Dried eggs	0.30	Pure dextrose	0.89
Gelatin	0.43-0.45	Pure sucrose	0.85
Hard candies	0.25-0.30	Maltose	0.92
Chocolate, plain	0.73	Sorbitol	0.55-0.65
Chocolate, milk	0.68	Dehydrated meat	0.72
Potato flakes	0.11	Dehydrated vegetables:	
Flour	0.65	Peas	0.25-0.45
Oatmeal	0.12-0.25	Beans	0.12
Dried skim milk	0.30	Dried fruits:	
Dry milk	0.20-0.30	Apples	0.70
Beef-tea granules	0.35	Apricots	0.65
Dried soups	0.60	Dates	0.60
Roast coffee	0.10-0.30	Peaches	0.70
Soluble coffee	0.45	Pears	0.65
Starch	0.60	Plums	0.60
Wheat preparations (macaroni, noodles, etc.)	0.60	Orange powder	0.10

From Ref. [23] with permission.

Several forms of mathematical functions have been proposed by researchers to describe the relative effect of water activity on the various types of quality changes. Quast et al. [24] reported that oxidation rate of potato chip was described by the following equation:

$$\text{Rate} \propto (C_1 + \frac{C_2}{\sqrt{a_w}}) \tag{16.37}$$

where a_w is water activity of food with C_1 and C_2 being constants.

Above monolayer value of the moisture content, a semilog plot of deteriorative reaction rate versus water activity generally results in straight line of increase before reaching a maximum level. Mizrahi et al. [25] correlated the nonenzymatic browning rate as a function of water activity by following equation:

$$\text{Rate} = C_1 \, a_w^{C_2} \tag{16.38}$$

For most dry foods, increase of water activity by 0.1 unit decrease the shelf life by two to three folds [26].

Table 16.7 Some useful mathematical descriptions of moisture sorption isotherm

Type	Function	Applicable a_w range
Linear	$M = C_1 a_w + C_2$	0.1-0.3
BET	$\dfrac{a_w}{(1-a_w)M} = \dfrac{1}{M_m C_1} + \dfrac{C_1 - 1}{M_m C_1}$	0.05-0.45
GAB	$\dfrac{M}{M_m} = \dfrac{C_1 C_2 a_w}{(1 - C_2 a_w)(1 - C_2 a_w + C_1 C_2 a_w)}$	0.1-0.9
Halsey	$a_w = \exp(-\dfrac{C_2}{M^{C_1}})$	0.1-0.8
Henderson	$a_w = 1 - \exp(-C_1 M^{C_2})$	0.1-0.8
Oswin	$M = C_1 (\dfrac{a_w}{1-a_w})^{C_2}$	0.1-0.85
Kuhn	$M = \dfrac{C_1}{\ln a_w} + C_2$	0.1-0.8
Iglesias and Chirife	$M = C_1 (\dfrac{a_w}{1-a_w}) + C_2$	0.1-0.6

M: moisture content (dry basis); a_w: water activity; C_1 and C_2: constant; M_m: constant as monolayer value.

It is generally accepted that maximum storage stability of the dehydrated foods is attained at the monolayer moisture content, where water molecules are adsorbed as mono molecular thickness on all the adsorption sites. Table 16.5 presents monolayer values for some dried foods. The monolayer generally amounts to a water activity of 0.2-0.4. Some dry food products lose suddenly their quality characteristics when the water activity increases beyond a certain limit, which is listed in Table 16.6. Textural property such as crispness in snack foods, for example, changes abruptly at a critical range of water activity.

Water activity is ruled by the moisture content in the food, which may change with time through the permeation process of water vapor across the packaging film. The equilibrium relationship between moisture content and water activity is explained by the sorption/desorption isotherm, which is obtained mostly by the experiment determining moisture contents of the food samples equilibrated under different relative humidity conditions. Many attempts were conducted to express the equilibrium relationship by mathematical forms. Some of the most widely used ones are given in Table 16.7.

16.3.6 Oxygen Dependence

Oxidation of lipid and pigments are strongly influenced by oxygen partial pressure in the package headspace, since the partial pressure determines the dissolved oxygen content in the food product. In many oxidation reactions, the rate of oxygen consumption R_{O2} follows the mixed order type equation [20, 27]:

$$R_{O2} = \frac{P_{O2}}{C_1 + C_2 P_{O2}} \qquad (16.39)$$

where P_{O2} is oxygen partial pressure, and C_1 and C_2 are constants. R_{O2} may also be regarded as rate of oxidation and its oxygen dependence for some food products is shown in Figure 16.11.

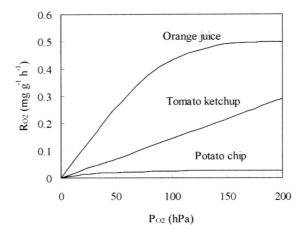

Figure 16.11 Oxygen dependence of oxidation rate for some foods. Drawn from data of Ref. [20].

For C_2P_{O2} much lower than C_1 (when P_{O2} is low and/or C_2 is very low), the consumption rate is directly proportional to P_{O2}:

$$R_{O2} = \frac{P_{O2}}{C_1} \qquad (16.40)$$

For C_2P_{O2} much higher than C_1 (for high P_{O2} and/or high C_2), the consumption rate is constant:

$$R_{O2} = \frac{1}{C_2} \qquad (16.41)$$

The partial oxygen in the package headspace may be influenced by the oxygen consumption of the oxidation reaction and by the permeation of oxygen through the package.

16.4 Environmental Factors Affecting Shelf Life

The environment that a food product will encounter is usually dictated by its food distribution channel whose pathways include storage, transportation, display, etc. (Figure 16.12). As discussed in Section 1.5, there are three different types of environments (ambient environment, physical environment, and human environment) which can greatly affect product shelf life and packaging requirements.

16.4.1 Ambient Environment

Temperature is usually the most important variable in the ambient environment, since it can greatly affect the rates of food deterioration and barrier properties of packaging materials. Good temperature control is important to ensure the quality and safety of foods, especially those frozen and chilled. Frozen food products are typically kept around -18°C. Some typical temperatures for chilled foods are as follows: 0-5°C for dairy products; -1 - +1°C for fresh and processed meat, poultry, and fish products; 0-4°C for leafy vegetable salads and soft fruits. However, temperature fluctuations are almost unavoidable during the storage and distribution of the food product. Figure 16.13 shows a typical temperature history encountered by chilled or frozen foods. Since the distribution temperature of shelf stable foods is not controlled, those food products experience a much wider range of temperature fluctuations depending on seasonal changes and the distribution environment.

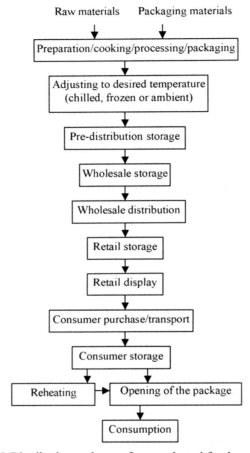

Figure 16.12 Distribution pathways for a packaged food product

Relative humidity is usually the second most important variable in the ambient environment, since many food products are moisture sensitive. It is expressed as:

$$H_R = \frac{p_w}{p_s} \tag{16.42}$$

where H_R is relative humidity, p_w is water vapor pressure in air, and p_s is saturation water vapor pressure. p_s increases with temperature as shown in Table 16.8 and may be estimated using the following equation [28]:

$$p_s = 611.2 \cdot \exp(53.479 - \frac{6808}{T} - 5.09 \cdot \ln T) \qquad (16.43)$$

where p_s is in Pa and T in K. Since relative humidity is seldom controlled in the distribution environment, the package is important for protecting food products which are moisture sensitive.

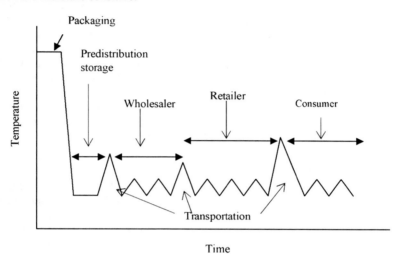

Figure 16.13 Temperature fluctuations encountered by chilled or frozen foods

Table 16.8 Saturation water vapor pressure as function of temperature

Temperature (°C)	Pressure (kPa)	Temperature (°C)	Pressure (kPa)	Temperature (°C)	Pressure (kPa)
-40	0.0129	-5	0.4018	30	4.2462
-35	0.0224	0	0.6112	35	5.6280
-30	0.0380	5	0.8725	40	7.3838
-25	0.0633	10	1.2280	45	9.5935
-20	0.1033	15	1.7055	50	12.3503
-15	0.1653	20	2.3389	55	15.7601
-10	0.2599	25	3.1693	60	19.9439

Gas composition is also an important variable in the ambient environment. Dry air normally consists of 20.9% oxygen, 78.1% nitrogen, 0.9% argon, and 0.03% carbon dioxide by volume; among them, oxygen is the most important because it can cause oxidation reaction in many food products. Packaging materials with good oxygen barrier are required to protect oxygen sensitive foods, especially those shelf stable foods stored at relatively high temperature.

Vacuuming or gas flushing is also useful packaging technique to remove oxygen inside the food package.

Lighting condition is another important variable. Illuminance is a common measure of light intensity that correlates well with human perception of brightness. Although illuminance is important to the appearance of the product on the display shelf, it does not take into account the ultraviolet component that plays a more important role in influencing deterioration reactions such as lipid oxidation and color discoloration. Therefore, the spectral distribution (particularly, ultraviolet component) and irradiance (W m^{-2}) should also be considered when analyzing the effects of lighting on food quality. The illuminance of sunlight is in the range of 1000-80,000 lux depending on the time of day, location, and season. Typical warehouses and retail stores in the food distribution channel have illuminance of 20-500 lux, while illuminated display shelves are typically under UVA irradiance of 0-20 mW m^{-2} with illuminance of 1000-3000 lux from fluorescent and/or halogen lamps.

16.4.2 Physical Environment

A food package may experience various kinds of mechanical stresses that can critically compromise the closure seal, hermetic condition, and appearance. These stresses are typically due to compression, shock, vibration, and handling encountered by the package during storage and transportation. Therefore, testing the package resistance to these stresses is an important part of shelf life study and whenever possible should be performed under actual conditions or conditions simulating the real world.

For sealed packages containing significant amount of headspace gas or residual gas, variations in temperature will cause the pressure and volume of the gas to change according to the Ideal Gas Law:

$$P \, V = n \, R \, T \tag{16.44}$$

where P is pressure inside the package, V is amount of gas, T is absolute temperature, n is number of moles, and R is the ideal gas constant. The increase or decrease in pressure sometimes induces mechanical stress causing damages to the package. The volume changes may also adversely affect the package.

Variations in the atmospheric pressure may disturb the balance between the pressures inside and outside flexible packages, inducing mechanical stresses to damage the package. The atmospheric pressure P(z) decreases with altitude according to the equation [28]:

$$P(z) = P_o \exp(-\frac{gz}{RT}) \tag{16.45}$$

where P_o is atmospheric pressure at sea level (101.325 kPa or 1 atm), g is gravitational acceleration (9.81 m s^{-2}), z is altitude (m), R is gas constant (287 J K^{-1} kg^{-1}), and T is absolute temperature (K). Variations in altitude of thousand meters are commonly encountered during transportation by air or over mountains, and Table 16.9 shows that the atmospheric pressure decreases greatly at these altitudes. It should be mentioned that variations in altitude affects only flexible packages containing significantly amount of headspace gas or residual gas. As the gas expands due to the lower atmospheric pressure, stress will be exerted onto the package and the volume expansion will depend on the pressure difference and the strength of the package.

Table 16.9 Changes in atmospheric pressure with altitude at 20°C

z (m)	P(z)/P$_o$	z (m)	P(z)/P$_o$
0	1.000	2000	0.792
500	0.943	2500	0.747
1000	0.890	3000	0.705
1500	0.840	3500	0.665

Example 16.1 Collapse of PET bottle due to temperature

Shredded cheese is packed inside PET containers in the production plant at 30°C. To inhibit mold growth in the cheese, CO_2 is flushed inside the containers. The containers are then transported to a storage area at around 4°C. Later it is observed that some containers collapsed at the corners or the walls in the storage area. Explain the reasons of this observation.

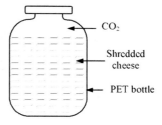

The collapse is due to pressure drop inside the package caused by the temperature decrease and absorption of CO_2 by cheese. The pressure drop creates an imbalance in pressure inside and outside the package. Hence a net force is exerted onto the bottle causing it to collapse.

Since the PET container is relatively rigid, its volume may be assumed to remain constant. Equation (16.44) may then be simplified as:

$$\frac{P_1}{P_2} = \frac{T_1}{T_2}$$

(16.46)

For temperature decrease from 30°C to 4°C, the pressure ratio may be calculated as:

$$\frac{P_{4C}}{P_{30C}} = \frac{4+273}{30+273} = 0.91$$

(16.47)

or 9% drop in pressure due to the temperature decrease. However, the drop is significantly higher than 9% when CO_2 absorption by the cheese is also taken into account.

16.4.3 Human Environment

Through the whole distribution channel of food products, human is involved in managing warehousing, transportation, retail purchase, and consumer handling. Marketing customs and logistic constraints such as pallet dimension restrict the package dimension adoptable. There would be variation of practices in storage time, storage period, and food consumption pattern for each stage of handling. Some outbreaks in cold chain and abnormal consumer behavior may result in unexpected fast food quality deterioration and shortening of the shelf life. Monitoring and checking systems can help the soundness of temperature and stock controls. More elaborate management of shelf life with consideration of variability in product itself and distribution chain can be possible with introduction of modern logistic system and intelligent devices such as time temperature indicator (TTI), electronic temperature data logger, and radio frequency identification tag. A shelf life management system considering variations of storage time and temperature in whole distribution chain has been proposed by Koutsoumanis et al. [29].

16.5 Package Factors Affecting Shelf Life

The package is important for protecting food products which are fragile or sensitive to oxygen, moisture, or light. In addition, packaging technologies such as modified atmosphere and aseptic packaging are also useful for enhancing the quality and safety of food products.

16.5.1 Packaging Parameters Important to Shelf Life

The important package parameters are those related to the package dimensions and type of materials used to construct the package. As described in Chapter 4, the permeation rate (Q) may be expressed by the equation:

$$Q = \frac{\bar{P}A}{L}\Delta p \tag{16.48}$$

where the package parameters are $\bar{P}$ (permeability coefficient), A (surface area), and L (package layer thickness). The equation includes information about the type of package materials through the permeability coefficient, as well as information about the package dimensions through the surface area and the thickness. The permeation rate may be manipulated by changing package parameter. According to the above equation, for example, the permeation rate may be reduced to ¼ of its original value if both the package surface is halved and thickness is doubled, while the partial pressure difference Δp is kept constant.

Permeation of a specific aroma compound may be used as a quality index for shelf life determination. For carbonated soft drinks, the amount of CO_2 permeated through the package wall is an important quality index. Sometimes, gas composition in package headspace may be used as a quality index; for example, oxygen concentration and relative humidity in the package headspace may be correlated to oxidation and textural change in the food product.

16.5.2 Permeation versus Reaction Controlled Shelf Life

The extent of influence of package permeation rate on product shelf life varies depending on the situation. Consider the hermetically sealed package containing an oxygen or moisture sensitive food as shown in Figure 16.14. Oxygen and water vapor from the external environment of the package can cause deterioration in the product by a series of steps as follows: (1) permeation through the package wall into the package headspace, (2) diffusion through the headspace to the food product surface, and (3) reaction with the food product until the extent of deterioration is no longer acceptable. The shelf life of the product is the total time required for these steps to complete. In most situations, the second step may be ignored since the package headspace is usually small and diffusion of oxygen or water vapor occurs quite quickly in the headspace.

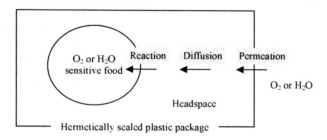

Figure 16.14 Permeation through package and reaction with food

The shelf life of the moisture or oxygen sensitive food package depends largely on the rates of permeation and reaction, which may occur quickly or slowly depending on the barrier property of the package and the stability of the food product. Below are some possible situations.

- Fast permeation, fast reaction: the package is a poor gas barrier and the food is unstable (i.e., highly sensitive to oxygen or moisture). Shelf life is short and in many cases unacceptable.

- Slow permeation, slow reaction: the package is a good gas barrier and the food is quite stable. Shelf life is not a problem.

- Slow permeation, fast reaction: the package is a good gas barrier and the food is unstable. Shelf life is controlled by permeation.

- Fast permeation, slow reaction: the package is a poor gas barrier and the food is quite stable. Shelf life is controlled by reaction.

The third situation (slow permeation and fast reaction) deserves particular attention because it applies to many commercial food products which need good gas barrier protection to meet shelf life requirement. The shelf life of the food package may be described as 'permeation controlled', since permeation is the rate limiting step controlling shelf life. The concept of permeation controlled shelf life is often an important assumption for most predictive shelf life models, where shelf life is equated only to the time required for permeation while considering the time for reaction to be relatively short and negligible.

The fourth situation (fast permeation and slow reaction) may be described as 'reaction controlled', since reaction is the rate limiting step controlling the shelf life of food package. This is equivalent to the situation where an unpackaged food product is exposed to the ambient environment, and the product shelf life is equated to the time required for the extent of reaction to become unacceptable.

16.5.3 Package Interactions

Figure 16.15 shows the common interactions associated with the package. Permeation involves the interactions between the permeating gas, package wall, package headspace, and food product. Migration involves the transfer of chemical additives from packaging materials into food which may lead to off-flavor development and safety issues. Conversely, flavor scalping involves the loss of flavor compounds from the food to the package. Detailed discussions on permeation may be found in Chapter 4, while those on migration and flavor scalping may be found in Chapter 5.

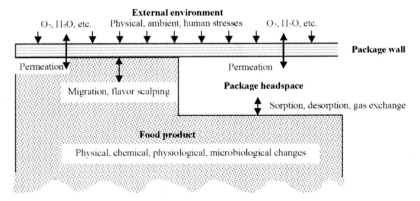

Figure 16.15 Interactions associated with package

16.6 Shelf Life Studies

Shelf life studies are important for determining sell-by-date, use-by-date, or other shelf life dates for package labels. They can also provide useful information for developing new products, minimizing product loss, and expanding distribution to distance markets. There are many ways to conduct shelf life studies depending on the food products, objectives, skills of the scientist or technologist, and other factors.

16.6.1 Testing Under Normal Conditions

The most accurate shelf life testing is to subject package samples to normal conditions encountered in the real world. This can be done by allowing the samples to go through the distribution channel, retrieving the samples at various points along the supply chain such as warehouse and retail store, and then evaluating the quality of samples. This type of testing is usually done after the food package has already been developed and is ready for the market.

During the product development process, prototype food packages are usually tested in the laboratory instead of the real world. Samples are placed inside environmental chambers where temperature and relative humidity are precisely controlled, and the quality indexes of the samples are measured at predetermined intervals. The selection of appropriate conditions requires a good understanding of the food product and its distribution and marketing environment. Temperature is usually the most important environmental variable since it affects the deterioration kinetics of all foods. Relative humidity and oxygen concentration are also important environmental variables for moisture sensitive foods and oxygen sensitive foods, respectively.

Preferably, test conditions in the laboratory should simulate the fluctuating temperature and relative humidity conditions in the real world. However, most shelf life studies are conducted at constant temperature and relatively humidity (typically representing the most severe conditions), since it is expedient to conduct experiments under constant conditions. Nevertheless, results from constant conditions are useful to compare the performance of different packages or to estimate the shelf lives of variable conditions. For chilled foods and frozen foods, temperatures are typically set between 0-10°C and around -18°C, respectively. For shelf stable foods, a wider range of temperatures should be tested, including high temperatures in the summer and subfreezing temperatures in the winter.

While shelf life testing under normal conditions provides the most accurate information, it also requires that the experiments last as long as the product shelf life, which may be several months to two years for commercial food products—a time too long to wait for timely development of new food products. Therefore it is desirable to shorten the time using accelerated shelf life testing.

16.6.2 Testing Under Accelerated Conditions

Accelerated shelf-life testing (ASTL) is a method for testing food products under accelerated conditions to shorten testing time or allow more tests to be done in a timely manner. This method is useful for products whose deterioration kinetics is known functions of certain accelerating factors, the most common of them being temperature.

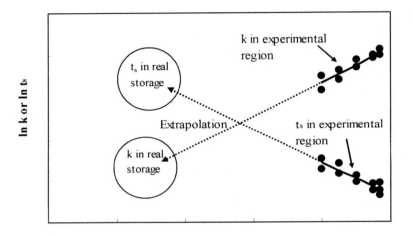

Temperature as accelerating factor

Figure 16.16 Concept of accelerated shelf life testing

The basic concept of ASLT involves conducting experiment to obtain data under accelerated conditions, and then extrapolating these data to estimate those data of normal conditions [30]. As illustrated in Figure 16.16, values of rate constant (k) and shelf life (t_s) measured experimentally under accelerated conditions (higher temperatures) are extrapolated to estimate k and t_s values of normal storage conditions. It is important to note that accurate extrapolation requires reliable kinetics models or other models. According to chemical kinetics, data extrapolation in Figure 16.16 should be performed using a plot of ln k or ln t_s versus T, not a plot of k or t versus T (see also Figure 16.9).

Temperature is the most frequently used accelerated factor and its application in ASTL will be illustrated in Section 16.6.3. Table 16.10 shows the recommended ASLT storage temperatures for some foods. For frozen foods, accelerated deterioration may be induced at -5 or -10°C. For shelf stable products, storage at 40°C can cause the faster quality deterioration such as flavor alteration and chemical oxidation. For some emulsion products, cycling between chilled and room temperatures will speed up the phase separation.

Table 16.10 Recommended storage temperatures for ASLT

Frozen foods	Dry and intermediate moisture foods	Canned foods
-40°C (control)	0°C (control)	5°C (control)
-15°C	23°C (room temperature)	23°C (room temperature)
-10°C	30°C	30°C
-5 °C	35°C	35°C
	40°C	40°C
	45°C	

From Ref. [31] with permission.

Relative humidity is another common accelerating factor for moisture sensitive foods. The high humidity condition provides high humidity differential, which enhance the rate of moisture gain through the packaging film following the permeation equation (Equation 16.48). By applying high humidity conditions, the time to reach certain critical moisture in a limited small extent of change can be shortened for the dry foods as follows:

$$\frac{t_{s1}}{t_{s2}} = \frac{Q_2}{Q_1} = \frac{\Delta p_{w.2}}{\Delta p_{w.1}} \qquad (16.49)$$

where t_{s1} and t_{s2} are the shelf life to reach a critical moisture limit under respective humidity difference potentials of $\Delta p_{w.1}$ and $\Delta p_{w.2}$ with respective moisture transfer rates of Q_1 and Q_2 (see also Section 16.7.2).

The ASLT results from available experimental data and kinetic models should be interpreted with caution when they are applied to the shelf life determination under normal conditions. Experience of similar products needs to be consulted and safety margins should be allowed. Agreement between the extrapolated prediction and actual behavior of the product quality under normal distribution conditions should be monitored even after the product is launched in the market.

16.6.3 Procedures for Shelf Life Studies

As mentioned earlier, there are many ways to conduct shelf life studies depending on the objective. For example, the food product may be tested either with or without the package, the test may be conducted in the laboratory or in the real world, and different accelerating factors may be used depending on the situations. During the storage test for shelf life determinations, a variety of physical, chemical, and microbiological quality attributes may be selected. When sensory attributes are measured, control samples may be presented to the sensory panel. The control samples may be prepared fresh every time when the test is made. Or the sample stored in frozen at -40°C may serve the purpose if the freezing does not change any adverse quality change. When frozen storage is not desired for control or reference samples, 0°C may be adopted for the purpose.

To illustrate the application of some of the concepts described in this chapter, a typical set of procedures for shelf life study in the laboratory is described below. It should be mentioned that these procedures assume that the food deterioration follows the chemical kinetics.

- Determine the primary deterioration modes, quality indexes, and critical limits of the food product or food package.

- Design an experimental plan and determine the number of samples required. Typically, 6-8 data points are required for each temperature. The time sampling intervals may be determined based on past experience with similar products or trial-and-error methods for new products. Daily examination may be required for perishable products such as fresh chilled foods and pasteurized drinks, while four month interval may be appropriate for shelf stable products such as canned foods.

- Store the samples at three different temperatures. In Figure 16.17, T_1 is usually the expected temperature to be encountered by the food product, and T_2 and T_3 are some higher temperatures for accelerated testing. Care should be taken to avoid excessive high temperatures so that the deterioration kinetics remains the same within the selected temperature range between T_1 and T_3. Excessive high temperatures may cause the deteriorative reaction to shift to other reactions.

- Take the samples out and measure the quality indexes at predetermined time intervals until the food product is no longer acceptable; that is, when the critical limit of one of its quality indexes is exceeded.

- As the experiment is progressing, not waiting until all the experiments are completed, plot B (quality index) versus t (storage time) as shown in Figure 16.17. Naturally, the data from T_3 (highest temperature) will arrive most early, followed by those from T_2 and then T_1.

- Determine the reaction order and rate constants from the plots of $[B]/[B_o]$ vs. t or ln $([B]/[B_o])$ vs. t. The reaction order will be obtained from the data of T_3 first and then later to be confirmed by the data obtained from T_2 and then T_1. While the reaction order should be the same, the rate constant is unique for each temperature. Once the reaction order and rate constant of a temperature are known, the shelf life may be predicted and the prediction can be verified later when the actual shelf life is obtained. An incentive of this procedure is that reasonable predictions can be made without waiting for the completion of experiment, which is useful for product development.

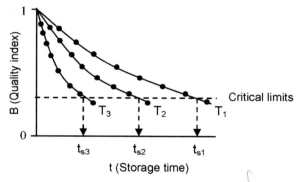

Figure 16.17 Kinetic data from shelf life studies

- Determine the product shelf life of each temperature based on the critical limit.

- Generate the shelf life plot by plotting ln t_s versus T (Figure 16.18). The shelf life t_{s3} will be obtained experimentally first, following by t_{s2} and then t_{s1}. Once t_{s3} and t_{s2} are known, t_{s1} can be estimated from the shelf life plot and later confirmed by actual data. Once t_{s3}, t_{s2}, and t_{s1} are known, the shelf life plot may be used to estimate the shelf life of other temperatures. Again, this procedure allows reasonable predictions before the completion of all experiments

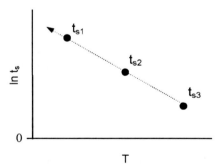

Figure 16.18 Generating shelf life plot from experimental data

Example 16.2 An example of shelf life study

Nonenzyamtic browning of nonhygroscopic whey powder (a_w = 0.44) was measured by Labuza and Kamman [12] in terms of optical density (OD/g solid) of aqueous extract solution at three temperatures. Assuming the critical limit for OD is 0.1, estimate the product shelf life at 15°C from the following data.

Time (day)	Browning at 25°C	Time (day)	Browning at 35°C	Time (day)	Browning at 45°C
0	0.018	0	0.016	0	0.017
30	0.042	10	0.051	2	0.052
60	0.062	20	0.079	4	0.071
90	0.075	30	0.106	7	0.224
120	0.097	40	0.138	11	0.253
150	0.119	50	0.164	18	0.317
180	0.126	60	0.202	28	0.443
210	0.147	70	0.233	35	0.508

As shown in Figure 16.19, the zero order reaction is attempted by plotting OD versus time. Since the linear plots fit the data reasonably well, it is concluded that the reaction is zero order.

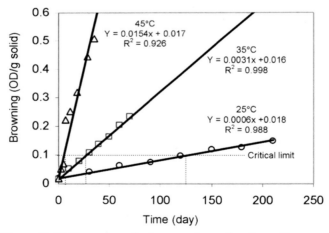

Figure 16.19 Browning of whey powder as function of temperature

The figure also shows the regression equations for the three temperatures. Since the critical limit for OD is 0.1, the shelf life at 25°C may be estimated as 137 days from the regression equation $0.1 = 0.0006\, t_{s,25} + 0.018$. Similarly, the shelf lives at 35°C and 45°C may be estimated as 27 days and 5.4 days, respectively.

Figure 16.20 shows the shelf life plot of ln t_s versus temperature, in which the data follow the expected linear behavior very well. The regression equation may be used to estimate shelf life at lower temperature. For example, the shelf life at 15°C may be estimated from ln $t_{s,15}$ = -0.1636 x 15 + 9.013, where $t_{s,15}$ is calculated to be 706 days.

Q_{10} may be estimated from the shelf life plot in Figure 16.20 using the relationship that its slope = -ln $Q_{10}/10$. The slope = -0.1636 = -ln $Q_{10}/10$, and thus Q_{10} = 5.1. Since the given temperatures 25, 35 and 45°C are exactly 10°C apart, Q_{10} can also be calculated from the shelf lives of these temperatures. According $Q_{10} = t_{s,45}/t_{s,35}$ = 137/27 = 5.1 and $Q_{10} = t_{s,35}/t_{s,25}$ = 27/5.4 = 5, which agree well with the Q_{10} value obtained from the slope.

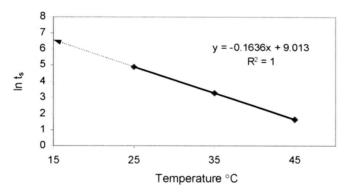

Figure 16.20 Shelf life plot for whey powder

It should be mentioned that the y-intercepts in the regression equations in Figure 16.19 are deliberately anchored at the OD values at time zero (the y-intercepts are 0.017, 0.016, and 0.018). Anchoring the y-intercepts is not a standard practice in regression analysis, but this is done in this case because the OD values at time zero are the average of more replicates than other data based on the original experimental design, and thus the initial OD are believed to be more accurate. If the y-intercepts are not anchored, Q_{10} values slightly different from the above will be obtained and those Q_{10} values also have wide variations among them.

16.7 Shelf Life Models

Shelf life models are mathematical equations which describe the relationship of the food, package, and environment for purposes such as predicting shelf lives of foods, designing food packages, and providing useful insights. Although experimental data are still needed, shelf life models can help to reduce the number of experiments required and provide a systematic way to extract useful meanings from the data.

16.7.1 Shelf Life of Chemical and Microbial Deteriorations

In case that chemical quality change dominates the product deterioration and its rate constant is known, the shelf life under the given conditions is described as given above by Equations 16.4 and 16.8:

$$t_s = \frac{[B]_o - [B]_c}{k} \text{ for zero order}$$

or

$$t_s = \frac{\ln\{[B]_o / [B]_c\}}{k} \text{ for first order}$$

If there are changes of storage conditions with time, reaction rate constant, k value may be evaluated for each conditions by using Arrhenius Equation 16.29 and then the cumulated quality changes may be obtained by stepwise calculation (Equations 16.3 or 16.7) until total shelf life is reached where the quality level arrives at a critical value. For determination of k at each condition, temperature or moisture dependence of the k may be used.

Prediction of microbial growth based shelf-life can be obtained by simple relationship using the maximum specific growth rate and lag time:

$$t_s = \lambda + \frac{1}{\mu} \cdot \ln \frac{N_c}{N_o} \tag{16.50}$$

where N_o and N_c are initial and critical microbial loads for the food. In Equation 16.50, it is assumed that N_c is located in exponential microbial growth phase whose rate is represented by μ. λ and μ can be obtained from experiment or from kinetics published. Intrinsic factors such as pH and other compositional data of the foods determine the λ and μ as well as extrinsic factors such as temperature and package atmosphere.

Software programs are available to predict the growth of pathogenic and spoilage microorganisms under various intrinsic and extrinsic conditions. Some notable programs include the Pathogen Modeling Program®, ComBase Predictor®, Sym'Previus®, and Seafood Spoilage Predictor®. These programs, some of them are freely available, can be easily found through Internet search.

Example 16.3 Estimation of shelf life based on nutrient loss

Kamman et al. [32] have reported that thiamin loss in a pasta product follows the first order reaction with rate constant of 1.99 x 10^{-4} day^{-1} at 25°C and activation energy of 129 kJ mol^{-1}. How long does it take for the retention of thiamin to reach 75% of the initial content at 25 and 35°C?

At 25°C, using the relationship of Equation 16.8,

$$t_{s,25} = \frac{\ln([B]_o/[B]_c)}{k} = \frac{\ln(1.00/0.75)}{1.99\times10^{-4}\ \text{day}^{-1}} = 1446\ \text{days}$$

At 35°C with the same conditions of quality limit, applying Arrhenius equation based shelf life relationship (Equation 16.32) to two temperature conditions yields

$$\ln(\frac{t_{s,35}}{t_{s,25}}) = \frac{E_a}{R}(\frac{1}{T_{35}} - \frac{1}{T_{25}}) = \frac{129000}{8.314}\times(\frac{1}{308} - \frac{1}{298}) = -1.690$$

$$\frac{t_{s,35}}{t_{s,25}} = 0.184$$

$t_{s,35} = 0.184$ x $t_{s,25} = 0.184$ x 1446 days = 266 days

It is shown that faster quality loss or shorter shelf life at higher temperature can be used in ASLT.

Example 16.4 Estimation of shelf life based on microbial growth

Hourly growth of the naturally occurring bacteria, pseudomonads on gilt-head seabream over 0 to 15°C, reported by Koutsoumanis and Nychas [33], can be represented by Belehradek (square root model) equations, $\sqrt{\mu} = 0.0211(T + 10.65)$ and $\sqrt{1/\lambda} = 0.0133(T+12.19)$ where T is temperature in °C. Estimate the shelf life for pseudomonads to multiply from initial level ($10^{4.45}$ cfu/g) to spoilage level ($10^{7.00}$ cfu/g) at 3 and 8°C.

At 3°C, from the square root model equation described above,

$$\mu = [0.0211 \times (3.0 + 10.65)]^2 = 0.0830\ h^{-1}$$

$$\lambda = 1/[0.0133 \times (3.0 + 12.19)]^2 = 24.501\ h$$

Using the Equation 16.50,

$$t_s = \lambda + \frac{1}{\mu}\cdot\ln\frac{N_c}{N_o} = 24.501 + \frac{1}{0.0830}\times\ln(\frac{10^{7.00}}{10^{4.45}}) = 95.2\ h$$

At 8°C, by the same way at 3°C,

$$\mu = [0.0211 \times (8.0 + 10.65)]^2 = 0.1549 \text{ h}^{-1}$$

$$\lambda = 1/[0.0133 \times (8.0 + 12.19)]^2 = 13.868 \text{ h}$$

Using the Equation 16.50 with parameters at 8°C,

$$t_s = 13.868 + \frac{1}{0.1549} \times \ln(\frac{10^{7.00}}{10^{4.45}}) = 53.8 \text{ h}$$

16.7.2 Shelf Life Models of Constant H₂O and O₂ Driving Forces

For oxygen or moisture sensitive packaged foods, it is convenient to define the terms 'maximum allowable O_2' and 'maximum allowable H_2O' beyond which the food products are no longer acceptable. Table 16.11 shows the maximum limits for some food products obtained from the literature. It should be mentioned that these values are for reference only, and experiments are recommended to determine more accurate maximum limits for specific food products of interest.

For moisture or oxygen sensitive foods contained in good barrier packages, the product shelf lives may be assumed to be permeation controlled as discussed in Section 16.5.2. The shelf lives may be estimated by the equations:

$$\text{Shelf life of moisture sensitive food} = \frac{\text{Maximum allowable moisture}}{\text{WVTR}} \quad (16.51)$$

$$\text{Shelf life of oxygen sensitive food} = \frac{\text{Maximum allowable oxygen}}{\text{OTR}} \quad (16.52)$$

The water vapor transmission (WVTR) may be estimated using the permeation equation:

$$\text{WVTR} = \frac{\overline{P_w} A}{L} (p_{w,out} - p_{w,in}) \quad (16.53)$$

where $P_{w,out}$ and $P_{w,in}$ are water vapor pressures outside and inside the package, respectively, $\overline{P_w}$ is water vapor permeability coefficient, A is package surface area, and L is package thickness. $p_{w,out}$ may be obtained by multiplying the saturated water vapor pressure and relative humidity. $p_{w,in}$ may be obtained by multiplying the saturated water vapor pressure and water activity of the food. If the change moisture content of the product is small, $p_{w,in}$ may be assumed as constant and thus the driving force ($p_{w,out} - p_{w,in}$) may be assumed to be constant.

Table 16.11 Maximum allowable limits for some food products

Food	Maximum allowable H_2O (%)	Maximum allowable O_2 (ppm)
Beer, wine	3, loss	1-5
UHT milk	3, loss	1-8
Soups, sauces	3, loss	1-5
Fruits juices, soft drinks	3, loss	10-40
Preserves, jams, pickles	3, loss	50-200
Dried foods	1, gain	5-15
Snack foods, nuts	5, gain	5-15
Instant coffee, tea	1, gain	15-50
Oils and fats, dressings	10, gain	50-200

Adapted from Ref. [20, 34, 35].

Similarly, the oxygen transmission rate (OTR) may be estimated using the permeation equation:

$$OTR = \frac{\overline{P}_{O2} \, A}{L}(p_{O2,out} - p_{O2,in}) \qquad (16.54)$$

where $\overline{P}_{O2}$ is oxygen permeability coefficient, $P_{O2,out}$ and $P_{O2,in}$ are oxygen partial pressures outside and inside the package. The driving force ($P_{O2,out}$ - $P_{O2,in}$) may be assumed to be constant.

Example 16.5 Package requirement to provide a shelf life

Dried green tea is known to be highly sensitive to moisture increase and is tolerable to only 2% increase from initial moisture of 3.0% in wet basis. The 3.0% moisture content corresponds to equilibrium relative humidity of 16%. The green tea of 100 g is going to be packaged in a plastic pouch of 11 x 19 cm. If the package has a shelf life of 60 days at 30°C and 80% relative humidity, what would be the required water vapor permeance of ($\overline{P}_W /L$)?

Allowable moisture gain is obtained as follows:

Dry solid weight = 100 g x (1-0.03) = 97 g

Initial moisture content in dry basis $= \dfrac{0.03}{1-0.03} = 0.031$

Final allowable moisture content in dry basis $= \dfrac{0.05}{1-0.05} = 0.053$

Maximum allowable moisture gain $= 97 \times (0.053-0.031) = 2.13$ g

By using Equation 16.51, WVTR $= \dfrac{\text{Maximum allowable moisture}}{\text{Shelf life}} = \dfrac{2.13}{60}$

$= 0.0355$ g day^{-1}

Now package conditions for the WVTR are specified:

Package surface area, A $= 0.11 \times 0.19 \times 2 = 0.0418$ m^2

Ambient saturation water vapor pressure at 30°C from Equation 16.43

$$P_s = 611.2 \cdot \exp(53.479 - \frac{6808}{T} - 5.09 \cdot \ln T)$$

$$= 611.2 \times \exp(53.479 - \frac{6808}{303} - 5.09 \times \ln 303)$$

$= 4200$ Pa $= 0.0415$ atm

Thus water vapor pressure of ambient air is given as:

$p_{w,out} = H_R \, p_s = 0.8 \times 0.0415 = 0.0332$ atm

Water vapor pressure inside the package is,

$p_{w,in} = a_w \, p_s = 0.16 \times 0.0415 = 6.64 \times 10^{-3}$ atm

From Equation 16.53

$$\frac{\bar{P}_W}{L} = \frac{\text{WVTR}}{A \cdot (P_{w,out} - P_{w,in})} = \frac{0.0355}{0.0418 \times (0.0332 - 6.64 \times 10^{-3})}$$

$= 31.98$ g day^{-1} m^{-2} atm^{-1} $= 0.042$ g day^{-1} m^{-2} mmHg^{-1}

The packaging film should have a water vapor permeance lower than 31.98 g day^{-1} m^{-2} atm^{-1}.

Example 16.6 Estimation of shelf life based on constant oxygen permeation

A certain spice is known to have the oxygen tolerance limit of 50 ppm. Estimate the shelf life of the vacuum packed spice. Product weight is 250 g and initial oxygen concentration inside the package is assumed as zero. The package surface area is 444 cm^2. The package film has the structure of 40 μm PET/10 µm PVDC/15 µm LDPE. O_2 permeability coefficients of PET, PVDC, and LDPE are 0.144, 0.035, and 36.1 cm^3 (STP) µm cm^{-2} day^{-1} atm^{-1}, respectively.

Maximum allowable O_2 in weight $= (250) \times (50 \times 10^{-6}) = 1.25 \times 10^{-2}$ g

Since molecular weight of O_2 is 32 g/mol and 1 mole of O_2 amounts to 22400 cm^3 (STP),

$$\text{Maximum allowable } O_2 \text{ in volume} = \frac{1.25 \times 10^{-2}}{32} \times 22400 = 8.75 \text{ cm}^3$$

(STP)Equation for calculating overall permeability of a multilayer film is obtained from Equation 4.21 in Section 4.3.4: $\sum \dfrac{L_i}{P_{O2,i}} = \dfrac{40}{0.144} + \dfrac{10}{0.035} + \dfrac{15}{36.1} =$

$$563.9 \text{ cm}^2 \text{ day atm cm}^{-3} \text{(STP)}$$

$$OTR = \left(\sum \frac{\overline{P}_{O2,i}}{L_i} \right) \times A \times \Delta p_{O2} = \frac{1}{563.9} \times 444 \times (0.21 - 0) = 0.165 \text{ cm}^3 \text{ (STP) day}^{-1}$$

$$t_s = \frac{\text{Maximum allowable amount of } O_2}{OTR} = \frac{8.75}{0.165} = 53.0 \text{ days}$$

16.7.3 Shelf Life Model of Variable H₂O Driving Force

The models in the previous section are relatively simple since the driving forces are assumed to be constant. If the moisture content of the food changes substantially with time, the constant driving force can no longer be assumed. Figure 16.21 depicts the typical state of moisture change for the food package. Described below is a model derived by Labuza et al [36].

The model is based on several assumptions:

1. Food is moisture sensitive and its shelf life can be determined by a critical moisture content M_c.

2. Package has good moisture barrier and shelf life of food is controlled by water vapor transmission (WVTR).

3. WVTR is accounted for entirely by the change in moisture content in food, assuming that the moisture change in package headspace is neglected except equilibration between headspace water vapor pressure and food moisture.

4. The region of interest in the moisture sorption isotherm (between M_i and M_c, see Figure 16.22) can be approximated using a straight line equation, $M = C_1 a_w + C_2$ (see also Table 16.7). This is a reasonable approximation in many real situations, which yields a simple mathematical solution for providing useful insight as illustrated below. However, it is not difficult to

incorporate nonlinear moisture sorption isotherms such as the GAB equation (see Table 16.7) in computer programs.

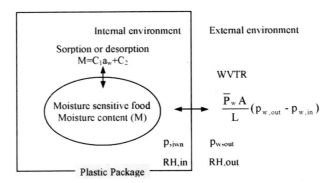

Figure 16.21 Interactions in moisture sensitive food package

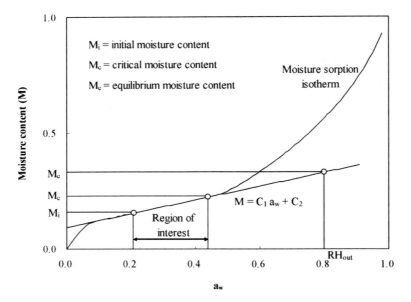

Figure 16.22 Straight line approximation of moisture sorption isotherm

Based on the above assumptions, model is formulated and solved as below.

Step 1: Incorporating mass balance and permeation equation

The mass balance tells that the change of moisture content in the food is determined by moisture permeated through the package by assuming the moisture change in the headspace is negligible:

Rate of moisture = Rate of moisture
gain or loss by food   permeated through package

This mass balance may be represented by the following equation which assumes moisture gain by the food (the situation for moisture loss can be derived similarly):

$$W_s \frac{dM}{dt} = \frac{\bar{P}_w A}{L}(p_{w,out} - p_{w,in}) \qquad (16.55)$$

where W_s is product dry weight, and M is moisture content of food (g moisture/g dry solid) at any time t.

Step 2: Incorporating moisture sorption equation

Note that $p_{w,in}$ the internal partial pressure of water, is changing since the food is absorbing moisture form the internal environment. This interaction between the food and the internal environment is assumed to be related by the linear moisture sorption isotherm as shown in Figure 16.22:

$$M = C_1 a_w + C_2 \qquad (16.56)$$

To incorporate this equation to $p_{w,in,}$ it is necessary to recall the definition (Equation 16.42),

$$H_R \equiv \frac{p_w}{p_s} \qquad \text{or} \qquad a_w \equiv \frac{p_w}{p_s}$$

Solving $p_{w,out}$ and $p_{w,in}$ from the above equations yields

$$p_{w,out} = \frac{M_e - C_2}{C_1} p_s$$

$$p_{w,in} = \frac{M - C_2}{C_1} p_s$$

Step 3: Mathematical operation to integrate to the shelf life with critical moisture content

$$W_s \frac{dM}{dt} = \frac{\bar{P}_w A}{L}(\frac{M_e - C_2}{C_1} - \frac{M - C_2}{C_1})p_s$$

$$\int_{M_i}^{M_c} \frac{dM}{(M_c - M)} = \frac{\overline{P}_w A p_s}{LC_1 W_s} \int_0^{t_s} dt$$

$$-\ln\left[\frac{(M_e - M_c)}{(M_e - M_i)}\right] = \frac{\overline{P}_w A p_s}{LC_1 W_s} t_s$$

$$t_s = \frac{\ln\left[\dfrac{(M_e - M_i)}{(M_e - M_c)}\right]}{\left[\dfrac{\overline{P}_w A p_s}{LC_1 W_s}\right]} \qquad (16.57)$$

From Equation 16.57 of shelf life model for moisture dependent shelf life, three factors (food, package, and environment) affecting shelf life can be reviewed as discussed in Section 16.1.2. In the equation, shelf life is determined by food variables of W_s, M_i, M_c, C_1; Package variables of $\overline{P}_w$, A, L; environmental factors of M_e, p_s. Another environmental variable temperature is not given explicit in the equation but is implicit through the variables of $\overline{P}_w$, C_1, and p_s.

Example 16.7 Estimation of shelf life based on moisture increase

100 g of dried gingers with moisture content of 7.0% (dry basis, monolayer in Table 16.5) are packaged in polypropylene film of thickness 0.08 mm (dimension 13 x 22 cm). Dried gingers need to maintain the moisture content below 9.8% dry basis for preventing nonenzymatic browning. What would be the potential shelf life for this product under the normal conditions of 25°C and 80% relative humidity? The polypropylene film has been reported to have a water vapor permeability of 2.69 g mm m^{-2} day^{-1} atm^{-1}.

The moisture sorption isotherm of the gingers is given as follows:

a_w	M (%)	a_w	M (%)	a_w	M (%)
0.11	5.89	0.41	10.71	0.75	18.00
0.23	7.15	0.51	11.19	0.84	25.09
0.32	8.86	0.57	12.60		

From initial moisture condition, dry solid weight of the packaged product is given by:

$$W_s = \text{weight x fraction of dry solid} = 100g \times \frac{100}{100 + 7} = 93.46 \text{ g}$$

Surface area of both sides in the package pouch,

$$A = 0.13 \times 0.20 \times 2 = 0.0572 \text{ m}^2$$

From the moisture sorption isotherm as shown in Figure 16.23, C_1=0.1464.

Hypothetical equilibrium moisture content on the linearized moisture sorption isotherm at RH of 80% is given by $M_e = 0.1464 \times 0.8 + 0.0415 = 0.1587$. At 25°C, saturation water vapor pressure is given from Table 16.8 (or Equation 16.43): $p_s = 3.169$ kPa $= 0.031$ atm.

$$t_s = \dfrac{\ln\left[\dfrac{(M_e - M_i)}{(M_e - M_c)}\right]}{\left[\dfrac{\overline{P}_w A p_s}{L C_1 W_s}\right]} = \dfrac{\ln\left[\dfrac{(0.1587 - 0.0700)}{(0.1587 - 0.0980)}\right]}{\left[\dfrac{2.69 \times 0.0572 \times 0.031}{0.08 \times 0.1464 \times 93.46}\right]} = 87.1 \text{ days}$$

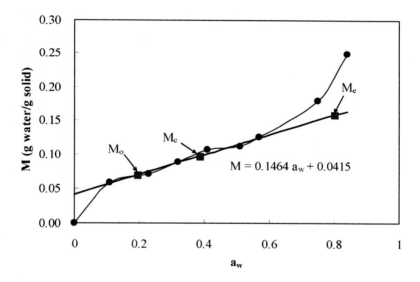

Figure 16.23 Moisture sorption isotherm of ginger at 25°C in Example 16.7

16.7.4 Shelf Life Model of Variable O_2 Driving Force

Many packaged foods deteriorate in the quality due to oxidation reactions with oxygen. The rate and degree of the oxidation depends on the concentration or partial pressure of oxygen, which has been initially present in the food and/or headspace, or has been permeated through the packaging material. When oxidation reactions determine the shelf life, the time to reach a certain level of oxidation is defined as shelf life. The tolerable level of oxidation depends on the sensitivity of the food to oxygen ranging from a few μg to a few mg per each gram of food (Table 16.11).

The oxygen balance on the package of the food undergoing oxidation is written as follows:

$$\frac{d[O_2]}{dt} = \frac{(\frac{\bar{P}_{O2}A \cdot (0.21-[O_2])}{L} - R_{O2}W)}{V} \tag{16.58}$$

where $[O_2]$ is oxygen concentration or partial pressure in atm, $\bar{P}_{O2}$ is oxygen permeability coefficient in mL μm h^{-1} m^{-2} atm^{-1}, R_{O2} is oxidation rate in mL kg^{-1}, W is the product weight (kg), and V is the package free volume (mL). The first term in the right side of the equation numerator amounts to the permeation through the packaging layer while the second term is the oxygen consumption by the food.

When the food is packaged in vacuum or gas-flushed with impermeable film, permeation is very limited with little change in oxygen concentration. In this case permeation becomes to be constant and balanced to the oxidation by the food, i.e., the permeated oxygen is instantly consumed by the food for the oxidation reaction. This simple case makes the shelf life determination easy, which has already been described in Section 16.7.2.

There would be another case with high permeable package, in which the permeation is much higher than the oxidation. If we disregard the oxygen consumption by oxidation process, the Equation 16.58 is simplified to:

$$\frac{d[O_2]}{dt} = \frac{\bar{P}_{O2} \cdot A \cdot (0.21-[O_2])}{L \cdot V} \tag{16.59}$$

The analytical solution of the differential equation is given as:

$$t_s = \frac{L \cdot V}{\bar{P}_{O2} \cdot A} \ln\left(\frac{0.21-[O_2]_i}{0.21-[O_2]_c}\right) \tag{16.60}$$

where the shelf life t_s is the time from initial oxygen concentration $[O_2]_i$ to critical oxygen concentration $[O_2]_c$. Similar derivation for general gas permeation was also given in Section 4.7.5.

Complete handling of Equation 16.58 can be found in Section 16.8 of the case with both oxygen permeation and oxidation reaction.

Example 16.8 Estimation of shelf life based on package oxygen concentration

It was thought that an active dry yeast product loses significant activity when headspace oxygen in the package exceeds 3.50% [37]. When 8 g of dry yeast is packaged with multilayered film pouch of 7.0 x 7.0 cm (film structure is same in Example 16.6), the oxygen consumption of the dry yeast is assumed negligible. What would be the shelf life of this package if the headspace has an initial oxygen concentration of 0.1% in 12 mL (STP) of headspace just after sealing?

From the Example 16.6 above, $\sum \dfrac{L_i}{P_{O2,i}} = 563.9$ cm^2 day atm cm^{-3}

Appling Equation 16.60, time to reach 3.50% oxygen concentration in the package is obtained as shelf life:

$$t_s = \frac{L \cdot V}{P_{O2} \cdot A} \cdot \ln\left(\frac{0.21 - [O_2]_i}{0.21 - [O_2]_c}\right) = 563.9 \times \frac{12}{2 \times 7.0 \times 7.0} \cdot \ln\left(\frac{0.21 - 0.001}{0.21 - 0.035}\right) = 12.3 \text{ days}$$

16.7.5 Package/Food Compatibility Dependent Shelf Life Model

As discussed in Chapter 5, migration of additive from the packaging material may limit the food shelf life because of concerns on the safety and disturbed flavor. Scalping of flavor compounds on the plastic packaging may cause the deterioration in sensory quality of the food products such as orange juice. The time to reach certain level of migration or adsorption of m_t can be obtained by the Equations 16.61 and 16.62, which were converted from Equations 16.26 and 16.27, respectively (see also Section 5.6.1)

$$t_s = \frac{\pi L^2}{4D}\left(\frac{m_t}{m_\infty}\right)^2 \tag{16.61}$$

$$t_s = \frac{1}{D}\left(\frac{V}{A} \cdot \frac{m_t}{m_\infty}\right)^2 \tag{16.62}$$

It should be noted that the above equation holds just for initial stage of migration or sorption with $\dfrac{m_t}{m_\infty} < 0.6$. For the more information on migration and scalping refer to Chapter 5.

Example 16.9 Estimation of shelf life based on migration from packaging

According to Baner [38], styrene monomer may lead to off-flavor at a concentration above 300 μg kg^{-1}. A coffee creamer of 7.5 g is packed in a polystyrene cylinder having a contact surface area of 11.3 cm^2 and thickness of 0.38 mm. The polystyrene cylinder is known to have an initial residual content of 800 mg kg^{-1} and a density of 1080 kg m^{-3}. Estimate the time to reach the critical level of off-flavor development. Assume that all the styrene monomers can be leached out with very long time of contact with diffusion coefficient of 3.10×10^{-18} m^2 s^{-1}.

Migration amount at critical limit is

$$m_t = 300 \ \mu g \ kg^{-1} \times 0.0075 \ kg = 2.25 \ \mu g$$

Total amount of styrene initially contained in the polystyrene cylinder is assumed to migrate in equilibrated state,

m_∞ = initial concentration x volume of the packaging material x density

= 800 mg kg^{-1} x 0.00113 x 0.38 x 10^{-3} m^3 x 1080 kg m^{-3} = 0.371 mg

$$\frac{m_t}{m_\infty} = \frac{2.25 \times 10^{-3}}{0.371} = 6.065 \times 10^{-3}$$

Using Equation 16.61

$$t_s = \frac{\pi L^2}{4D}\left(\frac{m_t}{m_\infty}\right)^2 = \frac{3.14 \times (0.38 \times 10^{-3})^2}{4 \times 3.10 \times 10^{-18}} \times (6.065 \times 10^{-3})^2 = 1.345 \times 10^6 \text{ s}$$

$$= 373.6 \text{ h} = 15.6 \text{ days}$$

or by approximate simple form of Equation 16.62,

$$t_s = \frac{1}{D}\left(\frac{V}{A} \cdot \frac{m_t}{m_\infty}\right)^2 = \frac{1}{3.10 \times 10^{-18}} \times \left(\frac{0.00113 \times 0.38 \times 10^{-3}}{11.3 \times 10^{-4}} \times 6.065 \times 10^{-3}\right)^2$$

$$= 1.713 \times 10^6 \text{ s} = 476.0 \text{ h} = 19.8 \text{ days}$$

Solution with more simplified approximation estimates a slightly longer shelf life.

16.8 Case Study: Shelf Life of Potato Chip with Two Interacting Quality Deterioration Mechanisms

Quast and Karel [27] studied the shelf life of a potato chip product packaged in plastic bags. The potato chip is moisture and oxygen sensitive and becomes unacceptable when its water activity exceeds 0.32 or it absorbs more than 1200 μL O$_2$ (STP) per gram of product. The relationship between moisture content and water activity of the product can be described by the Kuhn isotherm:

$$a_w = e^{\frac{-0.0387}{M + 0.0051}} \qquad (16.63)$$

The package is a good barrier to moisture and oxygen so that the deterioration of the food product may be assumed to be permeation controlled. That is, the product shelf life is limited by the rate of permeation through the package while moisture and oxygen are assumed to be absorbed instantly by the food.

The change of moisture content in the food is equated to the rate of water transmission rate through the package as follows:

$$\frac{dW}{dt} = \frac{dM \times W_s}{dt} = \frac{\overline{P}_w A \times (p_{w,out} - p_{w,in})}{L} = \frac{\overline{P}_w A \times (H_R p_s - a_w p_s)}{L} \qquad (16.64)$$

Substituting Equation (16.63) yields:

$$\frac{dM}{dt} = \frac{\overline{P}_w \times A \times p_s \times (H_R - e^{\frac{-0.0387}{M+0.0051}})}{L \times W_s}$$

(16.65)

The rate of oxygen concentration change is obtained from Equation 16.58 as follows:

$$\frac{d[O_2]}{dt} = \frac{(\frac{\overline{P}_{O2} A \cdot (0.21 - [O_2])}{L} - R_{O2}W)}{V}$$

(16.66)

in which oxidation rate is given as a function of $[O_2]$ (decimal) by:

$$R_{O2} = \left[\int R_{O2}dt + \frac{445.24 - 0.39111 \int R_{O2}dt}{100 \cdot a_w} \right] \cdot \left[\frac{[O_2]}{6.1924 + 818.36[O_2]} \right]$$

(16.67)

By solving the differential Equations 16.65-16.67, moisture content, oxygen concentration, and extent of oxidation can be obtained as reported by Quast and Karel [27]. Shelf life is terminated when the oxidation extent ($\int R_{O2}dt$) reaches 1200 μL O_2 (STP)/gram or moisture reaches 0.029 g/g dry solid corresponding to a_w of 0.32.

The common package conditions illustrated are as follows: temperature = 310K; outside environmental humidity = 0.4; initial moisture = 0.0005 g/g solid; initial oxidation extent =0; packaging film thickness = 1 mil (25.4 μm); surface area of packaging film = 0.0123 m^2; water vapor permeability coefficient of the film = 0.0010 g mil m^{-2} h^{-1} mm Hg^{-1}; oxygen permeability coefficient of the film = 9.1 cm^3 (STP) mil m^{-2} h^{-1} atm^{-1}.

Two experimental packages were simulated in moisture change, package atmosphere and oxidation extent by solving the differential equations above: one with product weight of 10 g and package headspace volume of 371.4 mL; another with 50 g and 337.0 mL. Figure 16.24 shows the calculation results obtained using a Mathcad$^®$ program, which can be found in the website given in the preface. For both cases spoilage occurred due to excessive oxidation rather than excessive moisture increase: time to oxidation quality limit is shorter than that to moisture limit as shown in the figure. The estimated shelf lives for the two experimental packages were 2200 and 2700 hours, respectively. As discussed by Quast and Karel [27], the computer simulation can make it easier to design and optimize the package.

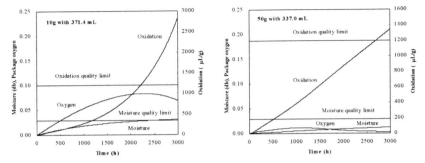

Figure 16.24 Progress of package oxygen concentration, moisture, and oxidation of the potato chip for two different packaging conditions

Discussion Questions and Problems

1. Provide some possible solutions to prevent the collapse of PET bottles in Example 16.1.

2. In Example 16.2 analyzing the data from Labuza and Kamman [12], the browning rate constant of nonhygroscopic whey was given as 0.0031 OD day^{-1} (g dry solid)$^{-1}$ at 35°C. The tolerable limit of acceptable quality is set as 0.1 OD/g dry solid. What would be the product shelf life at 28°C if the temperature dependence is described by an activation energy of 124.3 kJ/mol in Arrhenius equation? The initial browning level of the whey is 1.6 x 10^{-2} OD/g dry solid.

3. Assume an oxygen sensitive package food has a shelf life of 12 months. If the package thickness is decreased by 20% and the surface area is decreased by 30%, what is the new shelf life?

4. Below are the data obtained from a shelf life study for a food product:

T = 40°C		T = 30°C	
t (month)	B	t (month)	B
0	100.0	0	100.0
1	68.4	1	86.1
2	46.8	2	74.1
3	32.0	3	63.8
4	21.9	4	54.9
5	15.0	5	47.2

B is the primary quality factor of the food. A sensory panel has determined that the food is no longer acceptable when B ≤ 25.

a. Determine whether the data fit the first-order kinetics.

b. Estimate the rate constant and shelf life of the food at 40°C.

c.   Estimate the rate constant and shelf life of the food at 30°C.

d.   Estimate the rate constant and shelf life of the food at 20°C.

5.   The loss of the primary quality factor B of a food product is determined to follow the first order kinetics, and the rate constants at three different temperatures are given as follows:

T (°C)	10	25	45
Rate constant k (day^{-1})	0.0042	0.0094	0.0182

The food is stored at four different temperatures for durations shown as follows.

T (°C)	15	23	35	28
Time (days)	10	15	30	60

a.   Estimate the rate constants at 15, 23, 35, and 28°C.

b.   Plot $\ln(B/B_o)$ versus time for the entire duration (10 + 15 + 30 + 60 = 115 days). B_o is the value of B at time zero, and thus B/B_o is the fraction of quality remaining. The plot will consist of four lines, one line per duration, connecting together.

c.   Plot B/B_o versus time for the entire duration. Note that each duration will be represented by a curve instead of a straight line.

d.   If the maximum allowable quality loss is 40%, will the end of shelf life occur before the end of the 115 days?

6.   Prove the following relation between Q_{10} and E_a:

$$\log_{10} Q_{10} = \frac{0.522 E_a}{T(T+10)}$$

In this equation, T must be in K and E_a in J/mol. What is the value of E_a for $Q_{10} = 2$? T=25°C may be used in the calculation.

7.   The shelf life data for a food product are given as follows:

T (°C)	t_S (day)
12	216
25	83
33	49
45	23

As mentioned in Section 16.3.4, shelf life t_s can be estimated using (1) the Arrhenius plot, $\ln t_s$ versus 1/T, where T must be in absolute temperature, or

(2) the shelf life plot, ln t_s versus T, where T does not have to be absolute temperature. These plots may be done using the semi-log graph or regular scale graph.

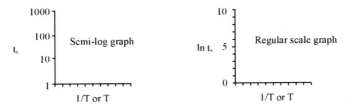

a. Plot the data using Excel to obtain an Arrhenius plot on semi-log graph and find t_s at 38°C.

b. Plot the data using Excel to obtain an Arrhenius plot on regular scale graph and find t_s at 38°C.

c. Plot the data using Excel to obtain a shelf life plot on semi-log graph and find t_s at 38°C.

d. Plot the data using Excel to obtain a shelf life plot on regular scale graph and find t_s at 38°C.

8. The shelf life of a food product at 25°C is 12 months. Construct a shelf life plot with a temperature range of 25°C to 45°C, and plot the lines for Q_{10} of 1, 2, 3, and 4. What is the shelf life of the food at 40°C if $Q_{10} = 3$?

9. A food has an 18 month shelf life at 15°C, and $Q_{10} = 3.5$. What are its shelf lives at 25°C and 30°C?

10. Migration of hazelnut oil into the chocolate from fat filling in the composite chocolate product is thought to cause hardening of the filling and softening of the chocolate. The migrated fat may form a eutectic mixture, which migrates to the chocolate surface with recrystallization of cocoa butter as bloom. According to Miquel et al. [19], diffusion coefficient of the hazelnut oil in the chocolate is 1.2×10^{-7} cm^2 s^{-1}. 4 mm thick filling is contacting a 6 mm thick chocolate layer in a composite product of 16 mm diameter. Initially the concentration of hazelnut oil in the fat filling and chocolate is 33 and 0%, respectively. What would be the time for the hazelnut oil concentration to reach 3% in the chocolate layer?

11. A freeze-dried *Lactococcus lactis* strain is a product used as a start culture for fermentation of yogurt. The *Lactococcus lactis* strain product has deactivation of k=0.111 day^{-1} at 25°C and activation energy of 34.62 kJ/mol in the range of 4-55°C [39]. How can you design the shelf life at 5°C based on the retention of 90% in its survival ratio?

12. 40 g of dry oatmeal initially at water activity of 0.52 (moisture wet basis: 5.3%) is to be packaged in flexible pouch of surface area 0.02 m². Shelf life of 9 months at tropical conditions (38°C, 90% RH) is required. Primary quality attribute determining shelf life is the moisture increase to 7.6% causing the lumpy structure [40]. Find the allowable moisture permeability of the packaging film whose thickness is 60 μm.

13. In a modified atmosphere packaging applying a high CO_2 concentration, loss of CO_2 may be a problem limiting its efficacy and shelf life. Derive a formula showing the package CO_2 concentration as a function of time. Assume that package volume is constant and CO_2 production from the food is negligible. CO_2 concentration in the ambient air may be assumed as zero. Please refer to Equations 16.59 and 16.60. You can also consult Section 4.7.5.

14. Karel and Nickerson [41] showed that ascorbic acid destruction and browning of dehydrated orange juice increase with water activity. Those two quality changes were significantly higher above water activity of 0.3. If 12 g of product at initial moisture of 1.5% (dry basis) is packaged in a laminated plastic film pouch of 4 x 10 cm and stored under relative humidity of 78% at 25°C, determine the shelf life for the product to reach a_w of 0.3. The laminated plastic film has a water vapor permeance of 5.3 g m⁻² atm⁻¹ day⁻¹ ($\overline{P}_w/L$). Water sorption isotherm at 25°C was given as following table:

a_w	M (g water/g dry solid)	a_w	M (g water/g dry solid)	a_w	M (g water/g dry solid)
0.01	0.005	0.22	0.032	0.58	0.182
0.06	0.007	0.32	0.074	0.75	0.340
0.11	0.011	0.44	0.110		

15. 200 g of dried anchovy at water activity of 0.6 was packed inside 50 μm thick low density polyethylene film pouch of 18 x 14 cm and showed constant oxygen concentration of 15.2% inside the package during 2 month storage at 25°C. Oxygen transmission rate of the film was 3700 cc m⁻² atm⁻¹ day⁻¹. What would be the oxygen consumption rate of the dried anchovy which is assumed indifferent of water activity change.

16. From the analysis of oxygen consumption data of a freeze-dried shrimp product reported by Simon et al. [42], its oxygen consumption in μL/(g day) is found to be described simply as $R_{O2}=46.0 [O_2]$, where $[O_2]$ is decimal oxygen concentration. 16 g of dry product is packaged with 2.9 mil thick plastic film pouch of 0.0075 m² surface area, which has oxygen permeance of 100 cc m⁻² atm⁻¹ day⁻¹. Immediately after storage steady state is known to be attained inside the package. What would be the steady state oxygen

concentration inside the package? The 'cut-off' value for the product quality is the extent of oxidation in 200 μL g^{-1}. Estimate the shelf life for this product.

Bibliography

1. TP Labuza. Shelf-life Dating of Foods. Westport, Conneticut: Food & Nutrition Press, 1982, pp. 1-87, 99-118.
2. Anonymous. Open shelf life dating of food. Food Technol 35(2):89-96, 1981.
3. MJ Ellis. The methodology of shelf life determination. In: CMD Man, AA Jones, ed. Shelf Life Evaluation of Foods. London: Blackie Academic & Professional, 1994, pp. 27-39.
4. RP Singh, Anderson, B.A. The major types of food spoilage. In: R Steele, ed. Understanding and measuring the shelf-life of food. Cambridge, UK: Woodhead Publishing, 2004, pp. 3-23.
5. D Kilcast. Sensory evaluation methods for shelf-life assessment. In: D Kilcast, P Subramaniam, ed. The Stability and Shelf-life of Food. Boca Raton: CRC Press, 2000, pp. 79-105.
6. MR Goddard. The storage of thermally processed foods in containers other than cans. In: CMD Man, AA Jones, ed. Shelf Life Evaluation of Foods. London: Blackie Academic & Professional, 1994, pp. 256-274.
7. T Martens. Harmonization of Safety Criteria for Minimally Processed Foods. Leuven, Belgium, 1998, pp. 37-39.
8. B Lund. The microbiological safety of prepared vegetables. In: A Turner, ed. Food Technology International Europe. London: Sterling Publication, 1993, pp. 196-200.
9. O Fennema. Chemical changes in food during processing-an overview. In: T Richardson, JW Finley, ed. Chemical Changes in Food during Processing. Westport: AVI Publishing, 1985, pp. 1-16.
10. JF Gregory III. Vitamins. In: O Fennema, ed. Food Chemistry. 3rd Ed. New York: Marcel Dekker, 1996, pp. 531-616.
11. LH Ababouch, L Souibri, K Rhaliby, O Quahdi, M Battal, FF Busta. Quality changes in sardines (*Sardina pilchardus*) stored in ice and at ambient temperature. Food Microbiol 13:123-132, 1996.
12. TP Labuza, JF Kamman. Reaction kinetics and accelerated tests simulation as a function of temperature. In: IS Saguy, ed. Computer-Aided Techniques in Food Technology. New York: Marcel Dekker, 1997, pp. 71-115.
13. RP Singh. Scientific principles of shelf life evaluation. In: CMD Man, AA Jones, ed. Shelf Life Evaluation of Foods. London: Blackie Academic & Professional, 1994, pp. 3-26.
14. DW Schaffner, TP Labuza. Predictive microbiology: where are we, and where are we going? Food Technol 51(4):95-99, 1997.
15. TA McMeekin, JN Olley, T Ross, DA Ratkowsky. Predictive Microbiolgy. Taunton, U.K.: Research Studies Press, 1993, pp. 11-164.

16. J Baranyi, TA Roberts. A dynamic approach to predicting bacterial growth in food. Int J Food Microbiol 23:277-294, 1994.
17. J Baranyi, TP Robinson, A Kaloti, BM Mackey. Predicting growth of *Brochothrix thermosphacta* at changing temperature. Int J Food Microbiol 27:61-75, 1995.
18. S Koseki, S Isobe. Prediction of pathogen growth on iceberg lettuce under real temperature history during distribution from farm to table. Int J Food Microbiol 104:239-248, 2005.
19. ME Miquel, S Carli, PJ Couzens, HJ Wille, LD Hall. Kinetics of the migration of lipids in composite chocolate measured by magnetic resonance imaging. Food Res Int 34:773-781, 2001.
20. K Eichner. The influence of water content and water activity on chemical changes in foods of low moisture content under packaging aspects. In: M Mathlouthi, ed. Food Packaging and Preservation. London: Elservier Applied Science, 1986, pp. 67-92.
21. TP Labuza, SR Tannenbaum, M Karel. Water activity and stability of low-moisture and intermediate moisture foods. Food Technol 24(5):35-42, 1970.
22. HA Iglesias, J Chirife. Handbook of Food Isotherms. New York: Academic Press, 1982, pp. 262-319.
23. SSH Rizvi. Thermodynamic properties of foods in dehydration. In: MA Rao, SSH Rizvi, ed. Engineering Properties of Foods. New York: Marcel Dekker, 1986, pp. 133-214.
24. DG Quast, M Karel, WM Land. Development of a mathematical model for oxidation of potato chips as a function of oxygen pressure, extent of oxidation and equilibrium relative humidity. J Food Sci 37:673-678, 1972.
25. S Mizrahi, TP Labuza, M Karel. Computer-aided predictions of extent of browning in dehydrated cabbage. J Food Sci 35:799-803, 1970.
26. MS Rahman, TP Labuza. Water activity and food preservation. In: MS Rahman, ed. Handbook of Food Preservation. New York: Marcel Dekker, 1999, pp. 339-382.
27. DG Quast, M Karel. Computer simulation of storage life of foods undergoing spoilage by two interacting mechanisms. J Food Sci 37:679-683, 1972.
28. CF Bohren, BA Albrecht. Atmospheric Thermodynamics. New York: Oxford University Press, 1998, pp. 55-56, 197-202.
29. K Koutsoumanis, PS Taoukis, GJE Nychas. Development of a safety monitoring and assurance system for chilled food products. Int J Food Microbiol 100:253-260, 2005.
30. S Mizrahi. Accelerated shelf-life tests. In: R Steele, ed. Understanding and Measuring the Shelf-Life of Food. Cambridge, UK: Woodhead Publishing, 2004, pp. 317-339.
31. GL Robertson. Food Packaging. New York: Marcel Dekker, 1993, pp. 338-380.

32. JF Kamman, TP Labuza, JJ Warthesen. Kinetics of thiamin and riboflavin loss in pasta as a function of constant and variable storage conditions. J Food Sci 46:1457-1461, 1981.
33. K Koutsoumanis, G-JE Nychas. Application of a systematic experimental procedure to develop a microbial model for rapid fish shelf life predictions. Int J Food Microbiol 60:171-184, 2000.
34. RJ Ashley. Permeability and plastics packaging. In: J Comyn, ed. Polymer Permeability. London: Elsevier Applied Science, 1985, pp. 269-308.
35. WE Brown. Selecting plastics and composite barrier systems for food packages. In: JI Gray, BR Harte, J Miltz, ed. Food Product-Package Compatibility. Lancaster: Technomic Publishing, 1987, pp. 200-228.
36. TP Labuza, S Mizrahi, M Karel. Mathematical models for the optimization of flexible film packaging of foods for storage. T ASAE 15:150-155, 1972.
37. WE Brown. Plastics in Food Packaging. New York: Marcel Dekker, 1992, pp. 358-395.
38. AL Baner. Case study: styrene monomer migration into dairy products in single portion packs. In: O-G Pringer, A Baner, ed. Plastic Packaging Materials for Food. Weinheim: Wiley-VCH, 2000, pp. 427-443.
39. M Achour, N Mtimet, C Cornelius, S Zgouli, A Mahjoub, P Thonart, M Hamdi. Application of the accelerated shelf life testing method (ASLT) to study the survival rates of freeze-dried *Lactococcus* starter cultures. J Chem Technol Biot 76:624-628, 2001.
40. SS Verma. A rapid method to determine the barrier requirements for dehydrated oatmeal product using flexible packaging films. Packag Technol Sci 10:291–294, 1997.
41. M Karel, JTR Nickerson. Effects of relative humidity, air, and vacuum on browning of dehydrated orange juice. Food Technol 18:1214-1218, 1964.
42. IB Simon, TP Labuza, M Karel. Computer-aided predictions of food storage stability: oxidative deterioration on a shrimp product. J Food Sci 36:280-286, 1971.

Chapter 17

Food Products Stability and Packaging Requirements

17.1 Introduction 543
17.2 Cereals and Bakery Products 545
    17.2.1 Cereal Grains and Flours ·· 545
    17.2.2 Ready-to-Eat Breakfast Cereals and Snacks ························· 547
    17.2.3 Fresh and Dried Pasta Products ··· 548
    17.2.4 Fresh Bakery Products ··· 550
17.3 Meat and fish products 553
    17.3.1 Fresh Meat and Poultry ·· 553
    17.3.2 Processed Meat Products ·· 556
    17.3.3 Fish Products ··· 557
17.4 Dairy Products 559
    17.4.1 Pasteurized and UHT Sterilized Milk and Cream ··············· 559
    17.4.2 Dried Milk Products ·· 561
    17.4.3 Cheese ··· 562
    17.4.4 Fermented Milks ·· 564
17.5 Confectionery Products 565
    17.5.1 Chocolate Products ··· 566
    17.5.2 Hard Boiled Sweets ·· 567
    17.5.3 Toffees and Other Confectioneries ···································· 568
17.6 Fats and Oils 570
17.7 Drinks 571
    17.7.1 Fruit Juices ·· 571
    17.7.2 Soft Drinks ·· 573
    17.7.3 Beer ·· 574
    17.7.4 Wine ··· 575
17.8 Fresh Fruits and Vegetables 576
17.9 Frozen Foods 581
Discussion Questions and Problems 585
Bibliography 585

17.1 Introduction

Foods being packaged include whole categories of cereal, horticultural, meat, fishery, dairy, and beverage products ranging from fresh to fully processed state. Each category food has its specificity in quality attributes, storage conditions, expected shelf life, and packaging tools applied. Quality deterioration mode differs with food type and the prevailing mode of a food may also be determined by the storage and packaging conditions. The packaging and distribution should consider the quality factors specific to the food category in its design. Effect of environmental factors on the quality factors should be evaluated and the way to protect them should be elaborated. The protection provided by package includes the measures against physical, chemical, and biological stresses, which are determined by the environmental conditions of the product's journey from food factory to consumers. Qualitative and quantitative information on quality deterioration depending on those stresses needs to be collected for proper design of food package.

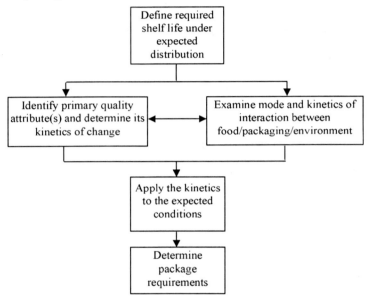

Figure 17.1 Flow diagram for designing package requirements

As discussed in Chapter 16, package requirements for the defined shelf life can be found from the knowledge of primary food quality factors interacting with package and environment (Figure 17.1). The required shelf life may be decided based on previous experience, consumer feedback, marketing, distribution, and so on. Once the required shelf life is determined, one needs to extract primary quality attribute and determine the kinetics of its deterioration. Next, the environmental conditions (temperature, humidity, and time related to the distribution system) must be determined. The effect of environment and packaging on quality change is characterized: interaction among food, packaging, and environment is looked up. The package requirements can then be estimated by applying the kinetics or shelf life model to the expected conditions.

The procedure may be iterated back and forth to find an optimum combination of shelf life, package design, and food supply conditions. This chapter analyzes the quality change aspects of different food types with regards to intrinsic and extrinsic factors, and then presents the way of designing the packaging, distribution, and shelf life of the packaged food.

Because of diversity of foods, it is impossible to deal with all the food types in a chapter. Some typical foods were chosen as examples showing the approach based on the scientific principles. So this chapter focuses on the interrelated scientific process of evaluating food stability and designing the food package system rather than comprehensive review on the whole category of food packaging.

17.2 Cereals and Bakery Products

17.2.1 Cereal Grains and Flours

Cereal grains still maintain the germination capability as seeds, therefore can sprout under proper conditions of temperature and humidity but after a certain period of time. However, they are very stable physiologically under dried conditions. In this context, packaging of these products normally plays a role of protecting from moisture. If the moisture content increases with poor protection, enzymatic and respiration activities increase and result in various kinds of quality degradation such as taste and nutrient contents. Cereal flour loses its physiological activity due to disintegration of tissues and removal of germs in the grinding process.

Main indexes for quality deterioration during storage of cereals and flours are increase of fatty acids from lipid hydrolysis by lipases, increase of reducing sugar from starch by amylases, increase of amino acids from protein, and decrease of vitamins [1]. Germinability and activity of succinic acid dehydrogenase, being index of life activity, decrease with storage time. Cereals may become harder with development of off flavor and off taste. With high moisture content, mold may develop on the surface. Sometimes, insect eggs may contaminate the cereal products and develop to adulthood when the appropriate conditions of humidity and temperature are met. Compared to whole grains the milled products are much more susceptible to lipid degradation by lipases because of their dispersion as consequence of milling [2]. The flour product may be added or fortified with enzymes, vitamins, or minerals, which may change the storage stability.

As mentioned above, the quality changes of cereal grains and flour depend on the moisture content. Temperature is another important factor affecting the shelf life of the products. Figures 17.2 shows the shelf lives of wheat grain as function of moisture and storage temperature. Usually cereals and flours dried to proper level of moisture content are stored and marketed under ambient temperature conditions. In this case the degree of water barrier property determines the quality deterioration and shelf life. Maintenance of low moisture content attains low level of fatty acid accumulation, better puffing, high germinability, and low fungi count. Mechanical strength of the film resistant to physical stresses is also required for protecting the film's integrity and barrier characteristics. For the wheat flour, moisture content needs to be less than 13.5%, desirably less than 12.0% for long period storage. Brown and milled rice with moisture less than 14.5% can preserve good quality with little changes in acidity, reducing sugar, and mold count at 25°C for 150 days [3].

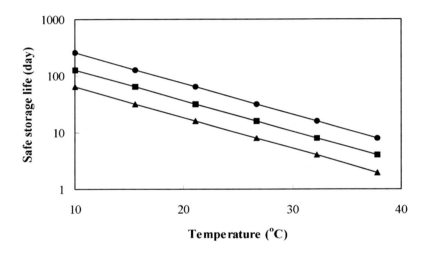

Figure 17.2 Safe storage life of grains as function of moisture and temperature. Moisture: ●, 14%; ■, 15.5%; ▲,17%. Drawn based on the data from Ref. [4].

When higher quality is desired for the cereals, they are stored at relatively low temperature. Brown rice is stored at the conditions of temperature below 15°C and relative humidity of 70-80% when long term storage is needed for better control of insects and microorganisms with high organoleptic quality after milling and cooking [1]. Sharp and Timme [5] reported better quality of brown rice stored at 3°C compared to those at 22 and 38°C.

Vacuum or modified atmosphere packaging (MAP) is effective somehow to control insect problems in cereals and flours [6]. Flushing the package with CO_2 or N_2 gas is reported to kill the insect eggs and larvae which had infested the wheat and wheat flour. Rice packaged under vacuum or modified atmospheres such as nitrogen and carbon dioxide undergoes slow changes in odor and flavor resulting in good sensory quality for the cooked one [7]. Shelled brown rice packaged with nylon-laminated plastic bag under CO_2 gas creates a tight vacuum-like appearance because of dissolution of headspace CO_2 into moisture and lipid of the rice, thus occupies less space in stack and shelf compared to regular bag, and provides better stability against lipid oxidation at refrigerated temperature [8].

17.2.2 Ready-to-Eat Breakfast Cereals and Snacks

Breakfast cereals cover corn flake, puffed rice, sugar-coated flakes, and flakes with dried fruits. Cereal snack includes tortilla, cracker, corn collet, and popcorn. Major quality attributes in ready-to-eat breakfast cereals and snacks with storage are organoleptic properties such as flavor and crispness, and analytical nutrient values when the product is fortified with vitamin/mineral and a claim is made [9-11]. Color is of less significance, since it is generally stable with age. Crispness is a strong function of moisture content. Puffed rice is extremely hygroscopic and loses its texture rapidly if the package does not provide a good moisture barrier. Puffed rice attains maximum crispness at 2% moisture content and loses crisp texture at 5% of moisture. Howarth [10] suggested critical moisture content as 6, 7, and 5% for flake, component-added cereal, and coated products, respectively. Figure 17.3 presents crispness of barley flake and sensory hedonic score as function of water activity, which decides critical water activity and moisture as 0.48 and 5.6%, respectively. Thus the liner or packaging film for these products should have good moisture barrier to ensure the required shelf life, which ranges from 6 to 16 months. Crispy nature of these products also requires a protection from physical damage.

The oxidative rancidity is another major mode of quality deterioration in breakfast cereals caused by the presence of unsaturated fatty acids. It is accelerated at moisture content below the monolayer value. For inhibiting the oxidation, antioxidants are added either directly into the formula or to the package material [9, 12, 13]. Butylated hydroxyanisole (BHA) and butylated hydroxytolune (BHT) are the most widely used ones even though α-tocopherol has been tried for incorporation in the plastic film by Wessling et al. [13]. Vitamins A and C fortified in the product decrease in the storage and thus the product needs to be overfortified in order to maintain the stated level on the label during the shelf life [10, 11]. As a means to protect the product from oxidation, light barrier material is applied with oxygen exclusion. For exclusion of oxygen, nitrogen flushing and/or inclusion of oxygen absorber may be applied. As practical choices, fiberboard box with antioxidant-incorporated liner of moisture barrier is used to exclude light and prevent the breakage. The liner usually has a structure consisting of aluminum foil, polyethylene (PE) waxed paper, polyvinylidene chloride (PVDC), and ethylene vinyl alcohol (EVOH). Foil-lined composite cans are used for certain brittle products on account of their excellent barrier property against moisture, oxygen, and light, and protection against physical stress.

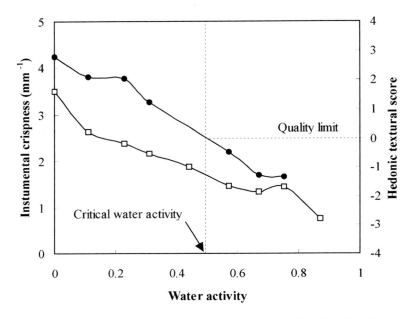

Figure 17.3 Crispness (□) and sensory hedonic score (●) of barley flake as
function of water activity. Hedonic score of 0 refers to 'neither like or nor
dislike', which is used as critical quality. Drawn based on the data of Ref. [14].

17.2.3 Fresh and Dried Pasta Products

Pasta products are prepared mainly by mixing the wheat flour and other
ingredients with water and extruding the mixed dough into desired shape. Most
products are dried by hot air, however, some products such as oriental instant
noodle are fried in oil. Dried pasta products include macaroni, spaghetti,
vermicelli, and noodles, and have moisture content less than 12%. The fat
content of the most dried pasta product ranges from 1.5% to 4.6%, while that of
fried instant noodle is about 18%. Fresh pasta products include plain wet noodle,
lasagna, and cannelloni, and have high moisture content ranging from 26 to 34%.
Some wet pasta such as ravioli and tortellini may be filled with cooked meat,
cheese, and vegetables inside.

Quality deterioration modes of the dried pasta products are moisture gain
(or loss), color loss, off-flavor development from lipid oxidation, and nutrient
loss [11]. Moisture gain may lead to mold growth and starch retrogradation,
while moisture loss may make the product too fragile and unacceptable for
certain products. Optimum moisture content of the dried pasta products for
storage is suggested as 10-12%. Oxidation of carotene pigments is catalyzed by

lipoxygenase from semolina flour and accelerated by exposure to light. Loss of nutrients such as lysine, riboflavin, and pyridoxin is also another quality deterioration mode and slowed down when packaged in opaque papers, films, and carton boxes [11, 15]. Packages with high oxygen and moisture barrier can help to preserve the quality of dried pastas. In case of oil-fried instant noodles, oxygen exclusion by inert gas flushing or oxygen absorbers can slow down oxidation of the products.

Microbial spoilage is a major problem in fresh pasta, and depends on the quality of raw materials, good manufacturing practices, and temperature control during production and distribution [16]; bacteria of concern include most cocci, lactobacilli, and aerobic sporeformers. Because of rapid quality deterioration at ambient temperature, the fresh pasta products are stored, distributed, and marketed under chilled conditions. Vacuum and modified atmosphere packagings are used extensively to extend the shelf life of fresh pasta products. The gas composition of modified atmosphere usually adopted for fresh pasta packaging covers 20-100% CO_2/0-80% N_2: shelf life can be extended from 1-2 weeks of air package to 3-4 weeks by appropriate MAP at 0-4°C. Addition of sauces or filling of meat reduces shelf life somewhat. As a further control, an oxygen absorber can be added inside the package to completely remove the headspace oxygen remaining originally or permeating from the outside. Figure 17.4 shows an effect of modified atmosphere packaging on the microbial quality of a fresh pasta product. Sinigaglia et al. [17] reported that vacuum packaging of fresh pasta could reduce both the growth rate and time to reach the maximum cell load for the mesophilic, psychrophilic, and coliform bacteria, but not those of *Staphylococcus aureus*; therefore initial hygienic condition is emphasized to be important for safety and long shelf life.

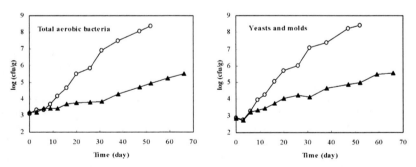

Figure 17.4 Microbial counts of fresh pasta in control and MA packages at 8°C. O: permeable air package; ▲: package with modified atmosphere of 22% CO_2/78% N_2. From Ref. [18] with kind permission of Korean Society of Food Science and Nutrition.

17.2.4 Fresh Bakery Products

Fresh bakery products include bread, cakes, pastries, waffles, crumpets, and doughnuts, whose water activity usually ranges from 0.78 to 0.98 [16]. Relatively high moisture content (>12%) makes these products have soft texture. Most of the bakery products are stored and distributed at ambient conditions, even though some products with cream or perishable components are chilled or frozen.

Table 17.1 Risk and chance of microorganism development related to water activity of bakery products

Type of product	Water activity	Microorganism likely to develop
Bread, cheese cake, custards, fruit pies,	0.86-1	Bacteria (*Bacillus cereus, B. subtilis, B. mesentericus, Serratia marcescens, Staphylococcus aureus*), yeasts, molds
Cake doughnuts, pound cakes	0.82-0.87	Yeasts, molds
Cakes filled with jam, fruit, or creams	0.78-0.80	Osmophilic bacteria, molds
Gingerbread	0.65-0.75	Xerophilic molds
Cakes with dried fruit, soft cookies	0.50-0.62	Osmophilic yeasts

Baking operations consisting of mixing, molding, proofing, and baking create attractive flavor specific to the product. Final baking step destroys some of enzymes and microbes, and improves the microbial stability of the products. Packaging has an important role to prevent the microbial contamination and retain the flavor. However, microbial spoilage, particularly of mold, stands as one of the main causes limiting shelf life because the baking does not destroy the microbial load completely and there is a chance of exposure to recontamination between baking and packaging. Chances of microbial deterioration depend on water activity and pH of the product, preservative addition, storage, and packaging conditions. Table 17.1 lists some of microbiological risks for the bakery products. Bacteria such as *Bacillus subtilis* and *Serratia marcescens* may survive the baking process and evoke roping and bleeding especially for the products filled with ingredients of high water activity and neutral pH. Yeasts such as *Saccharomyces cerevisiae* and *Saccharomyces rouxii* may cause the spoilage of fruit-filled products. Bacterial and yeast growth causes package swelling and off-flavors in the product [19].

Mold spoilage of bread and other bakery products comes mainly from post-baking contamination in cooling air, slicers, and other equipment surfaces. Some common spoilage strains of mold include *Rhizopus nigricans, Mucro mucedo, Aspergillus niger, Penicillium chrysogenum, Penicillium expansum, Geosmithia putterilli, Wallemia sebi,* and *Chrysonilia sitophila,* which produce visible white, yellow, or blue mycelium or conidia [20]. If the products are packed before complete cooling, moisture condensation will occur on the internal package and product surfaces, which promotes the growth of molds. Because mold spoilage is a predominant microbiological problem in the bakery products, mold-free shelf life has been a subject of designing preservation and package.

Moisture loss is a serious problem in bread and other bakery products when they are not wrapped in a moisture impermeable package. The extent of moisture loss is controlled by adopting the packaging of low moisture permeability and good seal integrity. Staling of bread is a major physical and chemical loss of quality in the storage, which includes flavor change and increased hardness [11]. It may be affected by moisture loss, but occurs even when there is no loss of moisture. Crystallization of amylopectin is mainly responsible for staling. Staling rate increases with temperature decrease up to 4°C, and decreases below that point.

Rancidity, oxidative or hydrolytic, can lead to off-flavors or bleaching of colors in the cakes [21]. Migration of moisture from one component to another, may lead to unpleasant texture in multi-component cakes. Matching equilibrium relative humidity of all the components or placing barrier between the components needs to be considered.

Table 17.2 Typical gas mixtures for packaging bakery products

Product	Gas (v/v %)		Mold free shelf life at ambient temperature
	CO_2	N_2	
Bread	100	-	1-6 week
Cakes	100	-	3-9 months
Croissants	100	-	15-25 days
Pastries	50	50	Up to 45 days

Adapted from Ref. [16].

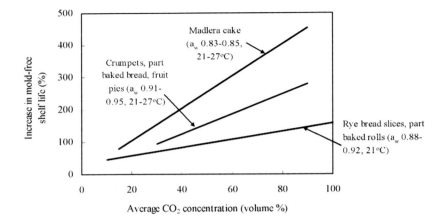

Figure 17.5 Extension effect of mold-free shelf-life of bakery products by CO_2 package atmosphere. Drawn according to the data from Ref. [22].

Operation of post-baking processes such as cooling, slicing, and wrapping under sterile conditions can reduce the contamination level of the product, and thus contribute to the shelf life extension. Treatment of the packaged product with ultraviolet light, infrared irradiation, or microwave heating can destroy some of the mold spores on the product surface, also extending shelf life [16]. MAP employing nitrogen and carbon dioxide gases has been applied successfully to the bakery products as a means to reducing the growth of mold and bacteria, moisture loss, and oxidation: 60% CO_2 and 40% N_2 is a typical gas composition of MAP for bakery products, while different optimum gas mixture may exist for each type of product. Table 17.2 presents some typical examples of the MAP conditions applied for the bakery products and Figure 17.5 shows the effect of CO_2 concentration on the shelf life extension. There are also some studies reporting the positive effect of CO_2 gas on retarding the staling of bread and fresh bakery products. Generally an ambient shelf life of at least 21 days can be applied for the category of bakery products if an appropriate MAP system and/or product modification is applied [21].

The effect of MAP can be strengthened by use of oxygen scavenger and ethanol emitter [23]: the anaerobic conditions created by oxygen scavenger can retard the oxidative quality changes and retard the growth of aerobic microorganisms; ethanol vapor generated from encapsulated ethanol sachet as well as sprayed into the package is useful to suppress the yeast growth on the bakery products of high a_w and low pH; in the cases of ethanol usage, the ethanol permeability of the packaging materials becomes an important issue for the packaging optimization. Soares et al. [24] recently tried packaging of bread by using propionate-incorporated cellulose acetate film sandwiched between the slices in order to reduce mold growth. Some minor problems of hardened texture, pale color, or slight taste change might be encountered with MAP of the bakery products.

17.3 Meat and Fish Products

Usually meat and fish products are distributed under chilled conditions below 8°C, preferably lower than 4°C. However sometimes the products are sold frozen for long time stability and convenience. This section deals only with chilled products and frozen products shall be handled in the later Section 17.9.

17.3.1 Fresh Meat and Poultry

Immediately after slaughter, meat carcass is cooled in air and aged for some time at low temperature. The aging time depends on the meat type and distribution method. The process of aging can be accelerated by electrical stimulation. Aging of meat reverses the toughening of the rigor mortis. The aged carcass is shipped whole or cut into the desired sizes and then packaged. Eviscerated poultry is cooled in iced water and then packaged with or without ice. The meat and poultry are stored, distributed, and marketed under chilled temperature. Optimum temperature for quality preservation is −1°C. Bacterial spoilage is the dominant factor causing quality deterioration. The characteristics of fresh meat make it an ideal growth medium for microorganisms: high moisture content, high nutrient content, nearly neutral pH, and the presence of glycogen, a fermentable carbohydrate. Spoilage can occur aerobically in the presence of oxygen, or anaerobically. Aerobic spoilage species are mostly composed of acinetobacteria, pseudomonads, and moraxellae [25], which give putrefactive flavor. In absence of oxygen the bacterial population is dominated by anaerobic lactobacilli for the meat with pH less than 5.8. Lactobacilli produce mild, acidic flavor without spoilage symptoms. However, the meat with pH greater than 5.8 will be spoiled by psychrotropic enterobacteria, *Brochothrix thermosphacta* and *Alteromonas puterfaciens*, imparting putrid flavor. It is generally known that putrid off-flavor, malodor, and slime can be detected when the bacterial count reaches about 10^7 to 10^8 organisms per cm^2 [11, 26, 27].

Even though microbial stability is of prime concern in fresh meat storage, storage life is, in many situations, limited by biochemical discoloration rather than excessive microbial growth [26, 28]. Color change of muscle pigment, myoglobin, often indicates deterioration of red meat such as beef and lamb. Oxygenated status of myoglobin determines the color of beef, lamb, and pork (Figure 17.6). Reduced form myoglobin has a dull purple color and is rapidly oxygenated to oxymyoglobin under oxygen environment. Oxymyoglobin has a bright red color which consumers associate with freshness. The oxygenation-deoxygenation process between myoglobin and oxymyoglobin is reversible. Oxymyoglobin is eventually oxidized with electron loss to metmyoglobin in long aerobic storage, which consumers regard as poor quality. On the other hand myoglobin with Fe^{-+} state is oxidized with electron loss slowly to metmyoglobin of Fe^{+++} state. Even though oxidation-reduction between myoglobin and metmyoglobin has some limited degree of reversibility, accumulated metmyoglobin on the surface layer limits the display time in the chilled showcase. Therefore oxygen availability on the meat surface, which is influenced by packaging conditions and flesh respiratory activity, determines its color status (Figure 17.6): under anaerobic condition reduced form of myoglobin dominates and under low oxygen environment of 1-3% metmyglobin of oxidized form prevails in brown color while the pigment exists in oxymyoglobin at higher oxygen concentration ($[O_2]$). In case of air permeable package, low temperature storage makes relatively higher amount of oxygen be supplied or diffused to the surface layer compared to the oxygen consumption by respiration activity of flesh cells, and thus maintains higher level of oxymyoglobin on the surface layer, which gives brighter color [25]. Carbon monoxide binds preferentially with myoglobin to form carboxymyoglobin which gives meat a stable cherry red color. On the other hand, oxygen permeable package may cause oxidative deterioration bringing about undesirable odor or flavor changes. However, color of poultry is affected little by oxygen availability because of its natural whitish appearance.

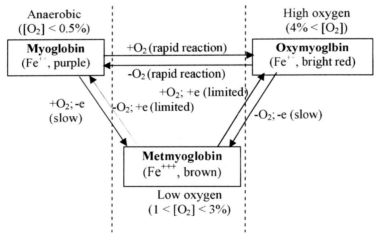

Figure 17.6 Status of myoglobin with state of iron dependent on the oxygen availability

Water-holding capacity is an important quality attribute of fresh meat. Loss of water-holding capacity is manifested by exudation of fluid known as 'weep', which is caused by shrinking of the myofibrils [29]. The 'weep' gives unpleasant appearance and may serve as a pool for microbial contamination. The amount of fluid exudation depends on the various physiological and environmental factors, and is not usually controlled by packaging. Minimized surface area of meat and not-too-tight shrunk wrap film help to attain less 'weep'. An absorbent pad is placed on the bottom of the package to absorb the exudated water.

Peptides and free amino acids released from breakdown of meat proteins can impart an objectionable flavor [25]. Tyrosine accumulated gives undesirable bitter flavor and may crystallize at the food surface.

Meat is frequently packaged under vacuum with heat-shrink film, nonshrinkable film or thermoformed tray. Vacuum packaging can aid in controlling oxidative rancidity and preventing the growth of normal spoilage bacteria [30]. When vacuum packaged, residual oxygen is consumed to produce carbon dioxide by the respiration process of the muscle tissue and microorganisms. Vacuum-packaged meat maintains the purple color, which is not favored by most consumers. Therefore vacuum packaging is usually applied to wholesale products in case of red meat. At retailers, the stored whole sale product package is opened for trimming, slicing, and repackaging in retail size trays permeable to oxygen. The bloom color can be maintained for up to 4 days of display life. Color is not so important in pork and poultry, and therefore retail vacuum pack is widely applied for these products. With vacuum packaging aerobic organisms such as pseudomonads are suspended, and lactic acid bacteria causing less offensive type of spoilage at high number are favored. Vacuum packaged beef with normal pH can be stored with extended shelf life up to 10-12 weeks at $0^{\circ}C$ and retail cuts with about 4 weeks at same temperature [26, 30]. In the vacuum packaging, the meat may be under pressure stress and increase the 'weep'.

Consumers' preference for bright red color originating from oxymyoglobin pushed the retail packages of red meat toward oxygen permeable tray. The tray is wrapped with cling films of polyethylene (PE) or polyvinyl chloride (PVC) which have low water permeability and high oxygen permeability. In this aerobic package, pseudomonads grow readily and lead to rapid spoilage of the product. Oxidative discoloration resulting in metmyoglobin is accelerated to limit the display life to 2-3 days.

MAP with gas of CO_2 concentration higher than 20% has a delayed and reduced growth of aerobic spoilage microorganisms on the meat. When bright red color is desired such as for retail meat cuts, high oxygen concentration is applied. An atmosphere of 60-80% O_2 and 20-40% CO_2 has been suggested as optimal for stabilizing the color and microbial conditions of red meat [25, 26]. High oxygen concentration keeps the meat color bright red. In order to maintain the desired level of gas concentration, higher level of CO_2 should be applied considering that O_2 is converted to CO_2 by respiration of the muscle tissue and microorganisms and headspace CO_2 is lost by dissolution in the meat. High volume of headspace is also desired for stabilized package atmosphere: a ratio of headspace to product of 3:1 is suggested for this purpose. Better sensory score in odor was obtained for the chicken leg with package atmosphere of higher CO_2 concentration and bigger volume [31]. However, high oxygen has a drawback of oxidizing meat lipids, and thus is not recommended for white meats [32]. O_2 concentration can also be very low or zero when meat color is not of concern such as for whole sale product. Anaerobic condition without oxygen prevents the oxidative quality deterioration. A centralized packaging concept called 'master-pack' places the overwrapped retail trays into the atmosphere of 100% CO_2 or other [28]. The master-pack is distributed at chilled conditions and opened for display at retailers. The separation of individual tray packs allows the oxygen supply to the meat surface through overwrapped film, which restores the bloom color. Master pack under CO_2 can have the extended shelf life up to 7 weeks and display life less than 4 day for retail packs. Addition of CO at 0.3-0.5% concentration can counteract undesirable color changes associated with vacuum or high CO_2 packaging, but has not been approved in most countries due to safety considerations.

It needs to be mentioned that all the packaging techniques are effective only when temperature is maintained at low temperature, preferably at -1°C. Because fresh meat quality and safety depend on temperature control, time temperature integrator can be put on the surface of the package: potential benefit would be greater with master pack. Irradiation has emerged as an effective tool to ensure hygienic quality of packaged fresh meat [26, 33]: microbial suppression can be attained without loss of enzyme activity, which helps tenderization of meat; trichinae can also be destructed for pork. The irradiation treatment under vacuum and modified atmosphere is reported to give generally better results of less rancidity and desired sensory quality. Irradiation may sometimes reduce water holding capacity of meat with resultant tough texture and irritant flavor. Dose level of 1-2.5 kGy is suggested as range for beneficial outcome without significant adversary effect.

17.3.2 Processed Meat Products

Meat processing covers a variety of procedures such as salted curing, grinding, emulsification, smoking, cooking, and retorting. The products include cured bacon, patty, cooked ham, luncheon meat, sausage, and hotdog. Sausage product may be stuffed into natural casings of gastrointestinal tract, collagen, and cellulosic materials. Plastic or metallic containers are widely used for packaging various products.

Storage stability depends on the level of curing agents, smoking, and heat processing. Salt added to the cured meat in 2-3% acts to keep the microbial quality, disperse the meat proteins, and add salty taste. Nitrite works to inhibit the growth of *Clostridium botulinum* and impart a characteristic flavor. Sodium nitrite is also combined with myoglobin to form nitrosylmyoglobin, a pink pigment. Heat processing such as cooking, pasteurization, and sterilization transforms nitrosylmyoglobin to nitrosohemochrome, another pink pigment. Oxygen can change nitrosohemochrome to oxidized porphyrins with green, yellow, or faded color. Antioxidants such as sodium ascorbate and sodium erythrobate stabilize the color development and reduce the rancidity [34].

Exposure to light can facilitate the lipid oxidation. Good hygienic conditions of packaging operation reduce the initial microbial contamination level. Usually the temperature below 10°C is suggested for storage and distribution of most processed meat products, even though some products commercially sterilized are stored at ambient conditions.

Vacuum packaging is mostly used for processed meat products [30]. Oxygen barrier of the packaging materials is of prime importance for preventing microbial deterioration, rancidity, and discoloration. Clipping may leave channels for oxygen exchange and give more chances of oxidative deterioration. O_2 scavenger may be applied to remove residual O_2 inside the package. MAP of 40-100% CO_2 with balance N_2 is useful for extending shelf life of some products such as ground beef patty, sliced ham, fried chicken, and sausages. Opaque films block the light and enhance the stability of color and lipid component. Transparent films, more appreciated by the consumers, are often obliged to include antifog additives in the plastic to guarantee clear vision through the package.

17.3.3 Fish Products

Fishes are considered more perishable than meats due to rapid autolysis by fish enzymes and fish oils are much more vulnerable to oxidation due to their high degree of unsaturation. Small amount of carbohydrate has usually been converted to lactic acid in the course of harvesting and endogenous catabolic reaction continues after death transforming adenosine triphosphate (ATP) to inosine monophosphate (IMP). Accumulated IMP is subsequently degraded to inosine and hypoxanthine (Hx), resulting in loss of fresh flavor [35]. After this stage, fresh fish held at refrigeration temperatures is spoiled by microbial flora on the surface and intestines. The most common groups responsible for microbial spoilage are *Pseudomonas/Alteromonas/Shewanella* and *Moraxella/Acinetobacter*. When the microbial spoilage is quite advanced, muscle tissue proteins are broken down to peptides and amino acids. Other components from microbial spoilage include free fatty acids, aldehydes, ketones, ammonia amines, and volatile sulfides.

Most fishes contain trimethylamine oxide (TMAO) as an osmoregulant. TMAO is reduced to trimethylamine (TMA) by Gram-negative bacteria, which imparts the odor of stale fish during storage [36]. CO_2 dissolved in tissue delays the TMA production by suppressing microbial growth and lowering the pH. Oxygen also has inhibitory effect on TMAO reductase. Fatty fish is also very prone to oxidation, because it has higher amounts of unsaturated fats. For elasmobranch fish containing urea, certain bacterial growth may produce ammonia and CO_2 by urease.

Odor development in fresh fish follows a progression beginning with a typical faint fish odor, then liberating a sickly sweet odor, finally developing into a distinctly fishy odor. There may also be putrid odors and ammonia-type odors upon development of hydrogen sulfide and indole compounds. The aerobic bacterial count of 10^6-10^7 usually corresponds to sensory quality limit in fish.

Physical deterioration of fish is apparent from fading color, development of color spots, and slime development on the skin. Eyes of fish will become sunken and cloudy as spoilage progresses. Finally, fish muscle loses integrity upon contact, and liquid leaks from the flesh if squeezed.

Traditional usual package of chilled fish are wooden or wax/plastic-coated corrugated boxes packed with ice [37]. Large master carton is used for bulk packaging in 4.5-9 kg (Figure 17.7). Fish is usually in direct contact with the ice and insulation panel of foamed polystyrene is located inside the carton board. Small retail-sized container may be located inside the master carton. However some delicate products such as fillets are contained in plastic trays or bags which are immersed in the ice or located in refrigerated chamber. Use of ice sets product temperature to 0°C, increases the package weight by 30-50%, and may cause leaking of melted water. Plastic inner lining may be used for preventing the leakage. Individual fish in a consumer package may be wrapped with water permeable plastic film sheet with humectant to absorb excess surface water (see Section 15.2.2).

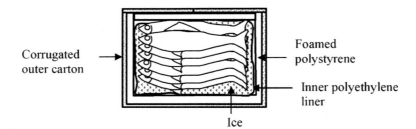

Corrugated outer carton

Foamed polystyrene

Inner polyethylene liner

Ice

Figure 17.7 Master carton for fish

Some fishes such as salmon and herring are salted and smoked for pleasant flavor and preservation hurdle. Salt content of the smoked fish is in the range of 2-4%. Smoking procedure reduces water activity of the product and gives some degree of storage stability. The chilled smoked products can be stored for 1-4 weeks at 0-5°C. For low salt product of smoked fish, maximum shelf life should be 10 days at temperature below 10°C in order to avoid botulism risk, unless aqueous phase salt content is higher than 3.5% or pH is below 4.5.

MAP can extend shelf life of fish products considerably by inhibiting growth of spoilage aerobic bacteria, reducing production of TMA and volatile bases, and retaining sensory quality [38]. CO_2 concentration of 40-100% with balance N_2 is usually applied. Sometimes O_2 concentration up to 30% is used for color retention of lean fish and reduced TMA level. Volume ratio of gas to product is important in attaining the desired effectiveness of MAP. Minimum ratio of 2 is suggested for preventing package collapse by dissolution of CO_2. Surface area exposed to CO_2 atmosphere is also another factor influencing the effectiveness of MAP and thus package design for maximized surface exposure such as dimpled bottom is applied. Vacuum packaging also has better quality retention of fish product compared to overwrapped or air one. The CO_2 produced by fish meat tissue and microorganisms may contribute to the improved stability. Active and intelligent packaging systems have been tried for fish packaging: one interesting approach is use of microperforated expanded polystyrene (PS) tray with liquid absorption and TMA scavenging capacity to delay microbial growth and attain low TMA accumulation inside the package [39]; Pacquit et al. [40] developed and tested a packaging device to detect volatile basic nitrogen as an indicator of fish spoilage.

Possible disadvantages of high CO_2 MAP are package collapse due to dissolution of CO_2 onto fish, increased exudate, discoloration, and increased toughness. Excessive drip may go along with the texture change of toughening. High CO_2 concentration can bleach cut surface due to low pH precipitation of sarcoplasmic proteins, and make eyes opaque and skin pigments pale. Reduction of volume by CO_2 dissolution can be overcome by slightly overpressured CO_2 flushing and pretreating the fish with CO_2 saturated water. Absorbent pad may be placed onto the bottom of the pack to absorb the drip or exudate. Bjerkeng et al. [41] incorporated an absorbent pad with sodium carbonate and citric acid in order to generate CO_2 gas for reducing volume deformation of 70% CO_2 package of cod fillets. Inclusion of N_2 filler gas or use of lower CO_2 level may alleviate the problems caused by high CO_2 MAP.

17.4 Dairy Products

17.4.1 Pasteurized and UHT Sterilized Milk and Cream

Pasteurization process originally based on thermal destruction of infectious *Mycobacterium tuberculosis* and *Coxiella burnetii*, kills Gram-negative psychrotrophic bacteria which cause spoilage of refrigerated milk products. Some typical examples of time/temperature combination for the pasteurization process are given in Table 17.3. Pasteurization also inactivates lipoprotein lipase which catalyses the breakdown of milk triglycerides to free fatty acids with development of off-odors and off-flavors. However thermoduric bacterial spores mainly consisting mostly of *Bacillus* spp. survive the pasteurization and can grow slowly at chilled temperature. They can cause spoilage of the pasteurized milk stored beyond shelf life and/or under temperature abuse conditions. Heat stable indigenous enzyme called plasmin and microbial extracelluar enzymes from outgrown psychrotrophic bacteria can also survive the pasteurization process and degrade milk fat and protein [42]. High quality milk with low microbial load should therefore be used to avoid the enzymatic spoilage problems. Provided the filling and packaging operations after pasteurization run aseptically, only heat resistant spore formers and thermostable enzymes grow or act to restrict the shelf life of the product. However, in usual practice, contamination after heat treatment occurs and contaminated spoilage bacteria such as *Pseudomonas* reduce the shelf life. Oxidized flavor development due to spontaneous lipid oxidation may limit the shelf life in the presence of oxygen.

Table 17.3 Typical time/temperature combinations with equivalent bactericidal effects in milk products

Product	Condition
Pasteurized milk and skimmed milk	30 minutes at 63°C (LTLT[a]) 15 seconds at 72°C (HTST[b]) 1-4 seconds at 85-90°C
Pasteurized cream 18% fat 35% fat	 15 seconds at 75°C 15 seconds at 80°C
Pasteurized concentrated milk	15 seconds at 80°C

[a]Low temperature long time; [b]high temperature short time. From Ref. [43] with kind permission of Springer Science and Business Media.

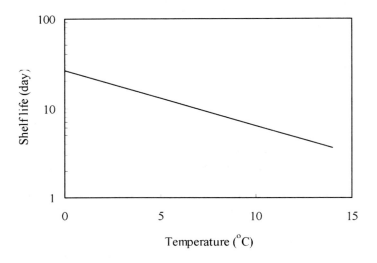

Figure 17.8 Shelf life of pasteurized milks at different temperatures based on sensory quality criterion. From Ref. [44] with kind permission of Elsevier.

Exclusion of oxygen by minimized headspace helps to have lower microbial count and higher flavor score of milk packaged in carton. Condition of exposure to light also influences oxidative degradation in ascorbic acid, riboflavin, and sensory flavor: opaque materials of glass, plastic, and paper board are desirable in quality retention. Storage stability of pasteurized milk is influenced by temperature greatly as shown in Figure 17.8. Typical commercial shelf life of the pasteurized milk is 7 days at temperature below 6°C [45].

When combined with aseptic filling and packaging, sterilization at ultra high temperature (UHT) at 135-150°C for 4-15 seconds destructs thermoduric spores and allows the milk and cream products to be stored at room temperature with extended shelf life of 90-150 days at ambient temperature [45]. In UHT milk, therefore, shelf life is dictated by proteolytic activity of plasmin and microbial enzymes from psychrotropic bacteria: the latter is more important and thus emphasizes use of low microbial load raw milk [42]. In UHT cream, calcium-induced aggregation called feathering limits the shelf life and cannot be overcome by packaging; sodium carbonate and tri-sodium citrate has been added to ameliorate this problem. It needs to be mentioned that UHT technology could be introduced mainly with thanks to the development of aseptic packages such as Tetra Pack® cartons.

Nitrogen flushed packaging has been reported to help to retain the flavor of milk, but not used in retail commercial packaging [46].

17.4.2 Dried Milk Products

Drying process of milk consists of pasteurization, evaporative concentration, and spray dehydration into hot air at 175-250°C for a few seconds. Even with high temperature condition at spray drying, indigenous and extracellular enzymes are not destroyed and accelerate fat oxidation. Autooxidation of unsaturated milk fat produces short chain aldehydes and ketones responsible for undesirable off-flavor. Nonenzymatic browning between lactose and milk protein induces off-flavor and color changes with nutritional loss of lysine. Its rate increases with moisture content whose critical limit for proper protection is about 4%. Dry milk at monolayer moisture content (water activity of about 0.3) remains acceptable for nine months at temperature below 40°C [11].

Consideration of these quality factors stresses the control of moisture content and protection against oxygen. Barrier against light helps to suppress the oxidative deterioration. Nitrogen-flushed, vacuum, or oxygen scavenging packaging is helpful for preventing fat oxidation and preserving flavor. Nitrogen flushing is particularly useful for bulk packaging of milk powder. Min et al. [47] also showed that flushing by 98% N_2 and 2% H_2 could reduce the browning and production of volatiles and CO_2 gas from dry whole milk. For more extended shelf life, storage at cold temperature (4-8°C) may be used [42].

17.4.3 Cheese

Cheese is manufactured by coagulation of milk by bacterial fermentation, rennet, or acid. The curd is separated with drainage of liquid, pressed, and ripened or fermented further with bacteria or molds. In case of process cheese, different types of cheese are combined or mixed into mold with addition of salt, emulsifier, and other ingredients. There are many cheese varieties and Table 17.4 shows a classification of them. Hard cheese has moisture content of 20-42%, semi-hard cheese are with that of 44-55%, and soft cheese about 55%.

Table 17.4 A classification of cheese in respect to ripening characteristics

Class of cheese	Example
Very hard	Parmesan
Hard (with no gas holes)	Cheddar
Hard (with eyeholes)	Emmental
Semi-hard	Port du Salut
Soft (unripened)	Cambridge
Soft (ripened)	Coulommier
Surface smear ripened	Limburg
Surface mold ripened	Camembert
Internal mold ripened	Roquefort
Acid coagulated	Cottage cheese
High fat (cream)	Cream cheese

From Ref. [48] with kind permission of Springer Science and Business Media.

Cheese has higher stability in nature compared to pasteurized or fermented milk. Its shelf life varies from a few days to several years being different with cheese variants [49]. Major quality changes during cheese storage are mold growth, color change, lipid oxidation, and sensory quality degradation, depending on the cheese type. High moisture fresh cheese may deteriorate by psychrotrophs, molds, and yeasts. Poor moisture barrier of the packaging film may cause weight loss. Storage temperature of cheese should be decided from consideration of microbial spoilage and physical structure or consistency: microbial stability is obtained at low temperature, but textural property may be harmed by low temperature. Optimal storage temperature for a cheese product can be selected between 4 and 13°C considering its quality characteristics depending on temperature.

Crystal formation on the surface of hard cheese such as Cheddar is also another factor aggravating the appearance. It is due to calcium lactate crystallization and can be alleviated by vacuum packaging maintaining tight contact between packaging film and product surface [50]. Some types of cheese such as mold ripened smear produce significant amount of carbon dioxide (Figure 17.9), which causes expansion of their packages. Some cheese products are often allowed to ripen after packaging, and thus needs proper level of CO_2 permeability [51-53].

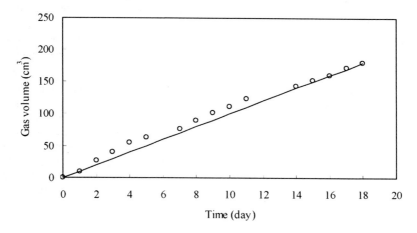

Figure 17.9 CO_2 production from 250g of Taleggio cheese at 6°C under 100% N_2. From Ref. [51] with kind permission of Italian Journal of Food Science.

MAP with CO_2 concentration of 20-100% and balance N_2 is widely used for packaging a variety of cheeses [46]. It suppresses microbial growths in psychrotropic bacteria such as *Pseudomonas*, molds, and yeasts. 100% CO_2-flushed pack collapses to zero headspace volume due to dissolution of CO_2 in cheese and shows the appearance similar to vacuum package. 100% CO_2 package preserves the microbial and sensory quality of cottage cheese [54, 55]. Alves et al. [56] also showed that 100% CO_2 package could effectively inhibit aerobic psychrotrophs and yeasts/molds with good keeping of sensory quality for sliced Mozzarella cheese. However in this case of sliced cheeses, high ratio of initial headspace to cheese is required to prevent total collapse of package due to CO_2 dissolution, which hampers ease of separation between slices [46]. However, in certain type of fresh cheese such as Cameros with high fat and moisture, 100% CO_2 atmosphere has negative effect on sensory quality even with excellent microbial inhibition: lower CO_2 concentration of 40-50% with 50-60% N_2 has been suggested for extending shelf life and retaining good sensory quality [57]. Vacuum packaging could extend the shelf life of

Stracchino, an Italian soft cheese by decreasing the growth rate of yeasts which cause the spoilage [58].

Protection against light and oxygen in MAP is necessary to prevent color degradation and lipid oxidation in cheese. Opaque or aluminum laminate films with low oxygen transmission rate retain yellow color with small oxidative quality changes [59, 60].

Packaging of some mold ripened cheeses should support the mold growth during the shelf life, which is possible under adequate level of oxygen and moisture. Film with high oxygen permeability is thus needed for these types of cheeses [45]. Desobry and Hardy [61] developed for Camembert cheese an active three-layer packaging material containing moisture absorber/desorber layer to prevent water condensation on internal packaging surface and maintain healthy *Penicillium camemberti*.

Processed cheese is normally stable microbiologically with shelf life of about 1 year at room temperature, because the cheese mixture is heat-processed at 70-140°C before packaging into pouches or polymer laminated aluminum foils [62]. Changes in appearance, structure, color, and flavor due to water loss, hydrolysis of polyphophates, crystal formation, oxidation, heat stable enzyme action, noenzymatic browning, and interaction with packaging are main factors of quality changes. In packaging aspects, high barrier packaging such as aluminum foil laminates and metal cans can help the quality preservation

17.4.4 Fermented Milks

Lactic fermentation of milk by *Lactococcus lactis*, *Leuconostoc cremoris*, *Streptococcus thermophilus*, *Lactobacillus delbrueckii*, *Lactobacillus acidophilus*, or *Bifidobacterium bifidum* gives the product low pH of 4.0-4.6, low lactose, improved digestibility, and many health functionalities [49]. The fermented milk group includes yogurt, sour milk, and buttermilk.

Even though low pH and dominating lactic acid bacteria of around 10^8/g stabilize the fermented milk products, microbiological, chemical, and physical spoilage can occur during storage and distribution. Yeasts and molds may grow and cause the microbial defects of the products. A flavor threshold is generally around at a yeast/mold count of 10^4 per mL [49]. When the products are stored for long time or at high temperature, enzymes from the starter bacteria may create offensive flavors bitter, acid, and cheesy. An excessive acid production may lead to an objectionable flavor, which also occurs with long time and/or high temperature storage. Contamination of metals like copper and iron and exposure to light may induce adverse effects on the product flavor. Flavor degradation also occurs with oxidation of milk fat. Interaction with packaging material may also cause off-flavor. Syneresis called whey-off may occur during storage and distribution, and is affected by storage temperature, vibration stress, and CO_2 gas included in the product. Stretch overwrapping of 227g yogurt waxed paper containers was reported to be effective secondary packaging in reducing the whey-off [63].

High gas barrier films are desirable for preventing oxidative changes and maintaining low level of dissolved O_2 in yogurt and fermented milk (Figure 17.10). Lowering dissolved O_2 level in yogurt is thought to help to obtain high viability of probiotic bacteria. High barrier pouches with high crystallinity film maintain high CO_2 concentration in headspace and give high sparking taste of the packaged yogurt [64]. Nitrogen flushing was reported in a study to give beneficial effect in extending the shelf life [46]. Normal shelf life limit for packaged yogurt is 24 days at storage temperature below 6°C [45].

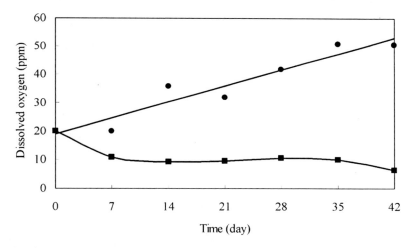

Figure 17.10 Dissolved O_2 in 200 mL yogurt packaged in tubs made of different materials and stored 4°C. ●: PS; ■: high barrier material. Reproduced from Ref. [65] with kind permission of John Wiley & Sons.

17.5 Confectionery Products

Confectionery products are mostly based on the high level of sugar in its composition and are thus generally stable microbiologically with a relatively long shelf life. However, they are prone to physical and chemical deteriorations such as textural and flavor changes. Different quality characteristics exist with a variety of products with different compositions and structures. Based on microstructure, confectionery products can be grouped into two categories: amorphous and crystalline types; amorphous one includes hard candy, caramel, toffee, and jelly; crystalline one covers chocolates, creams, fudge, fondant, pressed candy, marshmallow, and marzipan [66]. Water activity or equilibrium relative humidity (ERH) is another important factor determining product characteristics and quality deterioration mode as shown in Table 17.5. Porous structure of the aerated products can trap high amount of air and make it vulnerable to oxidation.

Table 17.5 Quality deterioration mode of confectionery products

Type of product	ERH (%)	Deterioration
Boiled sweets	20-30	Graining and stickiness
Toffees	<50	
Gums & pastilles	50-65	Stickiness, growth of molds and yeasts
Liquorice paste goods	53-66	
Turkish delight	60-70	
Jellies	65-75	Fairly stable at temperate conditions
Creampaste goods	65-70	
Marshmallow	65-75	
Marzipin	70-85	Drying out or mold growth
Fondant cream	75-85	
Jam	75-85	
Milk chocolate	68	Fat bloom, sugar bloom
Plain chocolate	70-72	

Adapted from Ref. [67-69].

17.5.1 Chocolate Products

Chocolate is consumed as straight chocolate or as coating on candy bars, cookies, and nuts. Plain chocolate is typically composed of cocoa mass, sugar, cocoa butter, lecithin, and vanillin. In milk chocolate, milk solid is added in about 8-10%. Sugar constitutes about 45-55% of the chocolate. Chocolate has a high fat content of 28-35% due to inclusion of cocoa butter.

Crystallization of fat on the chocolate surface, which is caused by fat migration and/or polymorphic transformation, yields a dull appearance and grayish spots called fat bloom. Moisture absorption on the chocolate surface or migration between components crystallizes the sugar crystal on the surface, which is called sugar bloom. Storage under fluctuating temperature conditions induces fat and sugar blooms. Exposure to melting temperature of cocoa butter (about 36°C) results in irreversible deterioration of chocolate. Storage in humid conditions or low temperature can cause moisture condensation on the surface and then result in sugar bloom. Plain and milk chocolates become soft and unacceptable at water activity above 0.75 [68]. Oxidative degradation of fat may create staling or 'cardboardy' flavor. Exposure to light can aggravate the oxidative rancidity particularly for white chocolate with low antioxidant content. Poor barrier packaging or wrapping of chocolate can pick up unacceptable flavor from the surroundings. There is also a concern for transfer of ink or solvent odor from packaging film. Chocolate products may attract insects when located in an infested environment. Special attention should be given to the storage and transit conditions. Packaging must be hermetically sealed with higher aroma barrier and resistant to creeping insects.

Therefore chocolate packaging is constructed based on the protection against heat, moisture, oxygen, and light [70]. Some degree of insulation against heat can be built into outer packaging by using foil-lined, metallized, and/or expanded polystyrene. Traditionally chocolate block has been packaged in aluminum foil or paper wrap, which provides adequate protection against moisture. When higher moisture protection is required, the foil is coated with heat-seal layer of polyethylene and heat-sealed. This also provides better protection against insects. Some products of irregular shape are pillow-packed with laminated plastic film by horizontal form-fill-seal machine. In this case sealing operation at lower temperature or pattern cold seal may be applied with use of proper sealant or adhesive for protection against heat [71]. In case of paperboard set-up box used for luxurious image, the outer box is wrapped in shrink film for extra moisture protection. Film coated with PVDC, EVOH, or metallized aluminum can give high protection against oxygen. In Japan oxygen scavenger is used for protecting chocolate confectionery from oxygen. Protection against light is achieved by foil, foil-laminated film, metallized film, or heavily printed film. Potential contamination of chocolates from taints, inks, or solvent in the packaging materials should also be carefully checked and prevented. The potential risk of migration is higher with the product having higher fat and lower crystallinity [72].

17.5.2 Hard Boiled Sweets

Hard boiled sweets in amorphous glassy structure are formed from cooling hot supersaturated sugar solution containing flavor, coloring, and acid. They have simple composition mostly consisting of sucrose and corn syrup with moisture less than 2%. At this very low level of moisture, the texture is hard and brittle. The glassy structure can change to a viscous state above glass transition temperature with concomitant sugar crystallization (graining). Glass transition temperature decreases with moisture and thus moisture pick up can cause graining and other textural changes at normal temperature. Texture is a very important quality factor for hard-boiled candy: adhesion, stickiness, and hardness are the main indexes [73]. As shown in Figure 17.11, the increased water activity decreases the candy hardness, and the product is acceptable up to water activity of 0.56 which corresponds to moisture of 1.2%. Therefore moisture protection is of utmost importance for candy packaging. Edible coatings and films such as pectinates, alginates, starch, and corn zein can be applied to minimize stickiness of the candy products. The desired volatile flavor may be lost through permeation or leakage through the packaging. Oxygen remained in package headspace or permeated from outside may change the characteristic flavor, produce off-flavor, and/or lead to rancid odor. Fading of bright color can occur by bleaching effect of light.

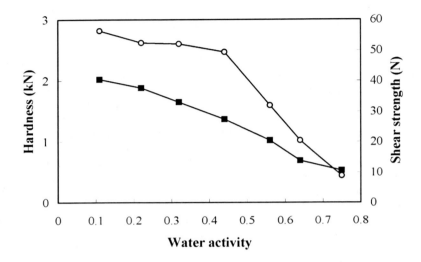

Figure 17.11 Textural properties of sugar confectionery at 27°C. ■ : hardness of hard-boiled candy; ○: shear strength of toffee. Drawn based on the data from Ref. [68].

The sweets may be packaged in sealed containers of metal cans, glass jars, and foil laminate bags [69, 71]. For consumers' convenience and prevention of sticking together, they may be individually packaged in twist wraps in film, foil/film bunch wraps, or bar wrappers in papers, films, and foils. The twist wrapping machines are the fastest equipments in packaging lines as they can reach a processing speed of up to 1200 units each minute. The individual wraps may be packed in plastic pouches, rigid composite cans, glass jars, and metal cans for further protection against moisture, light, or oxygen. Dead-fold properties are needed for tight twist-wrap for tiny sweets and are provided by the films such as metallized cellophane.

17.5.3 Toffees and Other Confectioneries

Toffees and caramels are made from blend of sucrose, corn syrup, milk ingredient, fat, emulsifier, and flavoring. The homogenized mix is cooked to a high solid content. Usual moisture content of toffees and caramels is 10-12%, higher than that of hard candy. The texture is thus plastic at ambient temperature. Maillard reaction between milk proteins and reducing sugars is responsible for characteristic flavor of toffees and caramels. Fat existing as droplets of various sizes plays an important role of reducing stickiness and contributing to mouth-feel. The major deteriorative quality changes during storage of toffees and caramels are shape loss, rancidity development, and softened sticky texture. Moisture pick up may induce graining on the surface and adhering to wrap paper or film; Figure 17.11 shows that shear strength of a toffee product stays almost constant at water activity of 0.11-0.44, but decreases drastically with further water activity increase. Waxed paper has traditionally been used for wrapping toffees and caramels offering some degree of moisture barrier. Outer packages such as metal containers, glass jars, and plastic pouches are applied for further protection from moisture and light.

Jellies and gums contain wide range of gelling agents besides sucrose and corn syrup to offer chewy texture. The gelling agents such as gelatin, starch, pectin, gum Arabic, and agar are frequently used ones. Texture of gum products becomes soft and sticky when they are located in humid conditions and absorb moisture. Conversely, warm and dry environmental conditions can lead to desiccation, especially in unpackaged or improperly packaged product. Artificial fruit flavors are also a common component of gummy candies. These compounds, which are often terpenes or related unsaturated compounds, are prone to degradation by oxygen. Gums and jellies have ERH in 50-75% range (Table 17.5), and thus can host many of the osmotic yeasts and molds. The high sugar and acid contents in gummy candy make microbial proliferation unlikely, although a processing defect or storage abuse may make the product suitable as microbe growth substrate. For example, gelatin is added as a gelling agent to the confectionery mixture, due to its heat sensitivity, after the high temperature boiling and cooling stages; if gelatin is not of proper microbial quality, there is danger of microbial spoilage.

Aerated confectionery such as nougat and marshmallow has quality characteristics similar to jellies and gums except that they have air dispersed as small bubbles in the structure. The incorporation of air in the matrix makes the product more susceptible to physical damage during handling, distribution, and storage. Increased surface area and included air can promote the oxidative changes. Collapse of air bubbles, drainage of the syrup, and shrinkage of the product are the common faults which limit shelf life of the aerated products and may be affected by moisture change [67].

Moisture control is a primary practical issue in packaging gums and jellies. These products are typically packaged by weight in transparent, heat-sealed heavy gauge pillow pouches. Flushing the package with an inert gas is an option, particularly for confectionery foams, although it is an expensive and temporary solution. Prior to sealing, the pouch can also be partially or totally evacuated to minimize oxygen. Higher levels of packaging, which prevent mechanical damage in shipping and retail, include paperboard boxes, corrugated boxes, and pallet stacking in warehouses.

The gums and jellies can be coated with sugar crystals or special glazing agents to prevent product sticking together during final processing and within the plastic packaging. This coating provides a desirable sheen to the product and retards sticking in heat and humidity.

Chewing gum typically consists of 25% gum base, 25% corn syrup, and 50% sugar [66]. Candy-covered chewing gum is usually coated by wax coating and well-protected against moisture. Packaging is done with convenience blister pack or in cans of metal or plastic for protection against physical damage. Stick gums require much more consideration to flavor loss and moisture change due to large surface area. Too low moisture content results in a hard and brittle texture, while too high moisture gives a soggy and tough texture [74]. The desirable sensory quality is maintained in the moisture content of 2.1-3.1% and the protection against moisture change is important. A typical commercial foiled paper package with 7 chewing gum sticks paper-wrapped was reported to have moisture transmission rate of 4.5×10^{-3} g day^{-1} mm Hg^{-1} [75]. These packs are distributed on whole sale in paperboard box and thus expected to have lower water permeation in practice.

17.6 Fats and Oils

A variety of fat and oil products are used for salad dressing, cooking, baking, confectionery, frying, and spreads. Physical form of the products may take solid fat or liquid oil: typical examples of solid form are butter, margarine, lard, and shortening; liquid type includes olive, soybean, corn, cotton seed, and sesame oils.

Fats and oils are generally stable microbiologically due to very limited amount of moisture in their compositions and thus have long shelf life with minor sensory quality change during storage. However, they are subject to chemical, microbial, and physical changes in some degree. Fats subjected to chemical and/or microbial changes may contain free fatty acids, glycerol, aldehydes, ketones, and lactones [69, 76]. The triglycerides, main constituent of the edible fats and oils, may be hydrolyzed to produce free fatty acids in the presence of moisture, which may develop unpleasant off-flavor defects called as hydrolytic rancidity. This hydrolysis of the triglycerides is catalyzed by lipase from animal, plant, or microbial sources (lipolytic rancidity), or proceeds spontaneously without enzyme action. The liberated fatty acids can be transformed to aldehydes and ketones with characteristic off-flavors. Unsaturated fatty acids in the backbone of triglycerides and free ones can readily be autoxidized with oxygen to produce hydroperoxides, which eventually become aldehydes, alcohols, and hydrocarbons. All these reaction products are volatile and responsible for rancid off-flavors. Oxidative rancidity may be activated or catalyzed by heat, light, lipoxygenases, and other catalysts. Some fat-soluble pigments may be produced from microorganisms and cause discoloration to yellow, red, purple, or brown. As mentioned above, there is generally no risk of microbial spoilage for the most pure oil and fat products because of the absence of water. Only margarine and spread have aqueous phase and thus have limited potential for microbial growth. However small droplet size of water in continuous oil phase suppresses microbial growth when the products are manufactured under proper conditions of quality control and hygiene.

Changes in crystal structure of solid fat may evoke loss of original texture and appearance [76]. Sandiness of a spread develops when large crystals are formed with exposure to high temperature. Oil migration resulting from short-term temperature rise may cause blemished surface on the margarine. Fats are very susceptible to absorption of foreign odors and must be protected from them [74]. Absorption of oil in the packaging material increases its permeability to aroma compounds by raising their solubility [77].

Controlled low temperature, nonfrozen storage of the fats and oils retards rancidity development, prevents transition of fat crystal type, and eliminates the risk of microbial spoilage. This is much more important in long storage with frequent use cycle in households or catering sector. Small package size may reduce the span of storage in use. However small package size may increase the ratio of surface area to oil volume and enhance the rate of oxygen ingress by diffusion through the package wall in case of plastic package. Active packaging incorporating oxygen scavenger in container wall has been shown to protect against the oxidation of oil [78]. Protection against light is also very important in preventing the oxidation, and should be considered in designing packaging and retail storage conditions. Therefore a typical consumer package of solid fat consists of foil-parchment laminate wrap located inside foil-laminated outer carton. Other package types of metal containers, composite cans, or plastic tubs are also used less frequently. For the liquid oils, tin cans, glass bottles, and plastic bottles usually from polyethylene terephthalate (PET), PS, or high density polyethylene (HDPE) are mostly used. In case of glass and plastic bottles, coloring of the containers may be applied for protection against light. El-Shattory et al. [79] reported that flavor of cottonseed and palm oils was preserved better in metal cans than in white plastic bottles, whereas the plastic containers were comparable to the metal ones in the oxidative changes of margarines due to the presence of water and salt. Recently aseptic paper-based carton came to use for packaging liquid oil. Devices for easy opening and dispensing are emphasized recently for convenience of usage. Snap-on caps, peelable lids, tear-away strips, and easy-to-pour spouts are examples.

17.7 Drinks

17.7.1 Fruit Juices

Fruit juices are manufactured from fruits by mechanical compression. Some other means of extraction and diffusion may be applied depending on the fruit type. Enzyme treatment may also be employed to break down the cellular structure and have better process yield. In many products, concentrated juice is diluted with water to have the state similar to original juice. As a variant of fruit juice, fruit nectars are obtained by adding water and sugar to fruit juice, concentrated fruit juice, fruit puree, and other fruit products. Usually fruit juices are pasteurized before or after packaging in order to destroy spoilage microorganisms and enzymes, and distributed at ambient conditions. In case of orange juice, 7 seconds at 91°C is required for inactivating pectic or pectinolytic enzymes, and 16 seconds at 74°C is sufficient for microbial reduction [80]. Temperature above 26°C needs to be avoided for the storage [81]. Volatile essence, which has been recovered in evaporation stage, is sometimes added to the juice diluted from concentrate for improving its flavor. However, fresh squeezed juice is directly packaged without pasteurization and distributed under chilled condition with short shelf life. Potential microbial hazards should be evaluated for the nonpasteurized juice. Most fruit juice products have sugar in range of 4-20 Brix degree and pH of 2.9-4.5. Lower pH and higher sugar content may improve the storage stability of fruit juices.

Major quality deterioration modes of fruit juices are destruction of ascorbic acid and carotenoids, nonenzymatic browning, and other oxidative reaction of flavors. Shelf life of nonpasteurized products may be restricted by microbial spoilage. Ascorbic acid content has been widely used as a quality index determining the shelf life of fruit juice. When a juice product is claimed on ascorbic acid content, it must be maintained throughout its shelf life. As with other oxidative reactions, ascorbic acid destruction is greatly influenced by initial dissolved oxygen, headspace oxygen, and oxygen supply through packaging material [80, 82]. The conditions with high oxygen supply and contact significantly accelerate the deterioration of ascorbic acid and flavor shortening the shelf life of fruit juices. Normal practices to exclude oxygen are initial deaeration of the product, flushing of headspace with inert gas, and package under high gas barrier material. Packages of glass bottle, metal can, plastic pouch, and paper carton laminated with aluminum foil provide an absolute barrier to oxygen. Cups and bottles of HDPE or polypropylene (PP) coextruded with EVOH or PVDC can also offer low oxygen permeation for fruit juice packaging. Nitrogen flushing is being used for added protection of fruit juices, particularly for bulk packaging unit such as bag-in-box. Liquid nitrogen is also sometimes injected to aluminum can package of fruit juices so as to provide rigidity and strength to the thin wall of the can [46]. The compressed liquid nitrogen injected in 0.12 to 0.25 g per can is evaporated to replace the air in the headspace before closing and maintain an internal pressure of 1 to 2 atm, which gives the can rigidity. Nitrogen flushed product is said to have other advantage of releasing an intense aroma as with liberation of internal pressure on opening the package.

Canned fruit juices of high acidity may suffer from corrosion of tin on the container wall limiting the shelf life and leading to leakage or swelling of the package in severe case. Dissolution of tin may also bleach the color of fruit juice. Selecting proper can lacquer can avoid the problem. Inside the package there may be competition for oxygen consumption among tin corrosion, ascorbic acid oxidation, and oxidation reactions of flavor and color changes [80]: ascorbic acid retention is higher for plain can than lacquered can because of oxygen consumption by tin dissolution in the former. In case of transparent packaging such as glass and plastic bottle, surface reaction stimulated by light may degrade the nutritional content, sensory quality, and pigment [83]; glass packed juices are inferior to canned products in terms of ascorbic acid retention. Most adequately processed and pasteurized fruit juices have a shelf life longer than 1 year at ambient temperature, but they are often stored and marketed in refrigerated conditions for better quality [81]. In case of some fruit juices packaged in the plastic containers, there is a concern about flavor adsorption on the plastic wall (flavor scalping), which may degrade their sensory quality (see Section 5.4). The inner layer contacting the product needs to be selected in consideration of adsorption capacity of a specific flavor component.

17.7.2 Soft Drinks

Soft drinks are generally recognized as sweetened water-based beverages with balanced acidity [84]. Sugars at concentration about 7-14% in the drink give sweetness in harmony with sourness coming from organic acids at 0.05-0.3% level. Flavors, colors, and fruit juices are added for desired organoleptic properties; other minor ingredients used in the formulation include carbon dioxide, high intensity sweetener (for diet drink), antioxidants, preservatives, emulsion, gums, vitamins, etc. Soft drinks may be divided into carbonated and still ones. Carbonation level expressed in volume ratio of CO_2 gas to liquid product ranges from 2.0 for a lightly carbonated beverage to 5.0 for a highly carbonated one. The 5 volume CO_2 produces 5 atm inside the container at ambient temperature and about 1% concentration in water [74]. Dissolved CO_2 provides a flavor that helps to compliment the refreshing carbonated beverage. Soft drinks also vary in the appearance: clear, cloudy, colored, or colorless. Natural fruit juices may be added for the flavoring. Carbonated soft drinks mostly command the market of soft drink and thus this section will mainly deal with carbonated beverage. Most carbonated soft drinks including regular cola, root beer, ginger ale, and lemon-lime are in low pH of 2.6-4.0 and contain sugars of 7-14%, which is helpful for their preservation. Many soft drinks also have some level of caffeine content (typically 120 mg kg^{-1}) while there are some decaffeinated products and noncaffeine drinks based on fruit juices and flavors. Diet colas and drinks contain little sugars to give negligible energy.

Choosing appropriate ingredients and formulation is the key elements for attaining stable product. Use of the stable flavor oils, clarified juices, emulsions, and antioxidants can give a significant extension of soft drink shelf life. Although there is no great microbial concern for carbonated soft drinks because of dissolved CO_2 and low pH, yeasts may cause spoilage problems such as off-flavors and sedimentation, and thus the product is pasteurized before the filling or under in-pack condition after packaging. The prepasteurized product is either directly filled into container or combined with other sterile ingredients such as water and CO_2 before filling. Preservatives may be added for some drink products with high microbial risk.

Flavor degradation and loss of CO_2 are the two major deterioration modes that would be observed in the carbonated soft drinks [85]. They both have a major impact on the sensory perception of the beverage by the consumer. Flavor degradation affects the overall sensory characteristics of the beverage by altering the flavor profile and/or producing off-odors. Reduction of dissolved CO_2 can cause flavor change and increase the risk of microbial growth. Decrease in CO_2 content of as little as 10% may cause changes in taste, making the product unacceptable [86]. Color loss and ascorbic acid degradation are also key quality factors to be considered in determining shelf life. Consumers often judge products by their overall appearance; therefore a faded color would give a negative connotation. Emulsions also contribute to the overall appearance of a beverage. Oil based flavors, such as lemon flavor, need an emulsifier to assist in mixing it with water. If the emulsion is not made and maintained properly, a phase separation and/or alteration in the cloudiness may occur to cause an undesirable appearance to consumers.

External factors such as storage temperature and light exposure affect greatly the flavor retention of soft drinks. Light and high temperature can degrade color and flavor. If some soft drinks are exposed to excessive light and high temperature for long time, their color tends to fade and flavor is oxidized to an unpleasant form. For improved protection of color and flavor, colored glass bottles of amber and green colors have been used for different types of soft drinks [74]. In case of plastic bottles, an additional packaging film/layer with ultraviolet ray barrier may be employed. High oxygen permeation of the packaging film also enhances the flavor deterioration. Deaeration of the syrup and inert gas flushing of the package help to reduce oxidative flavor degradation.

As mentioned above, degree of CO_2 loss affected by the package CO_2 permeability and integrity can determine the shelf life of the carbonated soft drink. Packaging with high CO_2 barrier and tight seal on the closure can assure the desired retention of CO_2. The CO_2 loss is also a function of the ratio between package surface area and volume for plastic packages with finite permeability to CO_2. Ideally a smaller ratio of surface area to volume is desired to minimize the CO_2 loss. It was emphasized by Del Nobile et al. [86] that proper temperature control is very important to attain desired retention of CO_2 inside the package; the decrease of CO_2 solubility and increase of plastic bottle's CO_2 permeability with temperature remarkably shorten the time to a given CO_2 loss when temperature fluctuation conditions are met.

17.7.3 Beer

Beer is produced by alcohol fermentation of malted barley and other carbohydrate sources. The yeast, *Saccharomyces cerevisiae* converts the substrate to alcohol, CO_2, and characteristic flavors. The CO_2 gas is collected and returned eventually to saturate the finished product. Typical beer products may be classified depending on fermentation method, color, and alcohol content: lager beer with bottom yeast fermentation, ale beer with top yeast fermentation, stout with caramelized malt for dark color, light beer with low alcohol below 3.9% v/v, and malt liquor with high alcohol above 6.0% v/v [87]. The product is pasteurized usually after packaging, but draught beer is packaged after microfilteration to remove the yeast. Pasteurization of 10 minutes at 60°C at slowest heating point is a usual heat treatment condition for bottled beer [85]. Flash pasteurization of 20 seconds at 70°C is a common pasteurization condition for bulk beer which is packed after heat treatment. Low pH of about 4.0 and some level of alcohol give the microbial stability to the product. Beer is similar to carbonated soft drinks in required properties for packaging, but is more sensitive to flavor deterioration stimulated by oxygen and light. This requirement has restricted beer packaging mostly to metal cans and amber glass bottles with absolute gas barrier. Minimization of headspace oxygen in the container can help to preserve the flavor [88]. The dissolved oxygen in the liquid and the amount of oxygen in the headspace of the package entering during packaging operation and diffused through bottle wall and closure degrade color, appearance, and flavor of beer shortening its shelf life [89]; current practice of 50 ppb oxygen restricts the shelf life to about 120 days. In order to protect the content from oxygen, closures lined with oxygen scavengers are introduced onto the bottled beer. An example is polyolefin liner incorporated with ascorbates and sodium sulfites [90]. The crown and aluminum roll-on closures with oxygen scavenging capacity can remove oxygen residing in headspace and permeating from outside. Recently multilayer PET bottle with an oxygen scavenger– incorporated inner layer has been introduced commercially for beer package. Polyethylene naphthalate (PEN) bottle usually in copolymer form with PET has also been considered as an alternative for beer packaging due to its excellent barrier (to oxygen, water vapor, and ultraviolet ray), mechanical, and temperature resistance properties. The other quality changes in the storage of beer are haze development and intensified color, which are also influenced by oxygen and presence of metal ions. Proper lacquer such as epoxy lining should be used as a coating on the inner surface of metal can to prevent development of metallic flavor and haze coming from metallic ions [88]. Ethylene concentration in the container headspace is coincided with oxidation and browning, and may be used as an index for quality deterioration.

Recently a small plastic insert called widget is placed inside a can package to produce foam similar to appearance of draught beer when poured into a glass [91]. The widget usually consisting of two piece plastic with hole(s) may be fixed in the can or floated. A small amount of liquid nitrogen is injected into the filled can just before seaming and migrated into the preflushed widget at equilibration. The widget works to jet a stream of nitrogen gas into the beer when the can is opened.

17.7.4 Wine

Wines are alcoholic beverages made from fermentation, mainly sustained by yeasts of *Saccharomyces* spp, of musts. Must is the grape juice obtained by mechanical compression and filtration of *Vitis vinifera* berries. The sugar amount of *Vitis viniferas* fruits can reach up to 20% and the extent of its fermentative transformation establishes the final ethanol content. The color of the wine depends on the pigments of the berry skins (mainly anthocyanins) and the winery technology used (the time of contact between must and skins, i.e., maceration). The wine flavors are related both to the grape cultivar and to the fermentation products. When other fruits and berries are fermented, the obtained beverages are normally named after the fruit like apple wine and elderberry wine and they are very different from the wine. Wine is one of the most ancient fermented beverages and the earliest evidence suggesting wine production are dated around the third millennium BC. Wines are not perishable products but their specific organoleptic properties (color, taste, and flavors) can be seriously affected by oxygen, light, and possible biological and mineral contaminants. Excessive access to oxygen may also allow bacterial growth to produce acetic acid from alcohol. Therefore, packaging is an important part of the product.

Wines are traditionally sold in glass bottles sealed using a cork, but other packages such as metal cans, bag- in-box, and plastic bottles are also used as innovative choices. Natural cork stoppers have been used mainly as closures of wine bottles. Screw caps and plastic corks have been introduced recently for wine closures. Synthetic plastic corks with an elastic property and low oxygen permeability are one of the most innovative developments of wine packaging. Oxygen permeation through wine bottle closures is a major concern for selecting a proper one: for natural cork oxygen contained initially in the stopper was found to diffuse slowly into the inside of the bottle [92]. In order to guarantee the same level of sensorial quality, this new packaging must ensure low oxygen permeability, low light transmittance, no chemical interaction with flavoring substances; a low permeability and no interaction are also required for retaining sulfur dioxide, which is still the only authorized additive in wine production for its antioxidant, color preserving, and antimicrobial properties.

17.8 Fresh Fruits and Vegetables

Fresh fruits and vegetables are biologically living tissues with biochemical metabolic activity: respiration, senescence, ethylene production, and organ development. While some of these activities are desirable and essential, most of them consume nutrients and energy sources with loss of freshness. Aerobic respiration process consumes sugar and O_2, and releases CO_2 and energy. Some degree of respiratory activity should be maintained for vitality of the fresh produce, but highly elevated respiration spends up the reserved energy and accelerates the process of senescence resulting in reduced sweetness and freshness. The physiological deterioration rate of the fresh produce generally increases with respiration rate. The pattern of respiration can be divided into two patterns: climacteric and nonclimacteric. Climacteric pattern of respiration, showing the respiration increased to a maximum with storage time, occurs with some fruits such as apples and bananas. Respiration of some fruits and most vegetables maintains relatively constant level through the storage after harvest, which is nonclimacteric. Ethylene (C_2H_4) is produced from most of the horticultural commodities and has functions of enhancing the rates of respiration, maturation, senescence, and other physiological metabolisms. Ethylene production becomes high with the produce damaged mechanically and/or microbiologically. The rates of respiration, senescence, and ethylene production generally decrease with lower oxygen and/or higher carbon dioxide concentrations, which can be attained more or less by packaging design.

Table 17.6 Fresh fruits and vegetables susceptible to chilling injury

Commodity	Approximate lowest safe temperature (°C)	Symptoms of chilling injury
Avocado	5-12	Pitting, grayish-brown discoloration of flesh
Banana	12	Brown streaking on skin
Cucumber	7	Pitting, water-soaked spots, decay
Eggplant	7	Surface scald, alternaria rot, blackening of seeds
Grapefruit	10	Scald, pitting, watery breakdown
Green pepper	7	Sheet pitting, alternaria rot on pods and calyxes, darkening of seeds
Lemon	10	Pitting, membrane staining, red blotch
Mango	5-12	Dull skin, brown areas
Okra	7	Discoloration, water-soaked areas, pitting, decay
Olive	7	Internal browning
Papaya	7	Pitting, water-soaked area
Pineapple	6-12	Brown, black flesh
Potato	3	Mahogany browning, sweetening
Pumpkin and hardshell squash	10	Decay, alternaria rot
Sweet potato	13	Decay, pitting, internal discoloration, hardcore when cooked
Tomato – mature green	13	Poor color when ripe, alternaria rot
Tomato – ripe	7-10	Water soaking and softening, decay

Adapted from Ref. [93, 94].

With progression of metabolic processes, fresh produce becomes soft and labile to bacterial and fungal infections in longer storage. Major fungal species for postharvest decay are *Alternaria, Botrytis, Diplodia, Monilinia, Penicillium,*

Phomopsis, Rhizopus, and *Sclerotinia* [93]. Bacterial decay is caused mainly by *Erwinia* and *Pseudomonas.* Fungal and bacterial infections are encouraged by high humidity conditions or free water on the produce surface. Physical damage also stimulates the microbial infection and decay.

Root, tuber, and bulb crops such as potato, onion, and garlic sprout after a dormancy period, which reduces the sale value of the produce. Continued growth of the organ may deteriorate palatability and marketability.

Transpiration, water loss takes place being proportional to water vapor pressure difference between produce surface and the surroundings. Water loss affects main qualities of fresh fruits and vegetables such as saleable weight, appearance, texture, and flavor. A loss in weight of only 5% often causes fresh produce to lose freshness and appear wilted. Too low humidity should be avoided in the storage and handling of the fresh fruits and vegetables. Packaging or wrapping the produce works to reduce the transpiration rate.

All of the physiological changes decrease with lower temperatures above freezing. Freezing injures the plant tissues making a mushy and water soaked texture, which is not desirable. Therefore chilled but nonfrozen storage is prerequisite for extended shelf life. However, too low chilled temperature is harmful for certain commodities mostly of tropical origin with occurrence of chilling injury. At low but nonfreezing temperature, various symptoms of chilling injury occur because of alteration in normal physiological and biochemical processes. Some symptoms are not apparent in chill storage but become evident when moved to a warmer temperature or cooked for consumption. Table 17.6 lists the fresh fruits and vegetables susceptible to chilling injury.

Relative humidity may affect transpiration rate, decay development, occurrence of physiological disorders, and uniformity of fruit ripening [95]. Condensation of water on the product over long period time enhances the rotting and decay. Proper relative humidity is 85-95% for most fruits and 90-100% for most vegetables. In summary the fresh produce can be stored and handled best under specified range of temperature and relative humidity, which is given in Table 17.7 for some commodities. More extensive lists can be found in reference [94] or other public websites (e.g., University of California Davis, Postharvest Technology Research and Information Center, http://postharvest.ucdavis.edu).

Basic packaging units for fresh produce distribution are mostly corrugated paperboard boxes, which can stand the stacking and handling stresses to protect the produce from mechanical damage [96]. Cushion pad on the box bottom, liners, wraps, and trays, mostly in expanded PS or corrugated paperboard, may be added as supplements for additional physical protection against drop and vibration stresses. Some holed bottoms are made for stable placement of fruits. Individual fruits inside the box may be wrapped in tissue, waxed paper, or expanded PS net for the protection. Vent openings on the container box are usually made for air flow to rapidly cool the packed produce to the desired temperature. There have been some efforts to standardize the boxes for efficient loading and transportation. For maintaining high humidity in the package and reducing the water loss from the produce, several means of moisture barriers are applied: polyethylene liners with small perforations, plastic film wraps with open end, and surface wax coating on the corrugated paperboard boxes. Surface wax coating of the carton board also gives water resistance which is needed when the packed product is hydrocooled or exposed to high humidity conditions for long period.

Table 17.7 Recommended storage temperature and humidity for some fresh fruits and vegetables with approximate storage life

Commodity	Temperature (℃)	Relative humidity (%)	Storage life	Freezing point (°C)
Fruits				
Apple	-1-4	90-95	1-12 months	-1.5
Banana, ripe	13-14	85-95	6-10 days	-0.7
Blueberry	-0.5-0	90-95	2 weeks	-1.2
Kiwifruit	-0.5-0	90-95	3-5 months	-1.6
Mango	13	85-90	2-3 weeks	-0.9
Olive	5-10	85-90	4-6 weeks	-1.4
Orange	0-9	85-90	3-12 weeks	-0.7
Peach	-0.5-0	90-95	2-4 weeks	-0.9
Persimmon	-1	90	3-4 months	-2.1
Pineapple	7-13	85-90	2-4 weeks	-1.1
Strawberry	0	90-95	5-7 days	-0.7
Vegetables				
Asparagus	0-2	95-100	2-3 weeks	-0.6
Broccoli	0	95-100	10-14 days	-0.6
Cabbage, late	0	98-100	5-6 months	-
Cauliflower	0	95-98	3-4 weeks	-0.8
Celery	0	98-100	2-3 months	-0.5
Cucumber	10-13	95	10-14 days	-0.5
Eggplant	8-12	90-95	1 week	-0.8
Garlic	-2	-	10 months	-3.5
Ginger	13	65	6 months	-
Green pepper	7-13	90-95	2-3 weeks	-0.7
Lettuce	0	98-100	2-3 weeks	-0.2
Mushroom	0	95	3-4 days	-0.9
Onion	0	65-70	1-8 months	-0.8
Potato	4	90-95	4-5 months	-0.6
Tomato, firm ripe	8-10	90-95	4-7 days	-0.5

Adapted from Ref. [94] and other sources.

Individual shrink packaging of the citrus fruits has been introduced to give the reduced water loss and increased shelf life [97]. The typical film used is HDPE 10 μm thick. This is also reported to prevent the spread of decay from infected fruits and promote the healing of the fruit skin injuries. Edible wax coatings are also applied on the produce surface for regulating water loss and delaying senescence [98]. Commercial uses are mainly for citrus, apples, mature green tomatoes, rutabagas, and cucumbers. The permeability of the coating layer should be matched to respiration or physiological requirement of the produce. The coating layer should allow the appropriate degree of gas exchange across the coated fruit peel in order to avoid anaerobic conditions inside the produce. The most widely used solutions for coating are beeswax, carnauba, paraffin, and polysaccharides of sucrose polyester.

Recently there has been significant growth in consumer prepackaging of fresh fruits and vegetables. The widely used forms of prepackaging are perforated polyethylene film bags and polystyrene trays overwrapped with cling film; they restrict the water loss from the produce and provide slight degree of modified atmosphere benefit. Hermetic polymeric package with resultantly lowered O_2 and elevated CO_2 can reduce respiration rate and ethylene production of the produce, and extend the shelf life. The permeability of the plastic film to O_2 and CO_2 should be tailored to obtain the optimal atmosphere specific to the commodity. The techniques of designing modified atmosphere package for the fresh produce have been dealt in Sections 13.3 and 13.4. Ethylene scavenger has been used partially for transport packaging of fruits to protect them from ethylene effect (see also Section 15.2.4). Recently controlled release of 1-methylcyclopropene, an ethylene inhibitor has been tried for the consistent inhibition of ethylene activity in the fresh produce packaging [99].

One popular sector in MAP of fresh produce is for fresh-cut fruits and vegetables. The fresh-cut fruits and vegetables provide both convenience and freshness to consumers, and attract the interests of many facets of the food industry including food manufacturers, retail food stores, restaurants, etc. These minimally processed products are prepared by a single or any number of appropriate unit operations such as peeling, cutting, shredding, etc. Compared to whole intact tissues, fresh-cut produce tissues have increased respiration and are more susceptible to oxidative browning and entrance of microbes. Consequently, the fresh-cut products suffer from fast quality deterioration and have a reduced shelf life. MAP of low oxygen and high carbon dioxide concentrations can thus effectively extend shelf life for minimally processed fresh products by delaying senescence. The modified atmosphere most commonly applied for the fresh-cut products are gas concentrations of 2-8 % O_2 and 5-12 % CO_2. Packaging with undesirable gas composition of too low O_2 or too high CO_2 concentrations may result in anaerobic respiration or CO_2 injury, which can do more harm than good.

Usually polyolefin films with high permeabilities to O_2 and CO_2 are used due to elevated respiration of the cut produce. The plastic bags may be flushed actively with the desired atmosphere or passively equilibrated to attain the wanted package atmosphere from the balance between the respiration and permeation. Sometimes the bags are vacuumized to give easy handling of the product with microatmosphere quickly modified: small headspace results in fast equilibration in the atmosphere. MAP designs are based on the balance between the produce respiration and permeation through the film (see Sections 13.3 and 13.4): permeability of the plastic film is very important to create and maintain an optimal atmosphere. To attain the required high permeation of the package some microperforations or diffusion leaks are made on the plastic film [100]. A specialized film has a capability to increase its gas permeability drastically at glass transition temperature ('switch point') adjusted deliberately for consideration of temperature abuse conditions. Detailed information on microperforation and selective permeation can be found in Section 15.4.3.

Recently MAP with high oxygen concentration has been proposed to allegedly slow down the microbial deterioration and inhibit enzymatic browning [101]: the high O_2 and high CO_2 modified atmosphere (50% O_2 + 30% CO_2) prolonged the shelf life of sliced carrots compared to storage in air at 8°C, and was comparable to the low oxygen modified atmosphere (1% O_2 + 10% CO_2).

17.9 Frozen Foods

Products sold in frozen state include fruits, vegetables, fishes, meats, bakery products, ice cream, and ready-to-eat multi-component products. Diversity of quality factors exists for frozen foods with different preparation and consumption patterns. Since microorganisms of hygienic concern will not grow at freezing temperatures and some will die upon freezing, microbiological issues are usually irrelevant for the stability of frozen foods, unless the products are manufactured under unsanitary condition and subjected to significant temperature abuse. However, physical and chemical deteriorations occur for long period storage at temperature below $-18°C$. Physical changes during frozen storage include recrystallization and freezer burn caused by sublimation of ice. Chemical changes proceeding during frozen storage are lipid oxidation, enzymatic browning, flavor deterioration, protein insolubilization, and degradation of chlorophyll and vitamins [102]. The freezing process itself can intensify enzymatic reactions and undesirable protein-starch interactions due to cell disruption and release of cellular components. This will result in a number of changes during storage as well as degraded sensory quality as the material is thawed. These negative effects are more pronounced if a slow freezing process is used; fast freezing tends to better preserve the integrity of the food structure because of the smaller size of ice crystal formed. Yet, there has to be a compromise between freezing speed and quality, since too high freezing speed may cause freeze cracks in the food product.

Moisture losses in frozen foods take place when they are stored without an adequate moisture barrier. The driving force for the moisture sublimation is the difference in water vapor pressure between the food surface and environment. The sublimation of ice on the surface results in surface desiccation and freezer burn with color change [103]. Moisture migration in packaged frozen foods leads to formation of ice or frost inside the package, which is converted directly to drip in thawing. The freezer burn will also impart changes in texture due to the misplacement of the water. The altered texture will also increase the specific surface for oxygen attack resulting in off-flavors. Moisture migration can be minimized by maintaining small storage temperature fluctuation and small temperature gradient inside the product [103, 104]. The use of moisture barrier packaging materials can reduce the moisture loss but cannot avoid this cascade of events. Vacuum packaging using materials with low permeabilities to water vapor and gas is often recommended to solve the problem effectively by eliminating any headspace that would allow undesirable sublimation-refreezing cycles. Components of different water activities produce moisture absorption and redistribution in multi-component frozen foods resulting in loss of textural characteristics. With minimized temperature change and difference in the product, inclusion of inner moisture barriers within the product also helps to reduce the damage due to moisture mobility.

Recrystallization with an increase of ice crystal size alters the texture of frozen foods. The increased ice crystal in the foods leads to redistribution of soluble solutes around tissue and denaturation of proteins, which in turn cause an increase of the drip after thawing.

Enzymatic activity can create a quality deterioration problem in raw and unheated frozen food. Even though enzyme activity is reduced at freezing temperatures with partial denaturation and low molecular mobility, there is often residual enzymatic activity that can be detrimental during the relatively long storage periods of frozen foods. Furthermore, the fact that enzymes and substrates are concentrated upon freezing is critical to food stability. When the substrate concentration upon freezing reaches a critical value for a specific enzyme, the reaction rate is substantially accelerated and the system will be subjected to increased deterioration. Loss of flavor due to lipoxygenase and peroxidase activities can cause rancidity in systems containing lipids and other susceptible ingredients. Lipolytic enzymes such as lipases and phospholipases hydrolyze fats and oils yielding glycerol and free fatty acids. The short-chain fatty acids (C_4-C_{10}) are associated with undesirable odors. Some enzymes of this group have been shown to be active in frozen food systems held as low as -29°C [103]. Lipolytic rancidity is a major concern for frozen fish products. In fruits and vegetables, endogenous enzymes such as pectin methyl esterase, chlorophyllase, anthocyanse, and polyphenoloxidase may degrade their texture, pigment, and color. Enzymatic browning can be accelerated in frozen plant tissue due to disruption of cells caused by ice crystals, which facilitates the contact between the enzyme and the substrate in presence of oxygen.

Oxidative rancidity reaction requires unsaturated oils/fats and oxygen, and can be initiated by enzymatic and nonenzymatic pathways. It is accelerated by light, pigments, and metals, including copper and iron. In lipid oxidation, the hydroperoxides produced from unsaturated fatty acids are eventually decomposed into short-chain aldehydes and ketones responsible for the off-flavors in the frozen food. Hexanal is often one of the major volatiles released from the lipid oxidation. The threshold values for these reaction products are very low, reaching the order of parts per billion (ppb, or μg/kg). To prevent the oxidation of lipids and other susceptible compounds, antioxidants can be added or combined with other methods (such as chelating agents, light proof packages, vacuum packages, or edible coatings) to obtain a synergistic effect.

Protein denaturation is caused by ice formation, recrystallization, dehydration, soluble solid concentration, and oxidation. The protein denaturation is commonly observed with decreased water holding capacity, textural deterioration, and loss of succulence. Protein changes of this type occur mostly in frozen flesh foods.

Nonenzymatic browning between the available amino groups and the aldo or keto groups, is often enhanced by freezing and subsequent storage, leading to a bitter off-flavor and color change [102]. Natural pigments present in frozen food materials are reactive and unstable under exposure to oxygen and light. If oxygen is available, myoglobin in meat products is oxidized to metmyoglobin, imparting an off-color to the meat surface even in the frozen stage. The photosynthetic pigments, chlorophylls a and b, responsible for the bright green color of many vegetables can be bleached irreversibly when the plant cells damaged due to freezing are exposed to light and oxygen.

All major components relevant to the nutritive quality of foods can be affected during freezing preservation. Heat labile water-soluble vitamins such as thiamin and ascorbic acid easily undergo oxidation in the presence of oxygen. Losses of the mentioned vitamins as well as tocopherols have also been observed during the shelf life of frozen food products. Degradation of fatty compounds may affect the availability and stability of liposoluble vitamins and essential fatty acids.

There may also be loss and/or transfer of volatile flavors, which reduces sensory quality of frozen foods, especially multi-component prepared products (84,118). As for most of quality changes, it is of prime importance to control the storage temperature so that frozen foods may stay as close as possible to a glassy state, which will reduce the molecular mobility required for undesirable chemical reactions and physical processes.

Packaging of frozen foods imposes certain special requirements in terms of temperature stability, barrier properties, insulation, compatibility with packaging machinery, and consumer appeal [105]. Suitable materials for packaging frozen foods should be stable physically and chemically over temperature range likely to be experienced by the product. The temperature applied for usual frozen food packages extends from –40°C to ambient temperature. However, high temperatures above 100°C may be encountered for 'boil-in-bag' frozen products. When microwave heating or cooking is expected, temperatures up to 300°C should be expected. As mentioned above, to protect the food quality from deterioration, the packaging material should be with adequate barriers to moisture, oxygen, and light depending on the product characteristics. Paperboard packaging blocking out light can obtain the well preserved vanilla flavor and low degree of oxidized flavor in ice cream [106]. They should not absorb oil, grease, or water from the packaged foods. Sometimes insulation property is required to maintain the product temperature consistently especially for bulk package from which a single portion is taken out intermittently. Examples are PS foam tubs or trays for ice cream and frozen cake. Metallized laminated foamed PS tray or bag is used for protecting against the radiation heat. High quality printing and graphics on the package surface is required for good consumer appeal. Recently antifog films with high transparency and gloss are available.

The packages for frozen foods cover direct film wraps, flexible bags including 'boil-in bag' packs, overwrapped trays, carton boxes, and rigid containers of plastics and metals [69, 107]. Popular method to package frozen foods is using polyethylene bags. If wrapped tightly around the food product, these materials help prevent freezer burn and several related problems discussed before. Higher degree of preservation is conveniently obtained using vacuum packaging in high barrier film containing EVOH, PVDC, or metallized layer. Vacuum packages are used primarily for shipping frozen food from packers to retail grocery, hotel, restaurant, and institution outlets. Exclusion of oxygen and light under moisture control helps to preserve the quality of frozen foods. Skin packaging is a modified form of vacuum packaging: a preheated film is dropped onto the product, which is supported on a lower web of the same film; the air between the two films is withdrawn, and an upper web forms around the food to produce a skintight package that is then heat-sealed in a vacuum chamber. It is used primarily for frozen meat products. Several types of material are used for skin packaging: most are a blend of ionomer material and low density polyethylene (LDPE). Gas flushed packaging is rarely used for frozen foods; however, Ray et al. [108] showed that frozen turkey strips formerly cooked retained higher organoleptic scores under nitrogen-flushed package of high oxygen barrier.

Edible coatings (applied directly on foods by dip or spray) and edible film wraps (preformed, then placed on foods or between food components) can function to prevent quality losses in frozen-food system in numerous ways. An edible coating can reduce the transmission rate of moisture and oxygen between the food and the surrounding atmosphere, preventing or slowing down quality changes associated with the oxidation of lipids, vitamins, pigments, and flavor compounds. Edible coatings can also be used to incorporate various food additives at specific locations. With regard to structural aspects, edible coatings can be used to improve the structural integrity and mechanical handling characteristics of products susceptible to shattering or fragmentation. The most common coatings are acetylated monoglyceride wax and corn amylopectin [107].

Discussion Questions and Problems

1. Choose a product in a supermarket and list the main quality deterioration modes. What protections are provided by packaging to ensure storage stability and shelf life for the product?

2. Piergiovanni et al. [51] suggested that modified atmosphere of 10% CO_2 and 90% N_2 is optimal for Taleggio cheese at 6°C. The cheese may be presumed to produce CO_2 gas at a rate of 1.69 x 10^{-3} cm^3 g^{-1} h^{-1} under similar condition (Figure 17.9). You are asked to package 250 g cheese in the plastic container with surface area of 600 cm^2. If you disregard the permeation of O_2 and N_2 through the packaging layer, which plastic films of reasonable thickness would you select for the appropriate packaging conditions? Please look over the gas permeability data given in Table 13.9 of Chapter 13 or other source. In case you have several choices in satisfying the required CO_2 permeability, also examine the possible alternatives based on the permeabilities to other gases.

3. What would be the benefits of vacuum packaging of frozen beef?

4. What kind of modified atmosphere packaging can be used for composite confectionery products? Give the reasoning for your selection.

5. Please find the optimal packaging condition for shredded lettuce from literature survey. How the packaging should be combined with several kinds of pretreatment and storage conditions?

6. According to Lee et al. [109, 110], red pepper powder, a condiment used for hot spicy taste, is degraded in quality by two modes of color loss, carotenoid destruction and nonenzymatic browning, whose rates are influenced by water activity and/or package atmosphere. Elaborate a packaging strategy to keep quality of red pepper powder for long time period of shelf life.

Bibliography

1. T Ishitani. Cereals. In: T Yano, ed. Handbook of Food Packaging (in Japanese). Tokyo: Japan Association of Packaging Professionals, 1988, pp. 1043-1046.
2. LF Murray, R Moss. Estimation of fat acidity in milled wheat products. Part II. Using colorimetry to ascertain the effect of storage time and conditions. J Cereal Sci 11:179-184, 1990.

3. S Yanai, S Tsubata, T Ishitani, S Kimura. Storability of rice stored in various plastic pouches. Nippon Shokuhin Kogyo Gakkaishi 25:563-569, 1978.
4. DA Fellers, MM Bean. The storage life of wheat based foods: a review. J Food Sci 42:1143-1147, 1977.
5. RN Sharp, LK Timme. Effects of storage time, storage temperature, and packaging method on shelf life of brown rice. Cereal Chem 63:247-251, 1986.
6. JH New, DP Rees. Laboratory studies on vacuum and inert packaging for the control of stored-product insects in foodstuffs. J Sci Food Agr 43:235-244, 1988.
7. S Yanai, T Ishitani, T Kojo. Influence of gaseous environment on the hermetic storage of milled rice. Nippon Shokuhin Kogyo Gakkaishi 26:145-150, 1979.
8. RL Ory, AJ Delucca, AJ St Angelo, HP Dupuy. Storage quality of brown rice as affected by packaging with and without carbon dioxide. J Food Protect 43:929-932, 1980.
9. SF Brockington, VJ Kelly. Rice breakfast cereals and infant foods. In: DF Houston, ed. Rice, Chemistry and Technology. St. Paul Minnesota: American Association of Cereal Chemists, 1972, pp. 400-418.
10. JAK Howarth. Ready-to-eat breakfast cereals. In: CMD Man, AA Jones, ed. Shelf Life Evaluation of Foods. London: Blackie Academic & Professional, 1994, pp. 235-255.
11. TP Labuza. Shelf-life Dating of Foods. Westport, Connecticut: Food & Nutrition Press, 1982, pp. 99-246.
12. J Miltz, P Hoojjat, JK Han, JR Giacin, BR Harte, IJ Gray. Loss of antioxidant from high-density polyethylene: Its effect on oatmeal cereal oxidation. In: JH Hotchkiss, ed. Food and Packaging Interactions. Washington DC: American Chemical Society, 1988, pp. 83-93.
13. C Wessling, T Nielsen, JR Giacin. Antioxidant ability of BHT- and α-tocopherol-impregnated LDPE film in packaging of oatmeal. J Sci Food Agr 81:194-201, 2001.
14. CK Mok, HY Lee, YJ Nam, KB Suh. Effects of water activity on crispness and brittleness, and determination of shelf-life of barley flake. Korean J Food Sci Technol 13:289-298, 1981.
15. EM Furuya, JJ Warthesen. Packaging effects on riboflavin content of pasta products in retail markets. Cereal Chem 61:399-402, 1984.
16. JP Smith, BK Simpson. Modified atmosphere packaging of bakery and pasta products. In: JM Farber, KL Dodds, ed. Principles of Modified-Atmosphere and Sous Vide Product Packaging. Lancaster: PA: Technomic Publishing, 1995, pp. 207-242.
17. M Sinigaglia, MR Corbo, GD Fabio, S Massa. Effect of under-vacuum packaging on microbiology of fresh 'home-made' pasta. Chem Mikrobiol Technol Lebensm 17:110-113, 1995.

18. DS Lee, HD Paik, GH Im, IH Yeo. Shelf life extension of Korean fresh pasta by modified atmosphere packaging. J Food Sci Nutri 6:240-243, 2001.
19. B Ooraikul. Modified atmosphere packaging of bakery products. In: B Ooraikul, ME Stiles, ed. Modified Atmosphere Packaging of Food. New York: Ellis Horwood, 1991, pp. 49-117.
20. JP Smith. Bakery products. In: RT Parry, ed. Principles and Applications of Modified Atmosphere Packaging of Foods. London: Blackie Academic & Professional, 1993, pp. 134-169.
21. HP Jones. Ambient packaged cakes. In: CMD Man, AA Jones, ed. Shelf Life Evaluation of Foods. London: Blackie Academic & Professional, 1994, pp. 179-201.
22. DAL Seiler. Modified atmosphere packaging of bakery products. In: AL Brody, ed. Controller/Modified Atmosphere/Vacuum Packaging of Foods. Trumbull, Connecticut: Food & Nutrition Press, 1989, pp. 119-133.
23. IS Kotsianis, V Giannou, C Tzia. Production and packaging of bakery products using MAP technology. Trends Food Sci Tech 13:319-324, 2002.
24. NFF Soares, DM Rutishauser, N Melo, RS Cruz, NJ Andrade. Inhibition of microbial growth in bread through active packaging. Packag Technol Sci 15:129-132, 2002.
25. CO Gill. MAP and CAP of fresh, red meats, poultry and offals. In: JM Farber, KL Dodds, ed. Principles of Modified-Atmosphere and Sous Vide Product Packaging. Lancaster, PA: Technomic Publishing, 1995, pp. 105-136.
26. DE Hood, GC Mead. Modified atmosphere storage of fresh meat and poultry. In: RT Parry, ed. Principles and Applications of Modified Atmosphere Packaging of Foods. London: Blackie Academic & Professional, 1993, pp. 269-298.
27. AA Kraft. Meat microbiology. In: PJ Bechtel, ed. Muscle as Food. London: Academic Press, 1986, pp. 239-279.
28. G Tewari, DS Jayas, RA Holley. Centralized packaging of retail meat cuts: a review. J Food Protect 62:418-425, 1999.
29. RA Lawrie. Meat Science. Oxford: Pergamon Press, 1991, pp. 101-224.
30. SA Muller. Packaging and meat quality. Can Inst Food Sci Technol 23:22-25, 1990.
31. M Eilamo, A Kinnunen, K Lavta-Kala, R Ahvenainen. Effects of packaging and storage conditions on volatile compounds in gas-packed poultry meat. Food Addit Contam 15:217-228, 1998.
32. PA Morrissey, JP Kerry. Lipid oxidation and the shelf-life of muscle foods. In: R Steele, ed. Understanding and measuring the shelf-life of food. Cambridge, UK: Woodhead Publishing, 2004, pp. 357-395.
33. M Lee, JG Sebranek, DG Olson, JS Dickson. Irradiation and packaging of fresh meat and poultry. J Food Protect 59:62-72, 1996.
34. AM Pearson, TA Gillett. Processed Meats. Gaithersburg, Maryland: Aspen Publishers, 1996, pp. 53-78.

35. HK Davis. Fish. In: RT Parry, ed. Principles and Applications of Modified Atmosphere Packaging of Foods. London: Blackie Academic & Professional, 1993, pp. 189-228.
36. DM Gibson, HK Davis. Fish and shell fish products in sous vide and modified atmosphere packs. In: JM Farber, KL Dodds, ed. Principles of Modified-Atmosphere and Sous Vide Product Packaging. Lancaster, PA: Technomic Publishing, 1995, pp. 153-174.
37. TC Lanier. Packaging. In: RE Martin, GJ Flick, ed. The Seafood Industry. New York: Van Nostrand Reinhold, 1990, pp. 194-204.
38. M Sivertsvik, WK Jeksrud, T Rosnes. A review of modified atmosphere packaging of fish and fishery products-significance of microbial growth, activities and safety. Int J Food Sci Tech 37:107-127, 2002.
39. L Franzetti, S Martinoli, L Piergiovanni, A Galli. Influence of active packaging of minimally processed fish products in a modified atmosphere. Packag Technol Sci 14:267-274, 2001.
40. A Pacquit, J Frisby, D Diamond, KT Lau, A Farrell, B Quilty, D Diamond. Development of a smart packaging for the monitoring of fish spoilage. Food Chem 102:466-470, 2007.
41. B Bjerkeng, M Sivertsvik, JT Rosnes, H Bergslien. Reducing package deformation and increasing filling degree in packages of cod fillets in CO_2-enriched atmospheres by adding sodium carbonate and citric acid to an exudate absorber. In: P Ackermann, M Jagerstad, T Ohlosson, ed. Foods and Packaging Materials-Chemical Interactions. Cambridge UK: The Royal Society of Chemistry, 1995, pp. 222-227.
42. DD Muir, JM Banks. Milk and milk products. In: D Kilcast, P Subramaniam, ed. The Stability and Shelf-life of Food. Cambridge, UK: Woodhead Publishing, 2000, pp. 197-219.
43. IJ Tuohy. Chilled storage and keeping quality of pasteurized fluid milk products. In: TR Gormley, ed. Chilled Foods. London: Elsevier Applied Science, 1990, pp. 37-63.
44. WS Duyvesteyn, E Shimoni, TP Labuza. Determination of the end of shelf-life for milk using Weibull hazard method. Lebensm Wiss Technol 34:143-148, 2001.
45. J Goursaud. Packaging of milk products. In: G Bureau, JL Multon, ed. Food Packaging Technology. Vol. 2. New York: VCH Publishers, 1996, pp. 251-276.
46. PJ Subramaniam. Miscellaneous applications. In: RT Parry, ed. Principles and Applications of Modified Atmosphere Packaging of Foods. London: Blackie Academic & Professional, 1993, pp. 170-188.
47. DB Min, SH Lee, JB Lindamood, KS Chang, GA Reineccius. Effects of packaging conditions on the flavor stability of dry whole milk. J Food Sci 54:1222-1224, 1989.
48. R Scott. Cheesemaking Practice. London: Applied Science Publishers, 1981, pp. 26-37.

49. P Walstra, TJ Geurts, A Noomen, A Jellema, MAJS van Boekel. Dairy Technology. New York: Marcel Dekker, 1999, pp. 517-708.
50. ME Johnson, BA Riester, C Chen, B Tricomi, NF Olson. Effect of packaging and storage conditions on calcium lactate crystallization on the surface of Cheddar cheese. J Dairy Sci 73:3033-3041, 1990.
51. L Piergionvanni, P Fava, M Moro. Shelf life extension of Taleggio cheese by modified atmosphere packaging. Ital J Food Sci 5:115-127, 1993.
52. L Giannuzzi, A Lombardi, NE Zaritzky. Microbial flora in hard and soft cheeses packed in flexible plastic films. Ital J Food Sci 10:57-65, 1998.
53. RS Topal. Effects of different packaging materials and techniques on the curing of Kashar cheese and on its surface moulds. Lebensm Wiss Technol 24:341-349, 1991.
54. CH Mannheim, T Soffer. Shelf-life extension of cottage cheese by modified atmosphere packaging. Lebensm Wiss Technol 29:767-771, 1996.
55. AB Maniar, JE Marcy, JR Bishop, SE Duncan. Modified atmosphere packaging to maintain direct-set cottage cheese quality. J Food Sci 59:1305-1308, 1994.
56. RMV Alves, CIGDL Sarantopoulos, AGF van Dender, JDAF Faria. Stability of sliced Mozzarella cheese in modified-atmosphere packaging. J Food Protect 59:838-844, 1996.
57. E Gonzalez-Fandos, S Sanz, C Olarte. Microbiological, physicochemical and sensory characteristics of Cameros cheese packaged under modified atmospheres. Food Microbiol 17:407-414, 2000.
58. I Sarais, D Piussi, V Aquili, ML Stecchini. The behavior of yeast populations in Stracchino cheese packaged under various conditions. J Food Protect 59:541-544, 1996.
59. CM Hong, WL Wendorff, RL Bradley Jr. Effects of packaging and lighting on pink discoloration and lipid oxidation of annatto-colored cheeses. J Dairy Sci 78:1896-1902, 1995.
60. G Mortensen, J Sorensen, H Stapelfeldt. Effect of light and oxygen transmission characteristics of packaging materials on photo-oxidative quality changes in semi-hard Havarti cheeses. Packag Technol Sci 15:121-127, 2002.
61. S Desobry, J Hardy. Camembert cheese water loss through absorbent packaging. J Food Sci 59:986-989, 1994.
62. W Schar, JO Bosset. Chemical and physico-chemical changes in processed cheese and ready-made fondue during storage. A review. Lebensm Wiss Technol 35:15-20, 2002.
63. ML Richmond, BR Harte, JR Gray, CM Stine. Physical damage of yogurt. The role of secondary packaging on stability of yogurt. J Food Protect 48:482-486, 1985.
64. SEA Jansson, CJ Edsman, UW Gedde, MS Hedenqvist. Packaging materials for fermented milk: effects of material crystallinity and polarity on food quality. Packag Technol Sci 14:119-127, 2001.

65. CW Miller, MH Nguyen, M Rooney, K Kailasapathy. The influence of packaging materials on the dissolved oxygen content of probiotic yoghurt. Packag Technol Sci 15:133-138, 2002.

66. RF Boutin. Confections. In: YH Hui, ed. Encyclopedia of Food Science and Technology. New York: John Wiley & Sons, 2000, pp. 465-476.

67. P Subramaniam. Confectionery products. In: D Kilcast, P Subramaniam, ed. The Stability and Shelf-life of Food. Cambridge UK: Woodhead Publishing, 2000, pp. 221-248.

68. P Ravichandran, KR Kumar. Moisture sorption and texture characteristics of sugar confectionery. Confect Product 63(11):33-34, 1997.

69. FA Paine, HY Paine. A Handbook of Food Packaging. Glasgow UK: Leonard Hill, 1983, pp. 199-254.

70. AV Martin. Chocolate confectionery. In: CMD Man, AA Jones, ed. Shelf Life Evaluation of Foods. London: Blackie Academic & Professional, 1994, pp. 216-234.

71. BW Minifie. Chocolate, Cocoa, and Confectionery: Science and Technology. 3rd ed. New York: Van Nostrand Reinhold, 1989, pp. 709-769.

72. DJ An, GW Halek. Partitioning of printing ink solvents on chocolate. J Food Sci 60:125-127, 1995.

73. D Schappil, F Escher. The importance of texture in hard boiled candy. Confect Product 68(5):13-15, 2002.

74. S Sacharow, RC Griffin. Principles of Food Packaging. 2nd ed. Westport, Conneticutt: AVI Publishing, 1980, pp. 276-402.

75. DH Chung, YH Lee, MS Yoo, YR Pyun. Prediction of shelf-life of chewing gum based on moisture gain and loss. Korean J Food Sci Technol 24:122-126, 1992.

76. J Kristott. Fats and oils. In: D Kilcast, P Subramaniam, ed. The Stability and Shelf-life of Food. Cambridge, UK: Woodhead Publishing, 2000, pp. 279-308.

77. P Hernandez-Munoz, R Catala, R Gavara. Effect of sorbed oil on food aroma loss through packaging materials. J Agr Food Chem 47:4370-4374, 1999.

78. MA Del Nobile, S Bove, E La Notte, R Sacchi. Influence of packaging geometry and material properties on the oxidation kinetic of bottled virgin olive oil. J Food Eng 57:189-197, 2002.

79. Y El-Shattory, MA Saadia, FH Said. Flavour changes due to effect of different packaging materials on storing of cottonseed oil, hydrogenated oil and margarine. Grasas Aceites 48(2):61-67, 1997.

80. S Nagy, RL Rouseff. Citrus fruit juices. In: G Charalambous, ed. Hanbook of Food and Beverage Stability. Orlando: Academic Press, 1986, pp. 719-743.

81. JG Woodroof. Storage life of canned, frozen, dehydrated, and preserved fruits. In: JG Woodroof, BS Ruh, ed. Westport, Connecticut: Academic Press, 1986, pp. 719-743.

82. NFF Soares, JH Hotchkiss. Comparative effects of de-aeration and package permeability on ascorbic acid loss in refrigerated orange juice. Packag Technol Sci 12:111-118, 1999.

83. S Limbo, L Torri, L Piergiovanni. Light-induced changes in an aqueous β-carotene system stored under halogen and fluorescent lamps, affected by two oxygen partial pressures. J Agr Food Chem 55(13):5238 - 5245, 2007.

84. RB Taylor. Ingredients. In: PR Ashurst, ed. The Chemistry and Technology of Soft Drinks and Fruit Juices. Sheffield, UK: Sheffield Academic Press, 1998, pp. 16-54.

85. GL Robertson. Food Packaging. New York: Marcel Dekker, 1993, pp. 338-380.

86. MA Del Nobile, G Mensitieri, L Nicolais, P Masi. The influence of thermal history on the shelf life of carbonated beverages bottled in plastic containers. J Food Eng 34:1-13, 1997.

87. P Jelen. Introduction to Food Processing. Reston, Virginia: Reston Publishing, 1985, pp. 175-198.

88. NAM Eskin. Biochemistry of Foods. San Diego: Academic Press, 1995, pp. 297-334.

89. L Kuchel, AL Brody, L Wicker. Oxygen and its reactions in beer. Packag Technol Sci 19:25-32, 2006.

90. FN Teumac. The history of oxygen scavenger bottle closures. In: ML Rooney, ed. Active Food Packaging. London: Blackie Academic & Professional, 1995, pp. 193-202.

91. R Miles. Packaging of beverages in cans. In: GA Giles, ed. Handbook of Beverage Packaging. Sheffield, UK: Sheffield Academic Press, 1999, pp. 16-52.

92. P Lopes, C Saucier, P-L Teissedre, Y Glories. Main routes of oxygen ingress through different closures into wine bottles. J Agr Food Chem 55:5167-5170, 2007.

93. RBH Wills, WB McGlasson, D Graham, TH Lee, EG Hall. Postharvest. New York: Van Nostrand Reinhold, 1989, pp. 17-143.

94. RE Hadenburg, AE Watada, CY Wang. The Commercial Storage of Fruits, Vegetables, and Florists and Nursery Stocks. Agricultural Handbook No. 66. Washington DC: United States Department of Agriculture, 1990, pp. 9-72.

95. AA Kader. Postharvest biology and technology: an overview. In: AA Kader, ed. Postharvest Technology of Horticultural Crops. Davis, CA: University of California, 1992, pp. 15-20.

96. FG Mitchell. Packages for horticultural crops. In: AA Kader, ed. Postharvest Technology of Horticultural Crops. Davis, CA: University of California, 1992, pp. 45-52.

97. S Ben-Yehoshua. Transpiration, water stress, and gas exchange. In: J Weichmann, ed. Postharvest Physiology of Vegetables. New York: Marcel Dekker, 1987, pp. 113-170.

98. EA Baldwin. Edible coatings for fresh fruits and vegetables: past, present, and future. In: JM Krochta, EA Baldwin, MO Nisperos-Carriedo, ed. Edible

Coatings and Films to Improve Food Quality. Lancaster, PA: Technomic, 1994, pp. 25-64.

99. AJ Macnish, DC Joyce, DE Irving, AH Wearing. A simple sustained release device for the ethylene binding inhibitor 1-methylcyclopropene. Postharvest Biol Tec 32(3):321-338, 2004.

100. AL Brody. High-gas-permeability packaging films for passive control of high-respiration-rate food content. In: AL Brody, ed. Modified Atmosphere Food Packaging. Herndon, Virginia: Institute of Packaging Professionals, 1994, pp. 161-172.

101. A Amanatidou, RA Slump, LGM Gorris, EJ Smid. High oxygen and high carbon dioxide modified atmospheres for shelf-life extension of minimally processed carrots. J Food Sci 65:61-66, 2000.

102. OR Fennema, WD Powrie, EH Marth. Low-temperature preservation of foods and living matter. New York: Marcel Dekker, 1973, pp. 150-239.

103. NE Zaritzky. Factors affecting the stability of frozen foods. In: CJ Kennedy, ed. Managing Frozen Foods. Cambridge, UK: Woodhead Publishing, 2000, pp. 111-135.

104. H Symons. Frozen foods. In: CMD Man, AA Jones, ed. Shelf Life Evaluation of Foods. London: Blackie Academic & Professional, 1994, pp. 296-316.

105. M George. Selecting packaging for frozen food products. In: CJ Kennedy, ed. Managing Frozen Foods. Cambridge, UK: Woodhead Publishing, 2000, pp. 195-211.

106. ML Suttles, RT Marshall. Interactions of packages and fluorescent light with flavor of ice cream. J Food Protect 56:622-624, 1993.

107. B Feinberg, RP Hartzell. Packaging materials and packaging of frozen foods. In: DK Tressler, WB Van Arsdel, MJ Copley, ed. The Freezing Preservation of Foods. Vol. 2. Westport, CT: AVI Publishing, 1986, pp. 260-288.

108. EE Ray, Y Olivas, M Vincent, D Smith. Effects of modified atmosphere packaging and types of pouches on product attributes of frozen (IQF) turkey strips. J Food Protect 56:626-629, 1993.

109. DS Lee, SK Chung, HK Kim, KL Yam. Nonenzymatic browning in dried red pepper products. J Food Quality 14:153-163, 1991.

110. DS Lee, SK Chung, KL Yam. Carotenoid loss in dried red pepper products. Int J Food Sci Technol 27:179-185, 1992.

Part Four

Packaging Sociology

The last two chapters are devoted to socioeconomic considerations of packaging. Chapter 18 introduces the important concept of sustainable packaging, a holistic approach to develop packaging systems by striking a balance between the needs of the society, the environment, and the economy. In fact, this book attempts to provide useful information and knowledge towards achieving the vision or goal of sustainable packaging. This chapter also proposes that sustainable packaging must also consider the functions of packaging, which is frequently ignored in the discussions of this topic. Chapter 19 discusses the human environment surrounding packaging including topics such as tamper evidence packaging, product liability, and label information.

594

Chapter 18

Sustainable Packaging

18.1 Introduction 595
18.2 Sustainable Packaging 596
    18.2.1 Basic Concepts ·· 596
    18.2.2 Development of Sustainable Packaging ·················· 597
18.3 Environmental Issues Relating to Packaging 598
    18.3.1 Solid Wastes ·· 598
    18.3.2 Hazardous Compounds ·· 598
    18.3.3 Ozone Depletion ··· 598
18.4 Packaging Waste Management 599
    18.4.1 Reduction ··· 599
    18.4.2 Reuse ··· 599
    18.4.3 Recycling ··· 600
    18.4.4 Composting ·· 601
    18.4.5 Incineration ·· 601
    18.4.6 Landfill ··· 602
18.5 Life Cycle Assessment 602
    18.5.1 Scope of LCA Studies ·· 602
    18.5.2 Methodology of LCA ··· 603
    18.5.3 LCA Studies ·· 604
18.6 Degradable Packaging Polymers 604
    18.6.1 Biodegradable Packaging ···································· 605
    18.6.2 Photodegradable Packaging ·································· 606
Discussion Questions and Problems 607
Bibliography 607

18.1 Introduction

In 1987, the World Commission on Environment and Development defined the term 'sustainable development' as the guiding principles for meeting the needs of the present without compromising the ability of future generations to meet their own needs. Similarly, the European Union defined sustainable development as a vision of progress that integrates immediate and longer-term needs, local and global needs, and regards social, economic, and environmental needs as inseparable and interdependent components of human progress. In these definitions, the general terms, guiding principles and vision of progress, are used because sustainable development is a complex process with no specific steps applicable to every situation.

Today there is a growing interest from organizations, governments, and companies around the world to give sustainable development useful and practical meanings. While the concept of sustainable development is important to virtually all industries, this chapter is focused on the development of sustainable food packaging.

18.2 Sustainable Packaging

18.2.1 Basic Concepts

The Sustainable Packaging Coalition, a leading organization for sustainable packaging, defines sustainable packaging as a target vision for creating "a world where all packaging is sourced responsibly, designed to be effective and safe throughout its life cycle, meets market criteria for performance and cost, is made entirely using renewable energy, and once used is recycled efficiently to provide a valuable resource for subsequent generations".

In this book, sustainable packaging is proposed as a holistic concept to developing packaging with three interrelated components, from left to right, as illustrated in the following framework.

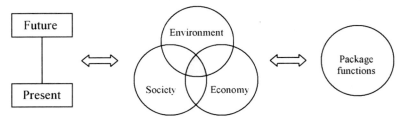

Figure 18.1 Conceptual framework of sustainable packaging

The first component is related to time and perpetuality. Sustainable packaging is aimed at not only meeting the needs of the present generation, but also meeting the needs of future generations. It requires a cradle-to-cradle flow of packaging materials in which the materials can be used repeatedly without depleting resource.

The second component is related to striking a health balance to meet the needs of the environment, society, and economy—a packaging system is not truly sustainable until all these are addressed in an equitable manner. For example, a packaging system with the sole purpose of maximizing profits is irresponsible, if it fails to address the needs of the environment and society. Similarly, a packaging system with the sole purpose of minimizing the negative impacts to the environment is unrealistic, if it fails to address the needs of the society and economy. In Figure 18.1, sustainability occurs only in the shaded area where the circles of environment, society, and economy meet.

The third component is related to the packaging functions (containment, protection, convenience, communication) described in Chapter 1. Unless a package can perform some useful functions, the justification of its existence is questionable. For example, a package may be considered environmentally friendly because it is biodegradable; however, if it fails to protect the product, the product will likely to be discarded and not used by the consumer. The packaging functions, unfortunately, are largely overlooked in the literature when sustainable packaging is discussed.

This book is written surrounding the conceptual framework in Figure 18.1. The socioeconomic factors are addressed in Chapters 1 and 19, while the impacts on the environment are addressed in this chapter. The packaging functions and related topics are addressed in the remaining chapters; for example, the chapters related to materials sciences are aimed at providing a better understanding to enable the production of packaging materials with certain useful functions, and the chapters related to packaging technologies are aimed at enabling the package or packaging system to perform certain useful functions.

18.2.2 Development of Sustainable Packaging

How best to develop sustainable packaging to meet the needs of the present without compromising the ability of future generations to meet their own needs? While there is no straightforward answer to this complex question, below are some useful guidelines:

- Avoid over-packaging by using minimal but adequate amount of materials to meet safety, quality, and market needs.

- Avoid toxic constituents and use energy efficient technologies whenever possible for manufacture and distribution of product.

- Whenever feasible, use packages that are made of renewable, environmentally friendly, or recycled materials, without compromising product safety and quality or greatly increasing cost.

- When developing packaging system or technologies, it is important to strike a healthy balance to meet the needs of the environment, society, and economy.

- Use methodologies such as life cycle assessment (LCA) to aid the development of sustainable packaging.

Chapter 1 has already discussed the socioeconomic factors relating to food packaging. Tables 1.1-1.3 in that chapter are useful tools to analyze package functions and technologies, which are also important for the development of sustainable packaging. Hence, this chapter will be focused more on the impacts of packaging on the environment.

18.3 Environmental Issues Relating to Packaging

18.3.1 Solid Wastes

According to the US Environmental Protection Agency, packaging constitutes approximately one-third of municipal solid waste stream by weight in many developed countries. As developing countries continue to improve their living standards, significant growth in packaging waste will also occur there. Many countries currently have laws or incentives to encourage the reduced use of packaging or recycling of packaging wastes.

As landfill space diminishes and waste disposal cost increases rapidly, solid waste has become a pressing socioeconomic and political issue. Packaging waste, being highly visible almost everywhere, is frequently a target of criticism. Much emphasis is now focused on how to effectively manage packaging wastes, a topic to be further discussed in Section 18.4.

18.3.2 Hazardous Compounds

Polyvinyl chloride (PVC) is a major packaging polymer under severe criticisms. The major concern is the fear that presence of PVC in incinerators would lead to increased production of dioxins, a family of highly toxic chlorinated hydrocarbons. Although PVC is a minor factor in dioxin production in incineration, some organizations including Greenpeace has called for elimination of chlorine-based chemical production processes.

Because of environmental concern, PVC containers are now frequently replaced by PET or PE containers in many countries. Polyvinylidene chloride (PVDC), although still widely used as gas barrier resin, is gradually being replaced by ethylene vinyl alcohol copolymer (EVOH), silicon oxide coated polymers, or aluminum oxide coated polymers. In recent years, a variety of phthalate plasticizers are also under attack as high levels of plasticizers have been found in a significant fraction of water samples in the environment.

18.3.3 Ozone Depletion

The ozone layer is a region of the earth's atmosphere, about 10 and 20 miles in altitude, containing relatively high concentrations of ozone (O_3). This layer prevents harmful ultraviolet and other high energy radiations penetrating from the sun to the earth. Hence depletion of this layer will lessen the protection of those harmful radiations, leading to higher skin cancer and cataract rates and crop damage.

In the 1970s, the loss of ozone was first noticed in the lower stratosphere over Antarctica. In the mid 1980s, the formation of an 'ozone hole' was discovered over Antarctica, and scientific evidence has since accumulated linking it to chlorofluorocarbon (CFCs) and similar compounds. In the past, CFCs were used in many commercial applications, including some relating to packaging such as aerosol propellants and foams. Today, CFCs are no longer produced and used in industrialized countries.

18.4 Packaging Waste Management

Generally there are six ways to manage packaging wastes. The first three ways (reduce, reuse, and recycle), collectively known as the 3 Rs, are aimed at minimizing the amount of post-consumer packaging materials into the waste stream. The remaining three ways are aimed at disposing packaging wastes in an efficient and environmentally friendly manner. In practice, combinations of these ways are used, since each has its strengths and weaknesses.

18.4.1 Reduction

Reduction, known also as source reduction, is an effective way to manage waste since it prevents the generation of waste in the first place. It conserves resources and reduces pollution which contributes to global warming. It can also reduce the costs for production, transportation, and waste disposal. However, careful considerations must be given to balance the benefits of source reduction and the functions of packaging. While over-packaging should be avoided to minimize the generation of packaging wastes, under-packaging should also be avoided to minimize product spoilage or damage.

At the process level, resource reduction may be achieved by ways such as selecting manufacturing processes which are environmentally friendly, optimizing distribution systems to conserve resource, and selecting suppliers and converters who are committed to good environmental practice. At the package level, resource reduction may be achieved by ways such as eliminating unnecessary materials in package design, substituting with thinner materials to reduce the amount of material, substituting with lighter materials to reduce the resource for transportation, using materials which are recyclable or require less resource for production, and packaging products in a single large pack instead of in smaller individual packs. For example, the weight of the 2-liter PET bottle has been reduced from 68 to 51 grams since its introduction in the 1970s.

18.4.2 Reuse

Reuse of packages is an old concept that extends the life of a package by refilling or reusing it multiple times at the consumer, retailer, or producer level. It decreases the demand of new materials, thereby contributing to source reduction and its associated benefits. Nevertheless, the practice of reuse is quite limited in food packaging.

Refillable bottles for milk, beer, and carbonated beverages were common in the past, but have steadily been replaced by disposable or recyclable bottles in recent years. The practice of refilling beverages is mostly limited to larger containers; for example, drinking water is often supplied in returnable 5-gallon polycarbonate bottles to offices and homes, and beers are often supplied in kegs (pressurized stainless steel barrels) at parties.

18.4.3 Recycling

Recycling involves converting materials that would otherwise become waste into new products, thereby diverting the materials from the waste stream to material recovery. While it is desirable to recycling material indefinitely, most recycling is actually down-cycling, meaning that materials are reused in less valuable applications in each cycle.

- **Recycling of Aluminium**

According to the Aluminum Can Association, aluminum cans are the most valuable container to recycle. They are also the most recycled consumer product in the United States, with over 50% recycling rate. Making cans from recycled aluminum requires only 5 percent of the energy for making cans from virgin ore. Recycling one kilogram of aluminum can save up to 8 kilograms of bauxite, 4 kilograms of chemical products, and 14 kilowatt hours of electricity. Once collected, post-consumer aluminum cans are remelted, rolled, manufactured, filled with the product, and returned to the retailer's shelf within two or three months.

However, recycling of aluminum from foil laminates is often not economical, since the amounts of aluminum in foil laminates are relatively small and recovering them is costly.

- **Recycling of Glass**

Like aluminum, glass may be recycled infinitely but its recycling rate is considerably lower. The use of cullet (recovered glass) in glass furnaces reduces the melting temperature of the glass. Using cullet allows the glass container industry to reduce energy input to its furnaces (see Section 7.4.1). Energy costs drop about 2-3% for every 10% cullet used in the manufacturing process.

Post-consumer glass containers are mostly collected through deposit systems and curbside recycling programs. The major market for recycled glass is container manufacturing which allows very little color contamination. While automated sorting equipment are available, sorting glass by color is often performed by workers. Mixed-color cullet may be used in the construction and related industries, but these low-value uses are often uneconomical.

- **Recycling of Paper and Paperboard**

According to the Paper Industry Association Council, corrugated board has a recycling rate over 70%, which is the highest among all packaging materials. Most of this material is collected from the retail stores and factories, rather than from individual consumers. The end-uses include new corrugated board, boxboard, and kraft paper for grocery sacks.

- **Recycling of Plastics**

HDPE bottles for milk and water and PET soft drink bottles are the two most widely recycled plastic packaging materials, because they are easily identified and available in large quantities. The major uses of recycled HDPE include plastic lumber, containers, pipes, etc. The major uses of recycled PET include fiberfill, carpet, clothing, containers, strapping, automotive parts, etc. Other polymers such as PP, PS, and PVC are also being recycled in small quantities.

Recycled plastics are allowed in food packaging applications such as egg cartons or vegetable trays where the potential for migration is limited. Sometimes, a recycled plastic layer is coextruded or laminated with a virgin plastic layer, where the virgin layer provides protection again potential contaminants from the recycled layer.

18.4.4 Composting

Composting involves the aerobic decomposition of biodegradable materials including most cellulosic packaging materials and biodegradable plastics. While composting is a disposal option for yard and food wastes, its application for managing packaging wastes is very limited. In practice, only cellulosic packaging materials have the potential to be composted. Compostable plastics are difficult to differentiate from noncompostable plastics and their availability and uses are still very limited.

18.4.5 Incineration

Incineration offers not only a substantial reduction in volume of solid waste but also the benefit of energy recovery. The major challenges of incineration are high cost and opposition from the public. Sophisticated and expensive pollution control equipment is needed to control pollution, since ash generated from noncombustible residues may contain harmful levels of heavy metals. While the EPA called for significantly increased use of incineration in the mid 1980s, the practice of incineration has not grown substantially since.

18.4.6 Landfill

Landfilling is the most common way to handle many types of wastes and can provide a source of methane as fuel. However, as existing landfills are depleting and new landfills are becoming more difficult to find and expensive to build, the other options for management wastes will become more attractive. Landfilling may also create the potential problem of groundwater contamination.

18.5 Life Cycle Assessment

Life cycle assessment or *life cycle analysis* (LCA) is a system approach for evaluating and minimizing the adverse impacts of packaging on the environment [1]. Below are two definitions of LCA:

- According to Society of Environmental Toxicology and Chemistry (SETAC), LCA is a process to evaluate environmental burdens related to products, processes, or activities, to identify potential impacts on the environment coming from energy or material consumptions, to identify, and to evaluate possible product improvements.

- According to International Organization of Standardization (ISO), LCA is a technique for compiling an inventory of relevant inputs and outputs of a product system, evaluating the potential environmental impacts associated with those inputs and outputs, and interpreting the results of the inventory and impact phases in relation to the objective of the study.

Although first introduced in the 1960s, the concept of LCA did not receive great attention until the late 1980s when solid waste became a worldwide issue. In 2002, the United Nations Environmental Program (UNEP) collaborated with SETAC to launch the Life Cycle Initiative, an international partnership aiming at putting life cycle thinking into practice and improving supporting tools, such as LCA, through better data and indicators.

18.5.1 Scope of LCA Studies

The interactions between a product and the environment may be assessed at various stages throughout the product life cycle:

- In *cradle-to-gate assessment*, the evaluation covers the stages from raw material to product manufacture. For example, the evaluation may involve extraction of crude oil, conversion of polymer resin from crude oil, and conversion of polymer resin to a product such as plastic bottle for consumer use.

- In *cradle-to-grave assessment*, the evaluation covers the stages from raw material through all stages in the supply chain until the product is finally disposed of. Taking the above example, the evaluation is extended beyond production of the plastic bottle to distribution, consumer use, and finally disposal in landfill.

- In *cradle-to-cradle assessment,* the evaluation is the same as cradle-to-grave assessment, except the product will be converted into a renewed resource instead of being disposed of at the end of its usable life.

LCA is typically associated with evaluating the environmental impacts of a product or process on the environment in cradle-to-grave mode. Strictly speaking, a product whose life cycle involves only cradle-to-grave does not meet the definition of being sustainable, since the product is not converted to a renewed resource at the end.

Sustainable packaging involves transforming packaging into a cradle-to-cradle system, in which the product is converted to a renewable source at the end of its life cycle. While the methodology of LCA is usually used for cradle-to-grave assessment, there is no reason why it cannot be used for cradle-to-cradle assessment of sustainable packaging systems.

18.5.2 Methodology of LCA

According to ISO, the implementation of LCA consists of four interrelated phases as illustrated in Figure 18.2.

LCA Framework

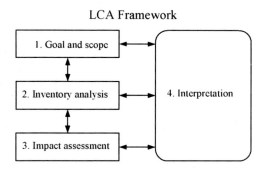

Figure 18.2 Four phases in life cycle assessment

1. Goal and scope definition. The first phase involves defining the package and related processes, establishing context in which assessment to be made, and identifying boundaries and environmental effects to be reviewed. Activities may also include determining what information is needed to facilitate decision making, what level of specificity is required, how data should be organized and result be displayed, and what ground rules for performing the work.

2. Inventory analysis. The second phase involves identifying and quantifying energy and raw material usage, solid waste disposal, and pollutant emission

during the product life cycle. Activities may also include developing process flow diagrams, designing data collection plans, collecting and evaluating data, and reporting results.

3. Impact assessment. The third phase involves evaluating potential human health and environmental impacts due to the material and energy consumption, solid waste disposal, and pollutant emission from the inventory analysis. Activities may also include selecting impact categories (such as solid waste, resource depletion, human health) to be considered as part of LCA, characterizing the magnitudes, and assessing the practical significances of potential impacts.

4. Interpretation. The fourth phase involves evaluating the results of the inventory analysis and impact assessment to select preferred products or processes with the least overall adverse environmental impacts. Activities may also include identifying significant issues (such as solid wastes and resource depletion) by reviewing data or information from the first three phases of the LCA process, establishing the confidence in and reliability of the results, reporting results, and making recommendations.

The double-headed arrows in Figure 18.2 denote that LCA is an iterative process, meaning that the four phases will repeat several times until optimum results are achieved. LCA should be an ongoing process since the society and environmental conditions are continuously changing with time.

18.5.3 LCA Studies

The methodology of LCA has been applied to food packaging and related studies. For example, LCA studies have been conducted on evaluating alternative coffee packaging [2]; comparing egg packages made of polystyrene and recycled paper [3]; comparing recycling, incinerating, or landfilling paper and cardboard wastes [4]; comparing environmental impact of polylactide (PLA) and petrochemical-based polymers [5]; evaluating performance of yogurt product delivery system [6].

18.6 Degradable Packaging Polymers

While plastics have many advantages, there are also growing public concerns about their negative impact on the environment. There has long been a great interest in developing low cost degradable polymers with properties suitable for packaging applications.

Degradation of plastics can be achieved by microorganisms (biodegradation) and ultraviolet (UV) light (photodegradation). Biodegradation is a natural process by which organic chemicals (such as natural polymers) in the environment are converted to simpler compounds, mineralized, and redistributed through elemental cycles such as the carbon, nitrogen, and sulfur cycles through the action of naturally occurring microorganisms [7].

Biodegradable packaging is gaining greater interest as composting is becoming a more attractive option for waste management. In an European directive, biodegradable packaging is defined as a material that must be capable of physical, chemical, thermal, and/or biological degradation such that this material used as compost ultimately decomposes completely into carbon dioxide and water [8]. Photodegradation of plastics involves the breakage of the polymer chains into smaller fragments by UV catalyzed oxidative reactions. After initiation by UV light, the degradation process may proceed without further UV light exposure to yield small molecular weight products desirously nontoxic and biodegradable.

Degradable packaging must satisfy the basic, but often conflicting, requirements of protection and degradability. The packaging material should remain stable to protect the contained food from the environment for the required time period, and then be readily degraded through proper mechanisms after use and disposal. Simplified general assumption is that the food package is not exposed to light and humid condition until opened and used for consumption, and it is then disposed to the outside light and/or moisturized conditions, which may trigger the degradation mechanism.

18.6.1 Biodegradable Packaging

Biodegradable plastic packaging materials important in food packaging applications are polylactic acid (PLA) and starch-based polymers [8]. Polylactic acid or polylactide is produced by condensation polymerization of lactic acid or intermediate monomer of lactide which is cyclic dimer of lactic acid. Raw material of L-lactic acid is produced from fermentation of sugars by lactic acid bacteria. Sugars for fermentation can be obtained from corn starch or other starch sources by enzymatic hydrolysis. PLA is thermoplastic and crystalline resin, and can be extruded, thermoformed, blow-molded, and injection-molded into sheets, films, trays, and bottles. The properties of PLA containers were reported to be superior or comparable to those of PET or polystyrene (PS). Commercialized PLA for food packaging is NatureWorks® PLA of Cargill Dow. Attaining GRAS (Generally Recognized As Safe) status, it is used now for direct food contact applications such as bottles for water, milk, fruit juice, and beer, trays for cheese and fresh salad, film bag for fresh salad, and twist-tie wraps for candies. With rising oil price and technology advancement, PLA may have the potential of competing with synthetic packaging plastics in the future.

Starch based polymers are produced by blending special kinds of modified starch with other biodegradable components in plasticizing solvent under heat. They are often called thermoplastic starch (TPS) or destructurized starch because of their thermoplastic nature. They can be formed into film bags and trays by existing plastics fabrication equipments. Blending of TPS with biodegradable synthetic aliphatic polyesters can produce the films and sheets with good water resistance and fabrication properties. One major application of starch based polymer is foamed cushion balls used for loose fill in protecting against physical shock and stress.

Polyethylene film simply blended with starch has also been produced to make it more vulnerable to biodegradation, but it is biodisintegrable rather than biodegradable. Only starch part is biodegradable with polyethylene part just losing its structural integrity [9].

Another biodegradable polyester attracting interest is polyhydroxybutyrate-valerate copolymer (PHBV) which is produced by microbial fermentation process. It has similar physical properties as polypropylene and can be blow-molded in bottles in commercial scale. However, the economical noncompetitiveness compared to synthetic plastics has stopped the commercial packaging application of PHBV since 1998. Synthetic aliphatic polyester synthesized by polycondensation of diol and dicarboxylic acid is considered to be a promising biodegradable plastic due to their property similar to low density polyethylene. Other biodegradable plastics of commercial interest are polycaprolactone and polyvinyl alcohol (PVOH). Both are used to a limited extent mostly in nonfood packaging applications such as medicine and agriculture.

There have been attempts to improve chemical and physical properties of biodegradable plastics such as thermal stability, gas barrier properties, moisture sensitivity, strength, melt viscosity, and biodegradation rate. One recent approach is nanoreinforcement of the biodegradable polymer by nanoclay particles (layered silicates) [8, 10].

For the biodegradation scheme to work efficiently, solid waste disposal needs to move to composting both on home and industrial level, which requires pretreatments such inorganic waste removal, size reduction, moisture addition, and mixing [8]. Without sound separate collection and treatment of biodegradable plastics, the management of composting system would not work properly and the use of biodegradable plastics in the packaging area would not help to relieve the environmental problem and concerns of packaging materials.

18.6.2 Photodegradable Packaging

Two methods are basically used for making polymer materials photochemically degradable: one method is to chemically incorporate photosensitive groups (usually carbonyls) into the polymer chains and another is to add photosensitizers such as metal salts, quinones, and peroxides. The former has found applications in polyethylene agricultural mulch films and refuse bags. The latter has in 6-pack loop ring for beverages, which is polyethylene-carbon monoxide copolymer. Desirable scenario of photodegradation should proceed with disintegration of plastics into powder form and the subsequent biodegradation [9]. However there are controversies about the validity of the proposed scheme of the degradation process.

Discussion Questions and Problems

1. Compare corrugated paperboard and expanded PS boxes for fresh produce packaging in technological and environmental perspectives. You may look for literatures of similar topics.

2. What kind of specific measures can be taken to improve sustainability in yogurt packaging and delivery system?

3. Compare reuse and recycle for glass bottle packaging of beer. Provide with advantages and disadvantages of each option.

4. Find any legislative laws and regulations which are enforced to reduce the environmental burden of food packaging in your country. Discuss their impact on everyday life and food industry.

Bibliography

1. MA Curran. Life Cycle Assessment: Principles and Practice. Cincinnati, Ohio: US Environmental Protection Agency, 2006
2. M De Monte, E Padoano, D Pozzetto. Alternative coffee packaging: an analysis from a life cycle point of view. J Food Eng 66(4):405-411, 2005.
3. A Zabaniotou, E Kassidi. Life cycle assessment applied to egg packaging made from polystyrene and recycled paper. J Clean Product 11(5):549-559, 2003.
4. A Villanueva, H Wenzel. Paper waste - Recycling, incineration or landfilling? A review of existing life cycle assessments. Waste Manage 27(8):S29-S46, 2007.
5. ETH Vink, KR Rabago, DA Glassner, PR Gruber. Applications of life cycle assessment to NatureWorks polylactide (PLA) production. Polymer Degrad Stabil 80(3):403-419, 2003.
6. AK Gregory, AW Phipps, T Dritz, D Brachfeld. Life cycle environmental performance and improvement of a yogurt product delivery system. Packag Technol Sci 17(2):85-103, 2004.
7. R Chandra, R Rustgi. Biodegrable polymers. Prog Polym Sci 23:1273-1335, 1998.
8. JJ de Vlieger. Green plastic for food packaging. In: R Ahvenainen, ed. Novel Food Packaging Techniques. Cambridge, UK: Woodhead Publishing Ltd., 2003, pp. 519-534.
9. SEM Selke. Packaging and the Environment. Lancaster: Technomic Publishing, 1990, pp. 141-151.
10. SS Ray, M Bousmina. Biodegradable polymers and their layered silicate nanocomposites: In greening the 21st century materials world. Prog Mat Sci 50:962-1079, 2005.

608

Chapter 19

Sociological and Legislative Considerations

19.1 Introduction 609
19.2 Tamper Evident Packaging...609
    19.2.1   Conventional Techniques Commonly Used ··························611
    19.2.2   Innovative Techniques···613
19.3 Product Liability..614
19.4 Labeling Information...616
Discussion Questions and Problems 618
Bibliography 618

19.1 Introduction

While food safety issues have been related to migration of potentially undesired chemicals from packaging materials, the consumer protection movement gave birth to many regulations affecting food packaging. At first strict control of product weight and labeling on the product information was enforced. It was advised to eliminate superfluous packaging [1]. Potential injury caused by product defect may be punished legally and thus appropriate provisions to prevent them should be established. There should also be deterrent actions or devices against malicious manipulation of food packaging for safe packaged foods. Recently concerns about bioterrorism underline the importance of secured food packaging without tamper. All of these troubles can be handled or avoided technically by wise design of food package. Food packaging technologists should actively respond to the questions and concerns raised by consumers. Expectations and demands of consumers in respect to packaging are long-term protection, no effect on taste, proper strength, easiness of transport or handling (opening), hygienic quality, recyclablity, and inexpensive cost [2]. There may be diversity of consumers in behavior of purchase and use of food products. Accounts on consumers' behavior are required for satisfaction from consumer side. Safe and satisfactory packaged foods can also be provided in consideration of social and economical environment.

19.2 Tamper Evident Packaging

Real psychological impact became to be felt broadly in society after malicious Tylenol® (analgesic medicine) contamination with potassium cyanide, which occurred in the Chicago area in Autumn 1982 by deliberate tampering of the capsules. The incident caused the death of 7 people bringing about total product recall and destruction of 31 million bottles. There have been numerous copycat incidents for drugs, eye drops, foods, and beverages [3]. Since then US FDA published regulations requiring tamper-resistant packaging for over-the-counter (OTC) drugs, cosmetic products, and contact lens solutions and tablets to make these solutions. Many cases of food package tampering have been reported in torrents [4]. The problem of product tempering is expected to escalate in the future and thus manufacturers have a responsibility to protect consumers against possible tampering acts by due diligence [5]. This means that even food products not required by regulations should be protected from possible tampering. Because food marketing has been moved to open display pattern in supermarket, tamper evident packaging has become more important these days. Food package design should consider the aspect of tamper-evidence. Consumers also like products that are tamper-evident.

Tamper evident package has been defined in US Code of Federal Regulations as one having one or more indicators or barriers to entry which, if breached or missing, can reasonably be expected to provide visible evidence to consumers that tampering has occurred. The FDA regulations require a labeling statement on most types of the containers to alert the consumers to the specific tamper-resistant feature(s) used. The labeling statement is also required to be placed so that it will be unaffected if a tamper evident packaging feature is breached or missing. US FDA suggests 12 features as possible examples of packaging technologies capable of meeting the tamper resistant packaging requirements [6]. The list includes (1) film wrappers, (2) blister or strip packs, (3) bubble packs, (4) heat shrink bands or wrappers, (5) pouches of foil, paper, or plastics, (6) container mouth inner seals, (7) tape seals, (8) breakable caps, (9) sealed metal tubes or plastic blind-end heat-sealed tubes, (10) sealed cartons, (11) aerosol containers, and (12) cans (both all-metal and composite). These 12 categories are merely examples and do not mean to be exclusive listing of acceptable devices. The use of one of these packaging technologies does not, by itself, constitute compliance with the requirements for a tamper-evident package. Packaging features must be properly designed and appropriately applied to be effective tamper-evident package.

Therefore tamper evident packaging should have some requirements for effectiveness: features immediately becoming apparent when moved or manipulated, readily understandable method of opening, easy-to-open feature, intactness of tamper evidence feature after sale, and aesthetics and positive essential image to consumers [7-9]. The effectiveness of tamper evident packaging can be measured based on the degree of difficulty in violating a specific package and restoring it to a near original condition [5]. Effective tamper evident features thus should provide greater difficulty in violating the product. However, they should not be the uncontrollable barrier to opening the package. The effectiveness of tamper evident packaging can be improved by better awareness of consumers about the feature.

In order to attain the objective of tamper-evidence to deter psychologically any attempt of manipulating the package, any existing packaging techniques can be used as features for tamper evidence [9]. Or an entirely new feature may be developed or deigned from the beginning for tamper evident packaging. Several features may also be combined to strengthen the effectiveness of the tamper evidence. Tamper evident packaging may take form of primary container/closure system, secondary container/closure system or a combination of both [7]. The system should provide the visible indication of package integrity when handled in a reasonable manner during manufacture, distribution, and retail display. Achieving tamper evident function without added cost is desired for easy application in practice.

19.2.1 Conventional Techniques Commonly Used

Tamper evidence is not an isolated single element of package protection. Overall task is product protection against contamination and possible safety risk. Therefore any packaging technique has more or less a functional feature of tamper evidence as a means of protection. Various packaging techniques commonly used can be applied or modified for designed function of protecting against tampering [8]. A shrink band or sleeve wrapper is frequently applied to a portion of the rigid containers of glass, metal, and plastics, usually at the juncture of the cap and container, but sometimes extends to far down the package. The band or sleeve is applied after container closing and heat processing, and then shrunk by heat to provide a tight fit (see also Section 11.7.3). Cellulosic wet shrink band is applied wet and shrink upon drying to provide tight fit on the product contour, but is not accepted as tamper evident by US FDA. The use of a perforated tear strip can enhance tamper-evidence allowing easy opening. The band film must employ an identifying graphics that cannot be readily duplicated.

Overwrapping individual packages tightly with shrinkable or nonshrinkable films can be applied as tamper evident protection. Printability of overwrapping film should be considered to carry the desired graphics for tamper evidence. The cellophane end flaps overlapping in the wrapper are not an effective tamper-evident feature because of the possibility that they can be opened and resealed without leaving visible evidence of entry [6].

Inner seal applied on mouth of glass or plastic container stays under the cap. The seal is usually made of waxed paper, plastic sheet, plastic film, foil, or a combination thereof, which readily separates from cap body on opening. The adherence to the cap is attained by cold seal or heat induction sealing (see Section 10.5). Seals applied by heat induction to plastic containers appear to offer a higher degree of tamper-evidence than those by an adhesive to create the bond. Typical induction liner under cap has a multiplayer consisting of a paper pulp, wax, aluminum foil, and polyethylene heat seal layers [9]. When the capped container passes under induction coil for the polyethylene layer to heat-seal onto its mouth, the conductive aluminum layer heats resulting in the absorption of the melted wax into the pulp and release of the film from the cap. Direct heat sealing may also be applied on wide-mouth containers. Pressure sensitive liners of foamed polystyrene are attached to the container mouth by the application torque, however, they are not considered as effective tamper-resistant features because of their vulnerable characteristics to entry with no visible evidence [6]. Container mouth seals must contain a distinctive design that cannot be readily duplicated.

Closure systems can be manufactured in breakable structure as with aluminum roll-on closure. Tear-off closure on glass bottles separates the body into two parts for opening. At the time of opening, metal or plastic caps of twist-off or snap-on types can be broken away into top cap and bottom rim on the bridges or membrane. Easy opening ends in the paper and metal cans are also a type of tear-apart closures. The button-style cap which pops up with breaking of vacuum on opening can indicate the previous opening by positioning at upper level. Numerous innovations in closure design are newly appearing as tamper evident. For the closure systems, refer to Section 11.4.

Carded packaging such as blister and skin packs can not be repaired once opened, and thus is tamper evident. Each food items are individually or in multiples sealed in clear plastic or plastic compartments with foil or paper backing. The individual compartment must be torn or destroyed to reach the product. Individual pack may also be located in a blister or compartment. The backing materials cannot be separated from the compartments or replaced without leaving visible evidence of entry. Empty or broken compartment notifies the signal of tampering. The carded packaging also offers space to carry the warning to consumers. For skin and blister packaging, refer to Sections 11.7.1 and 11.7.2, respectively.

Pouches or bags made of foil, paper, or plastics are sealed by heat or cold adhesive. The properly sealed pouch provides visible evidence of tearing or opening when the product is to be taken out. Cold seal is poorer in tamper-evidence while providing relatively easier opening. Vacuumizing pouch can shape close contact between food surface and packaging film. Loss of tight contact may be seen as a loss of package integrity and thus tampering.

Paperboard cartons sealed by hot-melt or cold glue can offer a visible sign of protection against tampering. Sealed cartons have widely been used for packaging crackers, breakfast cereals, and baking mixes. Even though the technique is sometimes taken as unable to solely satisfy the tamper evidence, it appears as an adjunct to the existing features.

Tape seals or labels are applied to join the closure onto the containers, or to seal a carton packs [8]. The concept of a breakable label can be applied to boxes and bottles as bridges between the container wall and closure. Breakable labels rely on the adhesive bond for their effectiveness, whose choice is important. Different adhesive types are often needed each for glass wall and metal closure for glass bottles (see Section 10.3).

Figure 19.1 shows proportional use of tamper evident tools in cultured dairy products. Even though the survey did not give any specific information on the tools used, conventional systems were shown to be commonly used for tamper evidence. Also about 10% of the products have used more than one tamper evident techniques [10].

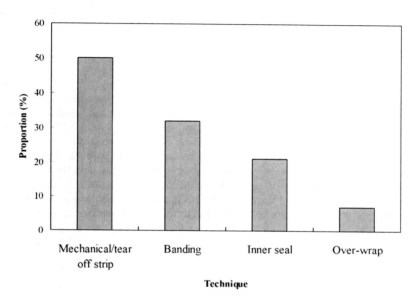

Figure 19.1 Proportion of systems used for tamper-evidence in cultured dairy products. Drawn based on the data of Ref. [10].

19.2.2 Innovative Techniques

Some innovative technologies have been developed and applied to tamper evident packaging. Photoelectronic scanners, microprocessor-controlled video cameras, and other inspectors can be used for on-line detection devices for seal defects or faults on surface and inside the containers [8]. Electronic discrimination of internal vacuum or pressure level can also alarm the defect in seal integrity. Oxygen indicator has been developed and used for detecting presence of oxygen inside the package [7]. The indicator changes color with oxygen presence and thus tell the possibility of tampering in case of transparent packs. A sheath of optical fibers which locates and shines on the side of a cap may indicate the tampering by loss of shining when the fibers are sheared with opening. Holographic image is being used for an identifying characteristic protecting against scanner and photo-copier.

As another innovative way, entire packaging system has been designed for high degree of tamper evidence. A method uses coding system incorporated into labels or magnetic strips on the package [7]. The system adopts two codes – one applied by a manufacturer on the date of delivery and the other by the store on the day or purchase. The applied information should be expected to agree with internal record, which is confirmed at check out. Other tracking systems can be used for protecting the product from tampering or counterfeiting. The intelligent packaging devices such as linear and two-dimensional bar codes, memory tags, and radio-frequency emitting tags, are used in distribution channels to keep track of pallet load of food products [11]. For the smart packaging devices, refer to Section 15.6.

19.3 Product Liability

Trend toward consumer protection has enforced wider responsibility on the food product manufactures and package makers. In many countries, product liability laws have been promulgated to protect consumers from defects of the product. Product liability refers to the liability of any or all parties along the chain of manufacture of any product for damage caused by that product. The damage covers personal injury or any loss of property. The responsible part takes the obligation to pay compensation for damage caused as a result of fault in certain legally defined situations [12]. The responsible part includes the manufacturer of component parts, an assembling manufacturer, the wholesaler, and the retail store owner. Therefore liability extends to the packaging: unsafe packaging is covered by the liability law [13]. A report from liability cases in Japan tells that there were 27 incidents in 1995 related with food packaging: 10 cases for cans, 9 for glass bottles, 6 for plastics, and 2 for paper containers [14]. Inappropriate handling and use of packages by children may cause unexpected injury to them.

Potential defects in food packaging may be caused by poor design such as weak strength and low heat stability, manufacturing error such as weak lamination and poor seal integrity, and improper instruction or warning for cooking and usage [14]. Prediction of consumer behavior, prevention of extraneous nonfood material, wise design of packaging, and appropriate instruction on the label can contribute to reducing the liability [14, 15]. Designing easier and safer opening, correct notice and information transfer, and dimensional design for safer handling units would help to reduce the liability.

Proper design of opening easiness should be important consideration for preventing consumers' injuries related to food packaging [16]. Too high opening torque of the closure makes it hard for common people to open the container, even though adequate torque is required for assuring safe and stable sealing during distribution and storage. Behavior and power of old people need to be considered in designing the opening torque [17]. Too difficult opening may force people to revert to unsuitable opening methods such as using knifes or pliers, which may result in accidents and injuries. Proper level of easy opening devices such as tear notch and tear tape for flexible pouches and plastic bottles can be helpful. In case of easy opening can end, the force level required for opening should be well controlled. Openability also needs to be balanced with tamper evident feature of food packaging: some tamper evident features of tear-off caps and crowns in beverage bottles sometimes cause incidence of cut-fingers [4]. For easy opening can end with full aperture, a torn surface with sharp edge may be left around score line after removal of the panel, endangering the safety of consumers. The designs with safety fold as shown in Figure 19.2 help to overcome the problem.

Figure 19.2 Examples of safety fold design on easy open end in food can. Arrow indicates score part.

Nonconformity of packaged foods due to packaging defects such as excessive tin can corrosion and leaking of paper packaging color into foods has also been the situation of civil responsibility [12].

A beer bottle may explode and hurt a consumer, which makes the processor liable [12, 13]. In 1976, an estimated 125,000 people have been reported to be injured by glass from carbonated soft-drink bottles that exploded on shelves in normal handling or when shocked [1]. Several measures have been used for avoiding the explosion of the glass containers. Standard levels were set for the pressure strength of bottle walls and for the ability of a bottle to withstand sudden temperature changes such as when moved from a refrigerator on a hot summer day. Usual standards state that a bottle's side wall must be able to withstand internal pressure of 1,500 kPa. Protective sleeve labels of foamed polystyrene or polyvinyl chloride have been used to protect from surface damage and scattering of glass shards potentially from explosion of pressurized bottle [18, 19]. Plastic coating combined with styrene butadiene rubber and polyurethane layers has also been used as with the purpose of coloring the bottle in Japan. The glass bottles must also be able to withstand the thermal shock of 42°C temperature difference when transferred from a hot-water bath of 63°C to a colder bath of 21°C [1, 19].

Relevant quality control management and technical auditing in the production line help to prevent the risk [12]. Responsibility of due diligence is on the sides of package manufacturers and food packers. Foreign bodies in the packages of food and beverage may cause potential harm to consumers, and thus detecting devices for all foreign bodies in the packages are installed to preventing them from reaching the consumers. Zhao et al. [20] proposed a method of scanning the flat bottom of the beverage container by ultrasound in order to detect foreign bodies such as glass, metal, and plastic pieces in the liquid food.

Regulatory issues about using active and intelligent packaging devices in food packages have been dealt formerly in Section 15.7.

19.4 Labeling Information

An important function of packaging is to give information to consumers, which may also be used as a marketing tool. Labels on food packages should contain correct information for consumers as guidance for the purchase, storage, handling, cooking, and preparation. Many countries enforce strict regulations on the contents and methods of food labeling. False statements of product on the label are subject to product liability [12]. Most typical mandatory data on food labels are product name, net contents, name and address of the manufacturer, packer, distributor, list of ingredients in descending order, quality grade, nutritional information, shelf life data, and method of usage [21]. There are some proposals for locations of printing and letter sizes on the specific information depending on the countries or states.

Some countries such as USA and Canada ask the food manufacturers that nutrition facts be printed on the label of most foods in order to help consumers plan a better diet. Regarding label information and nutritional claims, EU also has a regulation since 1990 that was recently updated (Regulation EC No 1924/2006): generally speaking the nutritional information is mandatory if a nutritional or health claim is reported in the label (general principles for all claims made on foods are established). These days there are growing needs from consumers for more nutritional information on the food label such as contents of trans fat and allergens. Health claims are permitted if based on scientific clues. Label claims summarize key health benefits.

Figure 19.3 An example of food label (Courtesy of Maeil Dairy Industry)

Other information of code dating, universal product bar code, legal symbols, and religious symbols may be added on food labels. Current regulations should be consulted for correct design of label information. The terms such as 'fresh', 'organic', 'sugar free', and 'fat free' should be used cautiously meeting the condition of definition. Figure 19.3 shows an example of food label. It needs to be mentioned that food labeling rules vary with product, package type, and country: some can be voluntary items and some can be mandatory ones; there may be exceptions and exemptions for some special cases. These days bar code system given also in Figure 19.3 as an example handles the product information. Recently two dimensional codes become available to carry on much more information on the packaged food. For the bar code system, refer to Section 10.4.2.

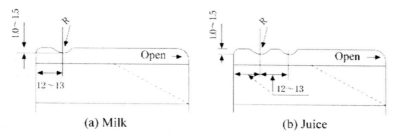

(a) Milk   (b) Juice

Figure 19.4 Examples of different shape concave cuts on top of liquid carton package indicating the content. Radius R is 2.5 or 6.5 mm. All the numeric values are in mm. Adapted from Ref. [22].

Figure 19.5 Braille characters on beer and beverage cans for blind people

Recently some special emphasis and regard have been given on the information for the old and disabled people [22]. Opening location in the package needs to be more clearly marked or informed by contrast color, instruction, or device for the old or blind people. Information on content is written obvious by shape or geometrical notation. Figure 19.4 shows examples tried for informing the opening mouth and contents of milk or juice to the sight-disabled people. Some beer or beverage cans manifest themselves by raised letters on the can ends for blind people (Figure 19.5). Specialized designs on the caps or package body are given for special category products such as sugar, salt, and pepper bottles for the similar purpose. Modern technologies such as sophisticated bar code or RFID tag are being adopted to help the disabled people to identify the package and use the product more easily.

Discussion Questions and Problems

1. Find an example of tamper-evident packaging from a supermarket shelf and evaluate the function and effectiveness of the applied feature.

2. Find a case of food product liability in past newspapers. Review and analyze the case. Show the way to avoid the problem by package design.

3. Examine the labels on the some food packages in supermarket and list the information items on them. If there are health claims on the label, examine them on scientific principles. Consult the rule or guideline of the corresponding country on food labeling.

4. As a conclusive remark on study on food packaging, holistic perspective on any packaging technique is desired in analyzing and developing a package. Select any three kinds of commercial beer packaging and then fit each into Tables 1.1-1.3 to see the adopted function relating with socioeconomics, technologies, and environments. Discuss the values and drawbacks of each packaging method in overall comprehensive aspects of society and technology development. Note that a packaging alternative of the material and function may have mutually interactive influence on the food preservation, human behavior, food supply system, etc.

Bibliography

1. S Sacharow. Packaging Regulations. Westport, Connecticut: AVI Publishing, 1979, pp. 78-101.
2. F Depledt. The consumer's point of view. In: G Bureau, JL Multon, ed. Food Packaging Technology. Vol. 1. New York: VCH Publishers, 1996, pp. 141-148.
3. HL Kaplan, RE Fawcett. Components of manufacturers' product liability based upon defec-tive packaging: foreseeability, superseding cause, and federal preemption. J Prod Liability 7:119-141, 1984.
4. JM Meister. Tamper evidence and closure innovations. Packaging 59(4):36-40, 1988.

5. JL Rosette. Tamper-evident packaging. In: AL Brody, KS Marsh, ed. The Wiley Encyclope-dia of Packaging Technology. 2nd Ed. New York: : John Wiley & Sons, 1997, pp. 879-882.

6. US FDA. Tamper-resistant packaging requirements for certain over-the-counter (OTC) human drug products. 1992.

7. B McKnight, A Cavalier. Tamper Evident Packaging. Symposium Proceedings. Chipping Campden: The Campden Food and Drink Research Association, 1990

8. Anonymous. Tamper evident packaging: technology. Packag Eng 28(6):109-171, 1983.

9. R Jansen, G Schelhove. Tamper evidence-a vital issue in packaging development. Packag Technol Sci 12:255-259, 1999.

10. RN Swing. What's hot, what's not with tamper evident packaging. Cultured Dairy Product J 23(4):10-13, 1988.

11. A Brody. What's active about intelligent packaging. Food Technol 55(6):75-78, 2001.

12. J Botrel. The agrofood business-legal liability. In: G Bureau, JL Multon, ed. Food Packaging Technology. Vol. 1. New York: VCH Publishers, 1996, pp. 113-128.

13. GW Williams. Small packaging - some product liability aspects. Ferment 5(1):76-78, 1992.

14. T Arita. PL law and quality management. In: The Society of Packaging Science and Technology Japan, ed. Dictionary of Packaging (in Japanese). Tokyo, Japan: Asakura, 2001, pp. 490-499.

15. JD Pagliaro. Exporting to the USA: product liability and how to avoid it. Eng Manage J 7:49-50, 1997.

16. N Theobald. Child resistance, tamper evidence and openability. In: N Theobald, B Winder, ed. Packaging Closures and Sealing Systems. Oxford, UK: Blackwell Publishing, 2006, pp. 231-259.

17. DA Carus, C Grant, R Wattie, MS Pridham. Development and validation of a technique to measure and compare the opening characteristics of tamper-evident bottle closures. Packag Technol Sci 19:105-118, 2006.

18. D Twede, R Goddard. Packaging Materials. 2nd ed. Leatherhead, UK: Pira International, 1998, pp. 49-54.

19. Y Yamato. Glass containers. In: T Kadoya, ed. Food Packaging. San Diego: Academic Press, 1990, pp. 105-116.

20. B Zhao, OA Basir, GS Mittal. Detection of metal, glass and plastic pieces in bottled beverages using ultrasound. Food Res Int 36:513-521, 2003.

21. ER Vieira. Elementary Food Science. 4th ed. Gaithersburg, MD: Aspen Publication, 1999, pp. 109-122.

22. Y Oki. Barrier-free packaging. In: The Society of Packaging Science and Technology Japan, ed. Dictionary of Packaging (in Japanese). Tokyo, Japan: Asakura, 2001, pp. 604-608.

Index

A

absorbance, 54
acid food, 363, 386, 392
acid polishing, 193
acrylic, 234-235
activation energy, 98, 415, 418, 499
active carbon, 448, 452
active packaging, 398, 446-462
additive, 254-255
adhesion, 30-31
adhesive, 293-296, 307
adsorption, 81-82, 112, 533, 571
aerobes, 406-407
aerobic bacterial count, 488
aerobic respiration, 400
aerobic spoilage microorganisms, 400
aerosol container, 231
air, 53
alkyd resin, 235
allyl isothiocyanate, 457
alpha-tocopherol, 459, 547
aluminum foil, 201, 258, 306-307, 436-437, 441, 567
aluminum tube, 230-231
aluminum, 45-48, 50, 64, 66, 70, 198-201, 218, 600
amino acid, 545
amorphous, 32-35, 48-49, 152-153, 462
anaerobes, 406-407
annealing, 190
antimicrobial effect, 412
antimicrobial packaging, 455, 456-459
antioxidant, 459, 547, 556
antioxidant releaser, 455, 459
argon, 400, 403
Arrhenius equation, 98, 129, 420, 499-500
ascorbic acid, 448-449, 572

aseptic packaging, 269, 379-391, 561
autolysis, 557
Avrami equation, 499

B

Bacillus cereus, 406-407, 489
bacteria, 503, 549-550
    psychrotrophic, 559, 561, 563
bacteriocin, 458
bag, 259-260, 303, 337-339, 612
bag-in-box, 339, 388
band sealing, 304
bar code, 297-301, 468, 472, 614
    two-dimensional (2D), 300, 434
bar sealing, 303-304
Baranyi model, 497-498
barrier, 9, 256, 379
base weight, 255
bead, 216, 223, 550
beating, 253-254
beef, 429
beer, 574-575, 615
Beer's law 54
Belehradek equation, 502
biodegradable packaging, 605-606
biodegradation, 38, 604-606
biodeterioration, 39
biofilm, 39-40, 219
biogenic amine, 468
bisphenol A, 234
blackplate, 204, 206, 208
blank, 264-266
bleaching, 37, 253
blister packaging, 345-346, 612
bloom, 566
blow and blow, 188
blow molding, 168-171
blown film extrusion, 166-167
blueberry, 415, 420
Braille character, 617
branched polymer, 143

bread, 411
breakfast cereal, 547
brittleness, 67
Brocothrix thermosphacta, 404, 408, 553
browning, 433-434, 441, 487
bulk pack, 388
burst strength, 72, 250, 254
burst test, 308-309
butylated hydroxytoluene (BHT), 455, 459, 547

C

C value, 362, 365, 367, 381
cake, 550
calcium hydroxide (Ca(OH)2), 448, 452
calcium oxide (CaO), 448, 451, 460
can, 220-226, 376, 379, 388
can end, 226-228, 615
candy, 567
canned food, 216, 221, 226-227, 517
caramel, 568-569
carbon dioxide (CO_2), 399-403, 574
carbon dioxide absorber, 448, 452
carbon dioxide emitter, 455-456
carbon monoxide (CO), 403-404, 554
carbonated soft drink, 513, 573, 615
carbonation, 317, 573
carboxymyoglobin, 403, 554
carded packaging, 345, 612
carton, 264-265, 267-268, 377, 387-388, 612
cast film extrusion, 165-166
cellophane, 32, 37, 48, 55, 59, 64, 270-272
cellulose, 50-51, 61-62, 66, 246-249
cellulose acetate, 271
cellulose ester, 61, 250
cellulose fiber, 246
cementing, 222-223
ceramic, 47, 194-195, 436
cereal, 545-546
cheese, 411, 503, 562-564

chemical bond, 19- 21
chemical kinetics, 493-497
chemical toughening, 192
chewing gum, 570
chilling injury, 577
chitosan, 458
chlorine dioxide, 458
chlorofluorocarbon (CFCs), 599
chocolate, 566-567
Clostridium botulinum, 360, 362, 364, 406-409, 411, 556
closure, 327-333, 612
    continuous thread (CT), 329-331
    design, 329
    dimension, 330
    press-on/twisted-off, 331
    roll-on, 332
    snap-fit, 332
    torque, 333, 614
CO_2 permeability, 93, 416, 418
CO_2 production, 415-416, 563
coating, 144, 156, 159, 161, 163, 171-172, 174, 206, 232-237, 257
code, 296-301
coefficient of friction, 64-66, 181, 190
coextrusion, 164-165
cohesion, 30
cold end treatment, 190
cold seal, 307-308
color change, 493, 553
combustion, 36-37
composite can, 266-267
composting, 601, 605-606
confectionery, 565-570
contour packaging, 344
converting, 256
copolymer, 144-145
cork, 45, 64, 273, 328, 332, 576
corrosion, 37-38, 109-110, 210-220, 615
corrugated board, 261-263, 601
corrugated box, 263
covalent bond, 21, 25-27

Coxelliae burnettii, 363-364, 559
cradle-to-cradle assessment, 603
cradle-to-gate assessment, 602
cradle-to-grave assessment, 602-603
cream, 411, 560-561
creep, 67, 72
creping, 256
crisping, 433, 441
crispness, 547
critical limit, 486
crosslinked polymer, 143
crown, 331
crystalline, 32-34, 48-49, 152-153,
    462
crystallinity, 34, 96
crystallization, 193, 493, 563, 566,
    567
cullet, 186, 600
curing, 236
cushioning, 74-75
cylinder machine, 255

D

D value, 359-361
Data Matrix, 300
degradable packaging, 604-606
density, 63-64, 270, 274
desorption, 81-82, 111, 492
deterioration, 483-485
    package dependent, 485
    product dependent, 485
dielectric constant, 428-429
dielectric loss factor, 428-429
dielectric sealing, 306
diffusion, 81, 111-114, 513
diffusion coefficient, 82, 113-116,
    129
dipole, 28-29, 427
dipole rotation, 427
double reduced plate, 206
double seam, 228-229, 334, 377, 388
draw and thin redraw (DTR), 226
drawn and redrawn (DRD) can, 225-
    226

drawn and wall-ironed (D&I, DWI)
    can, 224-225, 237
drip, 403, 559, 582
drum, 230, 267, 388
dry food, 503-504, 517
dry milk, 561
drying, 256
dual ovenable, 436, 439
ductility, 67
duty cycle, 433
dwell time, 303

E

EAN, 297-300
edible coating, 163, 567, 580, 584
edible film, 163, 567, 584
egg, 411
elastic modulus, 69-70, 181, 203, 209
elasticity, 67
electrolytic chromium coated steel
    (EECS), 207-208
electromagnetic radiation, 51-52
electromagnetic identification
    (EMID), 468
electronic article surveillance (EAS),
    468
electrophoretic coating, 236
elongation, 69-70, 272
environment, 11
    ambient, 7, 507-510
    human, 7, 512
    physical, 7, 510-511
environmental factor, 482, 485
epichlorohydrin, 234
epoxy resin, 234-235
equilibrium relative humidity (ERH),
    565-566
Escherichia coli, 406, 488-489
etching, 40
ethanol, 468
ethanol emitter, 457, 552
ethylene, 412, 453, 577
ethylene absorber, 448, 453
ethylene vinyl acetate (EVA), 61,
    155, 161, 295, 309, 328

ethylene vinyl alcohol (EVOH), 29, 93, 155, 160, 377-378, 412, 598
expansion ring, 227
extrusion, 164
extrusion blow molding, 168-169
extrusion coating, 171-172
extrusion lamination, 172-173

F

F value, 361, 365, 392
f value, 370
facultative organisms, 407
fat, 429, 566, 570-571
fatty acid, 545-546, 570
ferrous carbonate ($FeCO_3$), 455-456
fibril, 245-246
Fick's first law, 82, 113
Fick's second law, 84, 114, 498-499
filling, 317-327
    auger, 324-325
    bottom-up, 318
    count, 323-324
    diaphragm, 321
    gravity, 317-320
    horizontal, 342-344, 405
    level-sensing, 321
    liquid, 317-322
    piston, 321, 349
    pressure, 320-321
    solid, 322-327
    timed flow, 321
    vacuum, 319, 321, 324
    vertical, 340-342
    volume, 324-325
    voulme cup, 321, 324-325
    weight, 321, 326-327
fin seal, 338-339
first order kinetics, 495-496, 521
fish, 411, 557-559
flour, 545-546
fluoropolymer, 48
flute, 261-262
focusing, 441
folding carton, 264-265

food deterioration mode, 483-485
food safety, 4, 406-409
food simulant, 124, 135-136
form-fill-seal, 267, 339-344, 387
Fourdrinier machine, 255
Fourier's law, 45
freezer burn, 582
fresh produce, 576-581
freshness indicator, 467
friction 65
friction sealing, 306
frozen food, 432, 517, 581-585
fruit juice, 571-572
functional barrier, 118-120

G

gable top, 268
gas mixture, 412, 551
gas transmission rate, 90, 270
gas-permeable label, 461
germinability, 545-656
glass, 20, 32, 45-47, 50, 53, 55, 59, 64, 66, 70, 177-178, 600
glass jar, 189, 379
glass transition temperature, 49-50, 153, 181, 567
glassine, 259
gloss, 59-62
glucose, 467
gob, 187-188
Gompertz equation, 497
grammage, 64, 256, 262
Gram-negative bacteria, 400, 406, 408, 559
Gram-positive bacteria, 408
greaseproof paper, 258
growth rate (microbial), 497-498, 502, 522
GTIN, 300
gum, 569

H

haze, 58-59, 272
head box, 256
headspace, 9

heat capacity, 46
heat exchanger, 382
heat of combustion, 50
heat penetration, 369-371, 379, 392
heat seal, 272, 345, 348, 388
heat sealing, 301-307
heat stability, 378
helium, 335, 399
hemicelluloses, 245-248, 253
Henry's law, 82-83, 86, 401
hexanal, 492, 583
high density polyethylene (HDPE),
    48, 50, 59, 153, 155-156, 329,
    368, 378, 412, 418, 435, 439,
    601
high frequency sealing, 306
high temperature and short time
    (HTST), 380, 560
homopolymer, 144-145
hot air, 385
hot end treatment, 189-190
hot fill-hold-cool process, 391
hot filling, 367-368
hot gas sealing, 305-306
hot knife sealing, 305
hot melt, 161, 257, 293-295, 307-308
hot tack, 156, 309
hot water, 386
hot wire sealing, 305
humectant, 451, 558
hydrogen, 335, 399
hydrogen bonding, 29, 249
hydrogen peroxide, 385-386
hydroxymethylfurfural, 492
hyperbaric packaging, 398
hypobaric packaging, 398

I
ideal gas law, 103, 510
illuminance, 510
impact resistance, 73-74
impulse sealing, 305
incineration, 598, 601
induction sealing, 307, 611
infrared spectrum, 57

injection blow molding, 169-170
injection molding, 168, 329
injection stretch blow molding, 170-
    171
ink, 280-281
innovative technology, 4
insect, 567
intelligent packaging, 447, 462-464
interactions, 9
intermolecular force, 19, 20, 26, 151-
    152
intrinsic stability, 485
ionic bond, 21-23
ionic polarization, 427-428
ionizing radiation, 60-62, 386
ionomer, 223, 60, 93, 155, 160-161,
    309
iron, 32, 448-449
irradiance, 510

J
j value, 370
jelly, 569

K
keg, 229, 600
kraft paper, 70, 257, 259, 262
kraft process, 252

L
label, 291-295, 612, 616
    heat-sensitive, 293, 295
    holographic, 294
    in-mold, 294
    pressure-sensitive, 293
    leaflet, 294
labeling, 294-295
lacquer, 232-236
lactic acid, 557
lactic acid bacteria, 489, 564
lag time, 497-498, 502, 522
laminar flow, 383
lamination, 173-174, 208, 237, 257-
    258
land fill, 602

lap seal, 338-339
leak, 105, 334, 351
legislation, 7, 470
lethality, 365, 390, 392
life cycle analysis (LCA), 602-604
life style, 4
light, 510, 561, 566, 570, 575
light reflection, 59-60
light scattering, 58-59
lightweighting, 238
lignin, 245-248, 251-252
limestone, 186
linear low density polyethylene
    (LLDPE), 50, 155-156, 418,
    458
linear polymer, 143
lipase, 545, 570, 582
lipid, 503
lipid hydrolysis, 545
Listeria monocytogenes, 360, 364,
    402, 406-408, 488-489
load, 67
logistic equation, 497
low acid food, 363, 386, 392
low density polyethylene (LDPE), 45,
    46, 59, 66, 122, 153, 155-156,
    258, 378, 418, 435
low temperature and long time
    (LTLT), 560
low-tincoating steel (LTS), 206-207,
    237
lug cap, 331
lumen, 246
lysozyme, 458

M

machine direction, 255
mango, 419
master-pack, 556
meat, 411, 553-557
melting point, 48-50, 153, 203, 209,
    302, 311
membrane-filtered beer, 391
metal, 20, 24, 62, 429
metallic bond, 22-25

methylpentene, 435
metmyoglobin, 553-555
microbial contamination, 384
microbial count, 488
microbial inactivation, 358-362
microencapsulation, 460
microfilteration, 575
microperforation, 461-462, 581
microporous film, 418, 462
microwave, 62-63, 426
microwave heating, 430-431
microwave oven, 426, 434, 471
middle lamella, 245, 247
migration, 9, 110-121, 456-457, 498-
    499, 514, 533, 567
    modeling, 115-116, 125-132
    test, 123-125
milk, 559-561
modified atmosphere packaging
    (MAP), 391, 397-421, 546,
    549, 552, 555, 557-559, 563,
    580-581
    active, 398, 404
    passive, 398, 405
moisture, 256, 524, 534, 545-548,
    551, 562, 567, 569, 582
moisture absorber, 448, 451-
moisture sorption isotherm, 505,
    527-529
mold, 400, 411, 503, 545-546, 548-
    551, 562-564
molded cellulose, 270
molded pulp, 270
molecular orientation, 154
molecular weight, 149-151
monolayer moisture, 503
morphology, 149, 152-154, 245
municipal solid waste (MSW), 51,
    598
myoglobin, 400, 403, 553-554, 556

N

necking, 223
nisin, 458-459
nitrite, 556

nitrogen (N₂), 399-400, 403, 561, 572
nitrosylmyoglobin, 556
nitrous oxide, 399
nonenzymatic browning, 492, 503, 561, 572, 583
nonuniform heating, 432-433
nonwovens, 273-274
nutrition fact, 616
nylon, 32, 48, 53, 64, 93, 155, 162, 378, 412

O

O₂ consumption, 415-416, 506-507, 532
O₂ permeability, 93-94, 272, 416, 418, 564
off flavor, 136, 545, 570
oil, 429, 570-571
oleoresinous, 232, 235
one-way check-valve, 452
optical density, 54
optimal modified atmosphere, 413
organic acid, 468
organic type absorber, 449-450
organosol, 233
orientation, 154, 170-171
overall migration limit (OML), 133
oxidase, 404
oxidation, 400, 450, 502, 506-507, 532, 535, 561
oxygen (O₂), 399-400, 506, 524, 534-535, 556, 561, 572, 575
oxygen absorber, 448-451, 549
oxygen barrier, 378, 450
oxygen indicator, 466-467, 613
oxygen scavenger, 448-451, 552, 557, 567, 575
oxygen transmission rate, 90, 525, 527
oxymyoglobin, 399-400, 403, 553-555

P

package
flexible, 8
rigid, 8, 404
semirigid, 8
package development, 13
package integrity, 334-336
package permeability, 104
packaging function, 5, 446, 597
communication, 6, 447
convenience, 6
protection, 5, 446
packaging line, 313-316
packaging science, 3
packaging technology, 3
packaging waste, 598-599
paper, 20, 64, 66, 429, 436, 601
paper bag, 259-260
paperboard, 256, 380, 601
papermaking, 253
parison, 168-170, 188-189
partition coefficient, 114, 116, 129-130
passivation, 206
pasta, 411, 548-549
pasteurization, 194, 371-372, 559-560
pathogenic bacteria, 406-409, 467
PDF417, 300, 471
penetration depth, 429-430, 432-433
peracetic acid, 386
perforation, 418
permeability, 91-94, 97, 418
coefficient, 80
isostatic measurement, 99-100
quasi-isostatic measurement, 100-101
permeance, 91
permeation, 80-83, 416, 513-514, 524
container, 102-104
multilayer, 87-88
resistance, 86-87
permeation rate, 86, 512
phenolic, 234-235
photodegradation, 604-606
pigment, 490, 506

pillow pack, 338, 340-343, 405
pinhole, 105-106, 334, 461
plasmin, 559, 561
plastic, 20, 148
plasticity, 67
plasticizer, 255
plastisol, 233, 329
points, 255
polar polymer, 97
polarity, 25
polyacrylate, 451
polyamide, 32, 48, 50, 53, 55, 61-62,
    64, 162
    amorphous (AMPA), 378
polybutadiene, 412
polycaprolactone, 606
polycarbonate (PC), 53, 59, 61, 93,
    162, 364, 378, 439
polyester, 37, 47-48, 51, 53, 55, 59,
    61-62, 64, 66, 70
polyetherimide, 439
polyethylene (PE), 25, 32, 37, 61-62,
    64, 93, 143-146, 155, 267,
    274, 295, 329, 429, 598, 606
polyethylene naphthalate (PEN), 378,
    575
polyethylene terephthalate (PET), 32,
    93, 146-147, 155, 158-159,
    171, 208, 364, 412, 418, 435,
    450, 575, 598, 601
    crystallized (CPET), 378, 436,
    439
polyhydroxybutyrate-valerate
    copolymer (PHBV), 606
polylactic acid (PLA), 605
polymerization
    addition, 145-146
    condensation, 146-147
polymethyl methacrylate, 37, 50, 53
polyphenylene oxide, 439
polypropylene (PP), 37, 47, 50-51,
    53, 55, 59-61, 64, 66, 70, 93,
    155, 157, 166, 208, 274, 329,
    364, 377-378, 418, 435

oriented (OPP), 159, 166, 271-
    272, 418
    tacticity, 32
polystyrene (PS), 37, 50, 55, 59, 61,
    64, 66, 93, 155, 157-158, 306
polytetrafluoroethylene, 32, 55, 61
polyvinyl alcohol (PVOH), 606
polyvinyl chloride (PVC), 26, 29, 37,
    48, 50-51, 55, 60-61, 64, 93,
    155, 158, 329, 412, 598
polyvinylidene chloride (PVDC), 60-
    61, 66, 93, 155, 159, 271,
    377-378, 412, 598
pork, 411
potassium permanganate ($KMnO_4$),
    453
potato, 429, 503
pouch, 303-304, 335, 337-339, 377,
    388, 612
poultry, 411
powder coating, 236
preform, 168-171, 188
prelabelling, 192
press and blow, 188
pressing, 256
pressure, 181-182, 303, 317, 320,
    374-375, 510-511
pressure differential, 335, 351
pressure injection, 270
primary package, 7-8, 244
printing, 279-291
    dry offset, 286
    electrostatic, 290
    flexographic, 282-283
    gravure, 283-284
    inkjet, 289-290
    letterpress, 282
    lithographic, 285-286
    relief, 282
    screen, 287
    thermal transfer, 290
product liability, 614-616
Pseudomonas, 406, 557
pulp, 251-253
    chemical, 252

groundwood, 251-252
mechanical, 251-252
semichemical, 253

Q

Q_{10}, 501, 521
quality index, 486
quaternary package, 8

R

radio frequency identification (RFID), 297, 468-470
rancidity, 411, 492, 545, 551, 566, 570-571, 583
rate constant, 494-495
reaction, 513-514
reaction order, 494
recloseable, 238
recrystallization, 581-583
recycle, 238, 263, 379, 600-601
recycled plastics, 117-120
reducing sugar, 545-546
respiration, 414-416, 576
    model, 415
refining, 253-254
refractive index, 52-53, 181
regulation, 5, 7, 132-135, 470, 617
relative humidity, 250, 256, 508, 546, 578-579
relaxation, 67, 349
respiratory quotient (RQ), 414
retort pouch, 369, 376
retortable food, 268
retrogradation, 493, 499
reuse, 599-600
Reynolds number, 383
rice, 391, 546
roller coating, 236
rotary filler, 315
rubber, 47, 50

S

Salmonella, 406, 488-489
saturated steam, 384-386
scalping, 110, 121-122, 514, 533

screw cap, 329
seal distortion, 309-311
seal strength, 308-311
secondary package, 7, 244
selective permeation, 461-462
self cooling, 461
self heating, 460-461
self-venting, 441
sensory quality, 487
sensory tainting, 136
set-up box, 236
shaped can, 237
shelf life, 480-481, 551-552, 560-562, 565, 573
    accelerated testing, 500, 516-517
    model, 521-534
    testing, 515-521
shelf life dating, 481
shelf life plot, 500-501
shielding, 441-442
shrink band, 611
shrink packaging, 346-348, 580
side seam, 222
silica, 186
silica gel, 448, 451
silicone rubber, 418
silver zeolite, 458
sizing, 254, 256
skin packaging, 345, 612
smart packaging device, 464-465
snack, 411, 547
Snell's law, 52
socioeconomics, 9
soda ash, 186
sodium carbonate (Na_2CO_3), 448, 452, 455-456, 559
soft drink, 573-574
soldering, 222
solid board, 264, 266
solubility, 401-402
solubility coefficient, 83, 86
sorption, 492
source reduction, 599
sous vide, 363-364

specific migration limit (SML), 118,
    134-135
spinach, 429
square root model, 502
stainless steel, 45, 46, 48, 50, 209-
    210, 218
staling, 400, 493, 499, 551
Staphylococcus aureus, 406, 408,
    489, 549
steam-cooking microwavable
    package, 441
steel, 47, 64, 66, 203
sterile air, 387
sterilization, 194, 372-375
stiffness, 70
stoving, 236
strain, 67, 69
stress, 67, 69, 181, 320, 349
suction, 270
sulfate process, 252
sulfite process, 252
superabsorbent polymer, 451
superheated steam, 384-386
surface tension, 30-32
surrogate, 118-119
susceptor, 62-63, 120-121, 437-440
sustainable packaging, 595-598, 603
sweets, 567-568
symbology, 297
syneresis, 487, 564

T
tacticity, 32
tamper evident, 328, 332, 615
tamper evident packaging, 609-614,
tear strength, 71-72, 250, 254
temper, 205
temperature, 303, 499, 507, 546, 553,
    562, 566, 574, 577-579, 584
tensile strength, 67-70, 203, 209, 250,
    254, 272, 348
tensile test, 68, 308-309
tertiary package, 8
texture, 547, 551, 567-570
tetrahedron, 268

thermal capacity, 45, 181, 203, 209
thermal conductivity, 45, 181, 203,
    209
thermal expansion, 46-47, 181, 203,
    209
thermal toughening, 192
thermoforming, 174
thermoplastic, 144, 236, 304
thermoplastic starch (TPS), 605
thermoset, 144, 236, 329
thiobarbituric acid reactive
    substances, 492
three-piece can, 221-224
time-temperature indicators (TTI),
    465-466
tin, 50
tin free steel (TFS), 207-208, 217
tinplate, 64, 70, 204-207, 215
titanium oxide, 235
toffee, 568-569
tolerable thermal range, 47-48
toner, 281
toughness, 67, 70
transition temperature, 48-50
transmittance, 54-57, 181
transparency, 53-55, 183, 378
transpiration, 578
tray, 230, 264, 377, 388, 392
trimethylamine, 454, 467, 492, 557
tube, 230-231, 264, 267
turbulent flow, 383
two-piece can, 224-226

U
ultra high temperature (UHT), 561
ultraclean room, 391
ultrasonic sealing, 306
universal testing machine, 65
UPC, 297-300
UV light, 236, 386, 510, 605
UV transmission, 56-57

V
vacuum, 613
vacuum metallization, 174-175

vacuum packaging, 391, 397, 399, 404, 546, 549, 555, 557
vegetable parchment, 55, 258
vinyl, 233, 235
viscosity, 317, 350
vitamin, 490-491
volatile basic nitrogen (VBN), 492
volatile nitrogen, 468
volatile organic compounds (VOC), 137, 236-237, 281

W

water, 32, 53, 428-429
water activity, 363, 409, 502-506, 550
water-holding capacity, 554
water vapor permeability, 93-94, 104-105
water vapor pressure, 508
  saturation, 509
water vapor transmission rate (WVTR), 90-91, 104-105, 272, 524, 527-528
wax, 257, 306-307, 579
weathering, 40
welding, 222-223
wet glue, 294-295
wet-strength paper, 259
wetting, 31
wheat, 546
widget, 575
wine, 273, 575-576
wood, 45-47, 51, 64, 66, 245, 272-273
wrapping, 260, 336-337

X

X dimension, 301

Y

yeast, 399, 407, 503, 550, 562-563
yogurt, 564-565
Young's modulus, 69-70

Z

z value, 360-361
zeolite, 448, 452,
zero order kinetics, 494, 496, 522
zipper closure, 339

Lightning Source UK Ltd.
Milton Keynes UK
UKOW04n1544160914

238678UK00001B/25/P